ORGANIC CHEMISTRY SIMPLIFIED

ORGANIC CHEMISTRY SIMPLIFIED

BY

RUDOLPH MACY, PH. D.

CHEMICAL PUBLISHING CO., INC.
NEW YORK   1970

Preface

When Dr. Frank C. Whitmore was president of the American Chemical Society in 1938 and made the customary tour of local ACS sections, he used that occasion to spread the gospel of the electron theory of valence. At one of his lectures the author of this book sat in the audience among a mixed group of chemists consisting of technicians, students, and college graduates. The lack of familiarity of organic chemists with the electron was so obvious that it aroused in the author an urge to write an elementary introduction to organic chemistry in which the role of the electron would be emphasized.

This book is especially intended to serve two groups of readers: those engaged in work of a chemical nature who are not able to take a classroom course in organic chemistry, and those in a college course who find they have a need for a supplementary book to help clarify the approach to modern organic chemistry. In other words, the book was conceived as an integrated introduction to both electron-valence theory and organic chemistry at a level suitable for self-study.

The first edition of this book appeared in 1943 during World War II. A second edition, much enlarged, was published in 1955. For this third edition the book has been extensively rewritten, and more than enough material has been added so that it can serve as a textbook for a one-year college course. The novel arrangement of the subject matter in the earlier editions has been maintained. A teacher who prefers to lecture largely from his own notes should find no difficulty incorporating his material into the simple plan on which this book is based.

Baltimore, MarylandRudolph Macy

February 1969

Contents

PART 1

PART 2

CONTENTS

PART 3

PART 4

To the Reader

The plan of this book is different from that of standard textbooks of organic chemistry. It is divided into four parts, and the reader is advised to turn to pages 1, 55, 235, and 367 before proceeding. Special note should be made of the paragraph in italics on page 000.

Cross references are numerous in order to facilitate use of the book without formal instruction. The topics in each chapter are numbered, and for quick reference these numbers are given at the top of each page.

*The element carbon stands out as the central
luminary in the great mystery of life*

—H. E. ARMSTRONG

Frankland Memorial Lecture
Chemistry and Industry **53,** 459 (1934)

Part I

THE UNIQUE POSITION OF THE CARBON ATOM IN CHEMISTRY

Chemistry is the study of the properties of the elements and of the combinations that can be made from them. These elements are listed in the table on the inside front cover.

Most elements, when they enter into combination, form comparatively simple compounds, such as sodium chloride ($NaCl$) or potassium dichromate ($K_2Cr_2O_7$).

One quite exceptional element has the unique property of forming highly complex compounds, many of which are found in living things. This element is carbon.

Carbon tends to combine with itself to build up molecules with an apparently endless variety of chain and ring structures such as these:*

In combination with hydrogen, oxygen, and very few other elements, the carbon atom forms more than half the compounds known to science.

In Part 1 of this book we shall try to make clear why it is that carbon is so different from the other elements. This explanation requires an elementary study of the structures of atoms and molecules.

*See Tables 10.2 and 11.1 for the names of these two compounds.

I

The Nature of Organic Chemistry

1.1. Origin of Atoms and Molecules. Every schoolboy knows that two famous revolutions occurred in the later years of the eighteenth century. These are the American Revolution and the French Revolution.

In this same period of history, however, when Thomas Jefferson was writing the Declaration of Independence in Philadelphia and Marie Antoinette was losing her head in Paris, an equally important revolution was taking place in the scientific world. Chemists had finally begun to stage a revolution against methods that classed their work simply as the *art of alchemy*, and had begun to replace those methods by others that made it the *science of chemistry*.

The chemists discovered that the best way to determine the relationship of various materials to one another is to know their composition. They began to study substances on this basis as well as from their superficial nature.

Thus we find that Lavoisier, a French chemist, was devising methods of breaking down a uniform chemical substance into its constituent parts at the very same time Washington, in America, was struggling to build a uniform nation from an aggregation of states. Lavoisier was a leader of a revolutionary school of chemists who taught that we cannot truly understand the chemical behavior of substances until we know their composition.

He found out how to measure the composition of sugar and showed that it is composed of the three distinct substances—carbon, hydrogen, and oxygen—united in a definite ratio by weight. This ratio is indicated by the expression $C_{12}H_{22}O_{11}$.

The quantitative method of examining substances led John Dalton, an English schoolmaster, to the first attempt to picture the elementary particles of matter (known since ancient times as atoms) combined in definite ratios. This was in 1803.

Dalton's conception of the atom was that of a hard nucleus, which he represented on paper by a circle with certain signs indicating the nature of the atom, thus:

In this system the exact composition of a compound substance was easily pictured by placing the component atoms side by side. For example, the following shows the way Dalton wrote the formulas of the oxides of carbon:

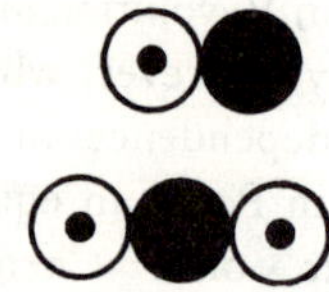

By means of these symbols Dalton was able to indicate how atoms unite not only in one definite ratio, but also in more than one ratio.

The symbols of atoms and formulas of molecules which we use in our modern chemistry seem so clear that we often lose sight of the fact that they have been in use for little more than 100 years, and that they are the result of many centuries of shrewd thinking of the most brilliant men.

1.2. The Birth of Organic Chemistry. In the old days men classified chemical compounds according to their origin; that is, they were either animal, vegetable, or mineral. Formic acid (from the distillation of red ants) and urea (from animal excretions) were called animal substances. Alcohol (from the fermentation of grain) was labeled a vegetable substance. Ordinary salt, of course, was placed in the mineral kingdom.

However, when chemists invented methods for prying into the composition of familiar substances, they discovered a surprising principle which is well illustrated by the data in Table 1.1. This table is a brief list of some substances commonly known about the year 1800, grouped according to the kingdom of nature in which they occur and with modern formulas.

These new adventures into the interior of molecules brought to light the remarkable fact that *carbon* atoms occur in all *living* things. The car-

TABLE 1.1.   Classification of Compounds According to Origin

Mineral compounds	*Compounds from vegetable life*		*Compounds from animal life*	
Sodium chloride $NaCl$	Sugar	$C_{12}H_{22}O_{11}$	Formic acid	CH_2O_2
Magnesium sulphate $MgSO_4$	Alcohol	C_2H_6O	Glycerin	$C_3H_8O_3$
Calcium carbonate $CaCO_3$	Prussic acid	HCN	Urea	CH_4N_2O
Iron sulphide FeS_2	Citric acid	$C_6H_8O_7$	Lactic acid	$C_3H_6O_3$
Magnesium carbonate $MgCO_3$	Tartaric acid	$C_4H_6O_6$		
	Lactic acid	$C_3H_6O_3$		
	Glycerin	$C_3H_8O_3$		

bon atom and life go hand in hand. (See page xii of this book.) The carbon atom is seldom a constituent of compounds obtained from the mineral kingdom; the carbonates listed in the table are exceptions to the general rule. Carbon compounds are not common in inanimate nature.

In the early 1800's it became evident that the science of chemistry could be divided into two branches: first, inorganic compounds, the substances from inanimate nature; second, organic compounds, the substances which always contain carbon atoms and are obtained from life. By 1850 it was further recognized that this definition of organic chemistry was not entirely accurate because certain carbon compounds isolated from living things can also be made in the laboratory. In fact, most of the carbon compounds recorded in our literature do not even occur in nature but have been made by synthetic processes.

Today we define organic chemistry as the *chemistry of carbon compounds*. The study of carbon compounds found in animal and vegetable life is often referred to as *biochemistry*.

The current literature on the chemistry of carbon is believed to contain data on nearly two million compounds. With the exception of hydrogen (Section 37.1) the carbon atom is the most prolific of the elements. This information on carbon compounds is collected in one library and called organic chemistry. The study of the compounds of all other elements constitutes inorganic chemistry.

2

The Organic Chemist Looks at
a Molecule

2.1. The Inorganic Point of View. To illustrate the difference between the inorganic and organic ways of looking at things we shall take as an example the simple compound, carbonic acid—H_2CO_3. This is an appropriate molecule to start with because it is not new to anyone who has studied inorganic chemistry and, since it contains a carbon atom, it may be classed as an organic as well as an inorganic compound.

Let us imagine that we are inorganic chemists and that we are describing in the usual chemical shorthand one of the chemical properties of carbonic acid, such as its reaction with an alkali:

$$H_2CO + 2NaOH \longrightarrow Na_2CO_3 + 2H_2O$$

Carbonic Sodium   Sodium   Water
acid   hydroxide   carbonate

We say to ourselves when we write this reaction,
"Carbonic acid is really ionized in solution, thus

$$H_2CO_3 \rightleftharpoons 2H^+ + CO_3^=$$

and sodium hydroxide also ionizes, thus

$$NaOH \rightleftharpoons Na^+ + OH^-$$

and if we evaporate the solution to dryness, the positive and negative particles change partners and we get sodium carbonate as a solid residue."

In general, inorganic chemistry deals with reactions like this one, between compounds that divide in water solution into two or more ions. These ions are electrically charged atoms or groups of atoms. (In Section 27.6 it will be shown that the ionization of acids is not so simple as pictured here for H_2CO_3.)

2.2. The Organic Point of View. An organic chemist does not generally write H_2CO_3 for carbonic acid. His experience leads him to

6

fix his attention on the molecule as a whole and to write out its structure as

$$HO-C-OH$$
$$\overset{\|}{\underset{}{O}}$$

He therefore looks at the reaction between carbonic acid and sodium hydroxide in this way:

$$Na\overline{OH + H}O{\diagdown \atop \diagup}C{=}C \quad \longrightarrow \quad {NaO \diagdown \atop NaO \diagup}C{=}O + 2H_2O$$
$$Na\overline{OH + H}O$$

Only if we have a clear picture of the detailed structure of most organic compounds can we understand how they go through their chemical changes. For example, the simple inorganic substance known as sulphurous acid is H_2SO_3 to an inorganic chemist, but is

$$HO{\diagdown \atop HO \diagup}S{=}O \quad or \quad {HO \diagdown \atop H \diagup}S{\diagup O \atop \diagdown O}$$

to an organic chemist. He thinks of this compound structurally, because he has found that the two hydrogen atoms in H_2SO_3 are not always equivalent. One of the hydrogen atoms sometimes acts as if it were linked to the sulphur atom; thus, by replacing each H atom with the organic group (C_2H_5) we can prepare two quite different compounds of the same formula, $(C_2H_5)_2SO_3$, and after considering their properties, the experts have assigned the following structures to them:

$$(C_2H_5){-}O{\diagdown \atop (C_2H_5){-}O \diagup}S{=}O \qquad (C_2H_5){-}O{\diagdown \atop (C_2H_5) \diagup}S{\diagup O \atop \diagdown O}$$

It is therefore apparent that if we picture the structure of sulphurous acid graphically we can understand its chemistry much better than by writing it H_2SO_3, even though we are not certain whether the acid itself exists entirely in one form or as a mixture of both. Such examples of dual structure are seldom found in inorganic chemistry but are common in organic chemistry.

It was only when chemists began to look at a molecule from the standpoint of its structure (as explained below) that the real science of organic chemistry had its beginning.

2.3. The Vital Force Theory. When a chemist looked at a molecule in the early days of organic chemistry, he could not possibly see it as plainly as we do today. Even with his limited vision, however, the early chemist was able to see that an inorganic compound is simplicity itself compared with the organic type. He was able to classify inorganic compounds into sulphates, nitrates, carbonates, and so on, but was completely lost when he tried to classify organic compounds:

Inorganic			*Organic*	
Magnesium sulphate	$MgSO_4$		Alcohol	C_2H_6O
Sodium sulphate	Na_2SO_4			
			Acetic acid	$C_2H_4O_2$
Magnesium carbonate	$MgCO_3$			
Sodium carbonate	Na_2CO_3		Glycerin	$C_3H_8O_3$
Magnesium chloride	$MgCl_2$		Urea	CH_4N_2O
Sodium chloride	$NaCl$			
			Sugar	$C_{12}H_{22}O_{11}$
etc.............			etc.............	

There seemed to be no logic to the way these elements (C, H, O and a few others) were able to combine in so many different ways *when present in organic compounds*. In those days, chemistry had only recently emerged from alchemy, and there was still a tendency toward ascribing unusual things to the agency of supernatural powers. For this reason, it was a common belief that when elements combined to form compounds in living nature they obeyed entirely different laws from those prevailing in inanimate nature. This idea was known as "vitalism." The extent to which the conception of a vital force influenced the thinking of the leaders in chemistry in the period 1800-1850 is said by some modern writers in chemistry to have been considerably exaggerated, but the following excerpt* from the May 1961 issue of *Scientific American* is illuminating:

> "The most remarkable scientific event of modern times is the publication of a treatise on chemistry proceeding on the same plan in organic chemistry as has been adopted for a century past in mineral chemistry; that is, forming organic substances synthetically by combining their elements with the aid of chemical forces only. The author who has performed demonstrations by this method is Berthelot, who has been occupied with organic synthesis since he first devoted himself to chemistry. Berthelot is not a vitalist. . . . . . ."

Eliminating the "Vital Force" theory was not nearly so important as the fact that the attention of chemists was directed to the structure of the molecule. For example, in the study of urea, it was found that a salt called ammonium cyanate has exactly the same composition as urea. A difficulty

---

*From "50 and 100 Years Ago". Copyright © May 1961 by *Scientific American*, Inc. All rights reserved.

like this could not be explained without an attempt at writing *structural* formulas. The organic chemist found he had to take a closer look at the molecules of the two compounds and wrote their formulas* as follows:

$$NH_4-O-C\equiv N \qquad\qquad NH_2-\underset{\underset{\displaystyle O}{\|}}{C}-NH_2$$

Ammonium cyanate   Urea

This, in brief, typifies the attitude we must adopt in trying to unravel the mysteries of organic chemistry. The structure of the molecule is all-important. We must consider the molecule as a piece of architecture, built up brick by brick. And just as model houses conform with a few, well-known, simple designs, so we shall find that organic compounds are constructed along lines that are beautifully simple.

* The formulas are given here in their modern versions.

3

Valence

3.1. The Meaning of Valence. In the preceding chapter we found that an organic chemist is really an architect who uses atoms to build up molecules. He is not content with a knowledge of how many atoms there are in a molecule, but also wants to know how they are put together. In other words, an organic chemist takes an extraordinary interest in the phenomenon of valence, which the reader will recall from inorganic chemistry is the word used to describe the attractive force between atoms.

The straight line used in written formulas to indicate which atoms are held together is called a valence bond. For most purposes this is all an organic chemist needs even today to express his thoughts on paper, or to show the structures of molecules.

The average student likes to read his chemistry from a background of definitions; therefore, before continuing on this topic we shall give the customary definition of valence as *the combining power of elements when they unite to form molecules*. The simplest of the elements is hydrogen, and here are some examples of hydrogen compounds:

$$H-O-H \qquad H-Cl \qquad H-H \qquad H-\underset{\displaystyle |}{\overset{\displaystyle H}{\underset{\displaystyle H}{C}}}-H$$

| Water | Hydrogen chloride | Hydrogen molecule | Methane |

In all hydrogen compounds it has been found that only one valence bond is ever necessary to show the linkage of a hydrogen atom to other atoms. Since the valence of hydrogen can be expressed by the number *one* it has been adopted as the standard, or unit, for assigning valence powers to other elements. For this reason, valence is also defined as *the number of H atoms with which an element can combine*.

From the preceding examples, it will be seen that oxygen has a valence of two in the water molecule, and carbon has a valence of four in the molecule of methane. The formula of hydrocyanic acid, $H-C\equiv N$, indicates that carbon as usual has a valence of four and that nitrogen has a valence of three.

3.2. Objections to the Old-fashioned Ideas on Valence. The scheme of using straight lines for valence bonds has been extensively employed by organic chemists. An example of this was given in the preceding chapter when writing the picture formula of sulphurous acid, H_2SO_3.

The objections to this conception of valence are:

1. It leaves in the mind of the student the impression that an atom is an object in space with a number of grappling hooks attached to it, by means of which it can hook on to other atoms.

2. It does not tell us why atoms have different valences, such as one for hydrogen, two for oxygen, four for carbon, and so on.

3. (Most important) it does not explain why the union between some elements is different from that between others.

This last factor is the principal complaint against the use of a straight line as a valence bond, and the importance of this criticism is apparent from the following illustration. The salt molecule and the chlorine molecule are given similar structures when using straight lines for valence bonds,

$$Na\text{---}Cl \qquad\qquad Cl\text{---}Cl$$

Sodium chloride   Chlorine

and there is nothing in these formulas to indicate that there is any difference between the way chlorine is united to sodium and the way the two chlorine atoms are united with each other. In each case, the valence (bonding power) is represented simply by a line.

However, we do know from inorganic chemistry that these two molecules differ a great deal in properties. When sodium chloride dissolves in water it breaks apart (ionizes) into Na^+ and Cl^- ions. Chlorine, however, dissolves only slightly in water and the dissolved molecules do not tend to separate in the same manner as do the sodium chloride molecules. Consequently, we think there must be something unusual about the valence bond, not expressed in our customary formulas, which permits two atoms to break apart so easily in one case yet holds them so tightly together in the other case.

For over a hundred years, physicists and chemists have been searching for a simple explanation of what holds atoms together. The chemists, naturally, could not wait for this mystery to be unraveled. While the search by physicists was going on, chemists decided to invent a symbol for themselves. In short, they decided to write the two atoms on paper and connect them with a straight line, this line being a symbol for the valence force, no matter whether that force may actually be like an ice tongs, a grappling hook, or an electromagnet.

Such symbols are common in science. When the average man is unable to set up a physical symbol for an idea, then the idea itself loses much of its meaning. For example, a meter is approximately a ten millionth of the earth's quadrant, but no one except a traveler in outer space has ever seen the earth's quadrant, so we commonly think of a meter as the length of our favorite meter-stick. The meter-stick is a physical symbol which we can see and feel, and without it the word meter would be quite vague to most of us.

Fortunately, the physicists have made a good beginning in their quest for the secret of atomic structure. On the basis of this new information, the chemists have developed a simple theory to explain the troublesome questions which we have just listed in connection with the problem of valence. The most fortunate thing about all this is that it does not upset any of the pet notions of the organic chemist. We still write our formulas with straight lines for valence bonds, but now all these lines mean much more than grappling hooks.

In the chapters that follow we shall describe briefly the structure of the atom, not from the standpoint of the physicist, but from that of the chemist who has waited so patiently for an explanation of valence. (A lazy reader could skip the next few chapters completely and pick up the story of organic chemistry again in Chapter 9. This proves that modern theories concerning the atom have not changed the fundamental nature of organic chemistry. This omission would leave the lazy reader, however, without a basic understanding of *modern* organic chemistry.)

4

New Ideas on Valence

4.1. The Nature of the Atom. We have just observed that the properties of sodium chloride, Na–Cl, are quite different from the properties of chlorine, Cl–Cl, and that the difference between the two compounds is presumably due to the difference in the way the atoms are bound together. If the atoms in these two compounds are linked differently, then there must be at least two kinds of valence.

Reasoning from what has been said about these two molecules, one might be led to suppose that a large variety of valence bonds is possible, depending on the particular atoms united in the compound, but this is not the case. We find it convenient to differentiate between two extremes in types of bonds, and we have selected Na–Cl and Cl–Cl as representative molecules because they are excellent examples of compounds in which these two general kinds of bonds are found. In these preliminary chapters we shall find that although the valence bond in Na–Cl is a type that occurs frequently in the architecture of certain inorganic compounds, the type of bond in Cl–Cl is more universal and is especially useful in explaining the structures of organic chemistry.

We know so much more about valence today because we know so much more about the structure of the atom. Since the reader may not be familiar with modern theories of the structure of the atom, a chapter on this topic will follow (Chapter 6). At that time we shall describe the atom as it appears to the chemist.

The atom is so complex that it is best to introduce it slowly. In the next few pages we shall consider the general nature and composition of an atom and the general nature of valence, before taking up the detailed study planned for a future chapter.

The evidence of modern science has led to the picture of the atom shown in Figure 4.1. The atom is a structure built up principally of positively charged particles called *protons*, negatively charged particles called *electrons*, and neutral particles, known as *neutrons*.

At the center of the atom is a nucleus which contains a certain number

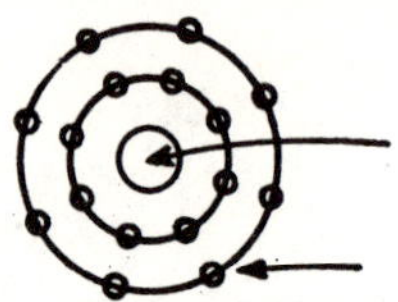

4.1.   The atom.

of neutrons and protons.   At various distances from the center of the atom a certain number of electrons are situated.   The electrons are supposed to bear the same relation to the nucleus as the planets bear to the sun at the center of our solar system.   In a normal atom there are an equal number of protons and electrons; this produces an electrical balance. When thus balanced electrically, the atom is said to be neutral.

The chemist is not particularly interested in the protons and neutrons at the center of the atom, but takes considerable interest in the activity of the external shell of electrons.   We have good reason to believe that most of the chemistry of the atom is associated with the behavior of these electrons.

Since we employ the Na–Cl and Cl–Cl *molecules* so frequently in the explanation of valence, let us examine the sodium and chlorine *atoms* on this basis of atomic structure.   Without any preliminary explanation we shall draw the pictures of these atoms with a certain number of electrons arranged in shells around the nucleus, as in Figure 4.2.

The reader will observe that we have constructed these shells with no more than eight electrons in each. This magic number eight is of great importance in the structure of the atom, as will be learned in Chapter 6 when atoms are considered in greater detail.   We do not yet need to know

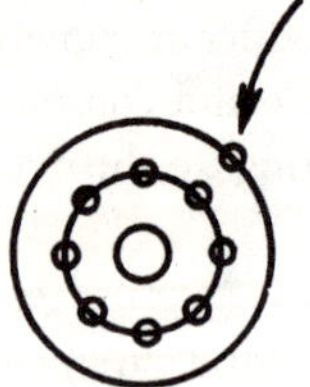 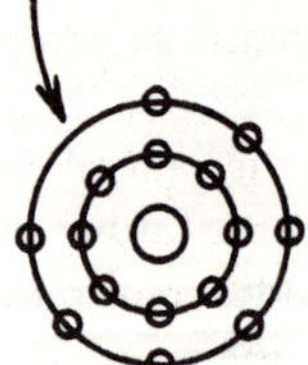

4.2.   Atoms of sodium and chlorine.

the reason for it in order to obtain a working knowledge of the nature of the valence bond.

4.2. Electrovalent (Ionic) Molecules. The sodium atom contains one electron that is attracted to a lesser extent by the positive nucleus than are all the other electrons. We have pictured this electron by itself at the edge of the atom. We know that this electron is held with only a slight attractive force and under certain conditions will completely depart from its parent, leaving one unbalanced proton in the nucleus.

The chlorine atom contains a nucleus which acts as if it were not completely satisfied by the number of electrons in the outer portion of the atom. If a loosely held electron comes within its radius of attraction, the nucleus of the chlorine atom will at once pull the stranger into its own territory to fill up its outer shell of eight.

When a molecule of sodium chloride is formed, the loosely held electron is drawn from the sodium atom by the greater attractive force of the chlorine atom and becomes a member of the chlorine family. The state of affairs in the resulting molecule may be pictured as shown in Figure 4.3.

Since the sodium atom has lost one electron in this process, one of the protons in its nucleus is left unbalanced. The resulting atom is therefore said to have a positive charge of one, and is a positive ion.

The chlorine atom, which has pulled away the loosely held electron from the sodium atom, now has one more negative particle than can be electrically balanced by its nucleus. The resulting atom is said to have a negative charge of one, and is a negative ion.

The sodium and chlorine atoms, as a result of this simple migration of an electron, thus become oppositely charged ions and hold on to each other by electrical attraction. Such a union between atoms is called *electrovalence* or *ionic valence*. This is the first of the two kinds of valence we started to discuss at the beginning of the chapter. (A more complete picture of the NaCl molecule will be found in Figure 8.1.)

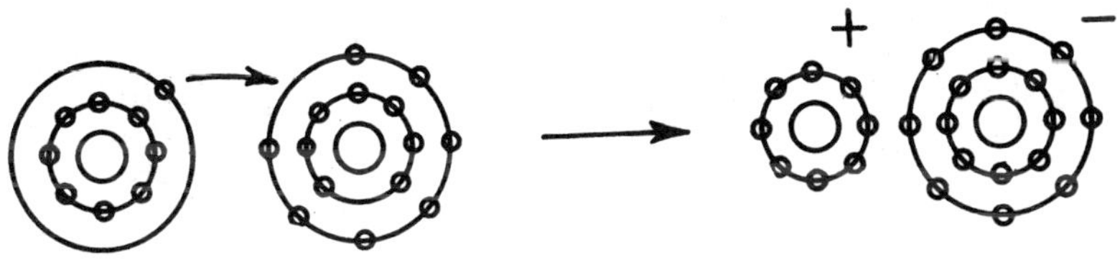

4.3.   Formation of an electrovalence bond. The molecule consists of ions, oppositely charged, electrically attracted.

4.3. Covalent Molecules. When two atoms of chlorine combine to form the Cl—Cl molecule the arrangement is much different from that in Na–Cl. As mentioned above the chlorine atom shows a marked tendency to pull an additional electron into its system. Two chlorine atoms can therefore help each other out by forming a close partnership as indicated in Figure 4.4.

In the chlorine molecule, each atom contains one more electron than it did before and thus its inherent tendency to add an additional electron is satisfied. The important fact, however, is that in the formation of the chlorine molecule no passage of an electron takes place from one atom to the other. It is a case of a mutual sharing of electrons. The new molecule is a complete unit in itself and does not consist of two oppositely charged parts.

To indicate the intimacy of this type of union between atoms and its spirit of partnership, we call it *covalence*. This is the second of the two kinds of valence. The student should constantly bear in mind that a *flow* of electric charge from one atom to another takes place in an electro-valence bond, whereas a mutual *sharing* of electric charge occurs in a co-valence bond.

4.4. Precursors of the New Valence Ideas. It is interesting to compare these two kinds of valence with the conception of valence as it developed from its earliest forms. The doctrine that compounds are composed of certain combinations of atoms was first stated in 1803 by Dalton (see Chapter 1). A few years later, Davy and Berzelius developed the first theory of a valence force between atoms, in which it was assumed that the atoms or parts of a molecule are held together by an electrostatic attraction. This, the reader will observe, is exactly in accord with our

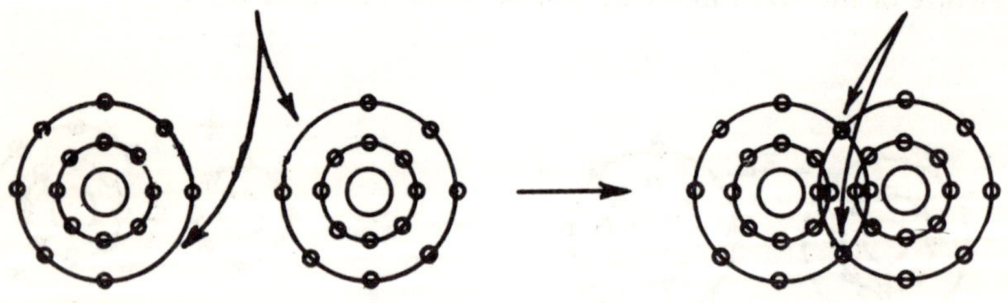

4.4.   Formation of a covalence bond. The molecule consists of atoms, sharing electrons.

modern idea of the electrovalence bond in Na^+Cl^-, for which we now have a good mechanical picture.

The difference between the old theory and the modern one is that in the the old one it was believed that certain elements were always positive and others always negative and they combined for that reason. Now it is believed they become positive or negative during the act of combining.

The electrical valence theory of Berzelius was satisfactory because most of the compounds for which the chemists of 1810 to 1830 had any detailed information were inorganic compounds, similar to sodium chloride in structure. They did not know enough about the composition of chlorine to get worried. If they had known, as we do now, that the chlorine molecule simply contains two atoms of chlorine (Cl–Cl), Berzelius would have been at a loss to explain how the two atoms are held together. He could not assume that one atom is positive and the other negative, since all atoms of chlorine are supposedly alike.

Later, when the composition of the chlorine molecule did become known, chemists realized that the electrical attraction theory of valence did not explain the structure of all compounds, and this became more evident after the period 1840 to 1850, when organic chemistry began its rapid expansion. It was found that most organic compounds (like the inorganic substance, chlorine) do not consist of oppositely charged atoms or groups; therefore, it was concluded that there must be some other kind of valence bond in addition to the simple attraction bond first announced by Berzelius. This second kind of valence is that which has just been described as covalence.

It is apparent, then, that our electrovalence bond is an old theory in modern dress; because of our knowledge of atomic structure, the actual mechanical process by which the atoms are linked together has become clearer. The covalence bond is of entirely modern origin; its presence in molecular structures is so universal that it has been called *the chemical bond*.

At this point, we must admit quite frankly that we are still content with a straight-line bond between atoms to indicate chemical combinations of all types. The straight-line bond was first used successfully in 1858 as a means of showing the positions of the various atoms in the molecule, as well as the combining powers of the atoms, and it is still a serviceable tool for the chemist.

5

The Unique Position of Carbon Among the Elements

5.1. The Periodic System of the Elements. In the preceding chapter we described how the chlorine molecule is built because we believe most organic compounds are covalent, like chlorine. But, the reader will observe, the chlorine molecule itself does not belong to organic chemistry. Organic chemistry, we learned in the very first chapter, is the chemistry of the carbon atom.

Let us see how carbon fits into the world of atoms and molecules. We shall consider the relationship of carbon to the other elements, and shall then be in a better position to explain why it is that carbon compounds are almost always *covalent* like the Cl–Cl molecule rather than *ionic* like the Na^+Cl^- molecule.

On the inside front cover of this book is the periodic table of the elements. To illustrate the usefulness of the table, notice the elements in the column labeled Group 1, where we find the closely related elements: lithium, sodium, and potassium. These have very similar chemical properties. For example, if we can prepare a compound like lithium nitrate, $LiNO_3$, then we know that there must also exist a potassium nitrate, KNO_3, with similar properties.

In Group 2, we find other families of related elements. At the bottom of the group the reader will notice the element radium. Now there are not many chemists who have handled either radium or its compounds, but any chemist with a working knowledge of the periodic system can give a fairly good account of its chemistry, simply from its similarity to the other elements in the family. Thus, the solubility of the sulphates in this family is in the following order:

$$MgSO_4 \qquad CaSO_4 \qquad SrSO_4 \qquad BaSO_4 \qquad RaSO_4$$

most soluble ←   → least soluble

Therefore, we know automatically that radium sulphate is the least soluble of all.

Uniform variations in properties, such as we have just described for the elements in Group 2, are also the rule for other columns of elements in the table.

5.2. Ionization Potential. The periodicity of the properties of the elements is shown strikingly by their ionization potentials. This property is the energy required to remove an electron from the outer shell of an atom, converting the atom to an ion as shown by the reaction $Na \rightarrow Na^+ + e^-$. This ionization is illustrated in Figure 4.3.

The relative ionization potentials of the elements are shown schematically in Figure 5.1, in which the vertical axis indicates the volts required to remove one electron from the outer shell. When we arrange the elements horizontally in the order of increasing atomic number, we periodically get back to elements with similar ionization potentials and, in general, similar properties. It will be seen that it is most difficult to ionize the noble gases in Group 0 of the periodic table, whereas the elements in Group 1 of the table, which we commonly call the alkali metals, are most easily ionized.

The relatively easy ionizability of metals like sodium is readily understood by referring once more to Figure 4.3. The positive charge in the

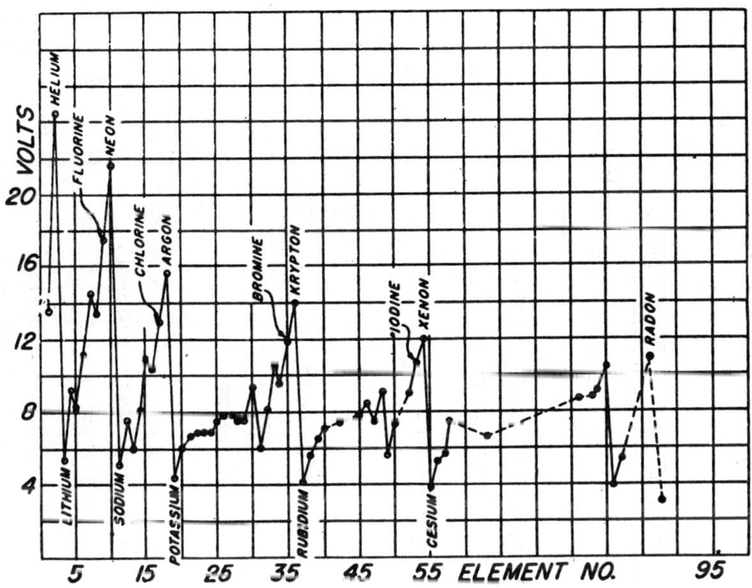

5.1 Ionization potentials of the elements.

nucleus pervades the entire atom, but this charge is so effectively screened by the inner shell of electrons that the effective positive charge at the surface of the atom is rather small, and the lone outer electron is easily lost.

Now let us look ahead a little and refer to Figure 6.1. Read across the row of elements from sodium to chlorine and note that the positive charge in the nucleus grows larger. The increase in positive charge will tend to pull in the shell of electrons and make the atom smaller, as illustrated in Figure 6.2. Since the positive charge of the chlorine nucleus is so much greater than that of the sodium nucleus, it will (even though screened by more electrons) exert a greater effect, so that there is a more positive charge on the surface of the chlorine atom. That is why we found in Figure 4.3 that the chlorine atom so readily attracts a negative electron, according to the reaction $Cl + e^- \rightarrow Cl^-$, to produce a chloride ion. The increased size of the chloride ion over that of the chlorine atom is apparent in Figure 6.2; this is the result of the expansion in the effective electron orbits when some of the positive charge due to the nucleus is offset by the one added electron.

The reasoning we applied to the sodium atom applies to the other elements in its group in the periodic table; they all have somewhat similar ionization potentials, as shown in Figure 5.1. In the same way, we can explain the similarity in behavior of the halogen family in Group 7 of the table. This will be commented on in greater detail in the next chapter.

5.3. Position of Carbon in the Periodic System. Now that we have reviewed the meaning of the periodic arrangement of the elements and

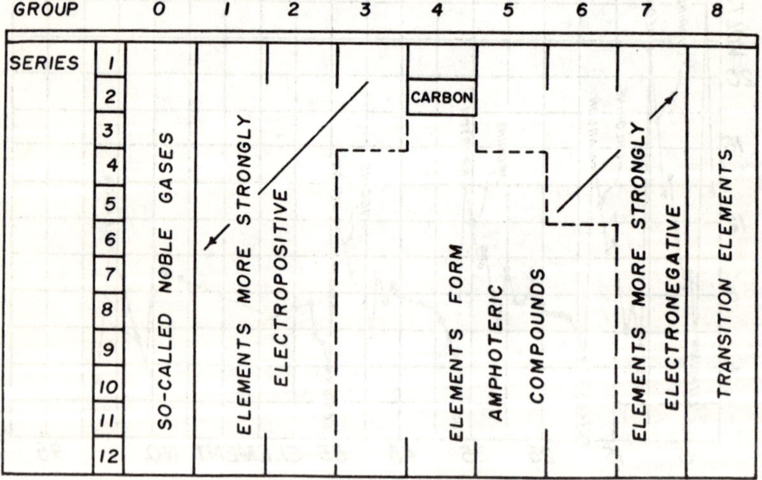

5.2. The periodic system outlined.

its usefulness in systematizing our knowledge, let us consider the position of the carbon atom in the table. To make the relationships of the elements more apparent we have constructed a skeleton outline of the periodic system as shown in Figure 5.2.

This brief outline indicates that with the exception of the noble gases in group O, which will be described in Section 6.4, the elements to the left of carbon are the electropositive elements. By this we mean that when they appear in molecules they tend to lose electrons to form the positive end of the molecule as explained in the case of sodium chloride in the preceding chapter. It was shown that in sodium chloride the sodium atom loses an electron easily and then has an excess of positive electricity:

$$Na^+ \qquad Cl^-$$

Positive end   Negative end

The more easily an element allows an electron to be taken from it by the opposing element in the molecule, the more electropositive it is said to be. The left-hand arrow in Figure 5.2 indicates the order in which the elements to the left of carbon become progressively more electropositive in character. These elements are said to be base-forming, as illustrated by sodium hydroxide, NaOH.

The other arrow indicates the variation in degree of electronegativity of the elements. The chlorine atom, we explained, has a positive charge on its surface, but we usually call it an electronegative element because it exerts a strong attractive force when pulling another electron into its atomic radius. Of course it does not actually become negative until it has acquired the electron, as shown in the formation of the molecule of sodium chloride. Just as the elements to the left of carbon in the periodic system of the elements are electropositive in character, the elements to the right of carbon are electronegative. The electronegative elements are also called acid-forming as illustrated by the molecule of hydrochloric acid, HCl.

The elements below carbon in the periodic table may be termed the weak-minded members of the family. Sometimes they have base-forming tendencies and enter into combination in the positive end of a molecule. At other times, under other conditions, they have acidic tendencies and appear in the negative portion of molecules. This is illustrated by aluminum, the oxide of which is *amphoteric* in that it reacts with either acids or bases:

$$Al_2O_3 + 6HCl \longrightarrow 3H_2O + 2AlCl_3$$

Aluminum
chloride

$$Al_2O_3 + 6NaOH \longrightarrow 3H_2O + 2Na_3AlO_3$$

Sodium
aluminate

The carbon atom cannot be definitely placed in any of the categories just described. It is considered neither electropositive nor electronegative, and its compounds rarely show amphoteric behavior (readers can profitably refer to Section 9.6 for a pertinent discussion of amphoterism.) Carbon occupies a key position among the elements. It is a transition substance, situated at the border line between the other recognized types. It has a set of distinctive properties all its own.

The outstanding tendency of the carbon atom is to *share* electrons with other atoms (covalence) in the formation of molecules. This will become more evident in the chapters that follow, especially when we apply our knowledge of covalence and electrovalence to the study of the polarity of molecules in Chapter 8.

6

The :Ö:C̈:T̈:Ë:T̈: in Chemistry

6.1. Historical Introduction. We have just seen how necessary it is to consider the elements as a big family in order to understand the position of the carbon atom among them. In the process of studying the big family we call the periodic system of the elements, we find that it is built up of a number of smaller family groups. These smaller families are set apart by a periodic arrangement in groups of eight, or *octet*. The reason for the octet of dots around each letter in the chapter heading will be clear when this chapter and the next have been read.

The periodic arrangement of elements in octets was first suggested about 1860. It has proved to be helpful for systematizing the knowledge and facts of chemistry as we have just indicated in the preceding chapter. It has also served as a basis for theorizing about the structure of the atom, as we shall explain in this chapter.

In 1897 the electron was discovered and it was shown to be a constituent of all atoms. Everything seemed bright and clear then to the scientific philosophers because all they would have to do was develop methods for determining how many electrons there were in each atom and how they were arranged in the atom. However, it was not evident how these electrons, or negative particles of electricity, were held together in the atom.

The answer to this last question was supplied in 1911 by Lord Rutherford. He proposed the theory that the atom is like the solar system in that it contains a positive nucleus corresponding to the sun and electrons which correspond to the planets. According to the modern version of this theory, the nucleus consists of protons and neutrons together with certain other entities that are being daily discovered and which act as a binding agent to condense the protons into a comparatively tiny space. The nucleus occupies only a very small space at the center of the atom, but serves to hold the electrons in their proper places.

This is the type of structure already introduced to the reader in Chapter 4, when we explained the new ideas on valence. In Figure 4 3 the models for the sodium and chlorine atoms were shown with positive nuclei, en-

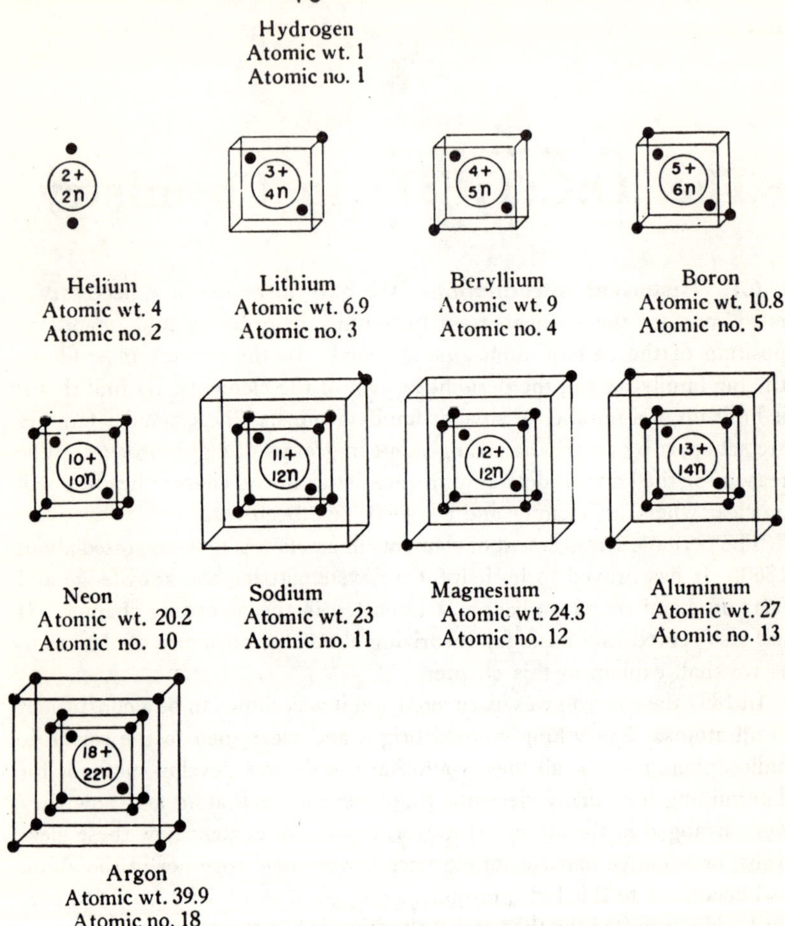

veloped by shells of negative electrons. We shall now consider this type of atom in detail.

6.2. The Cubic Atom. Determining the exact number of electrons in the atom of each element did not offer a very difficult problem. It was found that this quantity could be measured in several different ways. Finding out how the electrons are arranged in each element, however, proved to be much more difficult.

When the theoretical physicists and chemists got together to figure out what these arrangements are, it was quite natural for them to choose the periodic table of the elements as their guiding star. In short, they

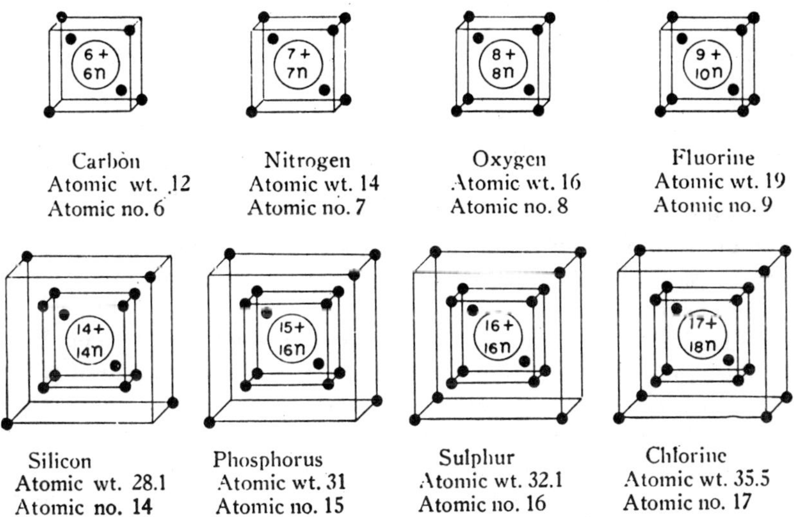

6.1. Cubic models of the first eighteen elements. Note the relation between the atomic weight of an element and the number of protons and neutrons in the nucleus.

The significance of the term atomic number will be discussed in detail in Chapter 37. It is apparent, however, that the atomic number is the number of electrons outside the nucleus, or the number of protons inside the nucleus.

decided that if every eighth element has similar properties then some outstanding principle may be discovered by arranging the electrons in the atom in such a way as to give paramount importance to a group of eight electrons. The noble gases of Group O until recently were all believed to be chemically inert; it was postulated they had zero valence because they had eight electrons in a peculiarly stable configuration, and that the elements of Group 7 tended to reach that stability of an octet by adding one electron, whereas the elements of Group 1 were assumed to reach the stable octet state by losing one electron (see Figure 4.3). This would explain why the Group 7 elements are electronegative and Group 1 elements are electropositive (see Figure 5.2).

A cube contains eight points symmetrically spaced. As far back as 1902 this figure suggested itself to Professor G. N. Lewis as a suitable hat rack on which to hang the electrons in "groups of eight." Cubic pictures like those in Figure 6.1 were finally arrived at in 1916. Figure 6.1 shows the first eighteen elements of the periodic system, which is as far as we can go in this introductory book.

The proton is indicated by a $+$ sign. The electron is generally indicated by a $-$ sign, but in these diagrams the electrons are more conveniently represented by black dots. The neutron is represented by the letter n and is a neutral particle whose weight is very close to that of the proton. The proton is nearly 2,000 times as heavy as an electron, so that the weight of an atom depends almost entirely on the number of protons and neutrons it contains.

The simplest atom is that of hydrogen; it is nothing more than one proton with one electron whirling around it. Hydrogen is the lightest of the elements, and in the tables of the elements it is assigned an atomic weight of 1. This tells us that the weight of the proton is one unit, since that is the only thing in the hydrogen atom with appreciable weight.

The second element, helium, is four times as heavy as hydrogen. Therefore, we know it must contain a total of four protons and neutrons. The protons and neutrons of the helium atom, as shown in the diagram, are massed at the center; this portion of the atom is called the nucleus. The nucleus of the helium atom contains two protons, which leaves two positive charges to be balanced by electrons outside the nucleus. These external electrons are called planetary electrons, because they (on paper at least) resemble the planets in motion around the sun.

Although hydrogen is the simplest of the elements, helium is the most stable. We shall refer to this again.

The third element, lithium, is seven times as heavy as hydrogen and contains three protons in the nucleus, which are balanced by external electrons. It appears that only two electrons can fit into the space immediately around the nucleus, so the third electron takes up a position slightly removed from the inner two. This third electron we have indicated in the diagram as placed at the corner of a cube, in accordance with our scheme of using a cube to represent the fundamental importance of the number eight.

Referring once more to the diagrams, we find that all the elements in the first horizontal series are constructed by drawing the outline of the *helium atom* and placing electrons at the corners of *the cube drawn around it.* For compactness the cubes are drawn around a helium structure rotated 45° from the position shown in the diagram of the helium atom itself. As we proceed from one element to the next, the nucleus constantly grows heavier, and there is a one-by-one increase in the number of electrons at the corners of the cube.

When the eight corners of the cube have been filled, that particular shell around the nucleus has its full quota. It will be observed that such a condition occurs when we reach the element neon, which is just below helium in the illustration and is the first element of the second horizontal series.

The next element in the second series is sodium, referred to in previous chapters. It contains one more electron than neon, but since there is no room for a ninth electron in the neon structure, this additional electron is found as the first member of still another cubic shell around the nucleus. The sodium atom contains three electronic shells—two electrons in the first, eight in the second, and one in the third.

The eighteenth element in the table is argon. This element is just below neon in the illustrations and is the first member of the next horizontal series. It has three filled shells.

Using the cubic atom, as we have done in this chapter, we get a much

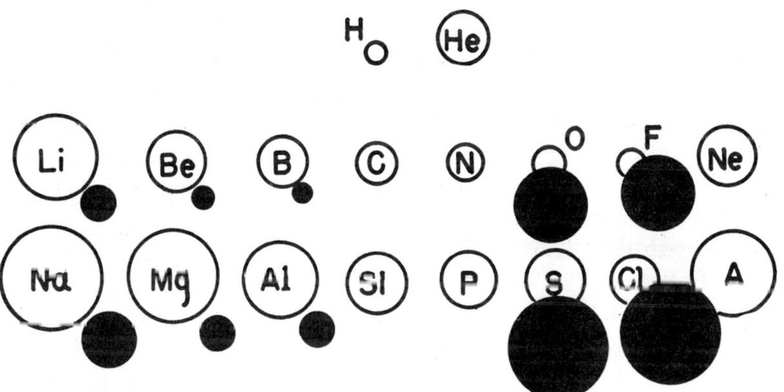

6.2. Relative sizes of atoms and ions. The ion is in black, below and to the right of the corresponding atom. [Adapted from a chart by J. A. Campbell, *J. Chemical Education*, **23**, 525 (1946) which shows most of the elements in the periodic system. A wall chart is available from W. M. Welch Scientific Co.]

better idea of the fundamental meaning of the periodic system. We see that beginning with helium the appearance of an additional shell occurs with rhythmic frequency at every eighth element. This is a periodic phenomenon, the periods in this case being octets. Also, if we inspect the vertical columns in the table we find that the external shells are similar and contain the same number of electrons. For example, lithium and sodium probably have similar chemical properties because they contain only one electron in the outermost shell. And the similar properties of fluorine and chlorine are due to the fact they contain seven electrons in the outer shell. The number of electrons in the outer shell is the same as the group number in the periodic system of the elements.

6.3. The Noble Gases.   We come now to the interesting group of elements, helium, neon, and argon, in the first vertical column of Figure 6.1. These are very symmetrical elements and very inactive. They are so inert, in fact, that they do not combine with themselves to form molecules like O_2, H_2, and Cl_2, but prefer to exist singly in the atomic state in the form of He, Ne, and A, which are gases. Because these elements are chemically inactive they have been called noble. This term was also applied to certain metals by ancient alchemists. For example, gold was called a noble metal because it does not tarnish (combine with oxygen) when heated in the air. Tin and copper tarnish readily and were called base metals.

Let us refer once more to the periodic table. There we find the noble gases all listed in the first vertical column with the heading Group 0 and assigned a zero valence because of their supposed lack of chemical reactivity. In Figure 5.1 we found the noble gases are the most difficult to ionize. If we glance again at the pictures of these elements in Figure 6.1, we can draw the inference (which is supported by many facts) that the inactivity must be due to their full quota of electrons. So far as electrons in the valence shell are concerned, these elements are completely satisfied. In 1962, however, it was found that some of the noble elements have "feet of clay." These are the larger, heavier elements below argon in the periodic table, with additional electron shells further removed from the nucleus. Xenon, for example, can form the compound XeF_4 with fluorine, the most electronegative of the elements; fluorine apparently is capable of causing at least partial removal of an electron from the valence shell of xenon. Reference to figure 5.1 shows that xenon has a considerably lower ionization potential than helium, neon, and argon. Krypton forms compounds, but much less readily than xenon.

6.4. The Octet Theory. Despite the fact that some noble elements have been found to be less noble than others, the basic conception in the electron theory of valence is that an atom is least active chemically when its external shell contains a full quota of electrons (an octet) and when the atom contains an equal number of protons and electrons. This is known as the octet theory, or the rule of eight.

We have indicated this theory in our chapter heading by the label :Ö:C̈:T̈:Ë:T̈: Each dot stands for an electron and each letter in the word is encased in a shell of eight electrons—the ideal state for an atom.

In the next chapter we shall find that atoms apparently go through chemical reactions with each other just for the sake of reaching that ideal state of the complete octet. It will also be seen that the process depends on a *pair* of electrons acting as a unit.

In the next chapter, therefore, in addition to the rule of eight, our attention will also be directed to the rule of two, or the :D:U:E:T:.

7

The :D:U:E:T: in Chemistry

7.1. Nucleus, Kernel, and Valence Shell. In the last chapter we emphasized the importance of the octet, or the rule of eight, in the chemistry of the atom, and the diagrams of the atoms in Figure 6.1 were given the shape of a cube in order to indicate a symmetrical arrangement of an octet of electrons in a valence shell. In this chapter we proceed to what is perhaps even more fundamental, namely, the duet of electrons, or the rule of two. It will be found that *the pair of electrons* is a unit referred to more frequently by the chemist, and *represents the modern idea of a chemical valence bond.*

As explained in the last chapter, the *nucleus* is the positively charged mass at the center of the atom, where practically all the weight of the atom is situated. In explaining the mechanism of chemical reactions, however, the chemist is less concerned with the nucleus than he is with the electrons around it. A moment's reflection will make it clear that the outermost electrons in an atom are probably those most intimately tied up with its chemistry. It is reasonable to suppose than when two atoms are brought into intimate contact their external electrons will be affected first.

Indeed, the modern theory of valence was originally based on the belief that the chemistry of an atom depends on the number and arrangement of the electrons in the shell farthest removed from the nucleus. For this reason, the external electron shell of an atom is called the *valence shell* and the rest of the atom beneath the *kernel.* Electrons in the valence shell are called the valence electrons; for example, there are three valence electrons in the atom of aluminum shown in Figure 6.1. Among the heavier elements, electrons of the inner shells may also sometimes engage in chemical combinations, but that need not concern us in this elementary study.

7.2. Valence Electrons as Dots. Since the chemist is primarily interested in the valence shell of the atom, let us examine first the valence shells of the noble gases. The noble elements which we have mentioned in this book are helium, neon, and argon, for which the cubic pictures may be written as follows, with special emphasis on the valence shell:

30

He 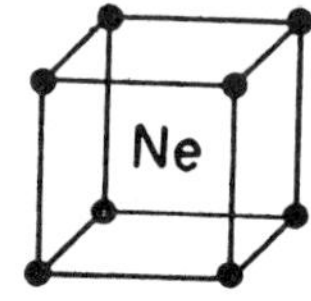 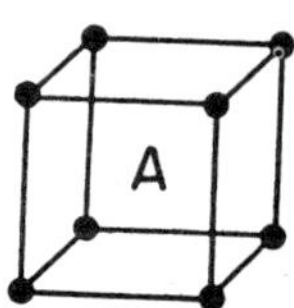

In these pictures, the eight electrons in the valence shell are placed at the corners of a cube. The rest of the atom inside the valence shell— that is, the nucleus and inner shells of electrons—which we call the kernel is represented by the chemical symbol of the element, such as Ne for neon and A for argon. The helium atom is remarkably inert (the two electrons shown in Figure 6.1 being nonreactive), so that we regard helium as having no valence shell at all. It is therefore simply indicated as a kernel by the symbol He. If the pictures we have drawn here are compared with those in Figure 6.1, it will be observed that the neon kernel looks exactly like the helium atom.

Although the electrons actually occupy planetary orbits, we have found it convenient to place them at the corners of a cube because in that way we correlate the electron theory with the periodic system (see Chapter 6). When atoms engage in chemical reactions, however, the electrons act in pairs, or groups of two, so that we find it even more convenient to place pairs of electrons at the corners of another symmetrical figure, the tetrahedron, as illustrated by the models in Figures 8.4 to 8.7. We can, however, indicate the electronic nature of the valence shell of an atom without giving the atom a definite shape at all. In short, the three noble elements just discussed can be represented in this simple manner:

$$\text{He} \qquad :\overset{..}{\underset{..}{\text{Ne}}}: \qquad :\overset{..}{\underset{..}{\text{A}}}:$$

Helium   Neon   Argon

The dots represent the electrons in the valence shell, and everything inside the valence shell is indicated by the chemical symbol.

If the first nineteen elements are written according to this scheme they appear as follows:

H·

He	Li·	Be.	B⫶	·Ċ·	·Ṅ⫶	:Ȯ:	:F̈:
:Ne̤.	Na·	Mg:	·Al:	·Ṡi·	·Ṗ:	:Ṡ:	:Cl̤:
·A̤·	K

In these symbols the pairs of dots represent the electrons that are "coupled" in certain subshells of the valence shell, a phenomenon which will be explained later in Section 7.8.

7.3. Tendency of Elements to Acquire Nobility. In the last chapter we stressed the fact that the noble gases have a *complete octet* of electrons in the valence shell. They are the only elements blessed with such a condition. If we assume that all the elements try to acquire an outer coat similar to that of the noble gases, we then have a very simple way of explaining chemical combination. The tendency for atoms to gain or lose electrons, to form a noble gas structure, is due largely to the charge on the kernel (Section 5.2 and Figure 6.1). The kernel charge of oxygen, for example, is 6^+ as compared with 4^+ for carbon. Therefore, oxygen has greater attraction for additional electrons by which to complete the octet and give the neon type of structure. Oxygen is said to be more electronegative than carbon. In fact, carbon stands in the unique position of either gaining four electrons to form the neon structure, or losing four electrons to yield the helium structure. It tends to share electrons rather than to acquire them or to lose them.

If the atom of chlorine gained one electron it would have a complete octet and would look like argon. In that way it would have the appearance of a noble gas, but of course it would still be only a chlorine atom with an extra electron, which is not balanced electrically by the nucleus. The resemblance is only skin-deep. Now examine the sodium atom. If it lost its lone valence electron, then its next lower electron shell would be uncovered and it would look like neon. This will be more evident if the reader consults the picture of the atom shown in Figure 6.1. Again the resemblance to a noble element is only skin-deep, and on the loss of the electron the atom of sodium becomes positively charged because of one unbalanced proton in its nucleus.

7.4. Electrovalence. The tendency for octet formation indicated in the preceding paragraphs for the two elements, chlorine and sodium, can be satisfied if they join each other in the formation of a molecule of sodium chloride:

$$\text{Na.} + \cdot \overset{\cdot\cdot}{\underset{\cdot\cdot}{\text{Cl}}}{:} \longrightarrow \text{Na}^+ : \overset{\cdot\cdot}{\underset{\cdot\cdot}{\text{Cl}}}{:}^-$$

In the resulting compound, both atoms assume the appearance of noble elements, but the transfer of the electron leaves one side of the molecule positively charged and causes the other end to be negatively charged.

This is the ionic type of molecule described in Chapter 4, and the bond between the atoms is an electrovalence bond.

Notice particularly that the valence bond consists of a pair (or doublet, or duplet, or duet) of electrons. The electron pair (:) takes the place of the straight line in the usual formula Na–Cl. It takes two electrons to constitute one valence bond. Each atom in this case contributes *one* of the electrons of the duet, and each of these atoms is said to have a valence of *one*.

Here is another example:

$$Cl\text{—}Mg\text{—}Cl \qquad\qquad :\ddot{C}l:^{-}\ Mg^{++}\ :\ddot{C}l:^{-}$$

Ordinary formula   Electron formula

In this salt the magnesium atom loses two electrons and is able to satisfy two chlorine atoms. The magnesium atom has a valence of two in magnesium chloride.

7.5. Covalence. It was difficult for a chemist to explain to a physicist how the two atoms of chlorine can combine to form a molecule of chlorine, Cl–Cl. Both atoms in this molecule are alike, and it does not seem likely that one becomes positive and the other negative just for the sake of mating. For this reason the covalent bond remained largely a matter of conjecture for several years after the usefulness of the electron theory of valence had become established through the efforts of chemists.

The physicists, however, took over with their calculations in quantum mechanics and showed that two like atoms can be held together by an electron pair with opposite spins. This spin is similar to that of the earth on its axis. If the two electrons in the pair are of the same directional spin the atoms repel each other. The formation of the covalent link is therefore as follows, the two electrons in the shared pair spinning in opposite directions:

$$.\ddot{C}l\cdot + .\ddot{C}l: \quad\longrightarrow\quad :\ddot{C}l:\ddot{C}l$$

In this case, no transfer of an electron is supposed to occur. The two atoms complete their octets by means of a close union in which the outer shells partially coalesce. The valence of chlorine is one, because each atom contributes one electron to the electron pair which is the valence bond. This type of union is covalent, as pointed out in Figure 4.4.

The formulas of a few other common molecules of the covalent type are as follows:

$$\begin{array}{ccc}
H & :\ddot{C}l: & :\ddot{C}l: \\
H:\ddot{N}:H & :\ddot{C}l:\ddot{B}:\ddot{C}l: & :\ddot{C}l:\ddot{C}:\ddot{C}l: \\
& & :\ddot{C}l:
\end{array}$$

Ammonia   Boron trichloride   Carbon tetrachloride

$(:\dot{N}\cdot + 3H\cdot)$ $(:\dot{B}\cdot + 3\cdot\ddot{C}l:)$ $(:\dot{C}\cdot + 4\cdot\ddot{C}l:)$

These compounds will be referred to again near the close of the next chapter.

7.6. Exceptions to the Rule of Eight. In the introduction to this chapter it was said that the rule of two is more important than the rule of eight. One of the numerous reasons for this statement is that there are many exceptions to the rule of eight (see index). An example is given in the preceding paragraph in the structure of boron trichloride, in which it will be seen that only six valence electrons on the boron atom are required to form the stable molecule. It is equally important to observe, however, that exceptions to the rule of eight are rare in the case of the element carbon among its nearly two million compounds.

7.7. Atomic Orbitals. To understand the mechanism of electron sharing, which was said in Section 7.5 to be the basis of the formation of molecules, and to understand why molecules have the characteristic shapes to be illustrated in Section 8.4, it is necessary to learn some of the details of electron habits that are not made apparent by Figure 6.1. We must also modify the concept that an electron is a "planetary" object rotating around the nucleus in an orbit like that of the earth around the sun. Instead of referring to an *orbit* we now introduce the term *orbital*, which means the region in space occupied by the electron.

The shape of the space (orbital) occupied by a so-called "electron cloud" can be calculated by wave mechanics; different electrons in a given electron shell may occupy orbitals of different shapes. The fundamental shapes of orbitals are familiar: a tennis ball (the spherical shape) and an hourglass (the nodal shape), as shown in Figure 7.1.

The crossover junction between the two halves of an hourglass is a node. In the atomic orbital we refer to a *nodal plane* which separates the two parts of the orbital and contains the nucleus of the atom. Thus, the nodal plane of the p_x orbital is the plane which contains the y and z axes in Figure 7.1. In the nodal plane, the probability of finding the electron is said to be zero.

The use of the word *node* in these discussions is similar to its use in phenomena familiar in everyday life. An example is the appearance of nodes in a vibrating string fixed at each end:

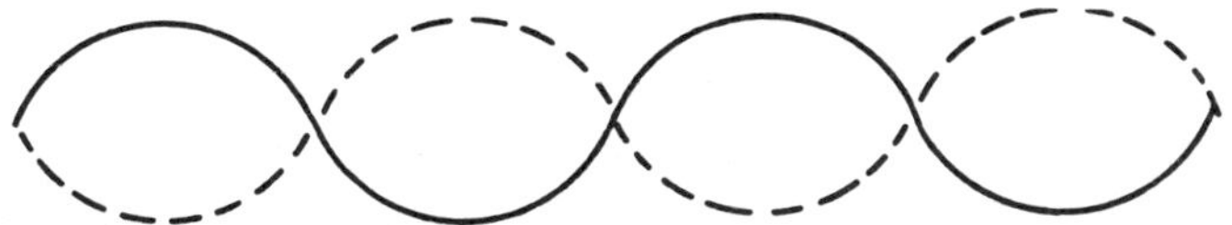

7.8. Energy Levels of Electrons. In the explanation of the pictures of the atoms in Figure 6.1 we referred to shells of electrons. We may also refer to those shells as energy levels. For example, the oxygen atom has two electrons in the lowest energy level (the first shell, principal quantum number 1), and six electrons in the next energy level (the second shell, principal quantum number 2). Larger atoms than those shown in Figure 6.1 have more electron shells, the principal quantum numbers ranging from 1 to 7. The shells may also be found labeled alphabetically from K to Q, a practice originally started by spectroscopists, who can follow by means of spectral lines the movements of electrons as they get bounced from one level to another when the atom is deliberately excited to an energy level higher than that of the normal "ground" state.

An important rule to remember is that the electrons in a given principal shell do not necessarily all have the same energy. There are subshells (sublevels of energy), the number of which is equal to the number of the shell. Thus, the second shell of the oxygen atom (principal quantum number 2) contains two subshells, but this was not indicated in Figure 6.1. The spectroscopists have labeled these two sublevels of energy with the letters s and p. An electron in the s energy level occupies a spherical orbital, as shown in Figure 7.1. The p energy level is higher than the s, and electrons in that level occupy a nodal orbital shown in Figure 7.1.

[Atoms larger than those we shall discuss have a wider variety of energy levels and orbitals. Thus, the fourth principal shell of a larger atom has four subshells of energy, labeled s, p, d, and f. The letter system was originated by spectroscopists to relate electronic energy levels to the characteristics of corresponding spectral lines, which are called *sharp, principal, diffuse, and fundamental.*]

Another important rule to remember is that the *square* of the principal quantum number tells us how many orbitals are permissible in an electron shell. The second shell of the oxygen atom has the principal quantum number 2; since $2^2=4$ there are, accordingly, four possible orbitals. The four orbitals are shown in Figure 7.1. Since the quantum number is 2 in the second shell, there are two levels of energy in that shell, according to the other rule cited previously; two of the electrons are in the s orbital and four are in p orbitals at a higher energy level.

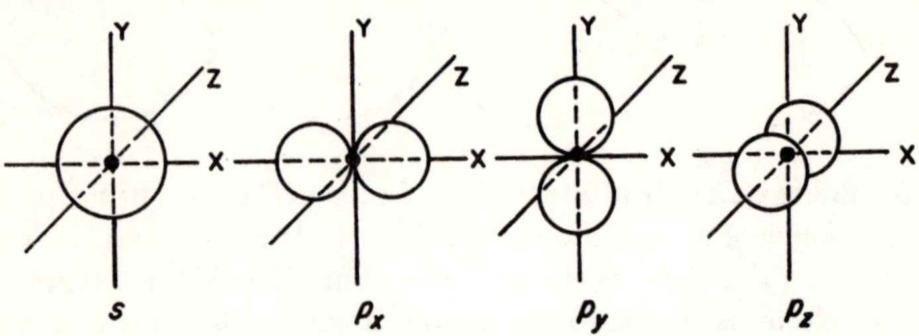

7.1. Representation of atomic orbitals. Black spot is a nucleus. The *s* orbital is spherical, without directional effect. The *p* orbitals are at right angles to one another—that is, 90° apart, on the conventional *x*, *y*, and *z* axes of solid geometry. Each of the *p* orbitals consists of two parts, with a nodal plane between them. All these orbitals can be occupied by a single electron or by two.

The nucleus is *enveloped* by an *s* orbital. It is *in a plane* separating the two parts of a *p* orbital. The two parts of a *p* orbital represent opposite phases of the wave function, calculated from quantum mechanics (compare Figure 15.9).

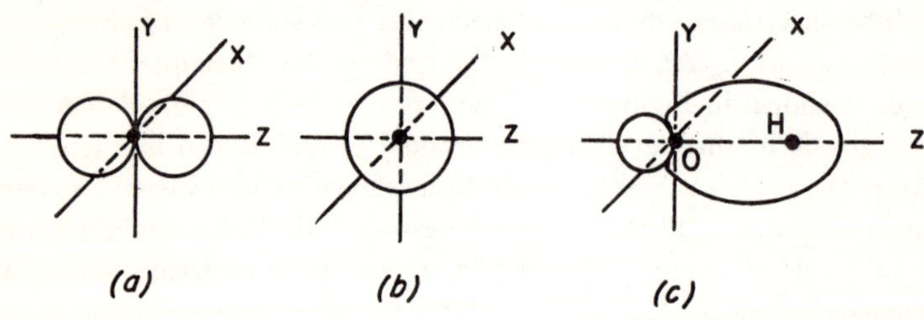

7.2. Formation of one of the $\cdot\ddot{O}:H$ bonds of water, $H:\ddot{O}:H$, through overlapping of *s* and *p* atomic orbitals. The black spot is a nucleus.

(*a*) The $2p_z$ atomic orbital of oxygen, occupied by one electron. See text (Section 7.9) for the reason why the z axis is drawn in the plane of the paper.

(*b*) The $1s$ atomic orbital of hydrogen, occupied by one electron.

(*c*) An *s-p* molecular orbital, occupied by two electrons with opposite spin. This is a σ type of bond.

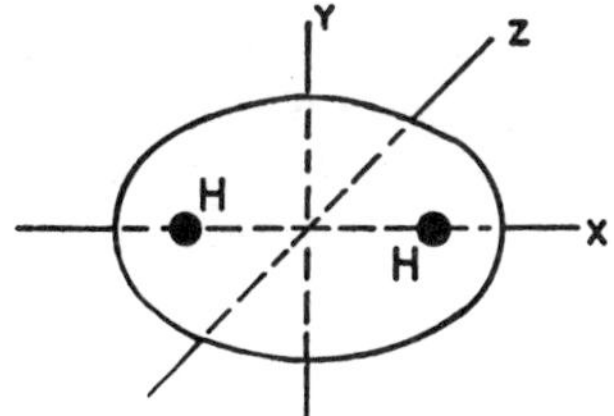

7.3. Formation of an *s-s* type molecular orbital, occupied by two electrons, through overlapping of two *s* type atomic orbitals (see Figure 7.2*b*) each with one electron. This is a σ bond.

H:H Molecule
(from H· + ·H)

We may now proceed to distribute the electrons of the oxygen atom, as follows:

$$1s^2; \; 2s^2, \; 2p_x{}^2, \; 2p_y, \; 2p_z$$

Boldface number is the principal quantum number. Letter shows type of orbital. Superscript shows number of electrons in orbital.

According to this shorthand expression the first shell of the oxygen atom has a pair of electrons; this shell has only one energy level (principal quantum number 1) and has only one orbital ($1^2=1$). That orbital is the *s* type in Figure 7.1 and contains both electrons. An orbital can hold no more than two electrons, and these must be of opposite spin—that is, affected oppositely by a magnetic field.

In the second shell, the *s* energy level fills up its quota of electrons before electrons can enter the higher energy level of the *p* orbitals. An electron already in a *p* orbital will not take a partner unless the other *p* orbitals have at least one each. All three of the **2***p* orbitals are occupied in the oxygen atom, with an electron pair in one of them.

To obtain a picture of the outer shell of the oxygen atom, the reader should visualize the six electrons distributed as just indicated and occupying orbitals as shown in Figure 7.1. The two *s* electrons completely encompass the nucleus. The four *p* electrons are in three orbitals at right angles to each other, and the probability of finding a *p* electron at the nucleus is zero.

[As an illustration of the complexity of larger atoms, the fourth electron shell of a large atom has four subshells (*s*, *p*, *d*, and *f*). Since $4^2=16$, it has sixteen orbitals (one *s*, three *p*, five *d*, and seven *f*). The space relations of the *d* orbitals are shown in Figure 30.1 Electrons in *f* orbitals do not take part in the formation of the simpler molecules.]

7.9. Molecular Orbitals. In reviewing the preceding remarks on *atomic* orbitals it should be understood from figure 7.1 that an electron in an *s* orbital has no directional effect; its motions are confined to a sphere encompassing the nucleus, and the probability of its occurrence is greatest at the nucleus and diminishes to a negligible value at the periphery of the sphere. An electron in a p atomic orbital is oriented in a definite direction, at an angle of 90° to other p orbitals; its occurrence probability is zero at the nucleus, and negligible at the boundary surface of the hourglass.

We can now use this information in building up molecules, and we start with the simplest—namely, hydrogen. The hydrogen atom has the electron distribution $1s$, which means that it has only one electron shell, that the orbital is s type, and that it has one electron in that orbital. Figure 7.2b shows the H atom and Figure 7.3 illustrates how two H atoms combine; the two nuclei act together and the spherical orbitals overlap to give a molecular orbital which is now eggshaped. The s-s bond is referred to as a σ type molecular orbital (σ, *sigma*, is the Greek letter s) because it resembles an s type atomic orbital which *encompasses* the nucleus.

In Section 7.2, the symbols of the first nineteen elements are written so as to show the paired and unpaired electrons in the outermost (valence) shell. Those symbols will be better understood by referring to a specific example, which again will be oxygen $:\overset{\cdot}{\underset{\cdot}{O}}:$ for which we have already used the electron distribution

$$1s^2;\ 2s^2,\ 2p_x{}^2,\ 2p_y,\ 2p_z$$

From this we see that in the **2** shell there are two orbitals already filled with electron pairs. These do not take part in reactions (except as described in Section 8.5). Two of the $2p$ orbitals, however, contain only one electron each and can take part in the formation of normal covalence bonds. Such bond formation is illustrated in Figure 7.2, where for the sake of clarity we have shown only one of the three $2p$ orbitals. We have also taken a liberty in that we have swung the z axis into the plane of the page in order to draw more clearly an orbital to which only one electron is allotted.

From the electron distribution in the oxygen atom we observe that the p_y orbital has only one electron, and can enter into bond formation just like the p_z orbital. The two OH bonds are shown in Figure 7.4, where it will be observed that the bond angle O—H should be 90°; deviations
$$\underset{\text{H}}{\overset{|}{}}$$
from this angle will be commented on in Section 8.4 Not shown in Figure 7.4 are the $1s$ orbital and the $2s$ orbital of the oxygen atom, which encompass the nucleus but contain satisfied electron pairs and do not take

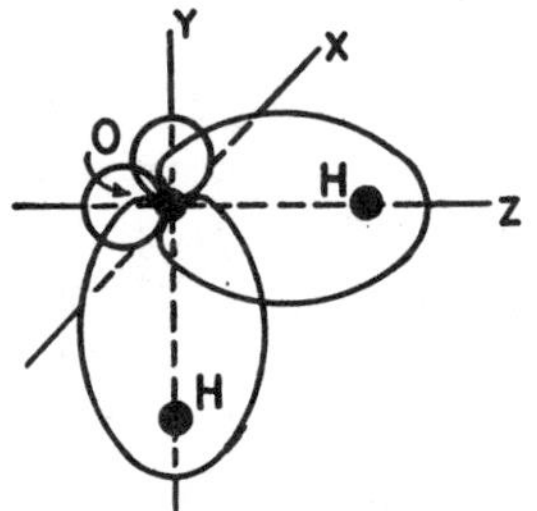

7.4. Formation of two *s-p* molecular bonds (see Figure 7.2*c*) at right angles. See text (Section 7.9) for position of third **2*p*** orbital and of the **2*s*** orbital. These are σ bonds.

$$\text{H:}\overset{..}{\text{O}}\text{:H Molecule}$$

$$(\text{from } \text{H} \cdot + \cdot \overset{..}{\text{O}} \cdot + \cdot \text{H})$$

part in this compound formation. We have also omitted the $2p_x$ orbital, because it already possesses an electron pair; it should be visualized, however, as a pure *p* type atomic orbital (hourglass) at right angles to the plane of the paper.

In this discussion we have illustrated how the lobes of atomic orbitals can blend to form a molecular orbital. The examples cited are called *bonding* orbitals; the electrons enter into a lower energy state in which there is greater stability. According to orbital theory, when atomic orbitals have the proper energy characteristics to permit a union they form not only the bonding orbital but also one that is *antibonding*. This orbital represents a higher energy level, and if an electron is in that orbital it contributes to instability of the molecule (see Section 15.6 and Figure 15.8). The electron will return to the more stable level, if permitted.

7.10. Hybridized Atomic Orbitals. The *s-p* molecular orbital in Figure 7.2*c* is made from two unlike orbitals (*s* and *p*), as contrasted with the *s-s* molecular orbital shown in Figure 7.3. The mixed *s-p* orbital has the hourglass feature of a *p* atomic orbital, with a nodal plane at the nucleus (in this case the plane of the *y-x* axes); it also shows the characteristic of an *s* atomic orbital, which encompasses the nucleus. This is illustrated in the diagram by placing both nuclei (O and H) inside the egg-shaped part of the orbital.

An important and very basic principle in orbital theory is that orbitals of *the same atom* can blend their energy and characteristics. This is called hybridization. The valence shell of the carbon atom (see Figure 6.1) is the second shell; the electron distribution for both shells according to the scheme outlined in Sections 7.8 and 7.9 is written

$$1s^2; \quad 2s^2, \; 2p_x, \; 2p_y$$

The pair of **1*s*** electrons can be disregarded in the chemistry of carbon. The valence shell has four electrons, but they are not equivalent; two of them are already paired and have energy corresponding to a $2s$ orbital, and the other two are in the somewhat higher energy level of a $2p$ orbital. The problem is to explain how the four electrons become equivalent, which is a condition we should expect on the basis of the chemistry of carbon; it nearly always has a valence of four, as in compounds such as methane, CH_4, carbon tetrachloride, CCl_4 etc. We shall use for illustration the methane molecule (Chapter 9).

The electron distribution just cited is for the *ground* state of the carbon atom—its most stable state. When a C atom reacts with four H atoms there is a considerable evolution of energy (Section 16.1), more than enough to promote the C atom to an "excited" state in which the pair of s electrons is "uncoupled" and one of them raised in energy to the p level. The electron distribution in the valence shell then becomes

$$2s, \ 2p_x, \ 2p_y, \ 2p_z$$

In the following discussion, in order to avoid confusion, we shall omit the use of the quantum number **2**.

The four electrons, of which there are two kinds (s and p), now pool their energy, with the creation of four electrons all of which are equivalent and are designated sp^3 to show they are hybridized from one s and three p electrons. In other words, there are now four sp^3 orbitals, all of which are equivalent. The hybrid sp^3 atomic orbital, as seen in Figure 7.5b, has the characteristics of the spherical s orbital of Figure 7.5a and of the

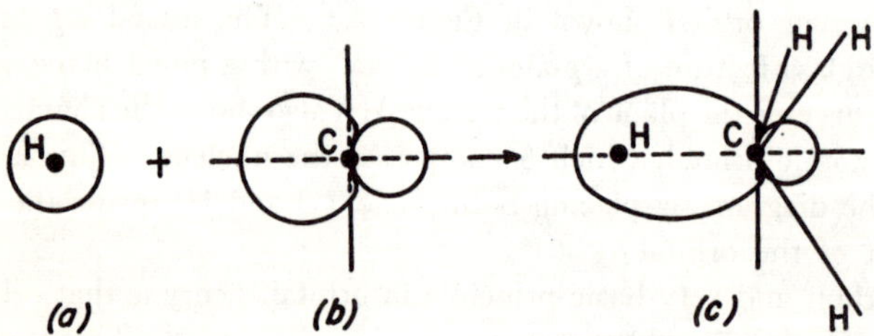

7.5. Formation of a $C{:}H$ orbital of methane, CH_4. Similar s-sp^3 molecular bond orbitals are formed with the three other H atoms shown. This is a σ type orbital.

(a) The hydrogen atom, with one electron in an s atomic orbital.
(b) One of the four sp^3 hybrid atomic orbitals of carbon, with one electron in the orbital.
(c) One of the four molecular orbitals constituting methane. It contains a pair of electrons.

TABLE 7.1. Hybridization of Orbitals in the Carbon Atom

The ground state	s^2	p_x	p_y		
Excited state	s	p_x	p_y	p_z	
The s orbital hybrid-ized with					**Example**
One p orbital	sp	sp	p	p	Ethyne (Section 15.5)
Two p orbitals*	sp^2	sp^2	sp^2	p	Ethene (Section 15.4)
Three p orbitals*	sp^3	sp^3	sp^3	sp^3	Methane (this section)

* The superscript shows how many of the p orbitals pool their energy with the s orbital.

hourglass p orbital shown in Figure 7.2*a*. The symbol sp^3 is read *ess-pee-three*.

Since all four of the sp^3 hybrid orbitals of the carbon atom are equivalent, in the molecule of methane (Figure 7.5) the four C:H bonds radiate from the nucleus of the carbon atom at equal angles of 109° 28′. The solid figure is a tetrahedron which will be described in greater detail in Section 12.2 and may be referred to at this time.

The types of hybrid orbitals possible in the carbon atom are summarized in Table 7.1. Hybridization in the atom is simply a process by which the several atomic orbitals reach an average energy level; for example, one s orbital and one p orbital will behave like two sp orbitals, and this kind of atomic orbital is present in certain carbon compounds.

7.11. Quantum Theory and Organic Chemistry. Some of the principles of quantum theory have been translated by able chemists into concepts that can be appreciated by lay chemists with little or none of the basic training needed to study the original formulations. The discussion in this chapter utilizes only enough of this elementary quantum theory terminology to explain the characteristics of some carbon compounds which will be described later in this book (for example, see benzene in Section 19.4).

8

North and South Poles

8.1. Nonpolar, Polar, and Ionic Molecules. This is the last chapter in our elementary study of the structure of atoms and molecules. In preceding chapters we gave most of our attention to the kinds of valence bonds that hold the atoms together in the molecule. In this chapter we shall consider the molecule as a whole and find out how the type of valence bond affects the properties of the molecule.

There are two kinds of valence—namely, electrovalence and covalence. There are three principal types of molecules, and their general nature can be illustrated by the following examples:

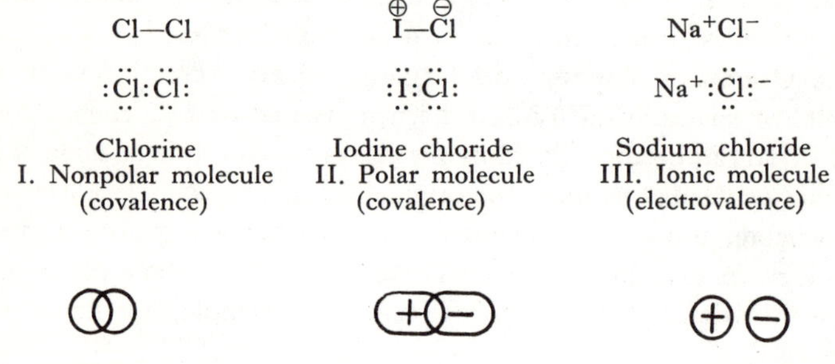

Chlorine	Iodine chloride	Sodium chloride
I. Nonpolar molecule	II. Polar molecule	III. Ionic molecule
(covalence)	(covalence)	(electrovalence)

In a covalence bond, as we have already learned, the atoms share electrons in such a way that there is no definite transfer of electrons from one atom to the other. The compound we have constantly used to illustrate this phenomenon is chlorine, Cl–Cl. This compound is said to be *nonpolar*.

The iodine chloride molecule, I–Cl, contains two atoms belonging to the same chemical family. Chlorine, however, is more electronegative than iodine. It exerts a stronger pull on the electron pair (:), which is the valence bond. If it is desired to emphasize this *polar* nature of a molecule like I–Cl, its formula is written in this book as shown in the preceding

42

examples, with $\oplus$ and $\ominus$ charges on the proper atoms ($\delta+$ and $\delta-$ signs are also sometimes used). The molecule may be regarded as having a north pole and a south pole, and consequently is also known as a *dipolar* molecule. The use of uncircled $+$ and $-$ signs is restricted in this book to ions, as in the formula of sodium chloride.

The third principal type of molecule is the *ionic* variety. In this case, there is a complete transfer of an electron from one atom to the other, as in the molecule of sodium chloride. The structure of this molecule is to be represented by two ions, Na^+ and Cl^-, held together by their opposite electrical attraction. In a crystal of salt there really are no NaCl molecules; instead there is a lattice-work of Na^+ and Cl^- ions, as shown in Figure 8.1. This is a model of a crystal of ordinary table salt, as determined by means of X-ray photographs. If the black balls are sodium ions the white balls are chlorine ions, or vice versa. The symmetrical way in which these simple ions arrange themselves at right angles to each other in space indicates why ordinary table-salt crystals are perfect cubes.

A crystal of salt is a giant molecule in which one Na^+ ion affects six Cl^- ions, or one Cl^- affects six Na^+ ions. Each of these ions is said to have a coordination number of six. The coordination number depends on the size of the ion—that is, on the number of oppositely charged ions with which it is surrounded.

Figure 8.2 is a model of silver bromide $Ag^+ Br^-$ which has the same type of crystal structure as sodium chloride. This model was built to scale, showing the relative sizes of the ions in the crystal (the small white balls represent silver ions and the larger, black balls are bromide ions).

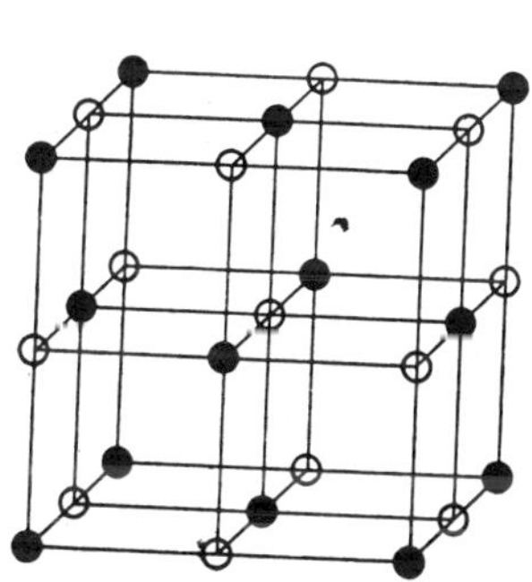

8.1. Arrangement of ions in the sodium chloride crystal.

8.2. Arrangement of ions in the silver bromide crystal [E. A. Hauser, *J. Chem. Education*, **18**, 165 (1941)].

Although the ions are mostly empty space as indicated by Figure 4.2, the sizes of the white and black balls show the relative effective distances from the positive nuclei of the ions to the outer electrons. Observe that each ion (as in Figure 8.1) has a coordination number of 6.

We have described the structures of ionic, polar, and nonpolar molecules before considering their properties; actually, the chemist has developed his knowledge of their structures from a study of their properties, some of which will now be explained.

8.2. Dielectric Constant. The dielectric constant is a simple physical property that has served as a useful tool in recent years for the study of the nature of molecules. In Figure 8.3, imagine first that the charged plates are separated by a vacuum, and that the strength of the electric field between the plates is E_{vac}. If a vaporized substance is placed between the plates, the value of E will fall because of the better conducting ability of the molecules than that of a vacuum. If polar molecules are employed they will swing into the positions shown in the diagram. It is said that the molecules are *oriented*. The electrical effect of the molecules is then opposite to that on the two charged plates and the strength of the electric field is reduced from E_{vac} to E_{mol}. The ratio E_{vac}/E_{mol} is called the dielectric constant.

The word dielectric conveys the idea of transmitting or holding an electric force without actually serving as a conductor. Polar compounds have a high reducing effect on the field strength between the plates and, consequently, have a fairly high dielectric constant. This is illustrated further in section 8.12.

Even nonpolar compounds have detectable dielectric constants. When such a compound as chlorine (see diagrams on p.42) is placed between oppositely charged plates, the two parts of the molecule are distorted by the

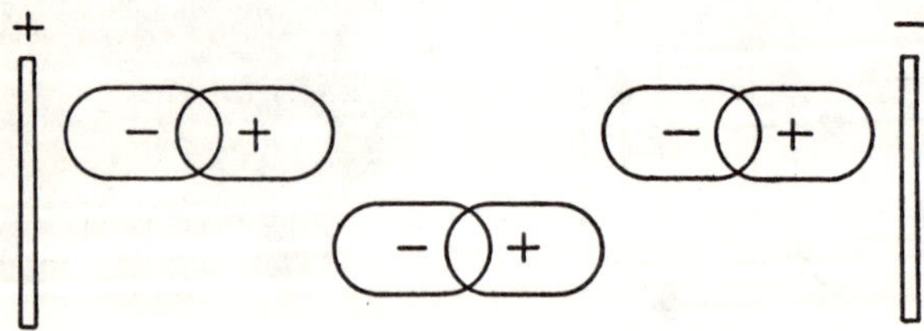

8.3.　Polar molecules between charged plates.

electric field and assume an "induced dipole" state. These molecules then become oriented as shown in figure 8.3, and reduce the field strength. Although the dielectric constant is appreciable for nonpolar compounds, it is of a very low order. It is only with difficulty that the component parts of such a molecule can be distorted to the state of an induced dipole.

The dielectric constant of a polar molecule is due to two principal factors: first, the *permanent dipoles* already present in the molecule, and second, the *induced dipoles* caused by distortion. It is the first factor that brings about the higher dielectric constant of polar compounds. Effects due to induced dipoles are very small compared to those caused by natural dipoles in the molecule.

Ionic compounds, of course, have permanent dipoles and should have high dielectric constants. Since most salts are nonvolatile, it is not possible to measure their dielectric constants by the vacuum method we have illustrated. Certain organic solvents, like benzene, however, are nonpolar and can often be used in place of the vacuum for dielectric measurements.

8.3. Dipole Moment. The dipole moment of a molecule is calculated from dielectric constant measurements. The meaning of dipole moment is quite simple although the method of calculation may be rather complex. In a molecule that consists of two poles the dipole moment is obtained by multiplying the distance d between the poles by the electric charge e on one pole. The electric charge, of course, is the same on each pole athough of opposite sign. The dipole moment, accordingly, is de.

The average value of d, the distance across a molecule, is 10^{-8} cm, and the charge on one electron is 4.80×10^{-10} esu (*esu* is the electrostatic unit of electricity, the quantity repelled by a force of one dyne when placed one centimeter from an equal charge in a vacuum). Accordingly, the dipole moment (de) of a highly polar molecule is in the vicinity of 4.80×10^{-18} esu-cm, whereas for a nonpolar molecule it is either zero or close to zero.

The dipole moments of several types of compounds are shown in Table 8.1, where the data are in Debye units (one such unit is equal to 1×10^{-18}

TABLE 8.1. Dipole Moments (in Debye Units)

Hydrogen, H_2	0.0	Hydrogen chloride, HCl	1.04
Carbon dioxide, CO_2	0.0	Hydrogen bromide, HBr	0.79
Carbon disulphide, CS_2	0.0	Hydrogen iodide, HI	0.36
Carbon tetrachloride, CCl_4	0.0	Iodine	0.0
Chloroform, $HCCl_3$	1.15	Bromine	0.0
Water, H_2O	1.85	Chlorine	0.0
Ammonia, NH_3	1.55	Silver perchlorate, $Ag\,ClO_4$	4.7

esu-cm).

When we properly interpret the data in this table, we get some idea of the immense value of dipole moments as a research tool in solving the problems of molecular structure.

8.4. The Shapes of Molecules. The H_2 molecule has a zero moment and is, therefore, a perfectly symmetrical compound, H:H, with no displacement of the electron pair toward either atom. If we inspect the data for the series of compounds HCl, HBr, and HI, it will be noticed that the moment is greatest for HCl and least for HI. This is qualitatively a measure of the relative electronegativity of chlorine, bromine, and iodine —that is, their displacement effect on the pair of electrons which bonds them to hydrogen.

In Section 7.8 it was shown that certain valence electrons are restricted in their movements to definite positions in space with respect to the nucleus. These space positions (p orbitals) are 90° apart on conventional x, y, and z axes. Valence-bond formation through p orbitals accounts for the fact that bond angles are not far from 90°, as shown in Figure 7.4. In Section 7.10, however, we learned that the CH_4 molecule has a highly symmetrical structure due to hybridization at the carbon atom; one s orbital and three p orbitals of the C atom pool their energy to form four sp^3 orbitals which radiate from the carbon nucleus at the tetrahedral angle of 109° as illustrated in Figure 12.2. If a molecule is found to have its valence bonds at close to the tetrahedral angle it is likely that the hybridization at the central atom is the sp^3 type.

For the ammonia molecule, NH_3, which is built up on the atom $:\overset{\cdot}{\underset{\cdot}{N}}\cdot$,

the student should be able to apply the principles of bond orbitals employed in Figure 7.4. The electron distribution in the valence shell of the nitrogen atom is

$$s^2, \; p_x, \; p_y, \; p_z$$

Single electrons are available in all three p orbitals, so that an N–H bond can be formed on all three axes of Figure 7.4. The measured H–N–H bond angles are about 107°, closer to the tetrahedral angle of 109° than to the right angle of 90° in Figure 7.4, so it is probable that the molecule is tetrahedral as in Figure 8.6, and that there is hybridization at the N atom to sp^3 orbitals, with the lone electron pair in an unbonded sp^3 orbital. If ammonia had the flat structure

$$\text{H}-\text{N}\Big\langle\begin{array}{l}\text{H}\\\text{H}\end{array}$$

the polar valences would balance to give a zero dipole moment. The high dipole moment which it actually has is in accord with Figure 8.6. The lone pair of electrons acts as a group (see Sections 29.6 and 30.6).

The water molecule appears to be a straight-line structure as normally written, and it also appears to be a right-angle molecule when pictured as in Figure 7.4. However, the H:O:H bond angle is about 105°, which is closer to the tetrahedral angle of 109° than to the right angle of 90°. The molecule has a tetrahedral structure as in Figure 8.5, with lone electron pairs occupying unbonded sp^3 orbitals, but the deviation from the theoretical 109° bond angle of H–O–H should be explained. One suggestion is that orbitals with H^+ nuclei at the ends are stretched due to electron attraction by the H^+, whereas the orbitals with lone pairs of electrons are puffed out and tend to squeeze the two OH orbitals to the smaller angle. Another suggestion is that when s and p orbitals hybridize the angle between the bonds gets larger as the percent of s character increases; for sp^3 orbitals the angle is the tetrahedral 109°, for sp^2 orbitals it is 120°, and for sp orbitals it is 180°. The angle in water may be partially accounted for on the basis that it has enough s character in its hybrid orbitals to raise the angle from the 90° in Figure 7.4 to the 105° as actually measured. A third factor to be taken into consideration is the repulsion by the positive H nuclei at the ends of the OH bonds; this would tend to widen the bond angle of H–O–H.

The valence bonds in carbon dioxide are polar, and on that basis it

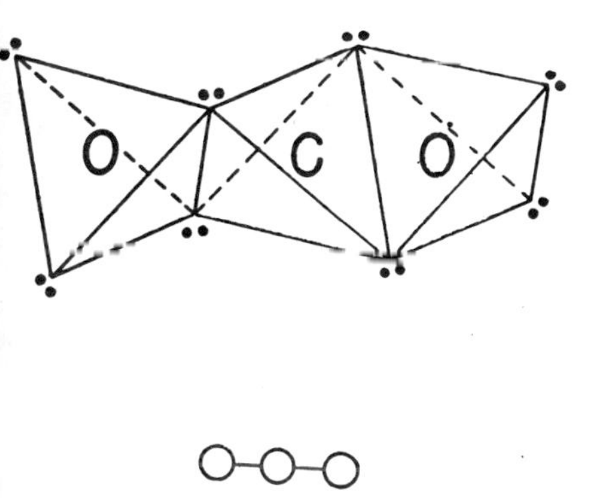

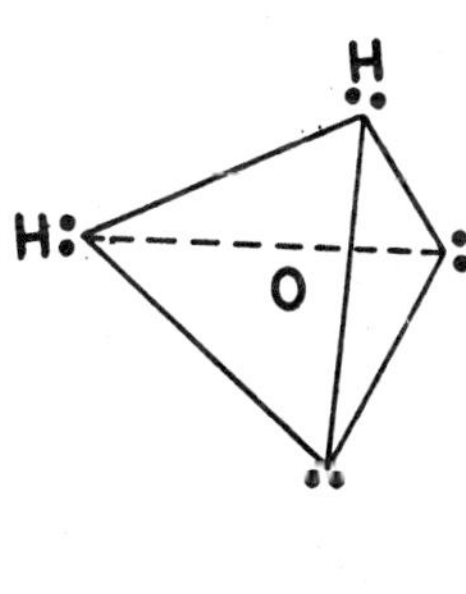

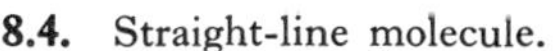

8.4.   Straight-line molecule.   **8.5.**   Angle molecule.

should be expected to have a fairly high dipole moment; however, the absence of dipole moment is proof that the three atoms are arranged in a straight line as shown in Figure 8.4. There are actually two dipoles, $O{=}C$ and $C{=}O$, but they act in opposite directions and the net dipole moment of the molecule is zero.

From the hybridization data in Table 7.1 and the method of orbitals described in Section 7.9, we can give a more convincing account of the straight-line structure of carbon dioxide. However, it is best to postpone that demonstration (see Section 16.6) and to employ the useful convention of the tetrahedral models shown in Figure 8.4. These models are based on the rules that (1) the atoms in a compound, generally, but not always, attain a cloak of eight electrons, and (2) the electrons act in pairs when forming compounds and are generally symmetrically situated.

The CCl_4 molecule shown in Figure 8.7 has a zero dipole moment, whereas the dipole moment of $CHCl_3$ is 1.15. From this information, it is safe to assume that the C–Cl bond is polar, but in CCl_4 the four polar

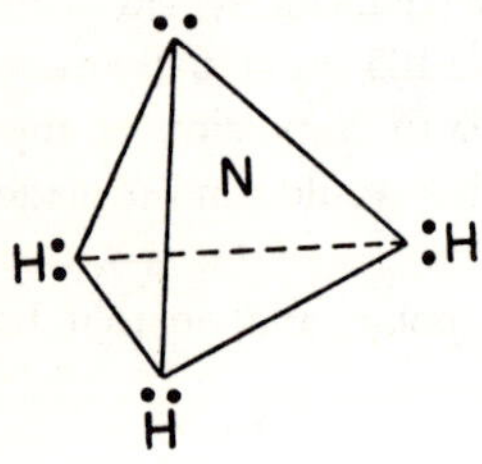

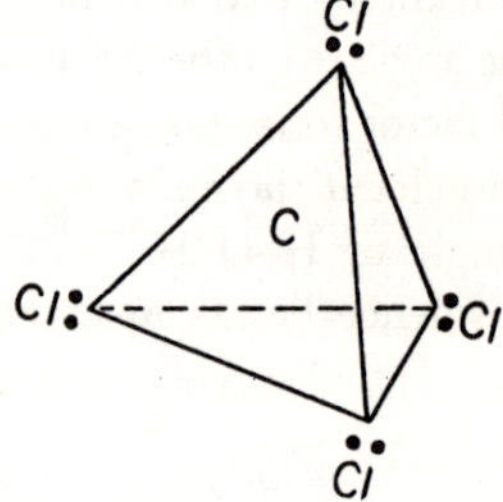

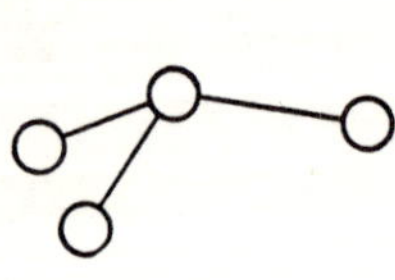

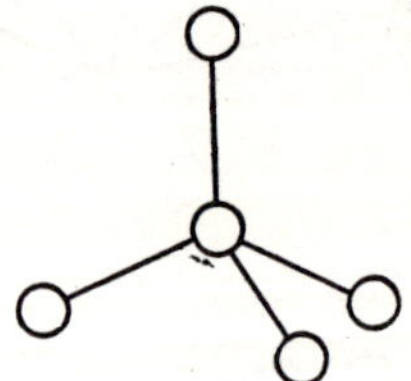

8.6. Pyramid molecule.   **8.7.** Tetrahedron molecule.

valences balance one another because of the symmetrical nature of the molecule. The $CHCl_3$ molecule is necessarily unsymmetrical and consequently has an appreciable moment; the Cl–C–Cl angles are close to 112°.

8.5. The Dative Bond. In Section 7.5 we wrote electronic structures for ammonia and boron trichloride. These two compounds unite easily to form a stable substance the structure of which is readily explained on the basis of electron theory.

$$
\begin{array}{ccc}
\text{H} & :\ddot{\text{C}}\text{l}: & \text{H} & :\ddot{\text{C}}\text{l}: \\
\text{H}:\ddot{\text{N}}: + & \text{B}:\ddot{\text{C}}\text{l}: \longrightarrow & \text{H}:\ddot{\text{N}}: \text{B}:\ddot{\text{C}}\text{l}: \\
\text{H} & :\ddot{\text{C}}\text{l}: & \text{H} & :\ddot{\text{C}}\text{l}:
\end{array}
$$

Ammonia Boron
trichloride

The N atom in NH_3 has two unshared electrons, whereas the B atom in BCl_3 has room for two more in its valence shell. The N atom donates the electron pair, and is called the *donor* atom. The B atom is the *acceptor*.

In a normal covalence bond, such as in Cl:Cl, one electron comes from each atom. In $H_3N:BCl_3$ both electrons in the bond come from the same atom; this is sometimes called coordinate valence and is a special kind of covalence. However, the bond is more often referred to as a dative bond, from the Latin verb for *give*, and the bond is designated by an arrow sign, thus, $H_3N \rightarrow BCl_3$, to show which atom furnishes the electron pair. Another scheme is to employ the usual straight line bond, with charges on the appropriate atoms, $H_3\overset{\oplus}{N}—\overset{\ominus}{B}Cl_3$. Phenomena explained by means of dative bonds will be found listed in the Index.

8.6. Formal Charge. The dative bond is polar; the molecule it builds is a dipole, but it must not be mistaken for an ionic compound. There is only a partial transfer of electric charge, not complete transfer. With polar bonds of this kind it is customary to speak of formal charges on the atoms concerned. For example, in NH_3 the N atom normally possesses five electrons (two in its unused pair, and three as its half-share of the electrons in the three N:H bonds). In the $H_3N \rightarrow BCl_3$ molecule, the N atom possesses only four electrons (its half-share of four electron pair bonds). A net loss of one electron gives the N atom a charge of $+1$, which is its formal charge. Similarly, the B atom can be calculated to have a formal charge of -1.

8.7. Mobile Electrons. The molecular structures described in Section 8.4 are due to pairing of electrons that become *localized*. Later we

Table 8.2. Electronegativity Scale of the Elements

Cs	0.7	H	2.1
Na	0.9	I	2.5
Li	1.0	S	2.5
Mg	1.2	C	2.5
Al	1.5	Br	2.8
Si	1.8	Cl	3.0
Pb	1.8	N	3.0
Hg	1.9	O	3.5
B	2.0	F	4.0

shall describe molecular systems in which certain of the electrons have considerable mobility and are called *delocalized*. This will be discussed principally in connection with the theory of resonance (Section 16.9). It will also be referred to in the study of the structure of benzene and of graphite (Section 19.7), where the mobile π electrons have much importance. The electrical conductivity of graphite is made possible by these mobile electrons.

The electron configurations in metals have also been made clear by the principles of wave mechanics that were briefly reviewed in Chapter 7. A metal is a lattice of positive ions. The electrons in the space around these ions can occupy certain energy levels. If the valence levels are not filled the electrons become mobile and the metal is an electrical conductor. If all the levels are filled the substance is an insulator.

8.8.   The Electronegativity Scale.   In Section 5.2 it was explained that the electric charge of the chlorine atom is really positive and for this reason strongly attracts electrons. It is therefore called strongly electronegative. Conversely, the sodium atom has only a weak positive surface charge, and since its attraction for electrons is feeble it permits electrons to be stolen rather easily. We should call sodium a weakly electronegative element, but in practice it is called strongly electropositive.

A roughly quantitative scale of electronegativity is shown in Table 8.2, which is based principally on the work of Pauling. The elements in this table should be looked up in the chart on the inside front cover of this book, and their positions studied with reference to Figure 5.2.

8.9.   Partial Ionic Character of Covalent Bonds.   We have already seen that HCl is covalent, as in I, and that its dipole moment shows it has a dipolar nature as depicted in II, but the molecule also has characteristics of the ionic molecule in III. This is not to be construed as meaning that the HCl molecule is occasionally ionic, but that the *covalent bond* in HCl has *partial ionic character*. The molecule acts as a hybrid which can vary in accordance with its environment (that is, with the solvent in which it is placed, etc.). This will be better understood when the phenomenon of resonance is discussed in Section 16.9.

$$\text{H}:\ddot{\text{C}}\text{l}: \qquad \overset{\oplus}{\text{H}}\!-\!\overset{\ominus}{\text{C}}\text{l} \qquad \text{H}^+\text{Cl}^-$$

$$\text{I} \qquad\qquad \text{II} \qquad\qquad \text{III}$$

The calculation of the degree of ionic character of HCl can be made as follows. From spectroscopic measurements, the distance between the atoms is 1.28 A, and if the molecule is completely ionic the dipole moment should then be 1.28 multiplied by 4.80×10^{-10}—that is, 6.14 Debye units. Actually, the dipole moment is found to be 1.04 Debye units. If we arbitrarily say that the covalent molecule H:Cl has a zero moment (whereas the ionic molecule H$^+$Cl$^-$ should have a moment of 6.14), then the molecule is actually $1.04/6.14 = 0.17$ or 17% ionic.

By means of calculations of this nature and other related methods, the ionic characters of many bond types have been tabulated, and these numbers compared with the data in Table 8.2. According to Pauling, a separation of two elements by 1.7 units in Table 8.2 corresponds to a bond between these elements with 50% ionic character. More than 1.7 separation units corresponds to ionic character, as in the case of Na$^+$ Cl$^-$, where $3.0 - 0.9 = 2.1$ units. The separation for H:Cl is $3.0 - 2.1 = 0.9$, and it is only 17% ionic. Apparently there is no such thing as 100% ionic character; even caesium fluoride (the extreme elements in Table 8.2) has some partial covalent character.

8.10. The Forces Between Ions and Between Molecules. Interionic forces are quite large, and are often called *Coulomb forces* after Coulomb's law, which states that the force of attraction is proportional to the product of the two charges and falls off with the square of the distance between them. Typical giant molecules built up from ions are shown in Figures 8.1 and 8.2, where it is seen that there is no diatomic NaCl molecule but a lattice of oppositely charged ions.

Intermolecular forces, by contrast, are relatively weak, and are generally referred to as *van der Waals forces*, after the man who developed the first satisfactory mathematical treatment of them. These attractive forces fall off very rapidly with the seventh power of the distance between molecules, so that they are not effective unless the molecules are within a few molecular diameters of each other. These are the forces that control most of the physical properties of covalent molecules like NH$_3$ and CCl$_4$ (Figures 8.6 and 8.7).

Since *interionic* forces are powerful, ionic compounds generally have high melting and boiling points. The process of melting means the break-up of the structure of a giant molecule, and boiling means complete rupture of the strong electrovalent bonds. Since *intermolecular* forces are weak, the simple covalent compounds have comparatively low melting and boiling points; it is relatively easy to separate the molecules of such a compound. These principles are illustrated by the following chlorides of elements from the first two horizontal series of the periodic system:

	Ionic		Covalent	
	$NaCl$	$MgCl_2$	BCl_3	CCl_4
Melting point, °C	804	712	—107	—23
Boiling point, °C	1413	1412	12.5	76.8

Although the simple covalent molecules of carbon tetrachloride (see Figure 8.7) are easy to separate, it is important to remember that it is possible to build giant molecules with covalence bonds (such as the diamond molecule in Figure 19.2 and the starch molecule in Figure 38.2) just as difficult to break up as the giant ionic salt molecule in Figure 8.1.

8.11. Electrolytes. When ionic compounds are melted, the ions are free to move about and are then able to carry an electric current. Ionic compounds are therefore called electrolytes. It is generally said that acids, bases, and salts are electrolytes, but this does not mean that they are necessarily ionic compounds. For example, boron trichloride (see preceding paragraph) is covalent although it is called a salt and an electrolyte; because it has considerable ionic character it reacts readily with water to form ions which make the resulting solution an electrolyte.

Similar reasoning applies to the electrolyte, hydrochloric acid, which is a "strong acid" although a covalent compound; it reacts with water to form an electrolyte as described in Section 27.6.

8.12. Ionic Dissociation. This is a term that can be applied to dissolution of compounds, like Na^+Cl^-, when placed in liquids, like water, which have high dielectric constants. Water is highly polar and has a dielectric constant of 80; this means that the electrostatic attraction between oppositely charged particles (Na^+ and Cl^-), when placed in water, is weakened eighty-fold over the force of attraction in air or in a vacuum. The act of dissolving sodium chloride in water simply means its dissociation into ions. It is analogous to raising the temperature of a crystal of Na^+Cl^- to its melting point.

8.13. Association and Related Factors. Just as certain types of molecules are readily torn apart into smaller units (ions), other types of molecules tend to combine with each other to form larger units by the process of association. At one time, this was vaguely attributed to their polar characteristics—that is, mutual attraction of oppositely charged poles on dipoles, as illustrated by the molecule of I–Cl in Section 8.1.

At present, the phenomenon of association can often be better explained on the basis of *hydrogen bonding*, as in the case of water (Section 25.10) and of acetic acid (Section 27.8).

Polar and ionic molecules usually dissolve readily in water, which is also polar. The solute and solvent molecules are surrounded by electric fields which exert a mutual attraction. A nonpolar molecule like Cl–Cl has a very weak field around it. Consequently, it is only slightly soluble in a polar liquid like water, because there is no opportunity for mutual attraction. The nonpolar substances dissolve more readily in nonpolar solvents, like benzene. Most organic compounds are nonpolar or at most only slightly polar and therefore dissolve best in nonpolar liquids. "Like dissolves like."

Part 2

THE ARCHITECTURE OF CARBON COMPOUNDS

Occasionally the student forgets that the carbon atom must be pictured with four bonds thus, $-\overset{|}{\underset{|}{C}}-$, and the process by which we can build molecules from these carbon atoms seems to lose its simplicity.

For this reason, in Part 1 of this book we did no more than try to develop a conception of the meaning of valence—that is, of the means by which atoms are joined together.

Now that we have made a study of the various ways in which atoms can be built into molecules we can proceed to the study of the organic chemist as an architect. What we *must* remember is that the carbon atom has a nearly constant valence of four.

"Structural organic chemistry, although developed without mathematics, except of the most elementary sort, is one of the very greatest of scientific achievements. . . It is this system of structural chemistry that served chiefly as my basis when I advanced my theory of valence."

G. N. Lewis, *J. Chem. Phys.*, **1**, 17, 1933.

9

Methane and the Structure Theory

9.1. Properties of Methane. We are now ready to consider a very simple compound of carbon called methane. The composition of this substance is expressed by the formula CH_4. It belongs to a family of compounds known as hydrocarbons, because they are composed of hydrogen and carbon, and nothing else.

Methane has several other names. It is called *marsh gas*, because it occurs in swamps as the result of the decomposition of organic matter induced by certain bacteria. Its mixture with oxygen or air is also called *firedamp* because its presence in coal mines often leads to serious explosions.

The principal sources of methane are natural gas, which contains about 90% methane, and coal gas, which contains between 30 and 40%. Methane has no color or odor, is only slightly soluble in water, and is only about one-half as heavy as air.

Since the molecular weight of methane has been found to be 16, in round numbers, it contains only one C atom and four H atoms ($C=12$ and $4H=4$). The structure of methane can therefore be represented by the following formulas, in which carbon is given its usual valence of four and hydrogen, its valence of one:

$$\begin{array}{ccc} & H & \\ & | & \\ H-\!\!&C&\!\!-H \\ & | & \\ & H & \end{array} \qquad or \qquad \begin{array}{c} H \\ H\!:\!\overset{\cdot\cdot}{\underset{\cdot\cdot}{C}}\!:\!H \\ H \end{array}$$

Methane is a very stable substance and is not even affected by chemical reagents which destroy most other substances. For example, it is not easily acted on when bubbled through sulphuric acid, nitric acid, strong alkalies, or powerful oxidizing agents.

Since methane is so inactive, we see at once that the union between carbon and hydrogen (the C:H or C–H union) is a very strong one. This is a fact which we shall refer to frequently during the rest of our study of carbon compounds. Occasionally, it is true, we may have to amend this rule slightly because we shall find that in larger molecules the presence and structural position of other atoms may weaken the C–H bond, but in general the rule of the great strength of the C–H bond is one that must always be kept in mind.

Although methane is so very stable, there are a few things we can do with it. In the first place, it will burn in oxygen, but this is a common property of many carbon compounds. The most noteworthy characteristic of the burning of methane is that its heat of combustion is very high and it is consequently a valuable fuel.

It is possible to make methane react with nitric acid to form a nitro compound (Section 20.3).

A more important property of methane and (from our elementary point of view in this book) its only chemical property of importance, is its reaction with chlorine or bromine (it does not react with iodine):

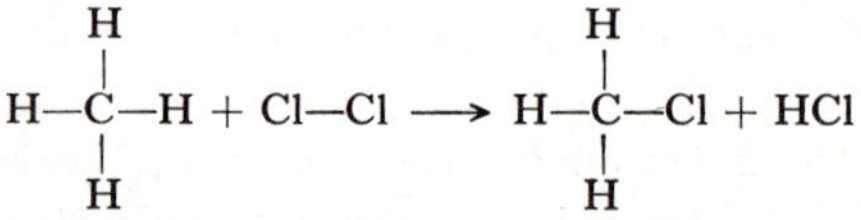

$$\underset{\text{Methane}}{\overset{\displaystyle H}{\underset{\displaystyle H}{H-C-H}}} + Cl-Cl \longrightarrow \underset{\substack{\text{Chloromethane} \\ \text{Methyl chloride}}}{\overset{\displaystyle H}{\underset{\displaystyle H}{H-C-Cl}}} + HCl$$

This is one of the very few methods at our disposal for breaking the C–H bond in methane.

The reaction between methane and chlorine takes place readily when the two gases are mixed and placed in diffused sunlight. Under properly controlled conditions other products may also be obtained, as follows:

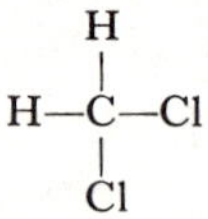

$$\underset{\substack{\text{Dichloromethane} \\ \text{Methylene chloride}}}{\overset{\displaystyle H}{\underset{\displaystyle Cl}{H-C-Cl}}} \qquad \underset{\substack{\text{Trichloromethane} \\ \text{Chloroform}}}{\overset{\displaystyle Cl}{\underset{\displaystyle Cl}{H-C-Cl}}} \qquad \underset{\substack{\text{Tetrachloromethane} \\ \text{Carbon tetrachloride}}}{\overset{\displaystyle Cl}{\underset{\displaystyle Cl}{Cl-C-Cl}}}$$

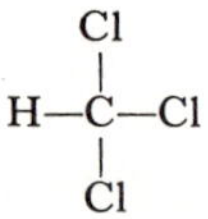

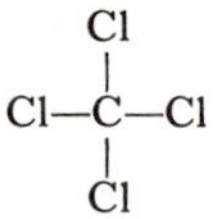

Each of these substances (made by other processes) is of commercial importance. Methyl chloride is a gas used in refrigeration machines, methylene chloride is a solvent, chloroform is an anesthetic as well as solvent, and carbon tetrachloride is used as a fire-extinguisher, solvent, and "dry-cleaner" for soiled clothing.

As stated before, iodine, the least active of the halogens, does not react with methane. Fluorine, the most active halogen (in fact the most active of all the elements), explosively decomposes methane, even at very low temperatures. In order to prepare fluorine compounds of methane, it is necessary to use indirect methods (Section 24.7). For example, the reaction between carbon tetrachloride, CCl_4, and antimony fluoride, SbF_3, yields dichlorodifluoromethane, CCl_2F_2, when SbF_5 is employed as a catalyst. This product is one of the nontoxic Freons used as refrigerants.

9.2. Structure Theory. The chlorine derivatives of methane, which we have just tabulated, illustrate a theory that has been of great importance in the development of organic chemistry. We wrote the molecule of methane with the various atoms in a certain space relationship. When we replaced the hydrogen atoms successively by chlorine atoms no appreciable change was indicated in the structure of the molecule.

The amazing progress in organic chemistry after 1850 was due largely to this belief among organic chemists that *the structures of carbon compounds are not appreciably altered by minor changes in composition.* This conception has generally been known as the structure theory.

9.3. Substitution. The direct replacement of one atom in a molecule with another atom is called substitution. As the reader has already observed, these substitution products are named by using the name of the parent molecule as the last name of the compound. Thus, dichloromethane means that two atoms of chlorine have been inserted in the methane structure in place of two hydrogen atoms. Besides these *systematic* names, certain *common* or *trivial* names are often employed. For example, trichloromethane is more often called chloroform.

9.4. Polar, Reactive Compounds Obtained by Substitution. The substitution process, by means of which we replace an H atom in a carbon compound with another element, is one of the most important features of organic chemistry. Before we have finished this chapter, it will be evident to the student that the simple molecule CH_4 can be used as the starting point for making a large number of more complex substances. In fact, methane is regarded as the mother substance of organic chemistry, in that all of the carbon compounds can be theoretically derived from methane.

As indicated in the last few pages, the nonpolar C–H bonds in the methane molecule are so stable that we cannot alter the molecule by any *simple* means except by the reaction with chlorine to give $CH_3–Cl$, chloromethane. (The reaction with nitric acid, described in Section 20.3, is much more drastic.) Since the CH_4 molecule is very stable, it does not lend itself for use in the building of the many carbon compounds that were just mentioned. The CH_3–Cl molecule, however, is polar and chemically active. It undergoes a large number of chemical changes, of which several examples will be given in the next section.

The substitution of a Cl atom for an H atom is therefore very often the first step in the preparation of more complex compounds. The Cl atom acts as a "hot spot" in the molecule, and the molecule can then be made to react with other substances.

When methyl chloride is shaken with a warm solution of potassium hydroxide in water the –Cl is slowly replaced by –OH:

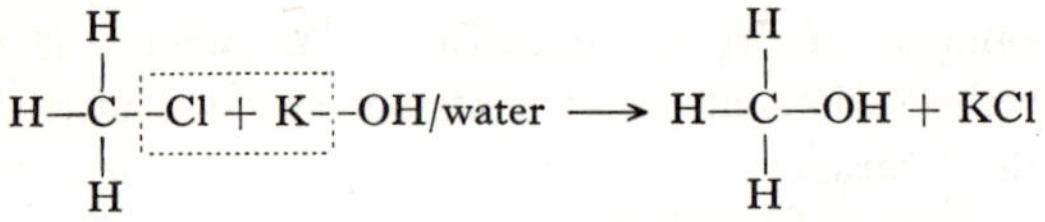

$$\underset{\substack{\text{Chloromethane} \\ \text{Methyl chloride}}}{H-\overset{\displaystyle H}{\underset{\displaystyle H}{C}}-Cl} + K-OH/\text{water} \longrightarrow \underset{\substack{\text{Hydroxymethane} \\ \text{Methyl alcohol}}}{H-\overset{\displaystyle H}{\underset{\displaystyle H}{C}}-OH} + KCl$$

This is a general method for replacing the halogen atom in a molecule by the hydroxyl group (–OH). Organic compounds containing the –OH group are called alcohols. The fact that the KOH is dissolved in water is so important that it is stressed in the equation by writing KOH/water. In Section 13.2 we shall find that in organic solvents a different type of product is obtained and that the reaction takes place more smoothly than this one.

If methyl chloride is dissolved in ether and pieces of metallic sodium are added to the solution, the following reaction takes place when the resulting mixture is heated for several hours:

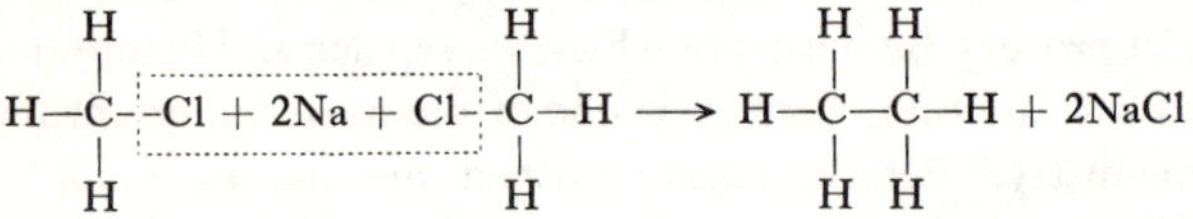

$$H-\overset{\displaystyle H}{\underset{\displaystyle H}{C}}-Cl + 2Na + Cl-\overset{\displaystyle H}{\underset{\displaystyle H}{C}}-H \longrightarrow H-\overset{\displaystyle H}{\underset{\displaystyle H}{C}}-\overset{\displaystyle H}{\underset{\displaystyle H}{C}}-H + 2NaCl$$

Dimethyl, or ethane

A significant feature of the organic compounds containing chlorine is the fact that they can usually be depended on to react with metals or metallic compounds. The reaction between CH_3Cl and Na was discovered a long time ago by a chemist named Wurtz and since then any reaction of this type is known as a *Wurtz synthesis.*

9.5. Reactivity of Chloromethane. Let us compare once more the formulas of methane and its halogen derivative:

$$
\begin{array}{cc}
\text{H} & \text{H} \\
\text{H:C:H} & \text{H:C :Cl:} \\
\text{H} & \text{H}
\end{array}
$$

Methane   Chloromethane
(nonpolar)   Methyl chloride
(polar)

In methane, we have nothing but C–H links which, as we have already pointed out, are nonpolar and are very stable. In the CH_3–Cl molecule there are three nonpolar C–H links and one C–Cl link that is polar because of an electron displacement toward the Cl atom.

In some of its reactions CH_3–Cl acts as if it breaks up into CH_3^+ and Cl^- ions, behaving just like Na^+ Cl^- the typical inorganic compound. For instance:

$$Na^+ \; Cl^- + Ag^+ \text{-} NO_3^- \longrightarrow AgCl + NaNO_3$$

$$CH_3 \text{-} Cl + Ag \text{-} NO_3 \longrightarrow AgCl + CH_3NO_3$$

The reaction between salt and silver nitrate takes place best in water, in which solvent they both ionize, and an instantaneous reaction takes place between the ions Ag^+ and Cl^- to form insoluble AgCl.

Methyl chloride is insoluble in water, but if it is dissolved in alcohol and mixed with an alcohol solution of silver nitrate, the corresponding reaction takes place, as we have just indicated. This is an instance in which a polar bond under the proper reaction conditions can become ionic; each CH_3^+ ion as it forms has only a very brief life.

9.6. Radicals. In the discussions of this chapter the reader will have noticed that we use the group of atoms CH_3– as a definite unit. It apparently remains intact during a chemical reaction. A group of atoms which goes through a series of chemical changes without any change in structure or composition is known as a radical or group.

The radical CH_3– is called *methyl* from its relationship to methane, CH_4. Here are the formulas of some representative compounds in which the methyl radical is present:

$$CH_3\text{—}H \qquad CH_3\text{—}Cl \qquad CH_3\text{—}CH_3 \qquad CH_3\text{—}OH \qquad CH_3\text{—}Na$$

Methane   Methyl chloride   Dimethyl, or ethane   Methyl alcohol   Methylsodium

The methyl radical, it will be seen, has a valence of *one*, in that it combines with only one atom of univalent elements, like Na or Cl, or with one uni-valent radical like –OH.

We mention some of these compounds derived from methane in order to stress once again the unique position of carbon among the elements. It was shown in Chapter 5 that the carbon atom is situated at the border line which divides electropositive from electronegative elements. To some extent, it shows amphoteric properties, since compounds are known in which it is combined with both types of elements. For example, in CH_3–Na the carbon atom is linked to a metallic (electropositive) element, whereas in CH_3–Cl it is linked to an acidic (electronegative) element. These compounds are pictured on the electron basis as follows:

Chloromethane
Methyl chloride   Methylsodium
Sodium methide

The sodium compound has properties that are salt-like; from the data in Section 8.2 it can be calculated that the C–Na bond has about 50% ionic character. Methylsodium is very reactive, igniting spontaneously when exposed to air. It is a relative of tetraethyllead, a substance used to improve gasoline for motor cars.

9.7. Ethane. This substance, just like methane, is a hydrocarbon and is a gas. It is found dissolved in crude petroleum and is also found in some natural gas wells. The structure of ethane has been proved to be, in most respects, similar to that of methane,

the larger part of the molecule consisting of C–H bonds.

The C–H union we have already explained in detail; it is a very stable union and all that can be done to it *readily* is to replace an H atom by halogen. Theoretically, then, we should be able to prepare the following compounds:

C_2H_5Cl $C_2H_3Cl_3$ C_2HCl_5

$C_2H_4Cl_2$ $C_2H_2Cl_4$ C_2Cl_6

and in fact all are known. Not all of them, however, can be made by the direct action of chlorine on ethane. The only part of the ethane molecule with which we are not familiar is C–C, the bond between the two carbon atoms. This valence bond, like C–H, is very stable, and all we need remember about it is that ordinarily it is not acted on by chemical reagents.

In these few paragraphs on ethane, the idea we wish to impress on the student is that the organic chemist, by inspection of the formula of the substance, can predict most of its chemical properties. Ethane contains the same elements as methane, linked together in the same way, and therefore should behave in the same way. Accordingly, we are not surprised to find that the following reactions take place:

$$
\begin{array}{c}
\underset{\text{Ethane}}{
\overset{\displaystyle H \quad H}{
\underset{\displaystyle H \quad H}{
H-\!\!\overset{|}{\underset{|}{C}}-\!\!\overset{|}{\underset{|}{C}}-H}}}
\; + \; Cl-Cl
\;\longrightarrow\;
\underset{\substack{\text{Ethyl chloride}\\\text{Chloroethane}}}{
\overset{\displaystyle H \quad H}{
\underset{\displaystyle H \quad H}{
H-\!\!\overset{|}{\underset{|}{C}}-\!\!\overset{|}{\underset{|}{C}}-Cl}}}
\; + \; HCl
\qquad (1)
$$

$$
\begin{array}{c}
\underset{\text{Chloroethane}}{
\overset{\displaystyle H \quad H}{
\underset{\displaystyle H \quad H}{
H-\!\!\overset{|}{\underset{|}{C}}-\!\!\overset{|}{\underset{|}{C}}-Cl}}}
\; + \; K-OH/water
\;\longrightarrow\;
\underset{\substack{\text{Hydroxyethane}\\\text{Ethyl alcohol}}}{
\overset{\displaystyle H \quad H}{
\underset{\displaystyle H \quad H}{
H-\!\!\overset{|}{\underset{|}{C}}-\!\!\overset{|}{\underset{|}{C}}-OH}}}
\; + \; KCl
\qquad (2)
$$

From the names given to these two products, it will be seen that the group CH_3–CH_2– is known as the *ethyl* radical. The condensed formulas of ethyl chloride and ethyl alcohol are C_2H_5Cl and C_2H_5OH, respectively.

$$
2C_2H_5-Cl + 2Na \;\longrightarrow\;
\underset{\text{Butane}}{
\overset{\displaystyle H \quad H \quad H \quad H}{
\underset{\displaystyle H \quad H \quad H \quad H}{
H-\!\!\overset{|}{\underset{|}{C}}-\!\!\overset{|}{\underset{|}{C}}-\!\!\overset{|}{\underset{|}{C}}-\!\!\overset{|}{\underset{|}{C}}-H}}}
\; + \; 2NaCl
\qquad (3)
$$

This last equation leads us directly to the subject matter of the next chapter.

10

Carbon Chains

10.1. Introduction. In this chapter the student becomes acquainted with the real spirit of organic chemistry. The organic chemist is to be described primarily as an architect. It will be found that the construction methods used by this builder of molecules are seldom more intricate than the simple principles we have learned from the study of methane and its related compounds.

In the last chapter a few simple compounds were described which consist, for the most part, of only carbon and hydrogen, and in some cases chlorine also:

$$
\begin{array}{ccc}
\text{H} & \text{H} & \text{H H} \\
\text{H:}\overset{..}{\underset{..}{\text{C}}}\text{:H} & \text{H:}\overset{..}{\underset{..}{\text{C}}}\text{ :}\overset{..}{\underset{..}{\text{Cl}}}\text{:} & \text{H:}\overset{..}{\text{C}}\text{:}\overset{..}{\text{C}}\text{:H} \\
\text{H} & \text{H} & \text{H H}
\end{array}
$$

Methane   Chloromethane   Ethane
(very stable)   (reactive)   (very stable)

From the properties of these compounds we arrived at the following rules:

1. The C–H bond is relatively inactive and is not affected by most reagents. An important property of the C–H union, however, is the replaceability of the hydrogen atom by chlorine and bromine.

2. The C–Cl bond is reactive. It is a polar bond in which the Cl is electronegative and enters into reactions especially with those reagents which contain electropositive elements, such as metals.

3. The C–C union is a very firm one. It is not broken by the action of most reagents, not even by chlorine which does disrupt the C–H bond. Later, some cases will be found in which the C–C bond is broken, but the rule of great stability can be safely applied to most compounds like ethane, in which two carbon atoms are connected by a single valence bond.

In this chapter a large number of compounds will be mentioned, the construction of which requires no more knowledge on our part than the rules given in the preceding paragraphs.

10.2. Synthesis. A dictionary will generally define synthesis as the building up of a substance from two or more constituents. From this standpoint the formation of a hydrate,

$$CuSO_4 + 5H_2O \longrightarrow CuSO_4 \cdot 5H_2O$$

would be a synthesis, since the resulting compound is more complex than its constituents. The organic chemist, however, looks at it differently. He limits his use of the word synthesis to the special case in which a more complex compound is formed containing *more carbon atoms* and in which the simpler compounds become directly linked to each other *through carbon atoms*. The material in this chapter amply illustrates this concept of synthesis.

10.3. Straight Chains. Perhaps the most remarkable thing about the carbon atom is the ease with which it unites with itself to form a chain, and in the Wurtz synthesis, described in section 9.4, we have an example of this phenomenon so characteristic of carbon.

We have already seen how a one-carbon compound, CH_4, can be changed into a compound with two carbon atoms, CH_3–CH_3, and by the same procedure we can build higher:

$$
\begin{array}{ccccc}
\text{H H} & & \text{H H} & & \text{H H H H} \\
\text{H--C--C--H} & \xrightarrow{Cl_2} & \text{H--C--C--Cl} & \xrightarrow{Na} & \text{H--C--C--C--C--H} \\
\text{H H} & & \text{H H} & & \text{H H H H} \\
\text{Ethane} & & \text{Chloroethane} & & \text{Butane}
\end{array}
$$

As before, this series of reactions indicates that ethane has been *chlorinated* in order to introduce a reactive halogen atom into the molecule. Two molecules of the resulting chloroethane, when treated with sodium under the proper conditions, will condense to form a straight-chain compound with four carbon atoms in the chain.

If we want, instead, a chain of three carbon atoms, all we need to do is couple a two-carbon chain with a one-carbon compound:

$$
\begin{array}{cccc}
\text{H H} & & \text{H} & \text{H H H} \\
\text{H--C--C--Cl} + 2Na + \text{Cl--C--H} & \longrightarrow & \text{H--C--C--C--H} + 2NaCl \\
\text{H H} & & \text{H} & \text{H H H} \\
\text{Chloroethane} & & \text{Chloromethane} & \text{Propane}
\end{array}
$$

And the following reactions (see paragraph in *italics*, Section 24.2) constitute the corresponding steps in the preparation of six-, seven-, and eight-carbon chains:

$$CH_3{-}CH_2{-}CH_2{-}CH_3 + Cl_2 \longrightarrow CH_3{-}CH_2{-}CH_2{-}CH_2Cl + HCl$$

$$CH_3{-}CH_2{-}CH_3 + Cl_2 \longrightarrow CH_3{-}CH_2{-}CH_2Cl + HCl$$

$$CH_3{-}CH_2{-}CH_2{-}Cl + 2Na + Cl{-}CH_2{-}CH_2{-}CH_3 \longrightarrow$$

$$2NaCl + CH_3{-}CH_2{-}CH_2{-}CH_2{-}CH_2{-}CH_3$$

Six-carbon chain

$$CH_3{-}CH_2{-}CH_2{-}CH_2{-}Cl + 2Na + Cl{-}CH_2{-}CH_2{-}CH_3 \longrightarrow$$

$$2NaCl + CH_3{-}CH_2{-}CH_2{-}CH_2{-}CH_2{-}CH_2{-}CH_3$$

Seven-carbon chain

$$CH_3{-}CH_2{-}CH_2{-}CH_2{-}Cl + 2Na + Cl{-}CH_2{-}CH_2{-}CH_2{-}CH_3 \longrightarrow$$

$$2NaCl + CH_3{-}CH_2{-}CH_2{-}CH_2{-}CH_2{-}CH_2{-}CH_2{-}CH_3$$

Eight-carbon chain

A hydrocarbon of this kind has been made with 100 carbon atoms in a straight chain.

In Table 10.1 there is a list of the first twelve of these straight-chain hydrocarbons, with some of their properties. The student should memorize the names of these members of the series of straight-chain compounds, for many compounds to be met with later are related to them and bear similar names. This memory task is rather simple, because, starting with the fifth compound, their names are derived from words in Greek and

TABLE 10.1. Straight-Chain Hydrocarbons (Paraffins)

Name	Molecular Formula	Structure	Melting Point, °C	Boiling Point, °C
Methane	CH_4	CH_4	-182.6	-161.58
Ethane	C_2H_6	$CH_3{-}CH_3$	-172.0	$-\ 88.5$
Propane	C_3H_8	$CH_3{-}CH_2{-}CH_3$	-187.1	$-\ 42.2$
Butane	C_4H_{10}	$CH_3{-}CH_2{-}CH_2{-}CH_3$	-135.0	$-\ 0.5$
Pentane	C_5H_{12}	$CH_3{-}(CH_2)_3{-}CH_3$	-129.7	36.08
Hexane	C_6H_{14}	$CH_3{-}(CH_2)_4{-}CH_3$	$-\ 94.0$	68.8
Heptane	C_7H_{16}	$CH_3{-}(CH_2)_5{-}CH_3$	$-\ 90.5$	98.4
Octane	C_8H_{18}	$CH_3{-}(CH_2)_6{-}CH_3$	$-\ 56.8$	125.6
Nonane	C_9H_{20}	$CH_3{-}(CH_2)_7{-}CH_3$	$-\ 53.69$	150.71
Decane	$C_{10}H_{22}$	$CH_3{-}(CH_2)_8{-}CH_3$	$-\ 29.72$	174.04
Hendecane	$C_{11}H_{24}$	$CH_3{-}(CH_2)_9{-}CH_3$	$-\ 25.61$	195.8
Dodecane	$C_{12}H_{26}$	$CH_3{-}(CH_2)_{10}{-}CH_3$	$-\ 9.65$	216.2

Latin (mostly Greek) which indicate how many carbon atoms there are in the chain. The first four members of the series (methane, ethane, propane, and butane) were not named in this systematic way because they were known before organic chemistry was systematized.

10.4. Isomerism. Poets often make use of words that do not fit the rules of grammar, and they excuse themselves with the phrase "poetic license." In the same way, teachers must often employ ideas which do not agree with their rules. For instance, in the preceding section we wrote this reaction:

$$CH_3\!-\!CH_2\!-\!CH_3 + Cl \longrightarrow CH_3\!-\!CH_2\!-\!CH_2Cl + HCl$$

Propane　　　　　　　　　　　　Chloropropane

This is a case of "teacher's license," for it does not tell all the truth.

It is practically impossible to control the experimental conditions so closely that the chlorine atom should enter the propane molecule only as indicated. There really are two possibilities:

```
    H  H  H              H  H  H
    |  |  |              |  |  |
H — C— C— C—Cl       H — C— C— C—H
    |  |  |              |  |  |
    H  H  H              H  Cl H
```

1-Chloropropane　　　　　2-Chloropropane

From the product of the reaction the two different compounds can be separated by distillation, for they have different boiling points. These two chloropropanes furnish an example of the phenomenon of isomerism.

Isomers are compounds of the same composition and molecular weight, but with different properties. The word isomer comes from the Greek, *isos*, meaning equal, and *meros*, meaning measure. In this example of isomerism, the two molecules have an equal measure of materials, namely C_3H_7Cl, but in one case the Cl atom is on an end carbon atom and in the other it is on the middle carbon atom. Because the atoms in the molecules are arranged differently, the two compounds have different densities, boil at different temperatures, and react at different rates of speed with other substances.

An elaborate system has been devised by which such compounds can be named so that no confusion (at least, very little confusion) arises when we refer to them. This system of nomenclature is too complicated for us to absorb all of it at one time, but we can take it piecemeal whenever occasion demands that part of it is explained

The two chloropropanes which have just been mentioned are named in such a way as to indicate that they are derived from a straight chain of three carbon atoms, in which the carbon atoms are numbered consecutively 1, 2, and 3.

1-Chloropropane is exactly the same as 3-chloropropane,

$$\overset{1}{C}H_2Cl-\overset{2}{C}H_2-\overset{3}{C}H_3 \qquad\qquad \overset{1}{C}H_3-\overset{2}{C}H_2-\overset{3}{C}H_2Cl$$

1-Chloropropane   3-Chloropropane

the only difference being the end of the molecule at which the numbering of the carbon atoms is begun. The convention is to *start at that end of the chain where a substituting element appears;* it is better to say 1-chloropropane than 3-chloropropane.

Here is another case of simple isomerism:

$$
\begin{array}{ccc}
 & H & H \\
 & | & | \\
Cl-& C-& C-Cl \\
 & | & | \\
 & H & H
\end{array}
\qquad\qquad
\begin{array}{ccc}
 & H & H \\
 & | & | \\
H-& C-& C-Cl \\
 & | & | \\
 & H & Cl
\end{array}
$$

1,2-Dichloroethane   1,1-Dichloroethane

These compounds are made by methods we cannot discuss at this point. From their methods of preparation, however, it is certain that in one case both Cl atoms are on the same carbon atom, whereas in the other case they are on separate carbon atoms. They have different properties, although the same composition, $C_2H_4Cl_2$.

10.5. Branched Chains. It is possible to employ the two chloropropanes in the Wurtz synthesis by reacting them with CH_3Cl; the products will be two different compounds with the same composition, C_4H_{10}:

$$CH_3-CH_2-CH_2-CH_3 \qquad\qquad \begin{array}{c} CH_3-CH-CH_3 \\ | \\ CH_3 \end{array}$$

Butane   2-Methylpropane

One of them is a straight-chain hydrocarbon, the other a branched-chain hydrocarbon. They are isomers with different properties as listed in Tables 10.1 and 10.2. The *last name* of a chain compound is obtained from the name of the *longest straight chain* which is present. In the preceding example of 2—methylpropane, it is clear that the longest chain contains three carbon atoms, C–C–C, and there is a methyl group, CH_3, on the second carbon atom in the chain.

We are now led to this observation on theoretical organic chemistry:

we are so certain the chemistry of carbon can be represented by the symbol $-\overset{\displaystyle|}{\underset{\displaystyle|}{C}}-$ that we feel we can arrange these atoms on paper at will to build up hydrocarbons that can be prepared in the laboratory. To illustrate, we can write out a string of six carbon atoms, thus,

$$-\overset{|}{\underset{|}{C}}-\overset{|}{\underset{|}{C}}-\overset{|}{\underset{|}{C}}-\overset{|}{\underset{|}{C}}-\overset{|}{\underset{|}{C}}-\overset{|}{\underset{|}{C}}- \quad -\overset{|}{\underset{|}{C}}-\overset{|}{\underset{|}{C}}-\overset{|}{\underset{|}{C}}-\overset{|}{\underset{|}{C}}-\overset{|}{\underset{|}{C}}- \quad -\overset{|}{\underset{|}{C}}-\overset{|}{\underset{|}{C}}-\overset{|}{\underset{|}{C}}-\overset{|}{\underset{|}{C}}-\overset{|}{\underset{|}{C}}-, \text{ etc.,}$$
$$\qquad\qquad\qquad -\overset{|}{\underset{|}{C}}- \qquad\qquad\qquad\qquad -\overset{|}{\underset{|}{C}}-$$

and arrange them in all the possible ways, filling in the necessary hydrogen atoms on the unused valence bonds. When the puzzle has been finished, we have no doubt that these new compounds we have worked out on paper can actually exist. In this example of the possible compounds containing six carbon atoms, it will be found that five arrangements are possible; their structures are listed in Tables 10.1 and 10.2 with some of their properties.

TABLE 10.2. Branched-Chain Hydrocarbons (Paraffins)
(Isomers of the straight-chain compounds in Table 10.1)

Molecular Formula	Name	Structure	Boiling Point, °C
*C_4H_{10}	2—Methylpropane	$CH_3-\overset{\displaystyle|}{\underset{\displaystyle CH_3}{CH}}-CH_3$	−12.2
C_5H_{12}	2—Methylbutane	$CH_3-\overset{\displaystyle|}{\underset{\displaystyle CH_3}{CH}}-CH_2-CH_3$	28.0
	2,2— Dimethylpropane	$CH_3-\overset{\displaystyle CH_3}{\underset{\displaystyle CH_3}{C}}-CH_3$	9.5
C_6H_{14}	2—Methylpentane	$CH_3-\overset{\displaystyle|}{\underset{\displaystyle CH_3}{CH}}-CH_2-CH_2-CH_3$	60.2
	3—Methylpentane	$CH_3-CH_2-\overset{\displaystyle|}{\underset{\displaystyle CH_3}{CH}}-CH_2-CH_3$	63.2
	2,3—Dimethylbutane	$CH_3-\overset{\displaystyle|}{\underset{\displaystyle CH_3}{CH}}-\overset{\displaystyle|}{\underset{\displaystyle CH_3}{CH}}-CH_3$	58.0
	2,2—Dimethylbutane	$CH_3-\overset{\displaystyle CH_3}{\underset{\displaystyle CH_3}{C}}-CH_2-CH_3$	49.7

* Branching cannot occur when there are less than four carbon atoms in the molecule.

It has been calculated that there are 355 isomers of the composition $C_{12}H_{26}$. Only a few of these isomers have been prepared, because organic chemists have been too busy with other problems to worry about this one. It is possible that some of these isomers cannot exist because of space considerations of the molecule, but it is certain that not more than 355 isomers are theoretically possible. It is of interest to note, however, that the nine possible C_7H_{16} compounds, the eighteen possible C_8H_{18} compounds and the thirty-five possible C_9H_{20} compounds have all been made. The straight- and branched-chain compounds are usually considered in one category as the open-chain compounds, to distinguish them from the ring structures (closed-chain compounds), described in the next chapter.

The general nature of the branched-chain hydrocarbons is similar to that of the analogous straight-chain compounds. However, if any specific property is considered, there are obvious variations in the magnitude of that property. For example, from the information in Tables 10.1 and 10.2 it is clear that for molecules of the same number of carbon atoms the boiling point is lowered by increasing the complexity of the molecule. With respect to chemical reactivity, the ease with which a C–H bond is broken depends on the position of the carbon atom in the molecule. It has been found convenient to label a C atom as primary ($1°$), secondary ($2°$), tertiary ($3°$), and (less frequently) quaternary ($4°$) on the basis of the number of other C atoms to which it is joined:

$$
\begin{array}{c}
CH_3 \\
\overset{4°}{|} \quad \overset{2°}{} \qquad \overset{3°}{} \quad \overset{1°}{} \\
CH_3\!-\!C\!-\!CH_2\!-\!CH\!-\!CH_3 \\
| \qquad\quad | \\
CH_3 \qquad CH_3
\end{array}
$$

The H atoms in this system follow the designations of the C atoms to which they are connected. In general, the activity is in the order $3° > 2° > 1°$, but in some types of reactions the order may be reversed. Chlorine will replace a tertiary H atom more readily than the others. (See Sections 24.2, 25.3 and 29.2).

10.6. Paraffin Hydrocarbons. All the hydrocarbons mentioned in this chapter belong to the same family. This is usually the first family studied in organic chemistry, because most other families of compounds can be built up by some ingenious trick of chemical architecture from these simple chain compounds.

1. All the hydrocarbons described in this chapter will, on examination, be found to agree with the general formula C_nH_{2n+2}. For example, in a compound of three carbon atoms, $n=3$, and the formula then is $C_3H_{2\times3+2}$

or C_3H_8. If there were sixty carbon atoms in a certain compound, its formula would be $C_{60}H_{122}$.

2. As already explained, these compounds are called hydrocarbons because they consist of nothing but hydrogen and carbon. They were originally called *paraffin* hydrocarbons in order to set them apart from the families of hydrocarbons which will be described in Chapter 13, as well as to indicate their general nature. The name was probably intended to mean slight reactivity from the Latin words *parum affinis*. The description of paraffin hydrocarbons as having only slight reactivity is not taken literally by the modern organic chemist. Although the two building units, C–C and C–H bonds, are relatively inert, we have already seen that the C–H bond is sensitive to the action of the halogens (Section 9.1) and later we shall find that it can be attacked through the processes called *nitration* (Section 20.3) and *sulphonation* (Section 20.4). An example of the reactivity of the C–C bond toward chemical reagents can be found in Section 11.2, and the effect of *pyrolysis* on this bond is described in Section 20.7.

In addition to being called the paraffin hydrocarbons, this family is also known as the *methane or marsh gas* series. Most families of carbon compounds, as we shall find out later, are named from the first or simplest member of the series, which in this case is methane (marsh gas) CH_4.

All the members of the paraffin hydrocarbons have names ending in *-ane*. This is the family symbol, the rest of the name simply indicating the number of carbon atoms in the molecule. For example, from the name *hexane*, we can at once tell that there are six carbon atoms in the compound (*hex*), and that it is a member of the paraffin series (*-ane*).

3. This family of compounds also constitutes what is known as a *homologous series*, a term applied to a series of compounds in which the successive members differ by a constant factor. For example,

$$CH_3{-}CH_2{-}CH_3 \ (C_3H_8) \qquad \text{and} \qquad CH_3{-}CH_2{-}CH_2{-}CH_3 \ (C_4H_{10})$$

Propane   Butane

differ by CH_2, and the same increment represents the difference between any other pair of successive members of the family. The members of a homologous series are said to be *homologs*.

4. In a homologous series, the physical and chemical properties generally show progressive variations. The gradual increase in boiling point of the successive members of the straight-chain series is evident in Table 10.1. The lowest members are gases; intermediate members, beginning with C_5H_{12}, are liquids, and members containing more than

sixteen carbon atoms (not shown in the table) are solid at room temperature.

Chemical reactivity, in general, decreases with an increase in molecular weight in this series. Thus, methane and chlorine react with explosive violence when placed in direct sunlight, but the succeeding members, with a higher carbon content, react much more slowly and quietly with chlorine.

10.7. Petroleum. This is an oil occurring widely in nature. About 90% of the crude petroleum is a complex mixture of several types of hydrocarbons. The open-chain hydrocarbons described in this chapter form a high percentage of the crude oil found in this country, especially in the Pennsylvania oil fields.

The many hydrocarbons in petroleum can be partially separated by careful distillation. The fraction which boils between room temperature and 70°C is used largely as a solvent and is called *ligroin* or *petroleum ether*. *Gasoline* is the mixture of compounds which distills from the crude oil between 70 and 200°C. and *kerosine* is the 200 to 300°C fraction. When the more volatile compounds, comprising gasoline and kerosine, are removed from Pennsylvania petroleum, a heavier liquid remains which can be further separated into *lubricating oils* and *paraffin wax*. The "mineral oil" sold by the druggist for use as a laxative costs more than that used in a motor car because it is more highly purified by removal of coloring matter and sulphur compounds.

Because of the great importance of gasoline in our national economy, much research has gone into improving its performance in engines. The efficiency of an engine depends largely on the extent to which the fuel used is free from knock. Long, straight-chain molecules knock badly, but the short, branched-chain compounds knock relatively little. The outline structures of the formulas of certain important gasoline compounds are shown here, with their common names in quotation marks:

$$
\begin{array}{ccc}
 & \overset{\textstyle C}{|} & \overset{\textstyle C}{|} \\
C-C-C-C-C-C-C \qquad & C-\underset{\underset{\textstyle C}{|}}{C}-C-\underset{\underset{\textstyle C}{|}}{C}-C \qquad & C-\underset{\underset{\textstyle C}{|}}{C}-\underset{\underset{\textstyle C}{|}}{C}-C \\
\end{array}
$$

Heptane	2,2,4-Trimethylpentane	2,2,3-Trimethylbutane
	"Isooctane"	"Triptane"

Heptane is one of the poorest fuels with respect to knock. It was once thought that isooctane was the perfect knockless fuel, but triptane is even better. Although isooctane is not useful alone as a fuel (because of its low volatility), it is used as an antiknock ingredient and as a standard. A gasoline equivalent in performance to isooctane has a 100-octane rating; if equivalent to heptane it has a zero-octane rating: if equivalent to a mixture containing 40% heptane and 60% isooctane it has a 60-octane rating, and so on. Addition of tetraethyllead to a gasoline acts like a high content of the antiknock hydrocarbons.

10.8. How to Name the Open-Chain Compounds. When an architect erects a modern skyscraper, he first builds a steel framework. This serves as an outline to which he later adds the necessary parts for a useful building. In the same way, the organic chemist regards a chain hydrocarbon as the skeleton framework on which organic compounds are are built. Take the following compound, for example,

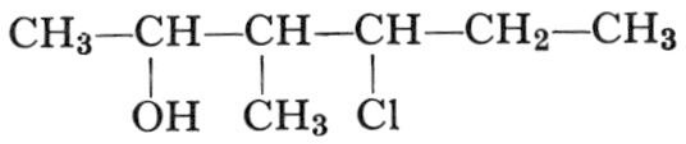

$$CH_3-CH-CH-CH-CH_2-CH_3$$
$$\qquad\ \ OH\ \ \ CH_3\ \ Cl$$

This is a formidable looking substance at first glance. To the uninitiated, it would seem to be a hopeless problem for anyone to write an intelligible name for such a substance.

To the initiated, however, this structure is very simple. In any office building or skyscraper the floors are numbered, and in order to visit a tenant it is merely necessary to whisper the name of the tenant to the elevator operator and he whisks us to the proper floor and lets us out.

We treat carbon structures the same way. Count the carbon atoms in the straight chain we have just written, starting from the left end. Then on the second floor we find –OH, on the third –CH$_3$, and on the fourth –Cl. And suppose we call it the *hexane* building. To locate the tenants, we just write out the list of tenants like this, 2-hydroxy–3–methyl–4–chlorohexane. Any organic chemist, given this name as the name of a compound, can sit down and write its complete structure. The first thing he would do is write out a chain of six carbon atoms, for the last name of the compound tells him it is a 6-story building.

$$C-C-C\ \ C\ \ C-C$$

Then he would number the atoms from 1 to 6 and place the tenants –OH, –CH$_3$, and –Cl in the proper positions.

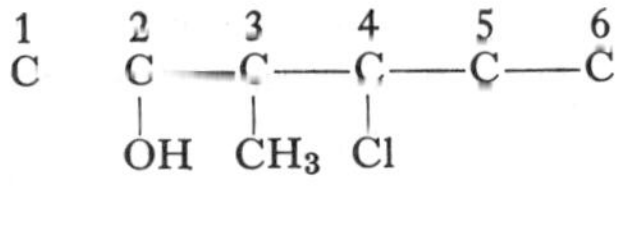

$$\begin{array}{cccccc} 1 & 2 & 3 & 4 & 5 & 6 \\ C & C & -C & -C & -C & -C \\ & | & | & | & & \\ & OH & CH_3 & Cl & & \end{array}$$

Finally, he would fix up the molecule with enough valence bonds and hydrogen atoms to give each carbon atom its usual valence of four.

$$H\text{---}\underset{\underset{H}{|}}{\overset{\overset{H}{|}}{C}}\text{---}\underset{\underset{OH}{|}}{\overset{\overset{H}{|}}{C}}\text{---}\underset{\underset{CH_3}{|}}{\overset{\overset{H}{|}}{C}}\text{---}\underset{\underset{Cl}{|}}{\overset{\overset{H}{|}}{C}}\text{---}\underset{\underset{H}{|}}{\overset{\overset{H}{|}}{C}}\text{---}\underset{\underset{H}{|}}{\overset{\overset{H}{|}}{C}}\text{---}H$$

This is a general rule for any chain compound. The last name is obtained by counting the number of carbon atoms in the longest straight chain. In this case, the longest chain is six and we consider the compound as a derivative of the paraffin hydrocarbon, hexane.

There are occasions when the longest chain in a compound is not apparent at first. These two formulas represent one compound in which the longest chain is five carbon atoms:

$$\overset{1}{C}H_3\text{---}\overset{2}{C}H\text{---}\overset{3}{C}H_2\text{---}\overset{4}{C}H_3 \qquad\qquad \overset{1}{C}H_3\text{---}\overset{2}{C}H_2\text{---}\overset{3}{C}H\text{---}\overset{4}{C}H_2\text{---}\overset{5}{C}H_3$$

$$\underset{\underset{CH_3}{|}}{\underset{|}{CH_2}} \qquad\qquad\qquad\qquad\qquad\qquad\qquad CH_3$$

$$\text{(I)} \qquad\qquad\qquad\qquad\qquad\qquad\qquad\qquad \text{(II)}$$

It should be named 3–methyl*pentane* and not 2–ethyl*butane*, although an organic chemist would recognize what is meant from either name. Structure (I) shows it as the butane derivative, while (II) shows it more correctly as a derivative of pentane.

10.9. Functional Groups. We have indicated that the general chemical chemical activity of a compound is due more to the substituents, such as –OH, than to the hydrocarbon part of the molecule. The organic chemist likes to refer to the –OH group as a "functional" group, meaning that it takes part, to a great extent, in the chemistry of the molecule in which it appears. If there are several such functional groups in the molecule, he picks out one of them and calls it the "chief function" and includes the name of this function in the last name of the compound.

The hydroxyl (–OH) group is called a chief function, and compounds with –OH groups attached to the carbon chain are called *alcohols*. The names of alcohols terminate in *-ol;* an example is ethanol (Section 25.1). The compound under discussion in Section 10.8 is considered to be an alcohol, more specifically as 2-*hexanol*, since the last name of the compound should correspond to the longest chain containing the functional group.

The numbering is begun at that end of the chain which gives the chief function the smallest number; on the basis of these rules the compound might be called 3-methyl–4–chloro–2–hexanol.

The student may argue that the Cl atom is quite active in such a molecule and may also properly be called a functional part of the compound. It has been agreed, however, to regard the Cl atom as a substituent in such a molecule. so the term *chloro* is usually seen as a prefix in the name of a compound.

The compound we are describing poses still another problem in nomenclature—namely, the order in which the substituents are to be listed. At present, it is the policy to follow the rules used by editors of chemical journals (see the next section) who prefer an alphabetical sequence in a case like this to assist them in their indexing. Our compound on this basis is 4–chloro–3–methyl–2–hexanol.

10.10. Old Nomenclature. Before the organic chemist became the efficient architect that he is today he did not look at structures in the systematic way explained in the preceding section. The early chemists had developed systems that are still in use for simple substances, but are not adequate for complex compounds. Molecules were called *normal* (represented by *n-*) when the carbon atoms were arranged in a straight chain. When the chain had a fork arrangement at one end, with a methyl group on a carbon atom adjacent to a *terminal* methyl group, it was called *iso*. If a carbon atom had four hydrocarbon radicals attached to it the resulting structure was *neo*. These designations can be illustrated by reference to the isomers of pentane which are listed in Table 10.2 where 2-methylbutane can be called *iso*pentane, and 2,2-dimethylpropane can be called *neo*pentane.

Another old system was to relate all structures to methane. Thus, in Table 10.2, the compound 2-methylbutane can also be called ethyldimethylmethane, with the substituents in alphabetical order.

10.11. New Nomenclature. For information on the naming of various types of compounds, the reader should refer to the index of this book under "nomenclature." The subject is too complex to be discussed in detail in a book of this kind. However, the system used in naming compounds follows that which is popularly called *systematic nomenclature*. This was first outlined by a committee meeting in 1892 at Geneva, Switzerland. It was modified in the period 1922-1930 by the International Union of Chemistry, whose recommendations are generally referred to as the I.U.C. Rules, and a meeting in 1949 of the International Union of Pure and Applied Chemistry resulted in what is called the I.U.P.A.C. system.

NOMENCLATURE*

"No phosphorus in phosgene gas,"
  The puzzled chemist said,
"There's no Cu in copperas,
  Plumbago has no lead.

"Things are not always what they
                seem,
  'Twas so in days of old.
There's no Au, in spite of gleam,
  In ore that's called 'fool's gold.'

"In selenite Se's not found
  And more the pity be.
In German silver, I'll be bound,
  One cannot find Ag.

"That magnetite has no Mg,
  Would not be hard to guess;
But platinite without Pt,
  Seems odd, I must confess."

The chemist hung his weary head.
  "Enough, enough," cried he,
"Now after this, I'll read instead
  Organic chemistry."

But to the patient man's dismay,
  Before he had gone far,
He found chromone, 'tis sad to say,
  Containing no Cr.

The chemist sighed, and said good-
                bye,
  "At least one thing I ken
That, when I find I want to die,
  There's death in KCN."

                    C. S. Powell

As intimated by the stanza in italics, the chemist feels that in organic chemistry compounds have names which tell him their chemical composition correctly. Nomenclature is supposedly more systematic in organic than in inorganic chemistry

But even in organic chemistry it is possible to be misled by the *common* names. For example, chromone (see Index) received its name from the Greek word *chroma*, meaning color, although this name might unfortunately lead us to believe that the compound contains chromium, Cr.

Ind. & Eng. Chemistry News Edition* **12, 161 (1934).

II

Carbon Rings

11.1. Formation of Closed Chains. Now that we have seen how the organic chemist can tie together a large number of carbon atoms into a chain, we have acquired a good insight into the methods he uses for building more intricate designs. The chain $-\overset{|}{\underset{|}{C}}-\overset{|}{\underset{|}{C}}-\overset{|}{\underset{|}{C}}-\overset{|}{\underset{|}{C}}-$ is his fundamental tool. The rest of this book is a description of the remarkable things he can do with it.

The first trick we shall explain is the formation of closed chains, or rings.

It is possible to make a dichloropropane in which the two Cl atoms are at the ends of the chain, as in the following illustration. Under the proper conditions, the effect of a metal on this compound is to remove both Cl atoms from the same molecule:

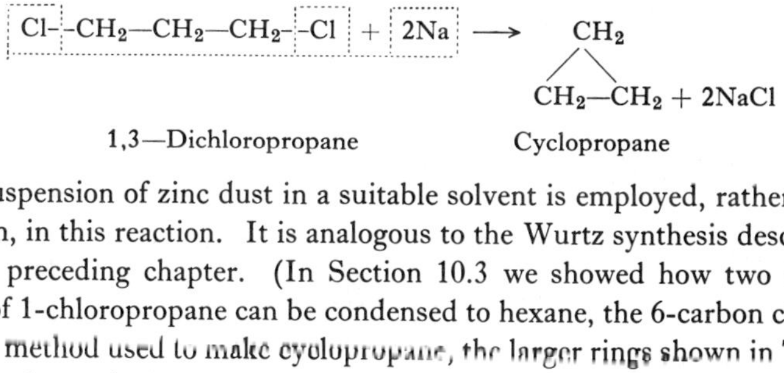

$$Cl-CH_2-CH_2-CH_2-Cl \; + \; 2Na \; \longrightarrow \quad \underset{\displaystyle CH_2-CH_2}{\overset{\displaystyle CH_2}{\diagup\diagdown}} + 2NaCl$$

1,3—Dichloropropane   Cyclopropane

A suspension of zinc dust in a suitable solvent is employed, rather than sodium, in this reaction. It is analogous to the Wurtz synthesis described in the preceding chapter. (In Section 10.3 we showed how two molecules of 1-chloropropane can be condensed to hexane, the 6-carbon chain.) By the method used to make cyclopropane, the larger rings shown in Table 11.1 can be made, but less direct methods may be employed to obtain good yields. Rings of this type have been reported containing more than thirty carbon atoms.

11.2. Cycloparaffin Hydrocarbons. It is easy to see that these ring structures are nothing more than the paraffin *straight*-chain compounds tied together at the ends of the chain. For this reason we call them the *cycloparaffin* hydrocarbons. In the process of tieing up the ends of a chain it is necessary to remove, indirectly, one H atom from each end.

77

The general formula of a paraffin hydrocarbon is C_nH_{2n+2}. Accordingly, the general formula of a cycloparaffin hydrocarbon is C_nH_{2n}. This is another family of hydrocarbons.

This ring family is also known as the *naphthenes*.

A third name for the family is the *polymethylenes*. This is probably the best of the family names, for it completely describes their structure. The $-CH_2-$ group occurs in many compounds and it has been named the *methylene* group, as a matter of convenience. This is a group with a valence of two, and must not be confused with methyl which has a valence

$$
\begin{array}{cc}
\text{H} & \text{H} \\
| & | \\
-\text{C}- & -\text{C}-\text{H} \\
| & | \\
\text{H} & \text{H} \\
\text{Methylene} & \text{Methyl}
\end{array}
$$

of one. Thus, the compound C_3H_6 can be written $(CH_2)_3$ and can be called trimethylene as well as cyclopropane. It is used in hospitals as an anesthetic, but requires considerable care because of the explosion hazard.

The cycloparaffins are found in large quantities in Russian petroleum and in the petroleum obtained in California and other western states.

The cyclic paraffin hydrocarbons in general have properties similar to those of the open-chain paraffins. The smallest member of the family, however, is under considerable strain simply because of its small size. As a result of the tendency of the C–C bond to open up to a more comfortable angle, cyclopropane behaves differently from the larger rings

TABLE 11.1. Closed-chain Hydrocarbons (Cycloparaffins)
(Compare the open-chain hydrocarbons in Tables 10.1 and 10.2)

Molecular formula	Name	Structural formula*	Boiling point, °C
C_3H_6	Cyclopropane	$\begin{array}{c} CH_2{-}CH_2{-}CH_2 \\ \mid\underline{\qquad\qquad}\mid \end{array}$	−34.4
C_4H_8	Cyclobutane	$\begin{array}{c} CH_2{-}CH_2 \\ \mid\quad\mid \\ CH_2{-}CH_2 \end{array}$	13
C_5H_{10}	Cyclopentane	$\begin{array}{c} CH_2{-}CH_2 \\ \mid\quad\mid \\ CH_2\quad CH_2 \\ \diagdown\diagup \\ CH_2 \end{array}$	49.5
C_6H_{12}	Cyclohexane	$\begin{array}{c} CH_2{-}CH_2{-}CH_2 \\ \mid\qquad\qquad\mid \\ CH_2{-}CH_2{-}CH_2 \end{array}$	81.4

* No attempt is made to portray the spatial configuration of these ring compounds, beyond indicating the cyclic structure.

toward some reagents. The H atoms in CH_2 groups are generally easily replaced by Cl or Br atoms; when cyclopropane is treated with chlorine the substitution takes place to some extent, but the ring opens to form an open-chain compound:

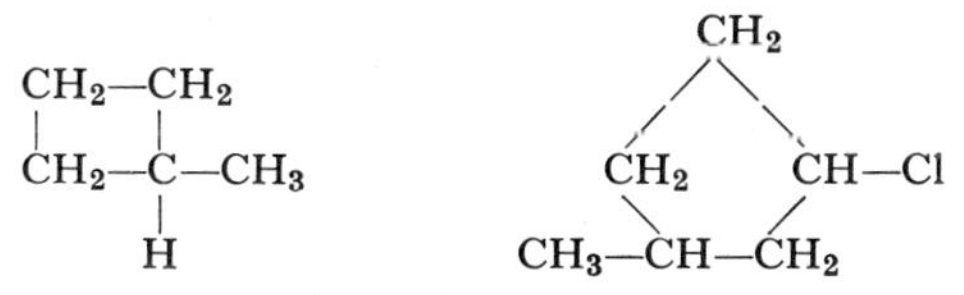

Cyclopropane   1,3-Dichloropropane

This is one of the few instances in which it is not difficult to break the C–C bond (see Section 10.1). Cyclobutane, the next member of the series of cyclic compounds does not behave this way.

11.3. How to Name the Closed-Chain Compounds. The naming of the cyclic compounds will be illustrated with a few examples:

CH₂—CH₂ etc.

Methylcyclobutane   1-Chloro-3-methylcyclopentane

In the case of compounds with several substituents, the convention is to number the atoms in the ring in a clockwise direction, with the 1 position assigned to the substituent with the highest alphabetical priority. Thus, chloro is 1 because *c* for chloro precedes *m* for methyl. Although this compound could be called 1-chloro-4-methylcyclopentane, the convention is to use the name with the smallest possible numbers.

11.4. The Different Kinds of Formulas. We have now collected so much information about the nature of carbon compounds that we must pause a moment to think a little more carefully about the formulas we have been using.

The *empirical* formula represents merely the ratio in which the atoms have combined to form a molecule; this ratio is found by a quantitative analysis. For example, C_nH_{2n} or $(CH_2)_n$ represents the ratio in which carbon and hydrogen are combined in the polymethylenes.

The *molecular* formula indicates not only the ratio of the atoms in the molecule, but also the total number of atoms in the molecule. Whereas CH_2 is the empirical formula for a series of compounds, the formula $(CH_2)_3$ is the molecular formula of just one compound in that series, the molecular weight of which was measured and found to be 42. (36 for three carbon atoms, and 6 for six hydrogen atoms). This molecular formula may be written either $(CH_2)_3$ or C_3H_6. It is the formula of cyclopropane.

Structural formulas are often preferred to molecular formulas in organic chemistry because they supply more information. For instance, $C_2H_4Cl_2$, at first glance, is simply the molecular formula of a compound derived from ethane, C_2H_6, but we know that there are really two compounds with the formula $C_2H_4Cl_2$. Their structural formulas are $CH_2Cl–CH_2Cl$ and $CH_3–CHCl_2$. These are sometimes also spoken of as *constitutional* formulas.

The *graphic* formula is one in which the atoms are written out completely, instead of only indicating the various groups in the molecule. The graphic formulas of the two compounds corresponding to the formula $C_2H_4Cl_2$ are

$$
\begin{array}{ccccc}
\text{H} & \text{H} & & \text{H} & \text{H} \\
| & | & & | & | \\
\text{H---C---C---H} & & \text{and} & \text{H---C---C---Cl} & \\
| & | & & | & | \\
\text{Cl} & \text{Cl} & & \text{H} & \text{Cl}
\end{array}
$$

1,2-Dichloroethane   1,1-Dichloroethane

It should be clear from these examples why structural or graphic formulas are used so extensively in organic chemistry, instead of the molecular formulas that serve the purpose so well for inorganic chemistry.

I2

Morphology of Chain and Ring Compounds

12.1. Introduction. Morpheus is the god of dreams in ancient mythology. His name is obtained from the Greek word *morphe*, which means shape. The Greeks were not the only people interested in dream shapes. One of the greatest achievements in the history of organic chemistry was the result of a dream in which the principal shape was a molecule of benzene (Section 19.1).

The word *morphe* is present in many technical terms. For example, lampblack is said to be amorphous, meaning that it is without shape. Snow is polymorphous; its crystals occur in many shapes. Morphology may be defined as the study of shapes or structures, and the shapes with which we are now concerned are those of the carbon compounds.

Up to this point, our carbon structures have been perfectly flat figures written in the plane of this sheet of paper in only two dimensions. Now we must begin to see the molecules in space in order to understand them. We must examine the molecule as a thing occupying space in three dimensions.

12.2. The Tetrahedral Carbon Atom. In Chapter 7 (see especially Section 7.9) we discussed the mechanism by which atoms build up to molecules through the process of electron sharing. Table 7.1 shows how the electron distribution around the carbon atom is modified in different types of compounds; in methane, all four valence electrons of carbon are in hybridized orbitals which are equivalent. The four orbitals radiate from the nucleus at equal angles of $109° 28'$. The symmetrical figure that results (Figures 12.1 and 12.2) is a tetrahedron, and the carbon atom in methane is referred to as a tetrahedral carbon atom.

The carbon atom retains this tetrahedral shape in all of its compounds of the methane type, and the valence angles vary only slightly with the nature of the attached atoms. For example, if a hydrogen atom in methane is replaced by a chlorine atom to give CH_3Cl, the valence angles will not

81

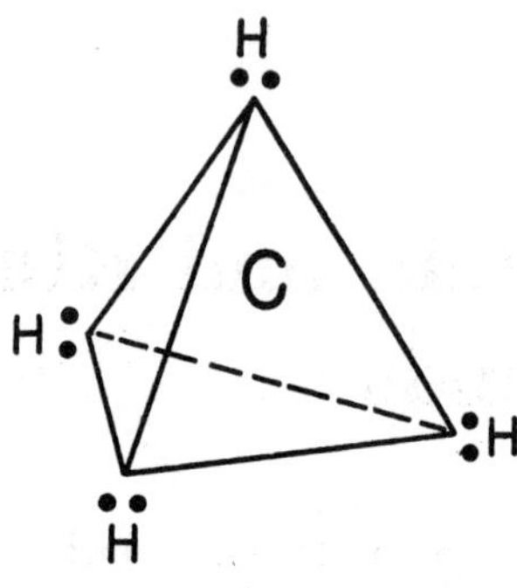

12.1 Methane. The four electron pairs at the corners of a tetrahedron.

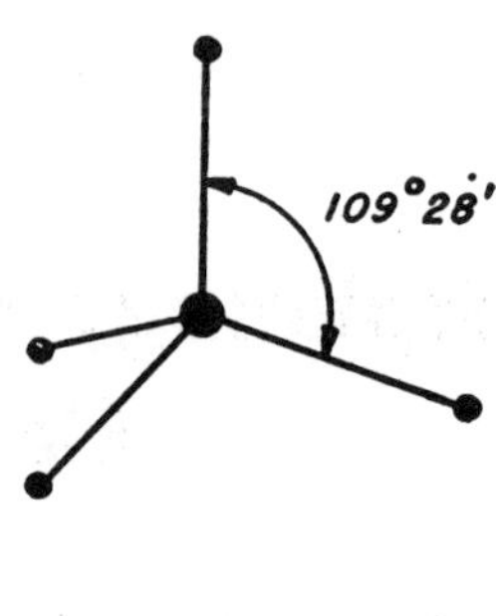

12.2. Methane. The four valence bonds directed to the corners of a tetrahedron.

be greatly changed even though the Cl atom is larger than the H atom and has a different pull on the electron pair.

The rest of this chapter will be devoted to a study of the various shapes of carbon compounds, in so far as these shapes are determined by the tetrahedral nature of the carbon atom as shown in Figure 12.2. At this point, the student may profitably review the subject matter associated with Figures 8.4 to 8.7.

12.3. The Conformations of Chain Compounds. In Table 10.1 are listed the first members of the paraffin series of hydrocarbons. What they look like when built up from the tetrahedral carbon atoms of Figure 12.2 is illustrated in Figures 12.3 to 12.7, in the molecules of ethane, propane, and pentane.

Pentane is a liquid; therefore the molecules have considerable freedom of motion. It is also true, within limits, that there is a freedom of rotation around the C–C bonds (the "limits" will be explained later when we study energy factors in Section 14.11). The various shapes a molecule can assume by permitting rotation around C–C bonds are called conformations, and any single representation is a *conformer*. These vary in stability. In the solid, crystalline state, pentane and other paraffin hydrocarbons are in the extended zigzag conformation. The ring pattern of pentane is probably its least stable conformer, but the ability of the 5-carbon chain to arrange itself in this pattern is shown by the ease with which cyclopentane (Table 11.1) is made.

The principles just mentioned have been known for some time, but

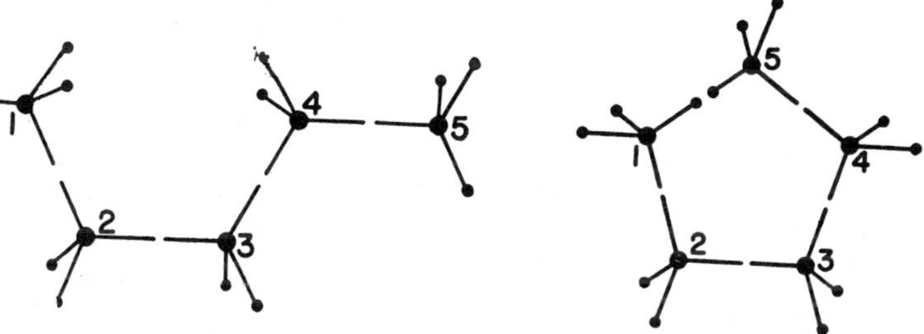

12.3   Ethane, CH_3—CH_3

12.4.   Propane, CH_3—CH_2—CH_3

12.5.   Pentane, zigzag pattern. CH_3—CH_2—CH_2—CH_2—CH_3

12.6.   Pentane, meander pattern.

12.7.   Pentane, ring pattern.

since about 1950 a school of chemists has found that the phenomenon
is important enough to have its own name (conformational analysis) and
its own terminology, some of which will now be defined. Figures 12.8
and 12.9 show two conformers of ethane—in each case a side view of the
molecule and a view from one end. The *eclipsed* conformer, in which all
the H atoms face each other, is less stable because of the greater repulsion effects of these atoms; this repulsion tends to cause rotation around
the C C bond to the *anti* or *trans* conformer (also see Section 14.11).
The *anti* conformer is also known as *staggered*.

The representation of conformers on the printed page can be further

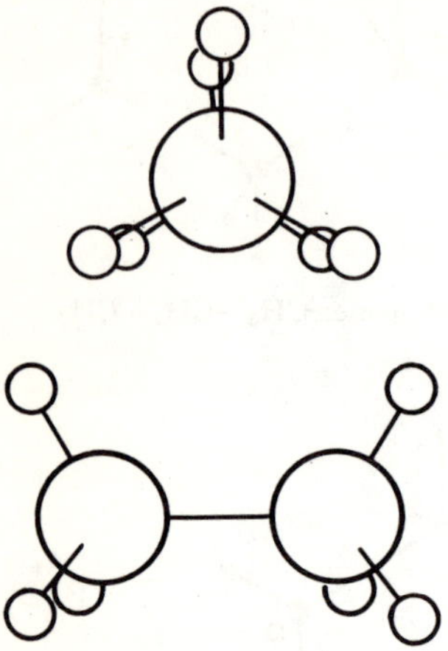

12.8. *Eclipsed* conformer of ethane. *Above:* View from either end. *Below:* side view.

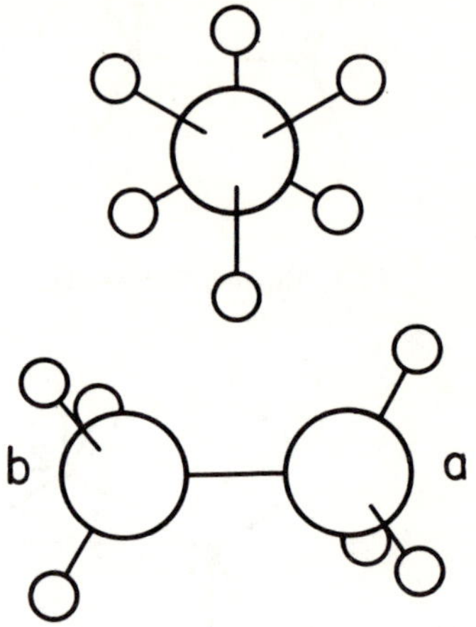

12.9. *Anti* or *trans* conformer of ethane. *Above:* View from end *a*. *Below:* Side view.

illustrated with the molecule of butane, which can be visualized as Figure 12.5 without the fifth carbon atom.   If we focus attention on the bond between carbon atoms 2 and 3, the side view of the *eclipsed* conformation is shown by Figure 12.10.   The end views by the so-called Newman convention are in Figure 12.11.   In this group of diagrams we introduce another type of conformer known as *gauche*, or *skew;* it represents all the positions

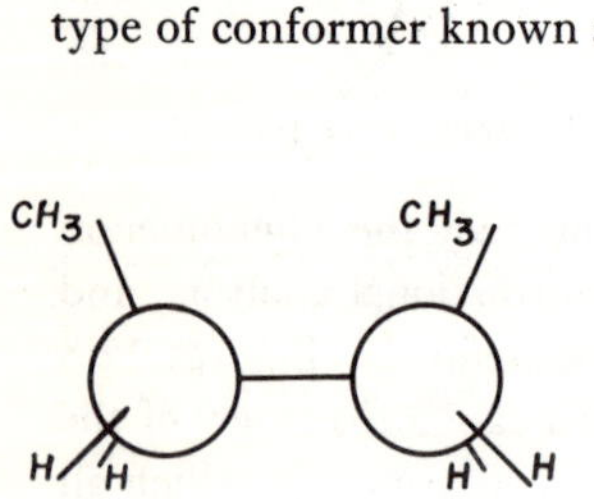

12.10. Side view of *eclipsed* conformer of butane.

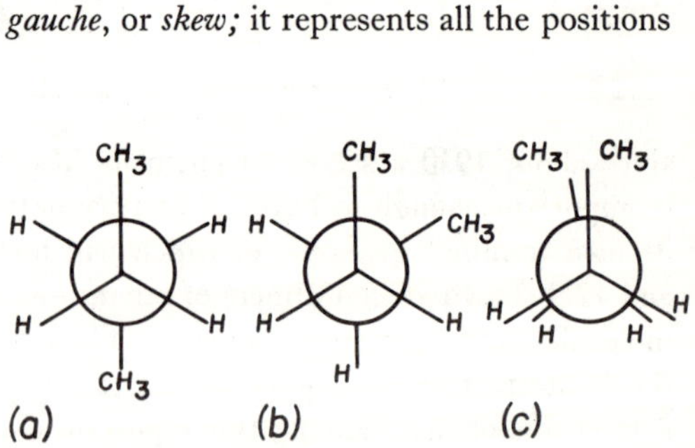

12.11. End views of butane conformers according to Newman convention.   (a) *anti.* (b) *gauche*, or *skew.*   (c) *eclipsed.*

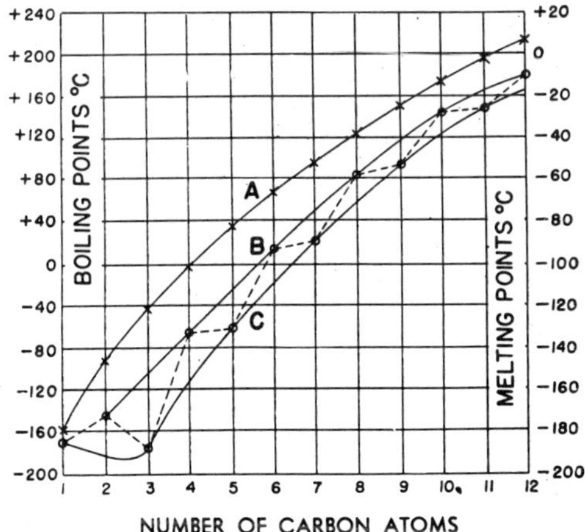

12.12. Melting points and boiling points of straight-chain hydro-carbons. Curve A: boiling points; curve B: melting points of even molecules; curve C: melting points of odd molecules.

intermediate between *eclipsed* and *anti*. The two CH_3 groups are the reference points on the basis of which we label these molecules as *eclipsed* or *anti*.

12.4. "Odd" and "Even" Molecules. The melting points and boiling points of the first twelve straight-chain hydrocarbons listed in Table 10.1 are plotted in Figure 12.12. The melting points form two smooth curves; the data for even molecules fall on curve B, whereas the data for odd molecules fall on curve C. The odd and even molecules form two different patterns when the zigzag, or staggered, chains are aligned in the crystal (see Figure 12.13). From X-ray analysis it has been found that these patterns are responsible for differences in crystal packing; the odd-

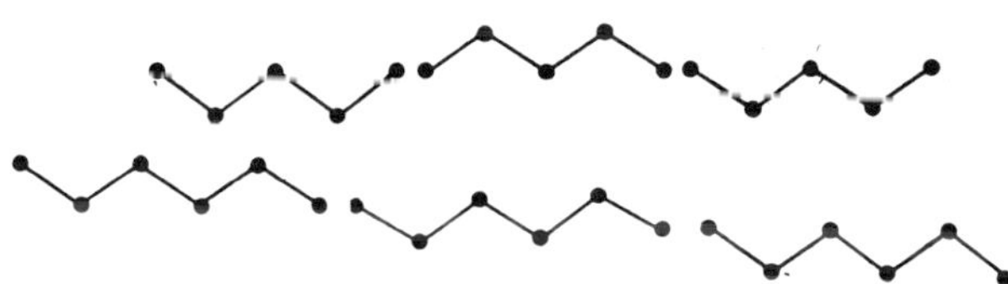

12.13. Alignment of odd (5-carbon) and even (6-carbon) chains in the solid state.

chain structures are slightly less stable, and melt at lower temperatures than the even-chain compounds.

The boiling points in this series of compounds (curve A) show no unusual behavior. They fall on a smooth curve because of the freedom of motion of molecules in the liquid state.

12.5. Conformations of Ring Compounds. Representative cyclic hydrocarbons with properties similar to those of the paraffin hydrocarbons are listed in Table 11.1. In the early 1880's the ring compounds smaller than 6-carbon had not yet been found in nature and had not been made in the laboratory. No less a famous chemist than Victor Meyer said they probably could not be made, but they were synthesized in that same period. In 1885 another famous chemist, Baeyer, announced his *strain theory* to account for the fact that the smaller rings were found to be chemically less stable.

The reasoning behind the Baeyer strain theory can be understood from Figures 12.2 and 12.7; if carbon atoms 1 and 5 of pentane in Figure 12.7 should join to form cyclopentane, the bond angles would be essentially the tetrahedral angle shown in Figure 12.2, and the angle strain would be almost nil (actually 0° 44′). It can be seen from Figure 12.7 that linking carbon atoms 1 and 3 to make cyclopropane introduces considerable strain (24° 44′), and similarly the strain in cyclobutane is 9° 44′. A 6-membered ring would have an angle strain in a negative direction (–5° 16′); in fact, the model in Figure 12.14 had to be assembled with springs instead of wood-pegs to link the carbon atoms. All this is based on the tetrahedral carbon atom, with the carbon atoms all in the same plane (that is, *planar*).

12.14. *Planar* conformation of cyclohexane. The H atoms are not shown.

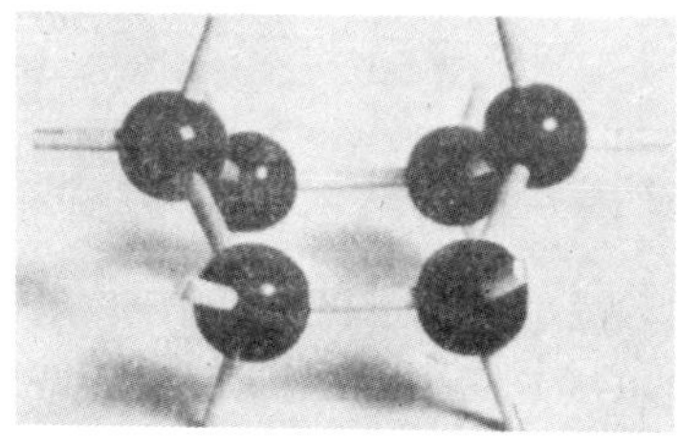

12.15.   The *eclipsed* conformation of cyclohexane; also known as *boat* or *tub* conformation.   The H atoms are not shown.

12.16.   The *anti* or *trans* conformation of cyclohexane; also known as *chair* or *staggered* conformation.   The H atoms are not shown.

These considerations would make cyclopentane one of the most stable in this series of compounds.

Modern methods of physical chemistry have shown that rotation around the C–C bond, just described for the open-chain compounds, occurs also in the cyclic series; in the ring compounds it is called *crumpling, puckering,* or *buckling*, as illustrated in Figures 12.15 and 12.16.

In connection with Figures 12.8 and 12.9 it was explained that rotation in the ethane molecule occurs in order that it may relieve the repulsion by the H atoms in the *eclipsed* conformation.   In the *anti* conformer the H atoms are as far apart as they can get.   Similarly, butane in Figure 12.11a prefers the *anti* conformation because then the CH_3 groups are further apart.   The strain that causes these rotations around the carbon-carbon bond is called *eclipsing strain*.   The strain we described in the Baeyer strain theory is called *angle strain*.

The crumpling of the cyclohexane ring in Figures 12.15 and 12.16 is due to eclipsing strain.   The C–C bonds in the resulting conformations have the angle shown in figure 12.2 and are therefore free from angle strain. The *anti* conformer is more stable than the *eclipsed* because it has less eclipsing strain.   It should be understood that in the course of random molecular motions many conformations intermediate between these two are possible.

The cyclopentane ring, at one time thought to be more stable than cyclohexane, is actually a little less stable; it crumples in order to offset the eclipsing strain which would prevail if it were planar like the molecule in Figure 12.7, but in its crumpled form the C–C bond angles are not as closely tetrahedral as in cyclohexane.   Cyclobutane likewise departs from

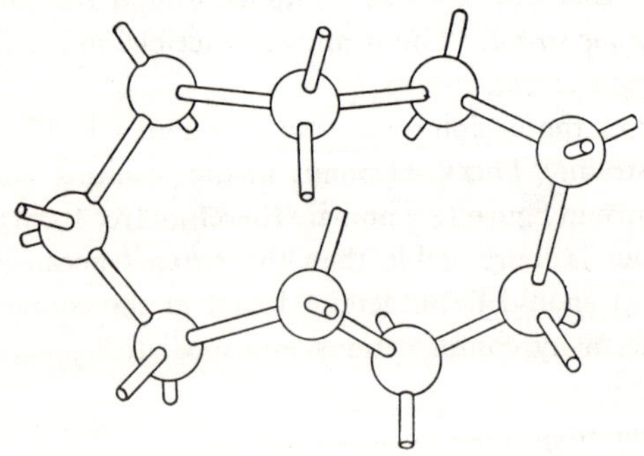

12.17. Chair form of cyclohexane, showing an axial bond *a* and an equatorial bond *e*. The H atoms are not indicated.

a planar form to reduce both eclipsing strain and angle strain as much as possible. Cyclopropane, however, cannot crumple, and has considerable angle strain as well as eclipsing strain; it consequently has greater chemical reactivity (see Section 11.2)

12.6. Axial and Equatorial Bonds. The importance of eclipsing strain to the stability of a molecule has led specialists in this field to give names to certain bond positions. Two of the more common terms are illustrated in Figure 12.17 in the chair model of cyclohexane. The *axial* bonds (a) are those at right angles to the general plane of the ring. *Equatorial* bonds (e) are those in the plane of the ring. (Only two of the bonds in the drawing are labeled.)

12.7. Large Rings. For many years no ring compounds containing more than six carbon atoms had been discovered in nature, and no ring compound containing more than nine carbon atoms had been prepared in the laboratory. It was felt that the absence of large rings was due to excessive strain. In 1920–1930, however, large ring compounds containing

12.18. Arrangement of carbon atoms in cyclononane, C_9H_{18}. One of several possible conformations, in all of which crumpling of the ring takes place.

seventeen and eighteen carbon atoms were recognized in nature (see Section 32.1), and in recent years cycloparaffins containing more than thirty carbon atoms in the ring have been prepared in the laboratory. These are quite stable, and it is believed they are crumpled rings.

It should not be supposed that the mere act of crumpling a ring of carbon atoms leads automatically to a structure free from strain. As already explained, we must also consider the eclipsing strain set up when similar atoms or groups are brought into opposition. That this is an important factor in large rings is illustrated by the model for the 9-carbon ring in Figure 12.18, which is shown with the H atoms omitted in order to make the picture clear, but with the valence bonds shown attached to the carbon atoms. It will be seen that when carbon atoms are tied into a **9-membered** ring some of the valence bonds on the carbon atoms are twisted in toward the center. The H atoms on these bonds have very little space available in the center of this crumpled ring and there is considerable mutual repulsion and thus a certain amount of strain.

Rings with about fifteen carbon atoms in the cycle are much roomier than the nine-membered ring shown in Figure 12.18 and the strain is quite small for these large rings. The very large rings probably have the shape of a circle that has been pulled out to resemble a flattened rubber band. A large ring behaves like two long zigzag chains lying parallel to each other and joined at the ends.

In concluding these remarks on ring compounds, it must be emphasized that the *relative ease of formation* of a carbon ring is not necessarily the same as its *relative stability*. Ease of formation of a ring depends largely on the probability of the end carbon atoms in a chain coming close together, and this probability is very great in a five-membered ring as we have already pointed out in Section 12.3.

Once a thirty-membered ring has been formed, however, it may be nearly as stable as a five-membered ring, even though it may have been harder to prepare because of the smaller probability of the end carbon atoms approaching each other.

12.8. Atom Models. These are indispensible tools in the study of molecules and their structures. At present a large variety of atom models can be purchased that serve various purposes. For lecture demonstrations, the ball and stick type in Figure 12.15 is useful. For research, elaborate sets are available which show distances between nuclei of atoms and the valence angles between them. The models in Figure 12.19 were assembled from a kit which has since been considerably improved with respect to the fasteners. These models are constructed on a scale of

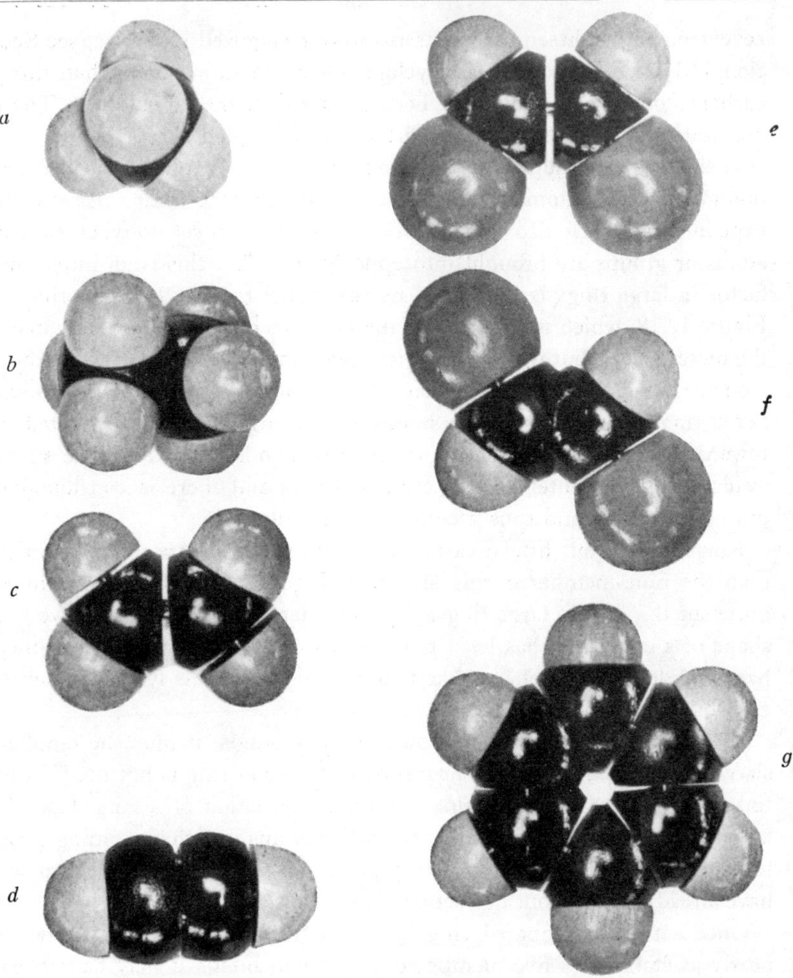

12.19.　Fisher-Hirschfelder molecular models.
(*a*)　Methane (Figure 12.1)
(*b*)　Ethane (Figure 12.3)
(*c*)　Ethene (Figure 15.1)
(*d*)　Ethyne (Section 13.5)
(*e*)　*cis*-Dichloroethene (Section 31.4)
(*f*)　*trans*-Dichloroethene (Section 31.4)
(*g*)　Benzene (Section 19.2)

100,000,000 to 1, so that 1 cm in the model represents 1 A (1 Angstrom unit is 0.000,000,01 cm) in the actual molecule. A more complex molecule built from these models is shown in Figure 38.3.

The models in Figure 12.19, which portray atoms linked by electron-pair bonds, should be compared with the model of silver bromide (Figure 8.2) which consists of ions linked by electrostatic attraction.

13

Double and Triple Bonds

13.1. Introduction. As explained in earlier chapters, the chain and ring hydrocarbons are relatively inactive. Since the chemist by definition is primarily interested in the transformations of substances, he would enjoy working with these structures to a much greater degree if he could increase their activity. This, we found, he can do by the simple trick of inserting a Cl atom into the molecule.

In this chapter we shall describe carbon compounds that do not depend on a foreign element for enhanced chemical reactivity. The greater activity in these compounds will be found to result from the way in which certain pairs of *adjacent* carbon atoms are linked in the chains and rings. These structural types have not yet been mentioned in this book.

Up to this point we have been careful to limit ourselves to those compounds in which each carbon atom uses only one of its valence bonds to combine with an adjacent carbon atom. This is called a *single* bond, and is represented as in Model I. Model II shows how the *double bond* is formed when each carbon atom uses two of its bonds, and Model III indicates how the *triple bond* is formed by the use of three bonds.

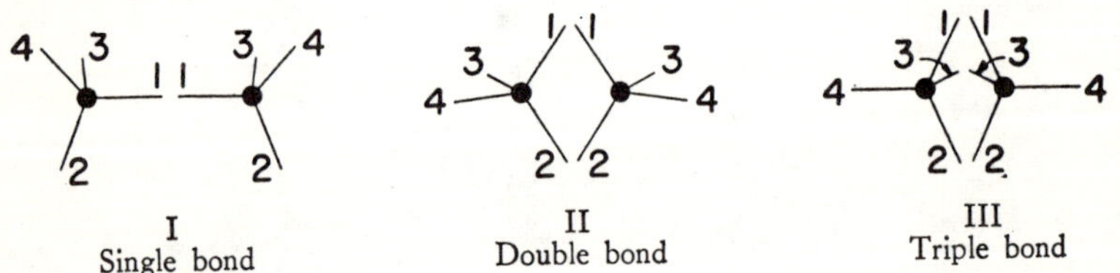

I	II	III
Single bond	Double bond	Triple bond

When projected on the plane of a printed page, these three kinds of carbon-to-carbon union are usually written in this way:

$$-\overset{|}{\underset{|}{C}}-\overset{|}{\underset{|}{C}}- \qquad -\overset{|}{C}=\overset{|}{C}- \qquad -C\equiv C-$$

Each carbon atom has four valences, and one, two, or three of these val-

92

ences may join with a second carbon atom. It should be observed in Model II that if four H atoms are attached to the bonds marked 3 and 4 they are in the *same plane* as the two carbon atoms. According to Model III the two H atoms attached to the bonds marked 4 are in the *same straight line* as the two carbon atoms.

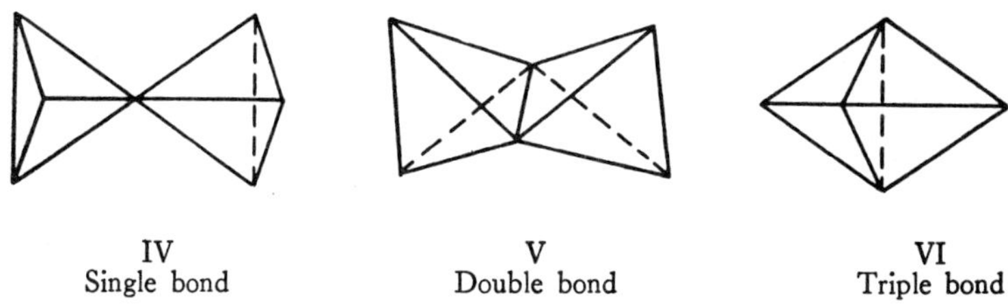

<table>
<tr><td align="center">IV
Single bond</td><td align="center">V
Double bond</td><td align="center">VI
Triple bond</td></tr>
</table>

If the full tetrahedral model of the carbon atom is used (Chapter 12), another interesting picture can be obtained of these three kinds of carbon-to-carbon bonds, shown in Models IV, V, and VI. A single bond consists of a point-to-point union, a double bond is an edge-to-edge union, and in the triple bond the union is face-to-face.

We should also consider the electronic formulas, in which a pair of electrons represents one bond:

$$\begin{array}{ccc} \text{H H} & \text{H H} & \\ \text{H:C:C:H} & \text{H:C::C:H} & \text{H:C:::C:H} \\ \text{H H} & & \end{array}$$

13.2. Preparation of Compounds with Double and Triple Bonds. It will be an easier task to explain the chemistry of molecules in which the carbon atoms are united by double and triple bonds if we first find out how they can be formed. We shall start with the simplest possible case, that is, a compound with only two carbon atoms.

$$\underset{\text{Ethane}}{\text{H}-\overset{\text{H}}{\underset{\text{H}}{\text{C}}}-\overset{\text{H}}{\underset{\text{H}}{\text{C}}}-\text{H}} \quad \xrightarrow{\text{Cl}_2} \quad \underset{\text{Chloroethane}}{\text{H}-\overset{\text{H}}{\underset{\text{H}}{\text{C}}}-\overset{\text{H}}{\underset{\text{Cl}}{\text{C}}}-\text{H}} \quad \xrightarrow[\text{alcohol}]{\text{K}-\text{OH}} \quad \underset{\text{Ethene}}{\text{H}-\overset{\text{H}}{\text{C}}=\overset{\text{H}}{\text{C}}-\text{H}} \quad + \quad \text{HCl}$$

The only two-carbon hydrocarbon we have yet described in this book is ethane, CH_3–CH_3. We found that ethane behaves like methane; it allows its H atoms to be replaced by chlorine or bromine atoms. After one H atom has been replaced by Cl, the compound has a greater reactivity.   If we treat the resulting chloroethane with an alkali under certain conditions, a molecule of HCl is split out from *adjacent* carbon atoms and a *double bond* is formed.   The KOH must be dissolved in alcohol (not in water) for the reaction to take place as indicated; the reaction also proceeds rather quickly.   In Section 9.4 it was shown that when the reagent used is a solution of KOH in water a *slow* reaction results in which –Cl is replaced by –OH.   The reader should compare the results obtained.

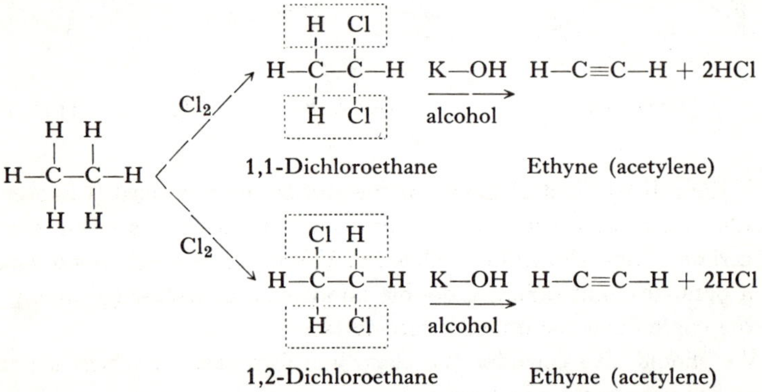

$$H-C\equiv C-H + 2HCl$$

1,1-Dichloroethane   Ethyne (acetylene)

$$H-C\equiv C-H + 2HCl$$

1,2-Dichloroethane   Ethyne (acetylene)

The preceding equations show a pair of isomers obtained when two Cl atoms are introduced into the ethane structure.   Each of the dichloroethanes yields a *triple bond* when treated with an alcoholic KOH solution; the compound formed is acetylene in each case.   Although these reactions by which multiple bonds can be formed may not be useful as production methods, they illustrate the structural relationship between the compounds with multiple bonds and the corresponding compounds with single bonds between carbon atoms.   Two atoms or radicals must be removed from adjacent carbon atoms to form an extra bond:

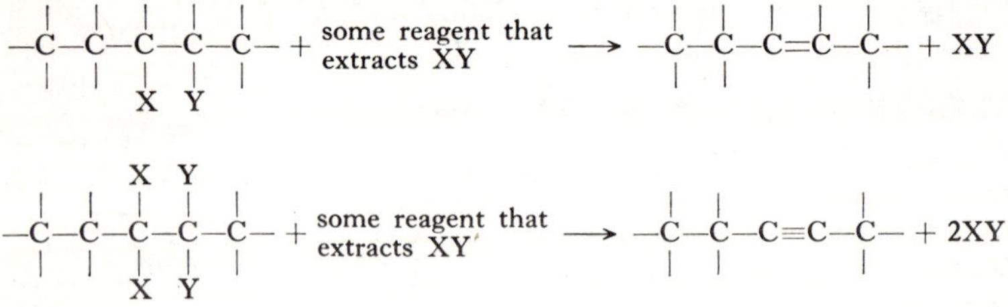

And this is a method of preparation for the double bond only:

$$\underset{\text{Ethyl alcohol}}{H-\overset{\overset{\displaystyle H}{|}}{\underset{\underset{\displaystyle H}{|}}{C}}-\overset{\overset{\displaystyle H}{|}}{\underset{\underset{\displaystyle OH}{|}}{C}}-H} + \text{heat, and certain dehydrating agents, like conc. } H_2SO_4 \longrightarrow \underset{\text{Ethene}}{CH_2{=}CH_2} + H_2O$$

13.3. How to Name Compounds with Double Bonds and Triple Bonds.

The double bond is indicated by the ending *ene* and the triple bond by *yne*, just as in preceding chapters the single bond was represented by *ane*. As usual, we consider the chain of carbon atoms as the general framework of the compound, and from the number of carbon atoms in the longest chain containing the multiple bond we derive the parent, or last name of the compound. If the 6-carbon chain compound described in Section 10.8 is altered by introducing a double bond at the fourth carbon atom from the left, the name and formula would be as follows:

$$H-\overset{\overset{H}{|}}{\underset{\underset{H}{|}}{C}}-\overset{\overset{H}{|}}{\underset{\underset{OH}{|}}{C}}-\overset{\overset{H}{|}}{\underset{\underset{CH_3}{|}}{C}}-\overset{}{\underset{\underset{Cl}{|}}{C}}{=}\overset{}{\underset{\underset{H}{|}}{C}}-\overset{\overset{H}{|}}{\underset{\underset{H}{|}}{C}}-H \qquad \text{4-Chloro-3-methyl-4-hexen-2-ol}$$

The $-C{=}C-$ group is regarded as one of the "functional" groups in a molecule (Section 10.9) and the presence of this group is, therefore, indicated in the main part of the name of the compound, thus, hex*ene*. The $-OH$ group takes precedence over the $-C{=}C-$ group, however, so the molecule is numbered from the left end in order to make the number associated with this chief function as small as possible. The compound is a *hexenol*—that is, an olefin (Section 13.4) and an alcohol.

Modifying the compound in such a way as to introduce a triple bond will change it to a *hexynol:*

$$H-\overset{\overset{H}{|}}{\underset{\underset{H}{|}}{C}}-\overset{\overset{H}{|}}{\underset{\underset{OH}{|}}{C}}-\overset{\overset{H}{|}}{\underset{\underset{CH_3}{|}}{C}}-\overset{\overset{H}{|}}{\underset{\underset{Cl}{|}}{C}}-\overset{}{C}{\equiv}\overset{}{C}-H \qquad \text{4-Chloro-3-methyl-5-hexyn-2-ol}$$

From what has been said it is apparent that double bonds or triple bonds can be situated in various parts of a molecule. Here are two compounds with a straight chain of five carbon atoms,

$$\underset{\text{2-Pentene}}{CH_3-\overset{\overset{H}{|}}{C}{=}\overset{\overset{H}{|}}{C}-CH_2-CH_3} \qquad \text{and} \qquad \underset{\text{1-Pentene}}{CH_2{=}\overset{\overset{H}{|}}{C}-CH_2-CH_2-CH_3}$$

which have the same molecular formula, C_5H_{10}, but different arrangements of the atoms. They therefore have different properties and are isomers (Section 10.4).

It is also possible for both double bonds and triple bonds to appear in the same molecule:

$$CH_2{=}CH{-}C{\equiv}C{-}CH_3$$

1-Pentene-3-yne
(Double bond gets the lower number)

Notice that when the $-C{\equiv}C-$ group is somewhere in the interior of a chain there are no H atoms on those two carbon atoms.

Many cyclic compounds are known in which double bonds are present —for example,

$$\begin{array}{c} CH_2 \\ \diagup\diagdown \\ CH{=}CH \end{array}\quad \text{Cyclopropene}$$

but relatively few have been prepared containing triple bonds. The smallest cyclic compound reported with a triple bond in the ring is cyclo-octyne, C_8H_{12}, the eight-membered ring, the first apparently unequivocal synthesis of which was published in 1953. (Also see the reference to benzyne in Section 19.8.)

13.4. The Olefin Hydrocarbons. We found in Section 10.6 that chain hydrocarbons formed exclusively with C–C units are called *paraffins*, have the general formula C_nH_{2n+2}, and have names ending in *ane*. The simpler members of the *olefin* family are shown in Table 13.1. Comparing the compounds in this table with the paraffins in Table 10.1, it will be found that these olefins can be regarded as paraffin hydrocarbons with *one* double bond somewhere in the chain. The olefins form a homol-

TABLE 13.1.   Olefin Hydrocarbons

Name	Molecular formula	Structural formula	Boiling point, °C
Ethene	C_2H_4	$CH_2{=}CH_2$	-102.4
Propene	C_3H_6	$CH_3{-}CH{=}CH_2$	$-\ \ 47.7$
1-Butene	C_4H_8	$CH_3{-}CH_2{-}CH{=}CH_2$	$-\ \ \ 6.47$
*2-Butene (cis)	C_4H_8	$CH_3{-}CH{=}CH{-}CH_3$	$+\ \ \ 3.73$
*2-Butene (trans)	C_4H_8	$CH_3{-}CH{=}CH{-}CH_3$	$+\ \ \ 0.5$
2-Methyl-1-propene	C_4H_8	$CH_3{-}C{=}CH_2$ $\quad\ \ \|$ $\quad\ \ CH_3$	$-\ \ \ 6.6$
†1-Pentene	C_5H_{10}	$CH_3{-}CH_2{-}CH_2{-}CH{=}CH_2$	$+\ \ 30.1$

* The difference between *cis-* and *trans-*2-butene will be explained in Section 22.3.

† Pentene is more often called amylene; four other isomers of the composition C_5H_{10} are known but are not included in this table.

ogous series of compounds, just like the paraffin series. The olefin series corresponds to C_nH_{2n}, since each compound contains two H atoms less than its corresponding paraffin compound.

It is also apparent from the examples given in the table that because of the double bond there are more isomers possible in the olefin series than in the paraffin series. If the meaning of isomerism has become hazy to the reader it should be reviewed in Section 10.4.

There are four isomers of the composition C_4H_8 in the olefin series, but only two of the composition C_4H_{10} in the paraffin family. Three of the four isomers of C_4H_8 are easy to figure out as being due to a straight-chain or branched-chain arrangement of the carbon atoms, or the position of the double bond. In the case of 2-butene, however, we find a kind of isomerism which we have not yet met and to which we shall devote a separate section (Section 22.3).

The early chemists who first studied the properties of ethene discovered that it forms oily compounds (this term probably referred to their insolubility in water) when it combines with chlorine or bromine. The reaction will be found written out in Section 13.9. Because of the oily nature of the addition product which is formed, ethene was known as the "oil-forming" or "olefiant" gas. This is why other hydrocarbons containing double bonds and related to ethene have been called, in general, *olefins*. Ethene has had a long history in which it was known as *ethylene*. This name probably will never become outmoded.

Ethylene and other olefins are obtained commercially by the cracking (breaking down) of hydrocarbons in petroleum. Ethylene is used in enormous quantities for making ethyl alcohol, polyethylene plastics, polystyrene, etc. Propylene is similarly used to make polypropylene, isopropyl ("rubbing") alcohol, and many other products.

13.5. The Acetylene Hydrocarbons. The simplest hydrocarbon containing a triple bond, H–C≡C–H, is *ethyne*, known more commonly as *acetylene*. Metal derivatives of acetylene are either *carbides* or *acetylides*

TABLE 13.2. Acetylene Hydrocarbons

Name	Molecular formula	Structural formula	Boiling point, °C
Ethyne	C_2H_2	CH≡CH	−83
Propyne	C_3H_4	CH_3—C≡CH	−23.3
1 Butyne	C_4H_6	CH_3—CH_2—C≡CH	+8.6
2—Butyne	C_4H_6	CH_3—C≡C—CH_3	+27.2
1—Pentyne*	C_5H_8	CH_3—CH_2—CH_2—C≡CH	+39.7

* The other possible isomers of pentyne have not been included in this table.

(Section 13.13) Acetylene is the first member of a family of hydrocarbons (which, as usual, takes its name from that of the simplest member) and is consequently known as the acetylene series. It is a relatively unimportant series of compounds except for the first member, acetylene itself.

The acetylene hydrocarbons can be regarded as the corresponding paraffin series (Table 10.1) with *one* triple bond somewhere in the chain of carbon atoms. Since each triple bond means the loss of four hydrogen atoms, the general formula for this series of compounds is C_nH_{2n-2} as compared with the formula C_nH_{2n+2} for the paraffins. The names of all the compounds in the acetylene family end in *-yne*.

It is of interest to observe that in Table 13.1 four isomeric olefin compounds were tabulated with the composition C_4H_8, but in Table 13.2, only two acetylene compounds can be written for the composition C_4H_6. The presence of the triple bond has in this case limited the number of possible arrangements of the carbon atoms.

Acetylene is made and used in large quantity. One source is calcium carbide (Section 13.13) which in turn is made from coke (carbon) and lime (calcium oxide). Acetylene is a gas, at one time used for illumination. A major use is in the oxy-acetylene flame for welding. It is explosive when subjected to shock, but is stable in acetone solution under pressure in tanks containing a porous absorbent.

13.6. Poly-enes and Poly-ynes. It is possible, of course, for a molecule to contain several (*poly*) double and triple bonds, as shown by the following examples:

$$CH_2{=}CH{-}CH{=}CH_2 \qquad\qquad HC{\equiv}C{-}CH_2{-}CH_2{-}C{\equiv}CH$$

$$\text{1,3-Butadiene} \qquad\qquad\qquad \text{1,5-Hexa-di-yne}$$

Butadiene will be discussed in Section 15.7, where it will be shown to have unusual properties because of its alternate arrangement of double and single bonds. The most interesting thing about hexadiyne is that it is an isomer of benzene (Section 19.1) the composition of which is C_6H_6.

13.7. Unsaturation, and Addition Reactions to Multiple Bonds. Olefins and acetylenes are said to be *unsaturated* because of their tendency to add on other complete molecules. This behavior is different from that described in Section 8.5 in connection with the dative-bond union of ammonia to boron trichloride. When adding to the multiple bond, the adding molecule in the course of the reaction divides into two fragments which attach to the two carbon atoms involved. The product is *saturated* in that the carbon chain now has only the chemically stable single bonds. In this chapter we shall illustrate the phenomenon with several examples in a qualitative way; a more detailed explanation on the basis of electron theory will be presented in Chapter 17.

13.8. Hydrogenation. Hydrogen can be added directly to compounds containing $-\overset{|}{C}=\overset{|}{C}-$ and $-C\equiv C-$ groups, usually without any difficulty:

$$CH_2{=}CH_2 + H_2 \longrightarrow CH_3{-}CH_3$$

Ethene   Ethane

$$CH_3{-}C\equiv CH + H_2 \longrightarrow CH_3{-}CH{=}CH_2 + H_2 \longrightarrow CH_3{-}CH_2{-}CH_3$$

Propyne   Propene   Propane

The customary procedure is to mix the vaporized hydrocarbon with the hydrogen and pass the mixture over powdered nickel at an elevated temperature. The process is called *hydrogenation*, and nickel is an excellent hydrogenation catalyst.

Addition reactions of this kind used to be referred to as "opening up" of the double bond and visualized as the breaking of one of the bonds in Figure 15.1. It should be observed that the addition always takes place *across the double bond,*

and that this is the reverse of what takes place in the formation of the double bond, as shown in Section 13.2. If there is a triple bond in the chain, the addition takes place in two stages, as indicated for the hydrogenation of propyne to propene and then to propane. In some instances, the reaction can be stopped at the first stage.

13.9. Addition of Halogen. Halogens, also, add directly to multiple bonds:

$$CH_2{=}CH_2 + Cl{-}Cl \longrightarrow \underset{\underset{Cl \quad\ Cl}{|\qquad |}}{CH_2{-}CH_2}$$

Ethene   Ethene dichloride
or, 1,2-Dichloroethane

The product of this reaction can be named in two ways. It can be named as an *addition* product of ethene or as a *substitution* product of ethane.

The addition of Cl_2 to a double bond is easier at a lower temperature, whereas substitution for hydrogen atoms in the chain is aided by a higher temperature. This is well illustrated by the action of chlorine on propene at temperatures of 250 to 500°C with yields of up to 90% of the substitution product:

$$CH_2{=}CH{-}CH_3 \;+\; Cl_2 \longrightarrow CH_2{=}CH{-}CH_2Cl \;+\; HCl$$

3-Chloro-1-propene
Allyl chloride

The addition reaction will be explained in Section 17.5 as an ionic reaction, whereas the substitution will be ascribed to a free-radical mechanism in Section 17.9.

13.10. Addition of Hydroxyl Groups. Certain oxidizing agents, such as potassium permanganate ($KMnO_4$) in aqueous solution, act on the double bond as follows:

$$CH_2{=}CH_2 \;+\; H_2O \;+\; O \longrightarrow \begin{matrix} CH_2{-}CH_2 \\ |\quad\; | \\ OH \;\; OH \end{matrix}$$

Ethene glycol

It will be seen that the action of the oxidizing agent is equivalent to the addition of a molecule of hydrogen peroxide (H_2O_2 or HO–OH) across the double bond. The product of this reaction, containing two –OH groups, is a member of the family of glycols (Section 25.7).

13.11. Unsymmetrical Addition. In the preceding examples of addition reactions, the added molecules are symmetrical substances, such as chlorine (Cl–Cl). An unsymmetrical added molecule is in general represented by H–X. Such unsymmetrical combinations exist in hydrochloric acid (H–Cl) and in sulphuric acid (H–SO_4H). These are inorganic substances which readily ionize by splitting up into a positive part and a negative part.

The addition of an unsymmetrical compound to a symmetrical compound containing a double bond does not offer anything unusual:

$$CH_3{-}CH{=}CH{-}CH_3 \;+\; HCl \longrightarrow \begin{matrix} CH_3{-}CH_2{-}CH{-}CH_3 \\ | \\ Cl \end{matrix}$$

Butene-2   2-Chlorobutane

However, if the addition takes place on an unsymmetrical double-bond compound, there are several possibilities:

$$CH_3-CH=CH_2 + HCl \longrightarrow CH_3-\underset{\underset{Cl}{|}}{CH}-CH_3 \text{ or } CH_3-CH_2-CH_2Cl$$

Propene   2-Chloropropane   1-Chloropropane

Experience has shown that the *negative half* of the adding molecule *usually* goes to the carbon atom with the least hydrogen. Therefore, 2-chloropropane is the predominant product in this reaction. This characteristic behavior of unsymmetrical compounds such as H–Cl, when added to a double bond, is spoken of as "directed addition" to the double bond and will be discussed further in Section 17.11. This is not, however, to be taken as an infallible rule.

According to this same principle of "directed addition," when HCl is added to triple-bond compounds, the negative half of the molecule tends to "pile up" on the same carbon atom:

$$CH_3-C\equiv C-H + HCl \longrightarrow CH_3-\underset{\underset{Cl}{|}}{C}=CH_2 + HCl \longrightarrow CH_3-\overset{\overset{Cl}{|}}{\underset{\underset{Cl}{|}}{C}}-CH_3$$

Propyne   2-Chloropropene   2,2-Dichloropropane

13.12. Effect of a Multiple Bond on Other Members of the Chain.
Although the most important characteristic of the multiple bond is that it readily adds on other substances, there are a few other properties to be considered which do not come in this category. Two of these properties are discussed in this section and the next. The presence of a double bond in a molecule makes a profound change in the behavior of the two carbon atoms at the ends of the bond, and often produces changes in other parts of the molecule. For example, in such a substance as this,

$$CH_3-\underset{\underset{(a)Cl}{|}}{C}=CH-\underset{\underset{Cl(b)}{|}}{CH_2}$$

it will be found that the two Cl atoms are very unlike in activity. The Cl atom marked (a) is very inactive, showing that the union $-\underset{\underset{Cl}{|}}{C}-$ is

stronger than $-\underset{\underset{Cl}{|}}{C}-$. It is difficult to make the (a) halogen atom go

through the reactions of the (b) halogen. In fact the (b) chlorine atom is even more reactive than it would be in a compound containing only single bonds (Section 17.3).

A good rule to remember is that in such a structure as this,

$$\ldots -CH_2-CH{=}C-CH_2- \ldots ,$$
$$\underset{\displaystyle OH}{|}$$

the compound is unstable, and rearrangements occur which will be described later (Sections 26.5 and 28.11). There are numerous exceptions to this rule, however; an example will be found in the vitamin C molecule (Section 35.8).

13.13. Acidity of the $-C{\equiv}C-H$ Group. In the triple-bond compounds an unusual characteristic arises when the triple bond is at the end of the chain, as in the following:

$$H-C{\equiv}C-H \qquad\qquad \text{and} \qquad\qquad CH_3-CH_2-C{\equiv}C-H$$

The hydrogen atom in the $H-C{\equiv}C-$ group has the properties of the H atom in acids. It can be replaced by metals:

$$Na^+[C{\equiv}C]^=Na^+ \quad \underset{200°}{\xleftarrow{\hspace{1cm}}} \quad \overset{Na}{H-C{\equiv}C-H} \quad \underset{100°}{\xrightarrow{\hspace{1cm}}} \quad [H-C{\equiv}C]^-Na^+$$

(Acetylene passed into melted sodium)

$$H-C{\equiv}C-H \quad + \quad 2AgOH \quad \longrightarrow \quad Ag-C{\equiv}C-Ag \quad + \quad 2H_2O$$

(Acetylene passed into ammoniacal silver nitrate)

The compounds formed with highly electropositive metals (review Section 8.8) are salts, but with certain of the heavier metals, like silver, the structure is believed to be covalent rather than ionic. Salts like the compounds of sodium and calcium are quite stable, but compounds of the heavier metals like silver are unstable and dangerously explosive when dry. Compounds in which one H atom is replaced by a metal atom are called *acetylides;* when both H atoms are replaced they are known as *carbides*. The most common of these compounds in commerce is calcium carbide, Ca^{++} $(C{\equiv}C)^=$. The crystal structures in this series of compounds are believed to be of various types, leading to different reactions with water. For example, calcium carbide gives acetylene when added to water, whereas aluminum carbide (Section 37.3) yields methane. The reason for the acidity of acetylene will be given in Section 15.5.

13.14. Oxidation. Oxidizing agents rupture a chain of carbon atoms more easily when the chain contains $C{=}C$ and $C{\equiv}C$ bonds. The initial stage in the reaction is presumably addition of $-OH$ groups across the multiple bond, as already described in Section 13.10. Continued oxidation then often breaks the chain at the point between the –OH groups.

An example of this behavior is the oxidation of oleic acid (Section 27.15), which is an eighteen-carbon straight-chain compound with one double bond in the chain. Shaking this acid with dilute potassium permanganate breaks it down into two lower acids, each with nine carbon atoms, proving that the $C=C$ bond of oleic acid is in the middle of the chain.

13.15. Detection of Double and Triple Bonds in a Molecule. By making use of our knowledge of the *chemical* activity of $C=C$ and $C\equiv C$ bonds, we can easily devise simple methods by which we can find out whether or not a molecule contains these groupings. The simplest method makes use of the ease with which $C=C$ and $C\equiv C$ can be oxidized. For example, assume we have obtained unlabeled containers with these three gases: ethane CH_3-CH_3, ethene $CH_2=CH_2$, and ethyne $CH\equiv CH$ and we are required to determine which is which. If we successively pass them into a dilute alkaline solution of potassium permanganate, the red color of the solution will rapidly disappear in the case of ethene and ethyne, but not in the case of ethane. In oxidizing the reactive double and triple bonds, the reagent loses its color.

Another method of detection of double and triple bonds is to treat the substances with a weak solution of bromine in carbon tetrachloride. This reagent has a pale red color, and the color gradually disappears as the Br_2 in solution adds on to the double and triple bonds.

A double bond can be usually distinguished from a triple bond by the fact that the triple bond adds twice as much bromine. This requires a quantitative analysis. When the $-C\equiv C-$ group is at the end of the chain, however, its presence can be detected easily by the formation of a metallic derivative, as explained in Section 13.13.

It is also possible to detect double and triple bonds in a molecule by *physical* methods, by applying the principles of infrared spectroscopy described in Section 14.17. The vibration frequency of the $C=C$ bond is about 1630 cm^{-1}. When using such data to identify multiple bonds in the spectrograms of compounds, it must be remembered that the intensity of absorption and the actual frequency at which absorption takes place will depend, to some extent, on the nature of the rest of the molecule. In Section 15.6, it is stated that absorption of light in the infrared region takes place only if the increased amplitude of the atomic oscillations results in a change in dipole moment; in a very symmetrical molecule, like that of ethene, the absorption is nearly nil. The effect of light on molecules containing double bonds will be considered in more detail in the study of dyes (Chapter 36).

13.16 Some Tricks of the Organic Chemist. In Section 9.4 we showed how the presence of Cl in a paraffin hydrocarbon gave the relatively inert molecule the ability to react with other substances. In this section we shall illustrate with some examples the corresponding importance of the multiple bond in the carbon chain.

1. Suppose we have the compound $CH_3-CH_2-CH_2Cl$, in which the Cl atom is at the end of the chain, but for some reason we want $CH_3-CHCl-CH_3$, in which the Cl is inside the chain. How can we make the change? The method is quite simple:

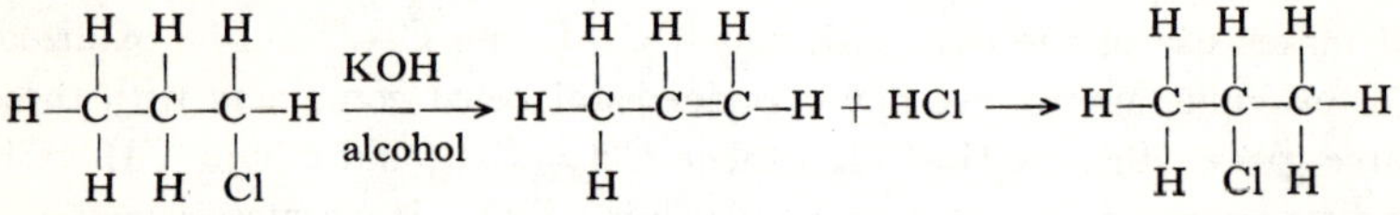

$$\underset{\text{1-Chloropropane}}{\begin{array}{c}H\ H\ H\\|\ \ |\ \ |\\H-C-C-C-H\\|\ \ |\ \ |\\H\ H\ Cl\end{array}} \xrightarrow[\text{alcohol}]{\text{KOH}} \underset{\text{Propene}}{\begin{array}{c}H\ H\ H\\|\ \ |\ \ |\\H-C-C=C-H\\|\\H\end{array}} + HCl \longrightarrow \underset{\text{2-Chloropropane}}{\begin{array}{c}H\ H\ H\\|\ \ |\ \ |\\H-C-C-C-H\\|\ \ |\ \ |\\H\ Cl\ H\end{array}}$$

We first remove the HCl from adjacent carbon atoms to form a double bond. Then, when we add a molecule of HCl back again, the Cl atom will go to the carbon atom with least hydrogen, as explained earlier in the chapter.

2. Suppose we have a quantity of CH_3-CH_2Cl and we wish to convert it to $CH_2=CHCl$. One of the possible series of reactions is as follows:

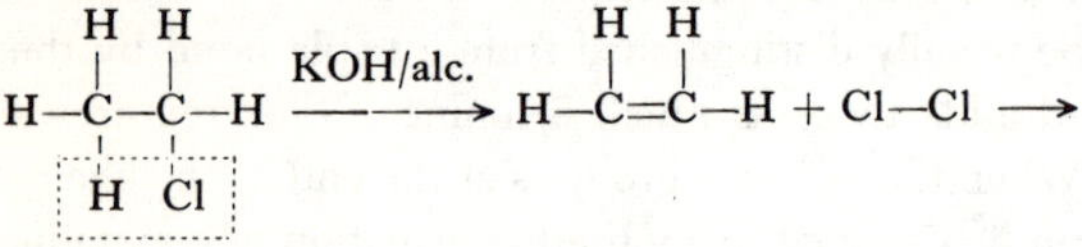

$$\underset{\text{Chloroethane}}{\begin{array}{c}H\ H\\|\ \ |\\H-C-C-H\\|\ \ |\\H\ Cl\end{array}} \xrightarrow{\text{KOH/alc.}} \underset{\text{Ethene}}{\begin{array}{c}H\ H\\|\ \ |\\H-C=C-H\end{array}} + Cl-Cl \longrightarrow$$

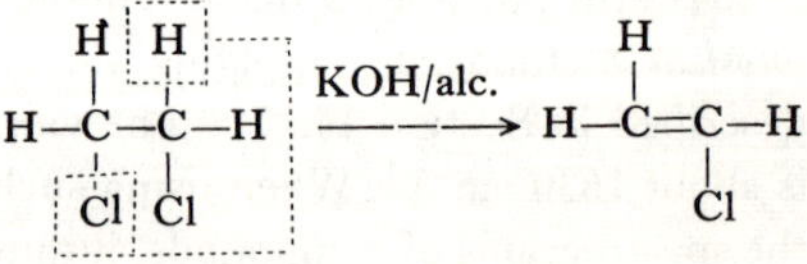

$$\underset{\text{1,2-Dichloroethane}}{\begin{array}{c}H\ H\\|\ \ |\\H-C-C-H\\|\ \ |\\Cl\ Cl\end{array}} \xrightarrow{\text{KOH/alc.}} \underset{\text{Chloroethene}}{\begin{array}{c}H\\|\\H-C=C-H\\|\\Cl\end{array}}$$

3. In the commercial preparation of gasoline at the oil refineries, it is often necessary to remove chain hydrocarbons which contain double bonds. The procedure followed is treatment with concentrated sulphuric acid:

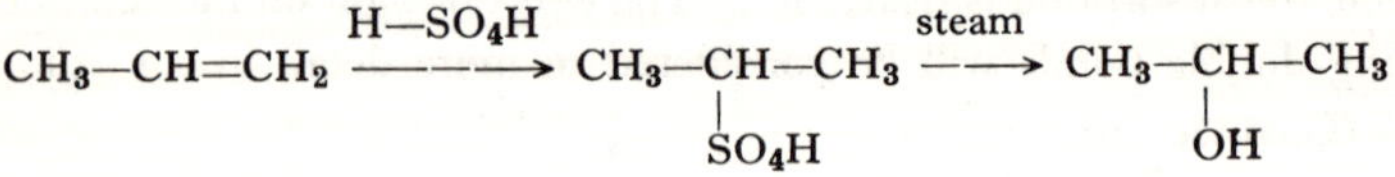

$$CH_3-CH=CH_2 \xrightarrow{H-SO_4H} \underset{\quad\ \ SO_4H}{CH_3-CH-CH_3} \xrightarrow{\text{steam}} \underset{\quad\ \ OH}{CH_3-CH-CH_3}$$

The acid adds on across the double bond when such compounds are present, and the resulting products, being insoluble in the rest of the mixture, are easily removed. When steam distilled, these acid addition products are converted into alcohols which have many commercial uses.

4. In the oil and fat industry the chemist has made extensive use of his knowledge of the double bond and its characteristics. This will be referred to in Sections 27.11 and 27.15.

14

Energy and Molecular Structure

14.1. Introduction. The inquiring reader will want to know how it was possible to arrive at some of the conclusions on molecular structure indicated by the models of molecules shown in the preceding chapters. Some of the reasoning was given in Chapter 8. The further explanations in this chapter belong, perhaps, in the domain of the physical chemist, but the organic chemist can easily afford to read a few pages on such an important aspect of his science. The continuity of this book will not be lost if the reader prefers to skip this chapter and remain aloof in the domain of "pure" organic chemistry. However, in the chapters that follow, references will frequently be made to sections in this chapter in which energy factors are described that are essential to an understanding of the nature of certain compounds.

In order to understand the theories involved we must review some of the elementary principles of physics. Fortunately, the principles we make use of are easy to comprehend and even the mathematical equations are quite simple, although it must be admitted that considerable genius was required to put these ideas into expressions that nonmathematical chemists can employ.

14.2. Energy. When a body is in motion it has a certain amount of energy due to its motion. This is called kinetic energy and is labeled E_k. A body, like a stretched rubber band, has potential energy E_p, equivalent to the kinetic energy required to put it in the configuration which prevails at a certain instant. This will be discussed later in Section 14.10.

Motion in a straight line is referred to in scientific literature as translational motion; consequently, kinetic energy due to straightline motion is called translational energy E_{tr}.

The body may also have kinetic energy due to its rotation around one or more axes, like a spinning top; this energy is E_r.

If the body is in constant vibration, like a tuning fork, the energy of vibration is designated E_v. It will be explained later (Section 14.7) that

106

energy of vibration consists not only of kinetic energy but also of potential energy.

The total energy E is obtained from the equation

$$E = E_{tr} + E_r + E_v$$

Every molecule has structure and occupies space. A molecule like hydrogen, for example, can be represented by a dumbbell like that in Figure 14.3b, which can move off into space in straight lines in any direction, can rotate on its axis, and can vibrate as if the atoms were held by a rubber band being alternately stretched and relaxed.

Much information with regard to the structure of a molecule can be found by a study of the distribution of its energy among the three types of energy. We shall first discuss E_{tr} rather fully in order to develop the basic line of thought with the simplest of the phenomena. The observations on E_r and E_v will be deferred until Sections 14.6 and 14.7.

14.3. Translational Energy. In order to determine the relation between molecular structure and energy of the molecule, we begin a study of gases in which the *hypothetical* molecules have no structure at all. The student will recall that this is the kind of molecule employed in developing the kinetic theory of gases, which assumes that molecules are simply "points" in space, that they have no attractive forces, are perfectly elastic, and have only translational motion. After studying this "ideal" gas, we shall return to gases in which the molecules are "real."

We may also recall the *ideal gas law*, a combination of the laws of Boyle and Charles,

$$\frac{PV}{T} = R \qquad \text{or} \qquad PV = RT \tag{14.1}$$

where R is a constant that shows the relationship of the volume V, the pressure P, and the temperature T for 1 mole of gas. The constant R can be evaluated in terms of heat energy and is 1.987 cal/mole/deg, but for most purposes, it can be said that $R = 2$ cal/mole/deg. The units of R will be defined in Section 14.4.

In any elementary text on physical chemistry we can find another simple equation involving the PV product; it is obtained from the fundamental assumptions given in the first paragraph of this section about the hypothetical point molecule, which has only translational energy E_{tr}. This "kinetic theory" equation is

$$PV = \tfrac{2}{3}E_{tr} \quad \text{per mole of ideal gas} \tag{14.2}$$

By combining Equations (14.1) and (14.2) we obtain for 1 mole of ideal gas

$$\tfrac{2}{3}E_{tr} = RT, \quad \text{or} \quad E_{tr} = 3 \times \tfrac{1}{2}RT \tag{14.3}$$

At the normal room temperature of 27°C ($T=300$°K), the kinetic energy of a mole of ideal gas due to its translational motion is

$$E_{tr} = 3 \times \tfrac{1}{2} \times 2 \times 300 = 900 \text{ calories}$$

The kinetic theory leads to the conclusion that at −273°C ($T=0$°K) the value of E_{tr} is zero—that is, there would be no translational motion at the absolute zero of temperature (but see remarks in Section 14.16 on the zero-point energy due to vibrational motion).

14.4. Molar Heat Capacity of Gases. When heat energy is added to a confined gas, various mechanical disturbances take place in the gas molecules. The nature and extent of these disturbances control the amount of heat energy that can be absorbed. Our hypothetical point molecule can move only in straight lines; it is therefore limited in the amount of heat energy it can absorb. More complex molecules, which have rotational and vibrational motions, can absorb more energy than a point molecule under the same conditions.

The amount of heat, in calories, required to raise the temperature of one molar weight of a gas through a temperature interval of 1°C (or 1°K) is called the molar heat capacity.

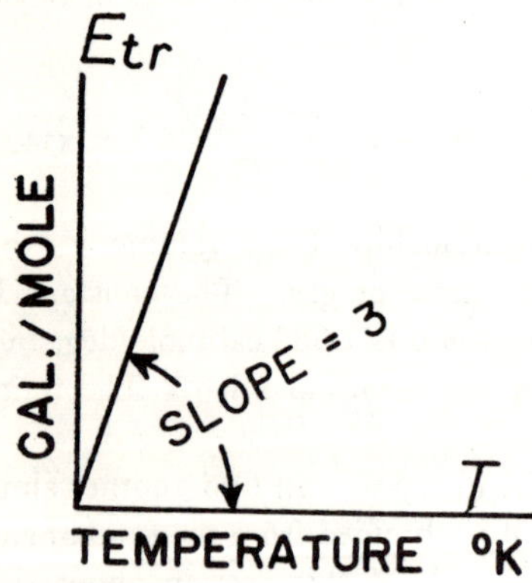

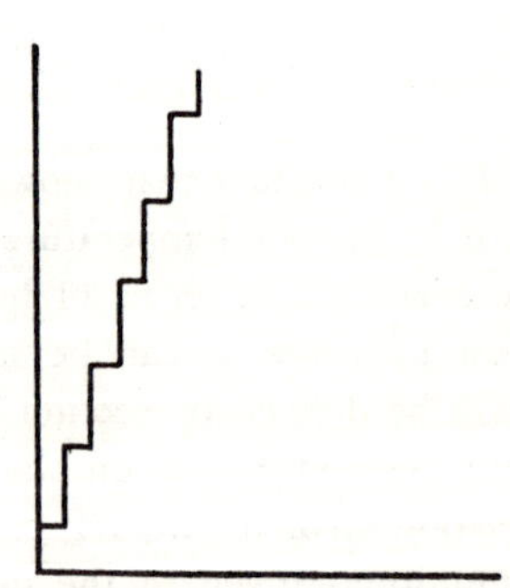

14.1.　Graph of equation 14.3, based on the kinetic theory.

14.2.　The equation in Figure 14.1, interpreted by the quantum theory (see Section 14.14). *The "steps" are greatly exaggerated.*

TABLE 14.1. Molar Heat Capacities of Gases at 15°C

Atoms per molecule	Name	Formula	C_v in cal/mole/deg	Degrees of freedom at 15°C
1	Helium	He	2.98	3
1	Mercury vapor	Hg	2.98	3
2	Hydrogen*	H—H	4.83	5
2	Hydrogen chloride*	H—Cl	4.99	5
3	Carbon dioxide**	O=C=O	6.71	7
8	Ethane**	H—C—C—H (with H H above and H H below)	9.51	—

* See Section 14.6.
** See Section 14.8.

If the experiment is done with the sample at constant volume the heat capacity is labeled C_v. If the gas is contained in a vessel maintained at constant pressure, so that the sample can expand while being heated, the resulting heat capacity will be greater and is called the molar heat capacity at constant pressure, C_p. In this book we shall consider only C_v.

Equation 14.3 applies to an ideal gas consisting of point molecules. If the temperature of 1 mole of such a gas is raised 1°, it can be calculated from Equation 14.3 that the energy due to its more rapid motion is increased to the extent of 3 calories (since R=2 and T=1). The relation is illustrated graphically in Figure 14.1, which shows that the translational energy E_{tr} is increased by 3 cal/mole for a rise in temperature of 1°. At the absolute zero of temperature the E_{tr} value would be zero, as already mentioned. The slope of the straight line is the heat capacity C_v.

This predicted value for an ideal gas of C_v=3 cal/mole/deg very closely approximates the measured values for gases which are composed of molecules that resemble point molecules. Such gases are helium and mercury vapor, which are monatomic (Table 14.1). It is apparent that energy supplied to these molecules serves only to increase their translational velocity. For a monatomic gas, therefore, we can substitute for the total energy E the energy of translation E_{tr} and apply Equation 14.3.

From Table 14.1 it will be seen that the values of C_v for molecules containing more than one atom are greater than predicted by the kinetic theory. The energy required to raise the temperature of 1 mole of polyatomic gases 1° is used not only to increase translational velocity (kinetic energy of translation) but also to raise the energy of molecular rotation and possibly the energy of intramolecular vibration. It was explained in Sec-

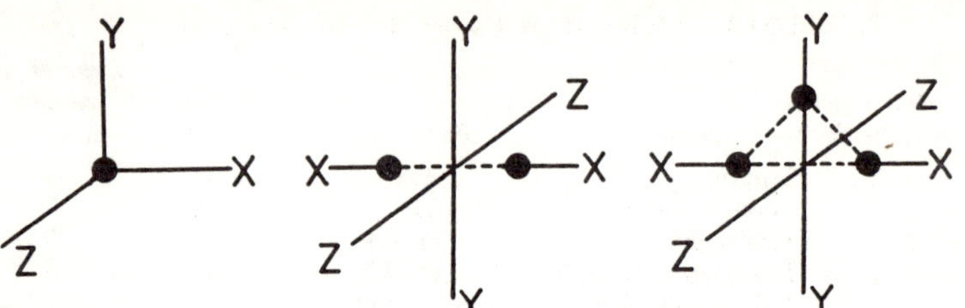

14.3a. Monatomic molecule, or "point" molecule, with three degrees of translational freedom.
14.3b. Two degrees of rotational freedom around axes YY and ZZ for a rigid diatomic molecule (or any *linear* molecule). Total degrees of translational and rotational freedom=5.
14.3c. Three degrees of rotational freedom around axes XX, YY, and ZZ for a rigid, nonlinear, triatomic molecule, such as water,

O
H   H. Total degrees of translational and rotational freedom=6.
[Adapted from A. A. Michelson, *Studies in Optics*, University of Chicago Press, 1927.)

tion 14.2 that $E=E_{tr}+E_r+E_v$, and we can substitute for E_{tr} its equivalent from Equation 14.3 and state that

$$E = (3 \times \tfrac{1}{2}RT + E_r + E_v) \text{ calories per mole, for any gas} \qquad (14.4)$$

The student should observe that we have completed one-third of our program. We have found out how to calculate what part of the total energy E of a mole of gas is energy E_{tr} due to motion of the molecules in straight lines. This motion does not depend on the structure. We shall discuss the E_r and E_v contributions to total energy in Sections 14.6 and 14.7, where it will be found that the structure of the molecules is important.

14.5 Equipartition of Energy. The motion of a point molecule, or of a monatomic molecule, can be represented by three components on axes at right angles to each other, as shown in Figure 14.3a. The average value of the X component is one-third of the total, and the same is true of the Y and Z components. The total energy of translational motion, from Equation 14.3, is $3 \times \tfrac{1}{2}RT$ calories per mole; each component is equivalent to one-third of this, or $\tfrac{1}{2}RT$ calories per mole.

A monatomic molecule is said to have three *degrees of freedom* because its motion can be defined by three variables. According to the principle of the equipartition of energy, the total energy is divided equally among these degrees of freedom. In the discussions that follow it should be kept in mind that when a molecule gains an additional degree of freedom its heat capacity rises by ½RT calories per mole.

14.6. Rotational Energy. A diatomic molecule, like H–Cl, is assumed in the discussion in this section to have a rigid dumbbell structure. Since the entire molecule is in translatory motion it has the three degrees of freedom associated with such motion, each of which contributes $\frac{1}{2}RT$ calories per mole to the total energy as just described in Section 14.5. In addition, the molecule has energy of rotation (E_r); it can rotate (tumble) about the axes YY or ZZ shown in Figure 14.3b, so that it has two additional degrees of freedom, a total of five. By the principle of equipartition of energy, each degree of freedom contributes $\frac{1}{2}RT$ calories per mole of energy:

$$E = E_{tr} + E_r = (3 \times \tfrac{1}{2}RT + 2 \times \tfrac{1}{2}RT) = 5 \times \tfrac{1}{2}RT \text{ calories per mole}$$

Now the molar heat capacity (C_v) is the change in energy per degree, and $R=2$, so that with T equal to unity, we have for the diatomic molecule

$$C_v(\text{trans.} + \text{rot.}) = (3 \times \tfrac{1}{2}R + 2 \times \tfrac{1}{2}R) = 5 \text{ cal/mole/deg}$$

This is the experimental value found for diatomic molecules as shown by the data in table 14.1. The possible rotation (spin) around an axis parallel to the bond joining the two atoms in a diatomic molecule is associated with negligible energy requirements.

14.7. Vibrational Energy. We can consider the atoms in a molecule as separate entities held together by an elastic bond. At a certain temperature (which depends on the molecule) the atoms are displaced with respect to each other, with a consequent restoring force, and vibration at characteristic frequencies sets in. This absorbs energy, so that C_v at a certain temperature rises above 5 cal/mole/degree.

The *act of displacement* of the atoms represents the kinetic energy factor in the oscillation and this constitutes one degree of freedom. The *degree of separation* of the atoms in an oscillation represents a potential energy change (see Section 14.10) which is equal to the change in kinetic energy, so that it is equivalent to a second degree of freedom. Each of these degrees of freedom absorbs $\frac{1}{2}RT$ calories per mole, so that vibrational energy is equal to twice this value, or RT cal. per mole. It is convenient to indicate a vibration by the symbol $\leftrightarrows$ in order to keep in mind that such vibration in any one direction is equal to two degrees of freedom. At high temperatures, then, we can have a total of seven degrees of freedom in a diatomic molecule,

$$E = E_{tr} + E_r + E_v = 3 \times \tfrac{1}{2}RT + 2 \times \tfrac{1}{2}RT + 2 \times \tfrac{1}{2}RT$$
$$= 7 \times \tfrac{1}{2}RT \text{ cal/mole}$$

and the molar heat capacity is

$$C_v(\text{trans.} + \text{rot.} + \text{vib.}) = 7 \times \tfrac{1}{2}R = 7 \text{ cal/mole/deg}$$

For hydrogen chloride and chlorine, the values of C_v at different temperatures are as follows:

	0°	100°	200°	500°	1200°	2000°C
HCl	5.00	5.09	5.27	5.46	6.13	6.9
Cl$_2$	5.95	6.3	6.7	6.9	7.1	7.2

14.8. Modes of Vibration. Any linear molecule should have the general characteristics described for a diatomic molecule, as to the three types of translational freedom and two types of rotational freedom. Such a linear molecule is carbon dioxide (Figure 8.4). When referring to vibration of the atoms, the possible motions are called modes of vibration.

Each atom in a molecule considered by itself has three degrees of freedom; if there are n atoms there are $3n$ degrees of freedom. However, if the molecule is diatomic or linear it has five degrees of freedom due to *translation* and *rotation*, so that if we subtract these five degrees of freedom from the total number possible, the modes of *vibrational* freedom are ($3n - 5$). Each mode of vibration is equivalent to RT cal/mole as explained in Section 14.7. For H–Cl, with its two atoms, there is only one mode of vibration ($3n - 5 = 6 - 5$); this can be pictured as the knobs of a dumbbell oscillating on the line which joins them. For O=C=O, we can similarly calculate that there are four modes of vibration, but actually there are only three. One is the symmetrical oscillation of the 0 atoms toward and away from the C atom, and the second mode is the corresponding vibration which is unsymmetrical (one O atom toward when the other is away). The third mode is oscillation of the C atom (a bending vibration) perpendicular to the long axis of the molecule. The fourth is oscillation in a plane at right angles to the third one just described, but at the same vibration frequency, so is disregarded; such vibrations, of the same frequency, are said to be *degenerate*, a term much used in spectroscopy and quantum mechanics.

Since a nonlinear molecule (Figure 14.3c) has six degrees of freedom due to translation and rotation, the possible modes of vibration it may have are $3n - 6$. For the nonlinear H–O–H molecule, which is shown in Figure 8.5, the formula $3n - 6$ yields $3 \times 3 - 6 = 3$ possible modes of vibration. The total energy of the gas at high temperatures can be calculated from

$$E_{tr} = 3 \times \tfrac{1}{2}RT \qquad E_r = 3 \times \tfrac{1}{2}RT \qquad E_v = 3 \times RT$$

which gives an E value (since $R=2$) of $12T$ calories. The maximum molar heat capacity C_v is then 12 cal/mole/deg.

Ethane is a nonlinear molecule (Table 14.1 and Figure 12.3); the maximum value for its heat capacity at high temperatures should be as high as 42 cal/mole/deg, based on the calculation procedure used in this section. However, in such more complex molecules many other factors, such as restricted rotation about a valence bond in the molecule (Sections 14.11 and 14.19), will alter the calculated value, and the molecule may even decompose at some high temperature before the maximum C_v is reached.

14.9. Heat Capacity and Temperature. Only energy that varies with temperature can affect the heat capacity; this follows from the definition in Section 14.4. In the case of translational energy, the variation with temperature is constant, as indicated in Figure 14.1 and in equation 14.3, which can be written $E_{tr}=3T$ since R is equal to 2 cal/mole/deg. In other words, for every rise in temperature of one degree the translational energy increases by three calories. This holds for the ideal gas and for monatomic gases. In Figure 14.4 this constancy is shown by the horizontal line labeled translation.

It has been shown in this chapter that when a molecule with more than one atom in it absorbs energy there is an increase not only in translational energy but also in rotational energy. However, according to old established theory, which is now called "classical" theory, diatomic gaseous

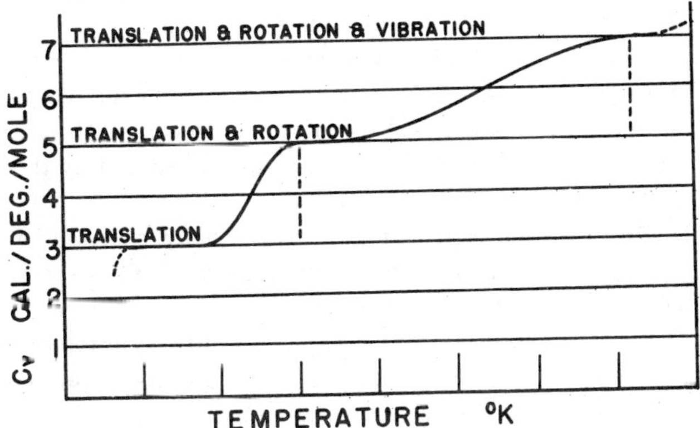

14.4. Effect of temperature on molar heat capacity at constant volume of simple gases. Vertical broken lines are required by classical theory. Slant curves are experimentally found, and predictable by quantum theory. The positions of the vertical lines and of the curves are purely schematic, not data for any one compound.

molecules, like H–H, should not rotate until they have acquired a certain amount of translational energy. At some temperature, represented by the first vertical dotted line in Figure 14.4, the molecules should have so much energy that they begin to rotate. The H_2 molecule in fact follows classical behavior up to about 50°K, showing a constant C_v of 3 cal/mole/deg characteristic of a *point* molecule, but instead of the C_v rising abruptly to 5 cal, experiment shows that there is a long, slow rise to that value. This slow change in C_v can be calculated by means of the "quantum" theory, and the ability to predict curves such as shown in Figure 14.4 was one of the early triumphs of the new theory.

The quantum theory will be discussed in Section 14.13, but it may be stated at this point that according to this theory a molecule absorbs energy in small units called *quanta*. This is shown schematically in Figure 14.2, where the step-like line indicates absorption of energy in small increments. These quantum increments, however, are so small in the case of translational energy (see Section 14.14) that for practical purposes translational energy is said to be absorbed continuously—that is, classically, as shown in Figure 14.1

Rotational energy quanta are larger than those of translational energy, but are still small enough so that at ordinary temperatures many molecules have such quanta; the result is the slant curve from the three to five levels in Figure 14.4. Vibrational energy quanta, however, are so large that at ordinary temperatures very few molecules have acquired them. For this reason, most diatomic molecules have C_v values of about 5 cal/mole/ degree at ordinary temperatures (translational and rotational motion energy only). Vibrational quanta are absorbed gradually as the temperature is raised (since each unit is so large) accounting for the gradual rise in Figure 14.4 from the five to seven level. This is also illustrated by the data for chlorine in Section 14.7.

14.10. Potential Energy. The energy a body possesses due to its position in space relative to a reference body is called potential energy E_p. The repulsive force that causes the separation of the two bodies in the first place is assigned a positive value; the attractive force that tends to bring the bodies together is negative. Physicists have adopted the concept that potential energy is zero at infinite separation of the body under consideration from the reference body; the energy becomes increasingly negative numerically as the body approaches the reference one.

14.11. Potential Energy Barrier. The literature of science contains many references to phenomena involving a potential energy barrier, or potential energy trough. This is illustrated in Figure 14.5, which shows

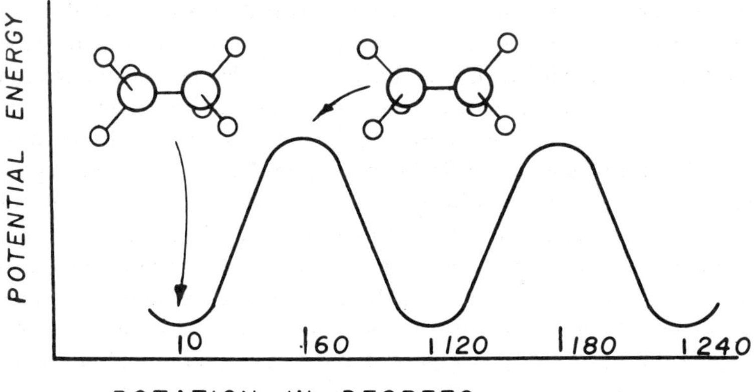

14.5. Potential energy barrier to the rotation of a methyl group in ethane, CH_3—CH_3, around the C—C bond. About 3,000 cal/mole are required to twist one of the CH_3 groups from its more stable *anti* conformation (in the trough) to the *eclipsed* conformation (at the crest). The two shapes occur alternately after each 60° of twist. To aid in visualizing the rotation, see Figs. 12.8 and 12.9.

an aspect of the energy relations prevailing in the ethane molecule. As explained in section 12.3 the *anti* conformation is the more stable of the two pictured and most of the ethane molecules are in that shape at ordinary temperatures. However, an appreciable fraction of the molecules are active enough at any instant to twist themselves into the *eclipsed* conformation; this is a 60° rotation of one of the methyl groups relative to the other methyl group, which is the reference body.

Rotation around the C–C bond is hardly "free"; the energy required to cause the rotation is about 3,000 cal/mole. This is rather high compared with the energy of translation of 900 cal/mole as calculated in Section 14.3. When an ethane molecule is in the fully *eclipsed* shape after a 60° rotation around the C–C bond, it is at maximum potential energy at a crest in the curve of Figure 14.5. The molecule releases this stored energy by flipping back to the *anti* conformation through a 60° rotation, and is then back in a trough of the curve. This is called a potential energy trough. The vertical distance on the curve from a trough to a crest is the *height of the barrier* to free rotation—in this case about 3,000 cal/mole, which explains why most of the molecules are in the potential energy trough. Once a molecule passes the *anti* conformation it runs into the barrier again, as shown by the succession of troughs in the curve. The method of estimating the height of the barrier is briefly described in Section 14.19.

In Section 12.5 it was explained that *anti*-cyclohexane is more stable than the *esclipsed* conformation. There is a restricting potential (energy barrier) that tends to keep the molecule *anti*, and the effect is large enough to be found through its contribution to the heat capacity of the substance.

14.12. Heat Capacity of Solids and Liquids. In order to complete the study of heat capacity which is summed up for gases in Figure 14.4, a few brief remarks are in order with respect to solids and liquids. The molecules in most crystals are restricted to only one type of molecular motion—that is, *vibration* about equilibrium positions of minimum potential energy. Consider crystals of metals, which consist of atomic molecules capable of vibratory motion in all directions. These motions can be resolved along three axes, in a manner similar to that shown in Figure 14.3*a*. The molecules in a crystal can therefore be said to have three modes of vibration.

Since it was shown in Section 14.7 that each mode of vibration is equal to RT cal/mole, the solid ideal element in crystal form should have an energy content at any temperature of 3 RT cal/gram-atom, and the value of C_v should be $3 \times 2 \times 1 = 6$ cal/gram-atom/deg. This agrees roughly with the empirical rule of Dulong and Petit that the atomic heat capacities of solid elements (with certain exceptions) are about 6.4 cal/degree at ordinary temperatures, the agreement being more satisfactory if the measured value is corrected to constant-volume conditions.

The principle of equipartition of energy, which predicts the heat capacity of solids at ordinary temperatures as explained in the preceding paragraph, does not hold at low temperatures, for the C_v of all solids approaches zero at the absolute zero (0°K.). The theory required to explain this was introduced by Einstein in 1907, who applied the quantum theory in the form of an equation involving a fixed vibration frequency (Section 14.13) of the atoms in the crystal lattice. Calculation of C_v by this quantum theory equation gives reasonable values at temperatures close to 0°K and the equation reduces to classical theory ($C_v = 3R$) at high temperatures.

The Einstein equation was improved in 1912 by Debye. The Debye calculation is used to extrapolate heat capacities to the absolute zero from measurements made below about 30°K.

As already stated, the molecules in most crystals can only vibrate. In some crystalline compounds, however, the molecules are in "swinging" oscillatory motion (a better word is *libration*) at low temperatures, but are not able to rotate freely because of the restrictive forces exerted by neighboring molecules. At a sharply defined temperature, sufficient rotational energy quanta are acquired by the molecules to enable them to begin free rotation in the solid phase. The melting point of 1,1,1-trichloroethane (CH_3–CCl_3) is 240°K, but from heat capacity measurements it was found that the molecules start to rotate in the solid phase at 224.3°K. Rotation of benzene derivatives (Table 19.1) and of a number of other compounds in the solid phase has been observed. The onset of rotation is accompanied by a decrease in density; the molecules need more room in which to rotate. From this review it is clear that vibrational and rotational motions are possible in solids, but there is no translatory motion.

Thermal effects on molecular motions of an ideal gas and of crystalline solids are now well understood, but this is not quite so true of the liquid state. At low temperatures the motion is mostly vibrational and rotational as in the solid state, and at high temperatures it also becomes translatory. Molar heats of liquids at high temperatures approximate those of corresponding gases.

14.13. Quantum Theory. In this chapter we have learned how to gain an insight into the structure of a molecule by determining the extent to which it is capable of absorbing three kinds of energy: translational, rotational, and vibrational. We found that this information can be determined rather easily by the measurement of the energy absorbed when the temperature is raised one degree—namely, the heat capacity. No mention was made of a fourth kind of energy: that due to motion of the electrons that make up the internal structure of the atoms and molecules. The energy required to alter electronic motions is so great that it does not appear in ordinary heat capacity measurements. However, we have already mentioned (Section 14.9) that quantum theory plays an important part in explaining absorption of energy by molecules; this theory is made easier to understand by looking at the electron arrangements in the atom and is briefly described in this section. From quantum theory we will be led to another way to explore the structure of a molecule—that is, through spectroscopy, which is even more direct than measuring heat capacity.

In section 7.1 it was stated that the electrons in an atom rotate around the nucleus in orbits corresponding to certain energy levels. When the electrons are in their assigned orbits the atom is in its "ground" state, but the atom is said to be in an "excited" state if an electron is sufficiently disturbed to rise to a larger orbit (higher energy level). The energy levels in an atom are fixed; the electron will not jump to its new orbit unless enough energy is supplied for it to reach the new level. This absorption of energy corresponds to a line in the absorption spectrum of the atom. If the electron falls back to a lower energy level, a corresponding line will appear in the emission spectrum. The energies of the various levels are proportional to $1/n^2$, where n is said to be the quantum number. The quantum numbers are small whole numbers—that is, 0,1,2,3,4, etc. In the level where $n=2$ the energy is one-fourth that in the level for which $n=1$.

When the cause of excitement has passed and the electron falls back into its original orbit, it liberates the energy (by radiation) it acquired when it became excited. It liberates a quantum of energy, the numerical value of which can be calculated from

$$\varepsilon = h\nu \tag{14.5}$$

where h is a universal constant known as Planck's constant and ν is the vibration frequency of the energy which is radiated. The wave length λ of the radiation is obtainable from the equation

$$\lambda = \frac{c}{\nu} \tag{14.6}$$

where c is the velocity of light (3×10^{10} cm/sec). If all the particles in a gram-atom or a gram-mole of substance are radiating energy, the total

TABLE 14.2. Relation Between Wavelength and Energy

Nature of radiation	Wavelength, A*	Radiation frequency, Hz**	Energy, cal/mole of quanta***	Effect on molecule
Radio	$>10^8$ (1 cm.)	3×10^{10}	2.858	Rotation
Infrared	2,000,000	1.5×10^{12}	142.9	Rotation
	500,000	6×10^{12}	571.6	Rotation and
	10,000	3×10^{14}	28,580	vibration
Visible	8,000			
	7,000	4.29×10^{14}	40,840	Outer electron
	4,000			displacement
Ultraviolet	3,000	1×10^{15}	95,270	Outer electron
	2,536.7	—	112,000	displacement
X-rays	100			
	1	3×10^{18}	2.86×10^8	Inner electron
	0.01			displacement
Nuclear	<0.01			

    * An average molecular diameter is about 3 to 10 A, where A is an Angstrom unit.
  ** The Hertz (Hz) 1 cycle/second. The symbol for frequency is ν, and for wavelength is λ. The frequency unit used by spectroscopists is generally $\bar{\nu}$, the wave-number in reciprocal centimeters, cm^{-1} ($\bar{\nu} = 1/\lambda$ and $\nu = 3 \times 10^{10} / \bar{\nu}$). Other units are 1 A $= 10^{-8}$ cm and $1\ \mu\ 10,000$ A. The wavelength region most thoroughly studied with respect to molecular motion (vibration) is the infrared.
 *** The amount of the quantum in calories can be obtained by dividing these data by the Avogadro number ($N\ 6.02 \times 10^{23}$), the number of molecules in one mole.

energy is equal to $N_{h\nu}$, where N is the Avogadro number (see footnotes to Table 14.2).

Since h is a constant, it is apparent from Equation 14.5 that the size of an energy quantum depends on the vibration frequency of the radiation (or on the wavelength). From Table 14.2, it will be seen that the quantum associated with comparatively long (radio) waves is small, but that the X-ray quantum is tremendously large. The outer electrons in an atom are relatively loosely held by the electrostatic attraction of the nucleus and are associated with relatively small quanta of energy. The inner electrons, which are tightly held by the nucleus, require quanta of high energy content to disturb them.

Chemical reactions take place through disturbance of the external (valence) electrons. The energy of visible light is sufficient, in many cases, to excite these electrons. The wavelength of 2536.7A in Table 14.2 is one of those radiated by incandescent mercury atoms. The energy of this ultraviolet radiation is more than enough to dissociate hydrogen molecules, in which the valence bond is exceptionally strong, as shown in Table 16.1

14.14. Quantum Theory and Translational Motion. Just as a person listening to music may be induced to tap his feet on the floor in unison with the music if the music is "right," a molecule in translatory motion if subjected to radiant energy may be induced to accept the energy. Since energy (according to modern theory) is always radiated in small packages (quanta) this is the way they will be absorbed by the molecule; however, the quanta will not be accepted unless they are of the right magnitude for the molecule to assimilate.

The ideal gas law, which we reviewed in Section 14.3, applies to a *gram-molecular weight* of substance, and we found that from the equation $PV = RT$ for one mole of gas we could proceed to an expression that indicates that the energy content is a function of RT. This is the classical-theory way of looking at things, and the RT product will be found in many equations derived by classical methods in textbooks of physical chemistry.

In quantum theory, however, attention is directed to the *individual* molecule, and in quantum-theory equations the product kT is more often found. The constant k is the Boltzmann constant, which can be calculated very simply from $k = R/N$ where N is the (Avogadro) number of molecules in a gram-mole.

The quantum level of energy of a molecule is generally represented by ε. Thus ε_0 is the lowest permitted state of energy, called the zero-point energy, and ε_1, ε_2, etc., are the energy states at quantum levels, 1, 2, etc. The translational energy at any quantum level n (in any one direction as in Figure 14.3a) is given by

$$\varepsilon_{tr} = \frac{n^2 h^2}{8ml^2} \tag{14.7}$$

where h is the Planck constant, m is the mass of the molecule, and l is the length of an edge of a cube holding the substance.

The total translational energy content ε_{tr} of a molecule at any temperature T is equal to $3 \times \frac{1}{2}kT$ (by analogy with Equation 14.3), but along any one axis, as shown in Figure 14.3a it is equal to $\frac{1}{2}kT$. We may compare this actual energy content of the molecule with the difference in energy levels from, say, quantum level 1 to quantum level 2. In Equation 14.7, this means calculating ε_{tr} when $n=1$ and when $n=2$. It will be found that ε_{tr} is about 10^{17} times as big as the difference between the two quantum levels of ε_{tr}. In other words, not only are the quanta of translational energy rather small, but also the quantum levels are so closely spaced that absorption of such energy is apparently continuous and "classical." The effect is shown by the continuous straight line in Figure 14.1, since the quantum steps in Figure 14.2 are actually infinitesimally small.

Translational energy levels cannot be determined spectroscopically; the energy levels are so close that they would form a continuous band in a spectrum. In Section 14.3 we found that the kinetic energy of translation of a mole of a gas at ordinary temperature is about 900 cal. Translational energy is sometimes called external energy because it can do work such as pushing a piston in an engine. It is also called thermal or heat energy. By contrast, the rotational and vibrational energies of a molecule are referred to as its internal energy.

14.15. Quantum Theory and Rotational Motion. Much larger quanta are required to affect the rotational motion than the translational motion of a molecule. This quantized energy is furnished by radiation in the infrared region (see Table 14.2); such energy causes a change in molecular rotation which, for many molecules, can be detected by the spectroscope in the form of simple absorption bands.

The rotational energy at any quantum level n for a diatomic molecule, such as H–H, or for any linear molecule, assuming it to be a rigid rotator as in Figure 14.3b, is given by

$$\varepsilon_r = \frac{n(n+1)h^2}{8\pi^2 I} \tag{14.8}$$

where h is the Planck constant and I is the moment of inertia of the molecule. For most simple gases, the calculation of total rotational energy by quantum theory is in agreement with that by classical theory, showing that the rotational quanta are quite small, even though they are larger than translational quanta. An exception is hydrogen, for which the moment of inertia I is so small and ε_r is proportionately so large (from Equation 14.8) that the quantum levels are fairly widely spaced. At 50°K all hydrogen molecules are in their zero rotational level of energy. The rotational quanta, being fairly large, are not completely acquired by hydrogen molecules in a gram-mole until the temperature of 300°K is reached, at which temperature it has a classical C_v of 5 cal. This is shown schematically in Figure 14.4.

14.16. Quantum Theory and Vibrational Motion. The general equation for vibrational energy for each vibrational mode of a molecule in terms of quantum levels is

$$\varepsilon_v = (n + \tfrac{1}{2})h\nu \tag{14.9}$$

where n has its usual whole number values of 0,1,2,3,. . .etc. In Figure 14.6, one of a pair of atoms is indicated as being in constant vibration (oscillation) about the equilibrium distance d_e. At absolute zero (0°K) there is no motion, and the system at point M is in the lowest vibrational state (that is, n is zero). From Equation 14.9, it is seen that the oscillator at 0°K has an energy of $(0+\tfrac{1}{2})\,h\nu$; in other words, it still retains a half quantum of energy. This is called the zero-point vibrational energy. The horizontal lines above point M in Figure 14.6 show the higher quantum levels that prevail at temperatures above 0°K. The vibration frequency is high (about 10^{13} to 10^{14} Hz), the same order of magnitude as the frequency of the infrared radiation in Table 14.2.

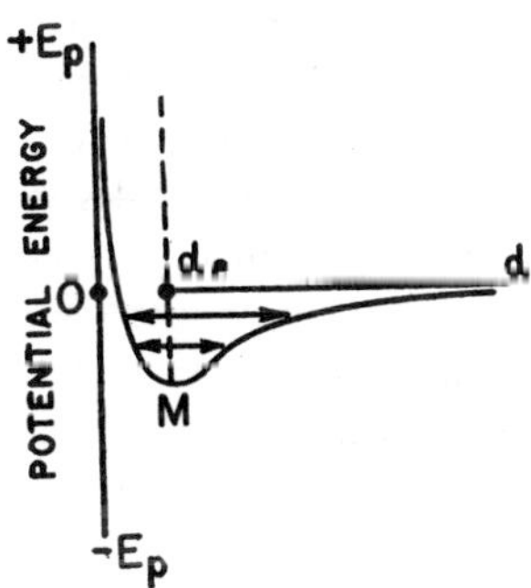

14.6.   A pair of atoms in a potential energy trough like that in Fig. 14.5. One atom is at 0, the other at equilibrium distance d$_e$.

Since translational quanta are small, all gases, even at very low temperatures, have acquired such quanta of energy. Rotational quanta are also small enough so that all the molecules of most gases at ordinary temperatures have acquired at least the first few levels of rotational energy. With respect to translation and rotation, therefore, most gases behave classically, as shown in figure 14.4; thus, for a diatomic molecule where $E=E_{tr}+E_r+E_v$ we can substitute for E_{tr} and E_r as follows:

$$E = (3 \times \tfrac{1}{2}RT) + (2 \times \tfrac{1}{2}RT) + E_v$$

and since $R=2$ cal/mole/deg, the molar heat capacity $C_v=5$ cal/deg.

At relatively high temperature, such as 1000 to 2000°C, the value of C_v goes up because vibrational motion occurs and the energy used to raise the temperature by 1° is absorbed not only by translational and rotational motion, but also by vibrational motion. The factor E_v which must be added to the last equation to account for the vibrational contribution to heat capacity can be calculated if the vibration frequency is known from spectroscopic measurements.

At ordinary temperatures, most diatomic molecules are in the quantum level ε_0 for vibrational energy. Their vibrational contribution to E_v and C_v is therefore zero. An exception is chlorine, which starts to acquire vibrational quanta at temperatures close to normal, as shown in Section 14.7.

14.17. Spectroscopy. The preceding discussions were concerned with quite simple gaseous molecules simply to show the basic nature of "quantized" energy. More complex molecules usually have vibrational modes as well as rotational motions, and these can be observed spectroscopically. As already stated, translational energy cannot be analyzed this way (Section 14.14). In this section we shall illustrate the spectroscopy procedure with substances that absorb infrared radiation.

If a sample is irradiated at a certain frequency, the energy of the incident radiation will be absorbed if the molecules in the sample can vibrate at two frequencies the *difference* of which is the same as the vibration frequency of the radiation. Another requirement is that a change must occur in the dipole moment of the molecule (Section 13.15). The percentage of radiation transmitted is plotted against vibration frequency or wavelength. Information on the nature of the atomic vibrations in a molecule can be gleaned from such spectrograms; thus, the two butanes in Figure 14.7 show absorption peaks at 1,450 cm^{-1} due to bending vibration of the C–H bond, and bands at 1,375 cm^{-1} which are characteristic of CH_3 group absorption. The double band (1,375 cm^{-1} and 1,360 cm^{-1}) in isobutane

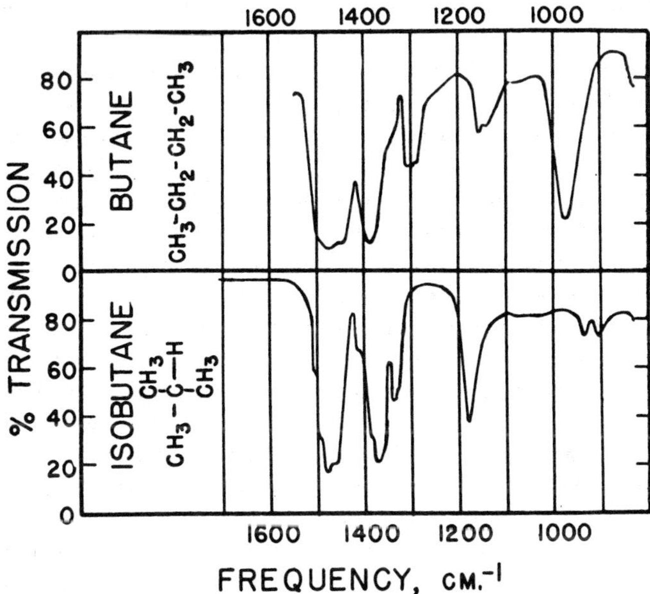

14.7. Infrared absorption spectra of isomeric butanes. [From R. B. Barnes, U. Liddel, and V. Z. Williams, *Ind. Eng. Chem. Anal. Ed.*, **15**, 659-709 (1943) which is the source of some of the material in this section]. See footnotes to Table 14.2 for the meaning of the frequency unit.

has been found to be characteristic of compounds with a terminal isopropyl group, CH₃–CH–CH₃.

We have already discussed the procedure for calculating the "fundamental" modes of vibration of linear and nonlinear molecules (Section 14.8). The chloroform molecule, H–CCl₃, has nine modes (from $3n-6 = 3 \times 5 - 6$) of which we may consider those modes associated with the C–H bond. A model can be prepared in which the atoms have their relative atomic weights (35.5 for Cl, 12 for C, and 1 for H) and the valence bonds are simulated by springs. Such a model shows clearly the *origin of characteristic vibrations* in the molecule. If the spring between C and H is stretched and released, these two atoms will oscillate at a certain characteristic frequency, but the three Cl atoms are relatively so heavy that they are not affected by this C–H "stretching" vibration. In other words, this mode of vibration takes place independently of other phenomena in the molecule, and study of the spectrograms of many compounds containing

C–H bonds shows that this stretching vibration has a frequency of 2,900 cm^{-1}.

The frequencies in Figure 14.7 do not go as high as 2,900 cm^{-1} to show the stretching vibration, but that figure does show a characteristic "bending" or "wagging" vibration of the C–H bond at 1,450 cm^{-1} which can be simulated on our chloroform model. This is done by displacing the H atom in order to bend the spring (without twisting the spring) and when it is released, the weighted atoms will again oscillate with practically no similar movement of the rest of the molecule. It may be mentioned that for any molecule there are $n-1$ stretching vibrations (where n, as usual, is the number of atoms) and the rest are bending vibrations. It is apparent that a catalog of infrared spectrograms of organic compounds is of immense value in elucidating the structure of an unknown compound, by comparison of its absorption bands with corresponding bands of compounds of known structure. The student should refer to the concluding paragraphs of Section 29.6 for a brief discussion of the ammonia molecule, NH_3, which at ordinary temperatures absorbs at a wave-length of about 1.25 cm, in what is known as the "microwave" region. The absorption is due to an intramolecular vibration which figuratively is equivalent to turning itself inside out at the rate of twenty-four thousand million times a second.

Structure analysis will be discussed again in Chapter 31.

14.18. Force Constants. In such a bond as C–C, the atoms oscillate rapidly about an equilibrium position. Force is required to change this equilibrium distance—that is, to push the atoms closer together or to move them further apart; these forces are generally referred to as *compression* and *stretching* forces, respectively. The phenomenon has been treated as a special case of Hooke's law for the simple harmonic vibrations of an elastic medium:

$$v = \frac{1}{2\pi}\sqrt{\varkappa/m} \tag{14.10}$$

where v is the vibration frequency in cm^{-1}, $\varkappa$ is the displacement, and m is the mass of the oscillating body. The following modification makes the equation adaptable to atomic systems:

$$v = \frac{\sqrt{10^5 N}}{2\pi c}\sqrt{f/\mu} = 1302\sqrt{f/\mu}\ \ \text{cm}^{-1}. \tag{14.11}$$

where c is the velocity of light (3×10^{10} cm/sec), μ is the reduced mass of the two atoms (explained below), and N is the Avogadro number (6.02

$\times 10^{23}$ molecules/gram-mole). The *force constant f* between the atoms is the restoring force per centimeter of displacement and is expressed in dynes/cm $\times 10^5$. The fixed constants (π, c, N) are all incorporated in the number 1302.

The reduced mass is obtained from the relation

$$\frac{1}{\mu} = \frac{1}{m_1} + \frac{1}{m_2}$$

where m_1 and m_2 are the relative weights of the vibrating atoms. Thus, when both atoms are carbon, as in the C–C bond, the reduced mass, μ is found to be 6.0, from the relation

$$\frac{1}{\mu} = \frac{1}{12} + \frac{1}{12}$$

With the proper spectroscopic data at hand, it is possible to calculate from Equation 14.11 the force constant f for various pairs of atoms. It has been found that for atoms united by a single bond (not only C–C, but also C–H, O–H, etc.), the value of f is between 4 and 6×10^5 dynes/cm. For atoms linked by a double bond, the value of f is twice as great (from 8 to 10×10^5 dynes/cm.) and for the triple bond it is three times as great (from 12 to 18×10^5 dynes/cm).

Since the same type of bond between any two atoms has roughly the same force constant, it is obvious that from this constant the approximate value of the vibration frequency of any pair of atoms can be calculated.

14.19. Calculated Heat Capacities. In the early pages of this chapter we defined heat capacity (Section 14.4) and indicated that it is a value obtainable by a calorimetric measurement. The data obtained by direct experiment were then evaluated in terms of the possible motions of the molecule.

In these closing pages of the chapter it has been shown that molecular gyrations can be recognized spectroscopically. Spectroscopic measurements can be made with a high degree of accuracy and the experts in quantum theory and statistical mechanics, by methods hinted at in this chapter, are able to use such data to obtain calculated values of the heat capacity which are more reliable than the calorimetric data. The discrepancy between a heat capacity calculated from spectroscopic (molecular) data and that determined calorimetrically sometimes leads to highly important discoveries. For example, the calculated heat capacity for ethane (Section 14.11) is a little higher than the calorimetric value. The calculation assumed that the compound is a free rotator about the C–C bond. The discrepancy indicates that the rotation is hindered, and the difference between the two values (about 3,000 cal/mole) is the energy equivalence of the hindrance to rotation.

15

π Electrons

15.1. Introduction. If we take a steel chain, like one used for lifting a ship's anchor, and weld two links to it where there had been only one before, we find that it is twice as strong at that one place. This is the picture:

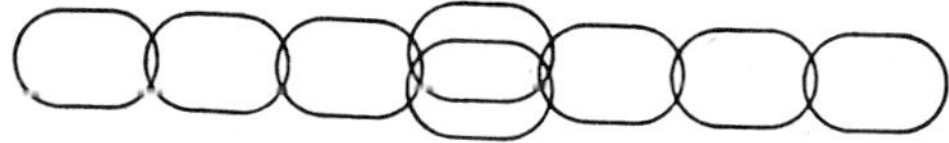

The double and triple links in the carbon chains, which we represent by C–C=C–C and C–C≡C–C, apparently have a similar strengthening effect on the physical properties of the molecule. Certain measurements, such as those of absorption spectra (Section 14.18), give evidence that the resistance to stretching and compression is about three times as great for a triple bond as for a single bond. Similarly, thermochemical data (Table 16.1) show that more heat energy is required to break the multiple bonds than to break the single bond. In the words of the physicist (or physical chemist) the multiple bonds are thermodynamically more stable.

Now all this may seem contrary to what we said about double bonds and triple bonds in Chapter 13 when we introduced these bonds to the reader. At that point, we stressed the fact that molecules containing these bonds are usually much more reactive than those with only single bonds. There is, however, no essential disagreement in these statements. Relative stability of substances when subjected to physical effects, such as change in temperature, may be quite different from the relative stability when treated with chemical reagents, such as chlorine. Both of these stability factors can be expressed precisely in mathematical terms, and in a later chapter we shall compare briefly thermochemical stability and chemical reactivity (Section 18.2) from the standpoint of the physical chemist.

For the present, however, we can use the term chemical reactivity in its usual loose sense—that is, the general willingness of a substance

127

(or of its valence bonds) to enter into chemical change.   Studying the chemical properties of compounds in which $C{=}C$ and $C{\equiv}C$ bonds are present, we shall find they are to be classed with the relatively reactive substances and that the double bond or triple bond serves as a reaction center and furnishes the chemist with a point of attack on the molecule.

We can obtain a helpful conception of the reason for the reactivity of the double bond by reviewing a method of its formation:

$$CH_3{-}CH_2Cl \xrightarrow{\text{KOH/alcohol}} CH_2{=}CH_2 + HCl$$

Chloroethane   Ethene

Much of the progress in the history of organic chemistry has resulted from the theory that the four valence bonds on carbon tend to act in straight lines, forming tetrahedral angles as shown in Figure 12.2. To explain the chemistry of ethene it was assumed the bonds were bent from the normal straight lines, as illustrated in Figure 15.1, where the double bond in the molecule is built from two coiled springs. In fact, ethene was regarded as the simplest of the cyclic hydrocarbons, and under considerable strain.   Although Figure 15.1 is still helpful in visualizing why compounds with double bonds are *unsaturated*—that is, *add on* other molecules as shown in Sections 13.7 to 13.11—electron theory offers a much more satisfying picture.   The double bond is formed so easily it does not appear likely it would yield the strained structure of Figure 15.1.

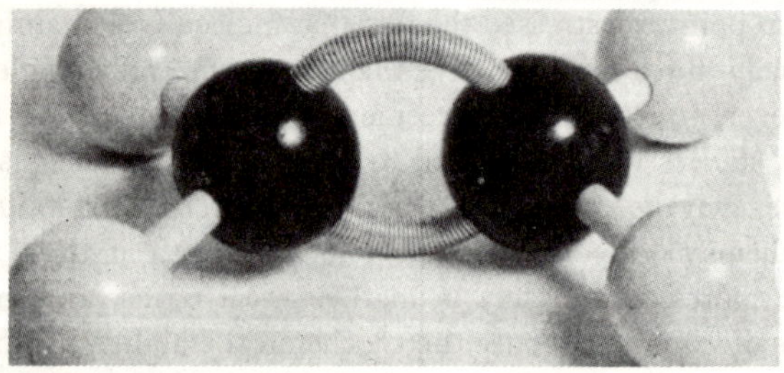

15.1 Model of double bond in ethene showing a supposed "strained" structure.

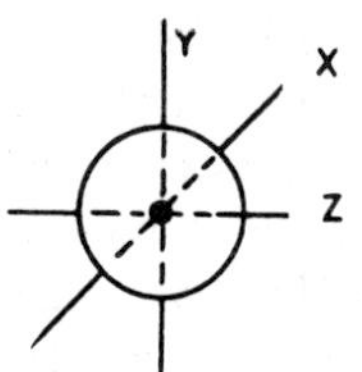

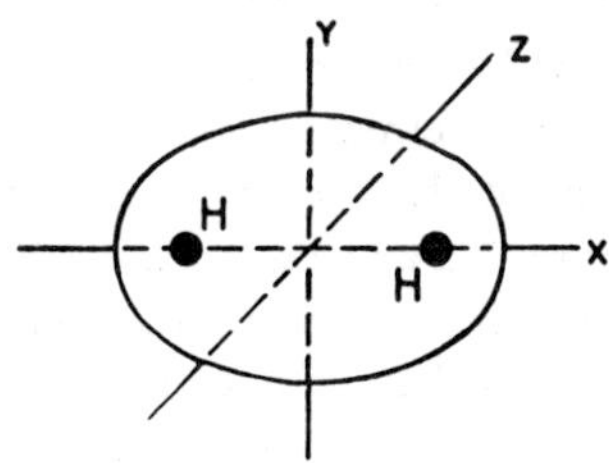

15.2. The 1*s* atomic orbital of the H atom.

15.3. The σ (or *s—s*) molecular orbital of the H—H molecule.

15.2. The Sigma Bond. When viewing the more or less elaborate pictures of the multiple-bond orbitals which will be developed in this chapter, it should not be forgotten that they are made by adding a second or third link to a bond with which we are already acquainted. This is the familiar *sigma* (σ) bond which was described in section 7.9. To review the nature of the *sigma* bond, figures 7.2b and 7.3 are reproduced in this chapter as figures 15.2 and 15.3.

One of the basic principles of physics is that all bodies prefer a greater freedom of motion; this greater freedom is associated with greater stability, which means a lower energy level. The electron in the H atom (Figure 15.2) is in an *atomic* orbital in which it encompasses a single proton. Two such hydrogen atoms readily combine to a molecule (Figure 15.3) because each electron then has greater freedom; it encompasses two protons instead of only one. The new *molecular* orbital is elliptical, contains both electrons, and is designated an *s-s* or σ orbital. When attaining this new freedom, the two H atoms give up about 103 kcal of energy (see Table 16.1). This is the energy that would be needed to separate them—that is, dissociate the hydrogen molecule.

Models I, II, and III in Figure 15.4 are based on corresponding diagrams in section 13.1. The straight line bond between carbon atoms, in

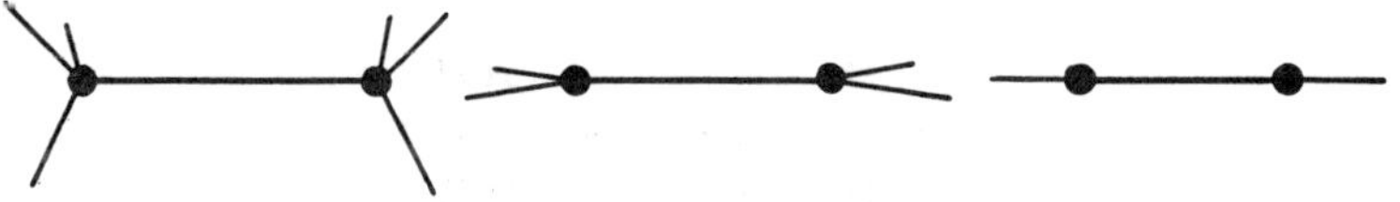

15.4 The valence-bond angles at the carbon atoms in ethane, ethene, and ethyne. Compare Section 13.1.

each case, stands for a σ bond, like the one in ethane, CH_3–CH_3. This should be remembered as we proceed to build on it the extra links which appear in ethene and ethyne. Models I, II, and III show the valence angles at the carbon atom. In I the carbon atoms are tetrahedral (see Figures 7.5 and 12.2) and all the H–C–H valence-bond angles are 109°28′. The carbon atoms in II are trigonal; the H–C–H valence angles are 360/3 =120° and are in one plane. In III the symmetry is digonal; the valence angles are 360/2=180° and are in a straight line.

15.3. The Ethane Orbitals. In Model I of Figure 15.4, all four valence bonds on each carbon atom are equivalent; this status is reached when three p atomic orbitals are hybridized with an s atomic orbital to give four orbitals of the sp^3 type as explained in Table 7.1. The overlapping of an sp^3 atomic orbital of one carbon atom with a corresponding orbital of the other carbon atom is illustrated in Figure 15.5; the resulting molecular orbital is sp^3–sp^3 and is σ type. The coalescence of a carbon atom sp^3 orbital with the s atomic orbital of a hydrogen atom to yield an s–sp^3 molecular orbital was shown in Figure 7.5 and should be referred to again. This also has σ character, and all the bonds in Model I are therefore of the σ type.

15.4. The Ethene Orbitals. In Table 7.1 it was shown that the atomic orbitals in the valence shell of the carbon atom can hybridize to a state in which only two of the p orbitals are involved and in which the four orbitals can be represented by sp^2, sp^2, sp^2, p. In Model II of Figure 15.4,

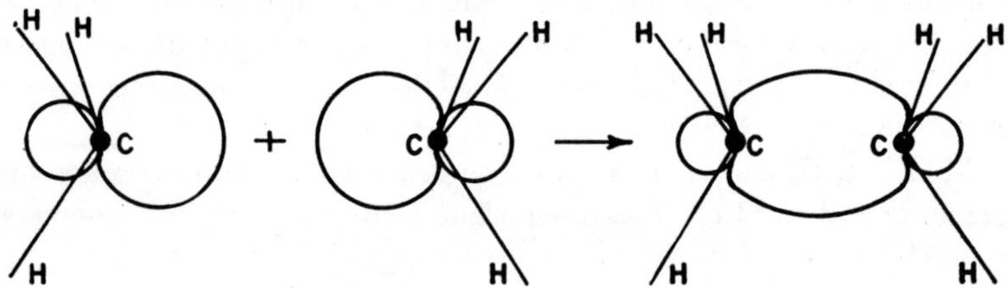

$$H:\overset{\displaystyle H}{\underset{\displaystyle H}{\ddot{C}}}\cdot \;+\; \cdot\overset{\displaystyle H}{\underset{\displaystyle H}{\ddot{C}}}:H \;\longrightarrow\; H:\overset{\displaystyle H\;H}{\underset{\displaystyle H\;H}{\ddot{C}:\ddot{C}}}:H$$

15.5. Formation of the C—C bond in ethane by overlapping of two sp^3 hybrid atomic orbitals. The C—C bond has σ character, both nuclei being encompassed by a molecular orbital containing two electrons.

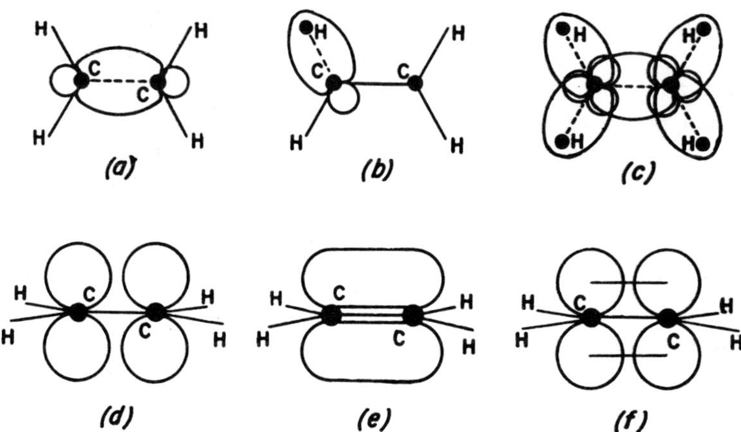

15.6. Disposition of the orbitals in ethene. Upper series of diagrams are top views of molecule, with all atoms in the plane of the paper. All the molecular orbitals are of the σ type. In the lower series, the molecule is viewed from the side (front); the *p* atomic orbitals are in plane of the paper, perpendicular to the rest of the molecule, and the molecular orbital is of the π type.

(*a*)   The C-C bond is an sp^2-sp^2 molecular orbital.

(*b*)   The C-H bond is an s-sp^2 molecular orbital.

(*c*)   The C-C bond and four C-H bonds are in the plane of the paper.

(*d*)   The hourglass-shaped *p atomic* orbitals on each carbon atom, before overlapping to form a *molecular* orbital.

(*e*)   Overlapping of the *p* atomic orbitals. The resulting molecular orbital, containing two electrons, is known as a π orbital.

(*f*)   A method of representing the status in (*e*)—that is, overlapping of two *p* orbitals, each of which has a single electron. The tie-lines are used to indicate overlapping. (See comments under Fig. 15.8.)

we use the three sp^2 orbitals on each carbon atom to create the three single bonds that are shown in trigonal symmetry. The C–C link (sp^2–sp^2 molecular orbital) is illustrated in Figure 15.6*a*; one of the C–H links (s–sp^2 molecular orbital) in Figure 15.6*b*; and the utilization of all three of the sp^2 atomic orbitals on each carbon atom is represented in Figure 15.6*c*.

Each of these molecular orbitals is of the σ type, in which a pair of nuclei is encompassed by a two-electron egg-shaped orbital.

The bottom series of diagrams in Figure 15.6 shows how the lone *p* atomic orbital on each carbon atom affects the molecule. Each hourglass orbital contains a single electron and is perpendicular to the plane that contains all the atomic nuclei in the molecule. The top row of diagrams shows that the two carbon atoms are bonded in the σ type, egg-shaped orbital of Figure 15.5, which is the normal single C–C bond. The second bond in the C=C bond of ethene is the orbital shown in Figure 15.6*e*; it is formed by coalescence of the *p* atomic orbitals in *d*, the driving force being the lower energy (greater stability) attained when the two electrons in the *p* orbitals can encompass both C nuclei in the same orbital. The new orbital can be likened to a squat hourglass or to a pair of fat sausages.

The resulting *molecular* orbital is labeled *p–p*; since it is very similar to a *p atomic* orbital with a nodal plane between the two parts, it is called a π (*pi*) orbital and the two electrons in it are called π electrons. To differentiate between the two types of carbon-carbon bonds in ethene we may write the structure $H_2C\!\cdot\!\cdot\!CH_2$, where the line is a σ bond and the dots are a π bond. It is apparent that the carbon atom in ethene has trigonal symmetry; it is connected to three atoms by σ bonds, and these bonds are in the same plane at equal angles of 120°. The electron pairs in these three bonds are said to be *localized*. The two electrons represented by dots are said to be *mobile* electrons. This will be discussed further in Section 16.4.

The σ type molecular orbital between carbon atoms in ethane (Figure 15.5) has cylindrical symmetry; it permits rotation of either CH_3 group about the C–C bond with a relatively small expenditure of energy of about 3 kcal mole, as described in Section 14.11. The double bond (Figure 15.6) between the carbon atoms in ethene, however, does not permit free rotation. It should be understood that both of the electron charge clouds in figure 15.6*e* belong to the same π orbital, and that the gap between them (the nodal plane containing the carbon nuclei) is free from electron charge. On the other hand, the two carbon nuclei in Figure 15.6*a* are encompassed by their σ orbital.

The C–C bond energy listed in Table 16.1 is 83 kcal/mole, and that of the C=C bond is 147. The difference, 64 kcal/mole, may be regarded as the stability of the π bond, which obviously is a big barrier to rotation; it is much higher than the 3 kcal needed for rotation about the single bond. The increased structural stability of compounds with $C\!\cdot\!\cdot\!C$ bonds is of considerable importance; it makes possible the existence of geometric isomers to be described in Section 22.3. The π bond is also important because the electron pair in that bond is much more exposed than the electron pair in the σ bond, and this leads to greater chemical activity. (Further remarks on the C=C bond energy will be found in Section 16.2)

15.5. The Ethyne Orbitals. The ethyne molecule is built up from carbon atoms in which hybridization (see Table 7.1) of only one *p* atomic orbital is involved, and the four orbitals of the carbon atom are then

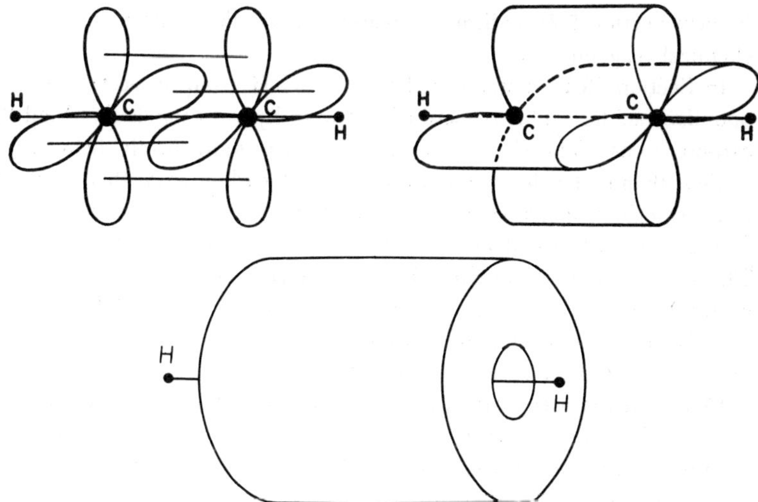

15.7. Molecular orbitals in ethyne. The σ orbital is not shown but is similar to that in Figure 15.6*a*.

Upper left: The four *p* atomic orbitals are shown; each has one electron.

Upper right: An aid to visualization of the overlapping atomic orbitals. The two π orbitals are perpendicular to each other and each has two electrons.

Lower: The π molecular orbital has this cylindrical symmetry.

represented by *sp, sp, p, p.* In Model III of Figure 15.4 the two *sp* orbitals on each carbon atom are used in making the H–C bond and the C–C bond. The two remaining *p* orbitals are perpendicular to the straight-line molecule shown in Model III; they are also perpendicular to each other. The hourglass *p* orbitals on the carbon atoms overlap just as illustrated in Figures 15.6*d* and *e*, except that in the case of ethyne there are four *p* orbitals. The overlapping results in two molecular π orbitals, as shown in perspective in Figure 15.7. This molecule may be represented by HC–CH, to indicate a σ bond between the carbon atoms enhanced by two sets of π electrons in two π bonds.

The C≡C bond of ethyne is even more stable structurally than the C=C bond in ethene. The energy required to separate the carbon atoms in ethyne (see Table 16.1) is greater than that needed to separate the carbon atoms in ethene. Chemically, however, ethyne is somewhat less active than ethene; the ethene molecule has a slightly lower ionization potential (review Section 5.2), indicating greater accessibility of its π electrons to chemical reagents.

In Section 13.13 it was stated that the C–H group in acetylene compounds is acidic. This is ascribed to the sp hybrid orbitals that bind the carbon atoms. The electron in an s orbital is more firmly held to the nucleus than is the electron in a p orbital. A carbon atom that gives rise to an sp orbital increases its electronegativity over that of a carbon atom in the usual sp^3 orbital because of its higher proportion of s character. The carbon atoms in ethyne, therefore, have a relatively high electronegativity, and the molecule may be represented by $H-C{\equiv}C^{\delta-}-H^{\delta+}$ to show the displacement of electrons toward carbon. This makes the H atom more detachable, or replaceable by metals.

15.6. Antibonding and Bonding Orbitals. The increased chemical activity of cthene and ethyne over that of ethane was just "explained" by simply stating that the electrons in the C=C and C≡C bonds are more "exposed" than the σ electrons which make up the C–C bond. Before amplifying on that statement let us look briefly at an aspect of molecular orbital theory needed even in this elementary account to fill out the picture.

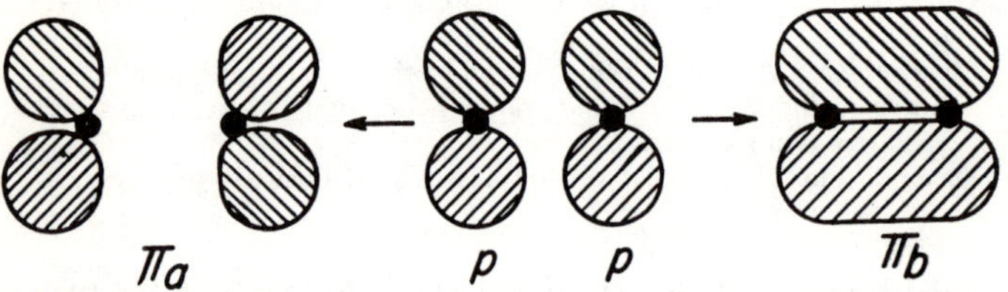

15.8. Interaction of a pair of p atomic orbitals from two similar atoms to give either antibonding (π_a) or bonding (π_b) molecular orbitals.

Observe from the shading that when the p orbitals overlap to a *bonding* orbital, they do so "in phase," whereas in the *antibonding* orbital the electrons are "out of phase." The nodal plane of π_b includes the nuclei of the two atoms. The nodal plane of π_a is at right angles to that of π_b.

The student is reminded that the hourglass (or dumbbell) used to represent a p atomic orbital is a convention. The electron density (wave function) in each half of an orbital extends to infinity, and the spheres simply indicate the boundary surface at which it is negligibly small. The two p orbitals should be drawn to show that at interatomic distance they overlap somewhat, but this would confuse the diagrams shown elsewhere in this book; for example, in Fig. 16.3, where Figure-8's are employed for the sake of clarity.

The slight gap between the two halves of a p or π orbital indicates the absence of electron density in the nodal plane.

The pair of p atomic orbitals in the center of Figure 15.8 are like those already described in Figure 15.6. In the ground state of ethene the p electrons coalesce to a bonding orbital labeled π_b, which encompasses the nuclei of both carbon atoms in the $C=C$ bond. However, it is possible to excite the molecule to such a degree that one electron is elevated to an energy state called antibonding, π_a. In this orbital the node plane does not include the nuclei of both carbon atoms; it is perpendicular to the line which joins the nuclei.

Ethene is a highly symmetrical molecule, so much so that it is quite resistant to the electron displacements illustrated in Figure 15.9, where c, d, and e show the probable changes that take place under excitation in certain chemical reactions or under the influence of radiation energy. In Section 14.17 it was stated that the structure of a molecule can be explored spectroscopically if a change in its dipole moment can be induced by the incident radiation. The stability of ethene toward excitation is indicated by the fact that radiation which it does absorb somewhat (1750A) corresponds to a rather high energy as shown in Table 14.2.

When the mechanisms by which ethene goes through its typical reactions are described (Chapter 17) it will not be found generally necessary to refer to such excited states as that illustrated in Figure 15.9. However, that set of drawings is helpful because it pictures the simplest compound with a double bond, and clarifies what is meant by an *electron shift* (see paragraph e under Figure 15.9). It is important to observe that an electron *pair* has not shifted; only *one* electron is involved, which is why there is only one plus sign and one minus sign. The orbital b has two electrons, one from each atom, and the completely polarized structure e has both; only one electron has shifted. Another way to explain this is to write the ethene structure as in Figure 15.10, where the crosses are electrons from H atoms, and circles are electrons from the C atoms.

In Structure I of Figure 15.10 each carbon atom has its quota of four valence electrons around it—that is, one-half of those it shares with other atoms (review section 8.6). In II, as a result of the electron shift, the C atom at the $\oplus$ end of the molecule has only three of its electrons, whereas the C atom at the $\ominus$ end has one more than its quota (two in an unshared pair and three in shared pairs). The customary way of writing such formulas is illustrated by III and IV. Facility in writing and comprehending such formulas as these is a necessity for studying modern organic chemistry.

15.7. Conjugate Systems. We have just observed that ethene is quite stable to electronic excitation and that high energy ultraviolet radiation

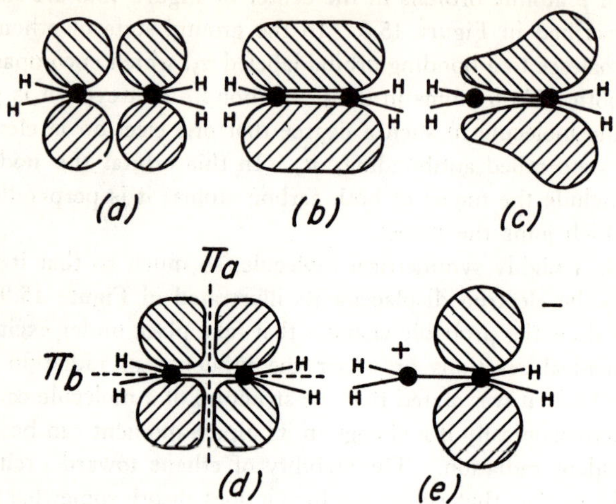

15.9.   Electron displacement in ethene.

(a)   The two p atomic orbitals, perpendicular to the $\mathrm{C-C}$ structure.   Each hourglass-shaped orbital contains one electron, which occupies both parts of the hourglass.

(b)   Overlapping of the p atomic orbitals to give the π molecular orbital (a pair of sausages).   The nodal plane contains the line which joins the two carbon nuclei.   The shading represents the opposite ($+$ and $-$) phases of the wave functions in the two parts of the orbital.   The orbital contains two electrons, both of which are called *bonding* (π_b) electrons.

(c)   Partial polarization caused by a reagent molecule which raises the effective (positive) nuclear charge at one end of structure b and increases the electron (negative) density at that end.   This *polarized* variation of the *ground* state of the ethene molecule should be distinguished from an electronically *excited* state shown in d.

(d)   Excitation (for example, by ultraviolet light at 1750 A) leaves one electron in the π_b orbital, but raises the other to a higher-energy orbital in which the electron is called *antibonding* (π_a). The π_a orbital has a nodal plane at right angles to that of the π_b orbital; in ethene it is midway between the nuclei of the two carbon atoms.   An excited structure such as d has a very short life of about 10^{-8} second.   Although much energy is required to create the π_a orbital in ethene, in more complex molecules (see Fig. 16.3), the π_a orbital may be present in the ground (stable) state.   With such a molecule, a reactant particle that raises the effective nuclear charge on one end of the molecule displaces the π_a nodal plane in that direction; this is another aspect of *polariza-tion*.   (*Continued on next page*).

is needed to create an excited state in the molecule. A more easily excited molecule is 1,3-butadiene, which has an alternate arrangement of single and double bonds. This

$$CH_2{=}CH{-}CH{=}CH_2 \qquad\qquad CH_2{=}CH{-}CH_2{-}CH{=}CH_2$$

1,3-Butadiene   1,4-Pentadiene

is a very important structural type in organic chemistry and is called a *conjugate system*. These double bonds are conjugated, whereas those in 1,4-pentadiene are said to be *isolated*. The isolated double bonds absorb ultraviolet radiation at about the same frequency as ethene does. The conjugate system, however, absorbs not only more strongly but also at a longer wavelength, although still in the ultraviolet. A longer wavelength corresponds to lower energy (Table 14.2). In the chapter on dyes (Section 36.14) it will be found that extended conjugate systems are an important factor in producing color, since they absorb the even longer wavelengths of visible radiation.

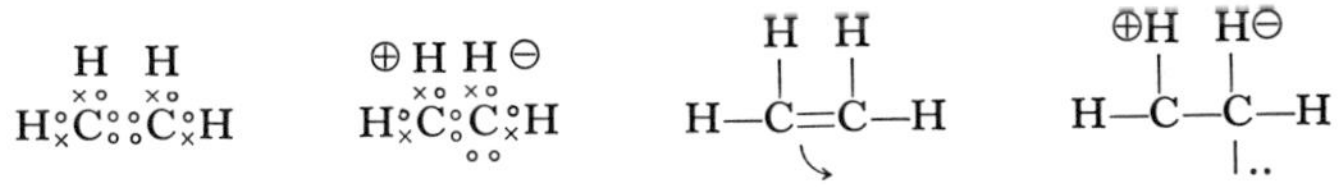

I   II   III   IV

15.10. Polarized structures of ethene.

The easy excitability of 1,3-butadiene can be ascribed to the fact that the conjugate system offers an easy path for π electron flow through the molecule; this is illustrated in Figure 15.11. The student should rework figure 15.11 with pencil and paper to learn how valence bonds can be flip-flopped through a molecule when it has a conjugate system. Each line in the formula represents an electron pair, and in a completely polarized structure a free electron pair (or bond) materializes at the terminus of the electron shift. As pointed out before, the result of the shift is the crea-

---

(*e*) Complete transfer of a π electron, consummating the process shown in *c*. Two electrons now occupy one orbital (—), and the other orbital is vacant (+). This diagram shows the equivalent of the formula $^{+}CH_2{-}CH_2^{-}$ in Fig. 15.10. It is apparent that the eletron *pair* now exposed to a reacting molecule is brought into position by transfer of a *single* π electron, although the phenomenon is generally referred to as the shift of a pair of π electrons.

(The author is indebted to Dr. R. S. Mulliken for suggestions regarding the layout of this figure.)

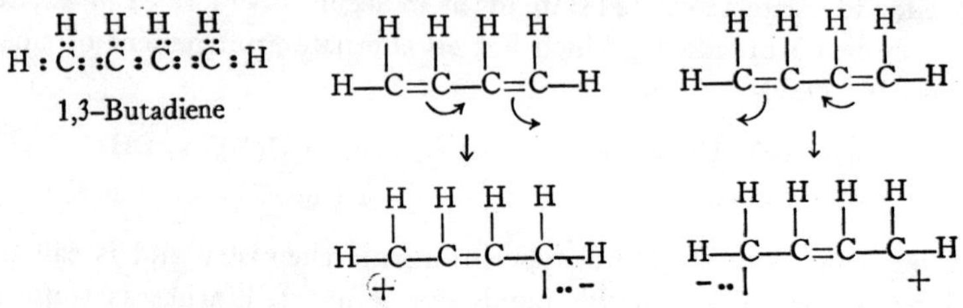

$$H : \ddot{C} : : \ddot{C} : \ddot{C} : : \ddot{C} : H$$

1,3–Butadiene

15.11. Flip-flop procedure for working out polarized structures of butadiene.

tion of a single $\oplus$ charge and a single $\ominus$ charge; it is, in its final effect, the displacement of a single electron.

Other important characteristics of butadiene will be described in Sections 16.4 and 16.5.

16

Bond Energies and Resonance

16.1. Heat of Formation. Before presenting the new principles in this chapter, we must once again briefly review some of the energy relationships among atoms and molecules. We have shown how the conversion of certain substances into other substances can be expressed by symbols such as these:

$$C + 2H_2 \longrightarrow CH_4$$

but in these reactions an essential factor has been omitted—namely, the extent of the energy or heat change. When the energy changes are considered, the reactions are more properly called equations:

$$C(s) + 2H_2(g) = CH_4(g) + 18.1 \text{ kcal}$$

or

$$C(s) + 2H_2(g) \longrightarrow CH_4(g) \quad [\triangle H_{298} = -18.1 \text{ kcal}]$$

These are accepted ways for writing a thermochemical equation. This one indicates that one atomic weight of carbon in the solid state (s) reacts with two molecular weights of hydrogen in the gaseous state (g) to yield one molecular weight of methane gas (g), with the liberation of 18.1 kcal of heat energy. The symbol $\triangle H_{298} = -18.1$ shows that this is the heat change at 298°K (or 25°C), with the negative sign indicating that heat is evolved in the reaction when it is read from left to right. The quantity 18.1 kcal/mole is the *heat of formation*, defined as the heat change when one mole of a substance is formed from its elements in their *standard states*—that is, their stable forms—at ordinary temperature and a pressure of one atmosphere.

Conversely, to decompose one mole of methane, i.e. to make the reaction go from right to left, requires the absorption of 18.1 kcal. This amount of energy is therefore taken to be a measure of the stability of the methane molecule. (The $\triangle H$ value is given a plus sign when the reaction is one that absorbs heat.)

139

16.2. Bond Energy. The CH_4 molecule (Figure 12.2) has four C–H bonds. We have just seen that energy to the extent of 18.1 kcal/mole is needed to break up these bonds when the reaction is the one described in the preceding section. However, if the reaction is one in which carbon is in the gas state the energy change is as follows:

$$C(g) + 2H_2(g) \longrightarrow CH_4(g) \quad [\triangle H_{298} = -395 \text{ kcal}]$$

One-fourth of this $\triangle H$ value is 98.8, which is the average bond energy of the C–H bond. For calculations in thermodynamics the experts have compiled extensive tables of such bond energy data, a sample of which is Table 16.1; these data are for complete decomposition into gaseous atoms in the ground state. The four C–H bonds of methane are not really equal in bond strength; it is estimated, for example, that the energy required to break up the first bond is 102 kcal/mole, and this is called the bond dissociation energy (BDE) of the first bond in methane.

The stability of a bond may vary with the nature of the adjacent atoms or structures in the molecule. The bond energy for the C=C bond in Table 16.1 is 147 kcal/mole, but this value may be quite different depending on the molecule in which the C=C bond is present. For the specific case of ethene (Section 15.4, last paragraph) the BDE is 125 kcal/mole rather than the 147 kcal/mole listed in Table 16.1. The bond energies in the first column of the table are often called *empirical* bond energies (BE). When they are values calculated by the same expert they are self-consistent when used for other thermochemical calculations.

The reader is also warned as to the ambiguity in certain of the definitions in thermochemistry. In the equation

$$C(s) + 2H_2(g) = CH_4(g) + 18.1 \text{ kcal}$$

we indicate that the elements carbon and hydrogen lose 18.1 kcal of heat energy to the surroundings when they form one mole of methane. Methane, therefore, has less energy than its elements and energy must be supplied to it in order to decompose the molecule. It is called a stable molecule because of its relatively low energy content. In the chemical literature, the term heat of *formation* is often used interchangeably with heat *content* and this sometimes leads to confusion; for example, 18.1 kcal is simply the energy associated with the formation or decomposition of methane, but is not the energy in the molecule.

In the table of bond energies, likewise, we do not mean that the H–H bond has an energy of 103.4 kcal. The actual energy content of this bond is relatively small as shown by the fact that considerable energy

TABLE 16.1. Some Bond Energies* st 298°K (25°C)

Bond energy (kcal/mole)

Nature of bond	*Empirical (BE)*	*Dissociation (BDE)*
H—H	104.2	103.4
H—Cl	103.2	102
H—Br	87.5	87
H—I	71.4	71
C—C	83.1	84 (CH_3—CH_3)
C=C	147	125 (CH_2=CH_2)
C≡C	194	230 (CH≡CH)
C—O	84.0	
C=O**	174	
C—H	98.8	
O=O	118	118
C—Cl	78.5	81 (CH_3—Cl)
C—F	105.4	108 (CH_3—F)

* For complete dissociation into gaseous atoms.
** When the bond is in compounds called ketones (Section 26.1)

(103.4 kcal) must be added to it in order to destroy it.   The data in the table of bond energies indicate relative stabilities; a high bond energy means high heat of formation and great stability, but low actual energy content.

16.3. Some Uses of Bond Energy Data.   In the first place, the data can be employed to estimate the heat change that occurs in a reaction when the information is not available in the literature.   To illustrate this principle with a simple example, consider the equation for the hydrogenation of 1-butene:

$$\underset{\text{1-Butene}}{H-\overset{\overset{\displaystyle H}{|}}{C}=\overset{\overset{\displaystyle H}{|}}{C}-CH_2-CH_3(g)} + H_2(g) = \underset{\text{Butane}}{CH_3-CH_2-CH_2-CH_3(g)} + 30.3 \text{ kcal}$$

The heat evolved in this hydrogenation reaction has been accurately determined, with all the substances gaseous at 1 atm. and at 355°K (82°C), and found to be 30.3 kcal/mole as indicated in the equation.   If the data were not available it could be calculated from the data in Table 16.1 as follows:

Bonds broken	*Heat absorbed*
H—H	−104.2
C=C	−147

Bonds formed	*Heat evolved*
C—C	+ 83.1
2 × C—H	+197.6

Net heat evolved
+29.5 kcal/mole

The heats of reaction and formation vary to some extent with temperature. The current tendency is to record such data for measurements made uniformly at 25°C (298°K). Much of the data in the handbooks are for a temperature of 18°C (291°K), and some data, as shown in the preceding example, are obtained at such unusual temperatures as 82°C. The variation over a range of as much as 50° is usually less than 0.1 kcal /mole, so that data from various sources can generally be used for approximate calculations without temperature correction.

Another application that may be mentioned briefly is the calculation of the relative electronegativities in Table 8.2. With data available for the bond energy of such *normal covalent* bonds as H:H and C:C, it may be assumed that half the bond energy of each of these, when added together, should give the bond energy of C:H. From the data for H–H and C–C in Table 16.1 the energy of the bond C–H calculated in this way is ½ (103.4+83) or 93.2 kcal. The deviation from the listed value of 98.8 kcal. for this bond is interpreted as being due to deviation from *normal covalent* behavior. The C:H bond has partial ionic character due to the difference in the relative electronegativities of the two atoms, that is, difference in attractive power for the electron pair. By ingenious arithmetic the experts have translated such deviations into the set of values shown in Table 8.2.

16.4. Hydrogenation of a Conjugate System. We have just seen that addition of hydrogen to the C=C bond is accompanied by evolution of 30.3 kcal of heat when one mole of reagent is used. Experiment has shown that this is the usual extent of the heat change when a double bond in a compound like butene is hydrogenated. If there are two double bonds in the molecule, the heat evolved on hydrogenation is almost exactly twice as great, or 60.6 kcal/mole. However, if the double bonds are conjugated as in 1,3-butadiene (Section 15.7), the rule is not followed. This is illustrated by the following examples:

CH_2=CH—CH=CH_2 Heat of hydrogenation to butane = 57.1 kcal

     1,3-Butadiene

CH_2=CH—CH_2—CH=CH_2 Heat of hydrogenation to pentane = 60.8 kcal

     1,4-Pentadiene

The heat of hydrogenation of 1,3-butadiene is 3.5 lower than one would expect it to be. This is a quantitative measure of the increased stability of the molecule due to its conjugated system (alternating double and single bonds). 1,4-Pentadiene gives the expected heat of hydrogenation.

16.5. Stability of a Conjugate System. In Section 15.7 we found that 1,3-butadiene is an easily excited molecule and that plausible structures of this compound which should be chemically reactive can be worked out readily on paper by simply flip-flopping the valence bonds. This flip-flop procedure was said to be equivalent to movement of a π electron through the molecule. Now we shall examine the phenomenon without use of valence bonds but with the concept of molecular orbitals.

The better than theoretical stability of butadiene which is indicated by its heat of hydrogenation can be largely accounted for by the mobility of its π electrons. It was explained in Section 15.2 that all bodies prefer greater freedom of motion, which is associated with greater stability. In butadiene there are four π electrons; in Figure 16.1 it is clear that each of these π electrons, instead of being restricted to two carbon atoms, can distribute its energy over four carbon atoms if these atoms can arrange themselves as shown.

The normal bonds between the atoms are of the σ type, just as shown for ethene in Section 15.4 and Figure 15.6. On each of the four carbon atoms there is left one p atomic orbital each of which has one electron; they can attain a system of lower energy if they overlap. In seeking this lower level of energy the molecule becomes planar—that is, the H atoms and C atoms are in a plane that is also the nodal plane of the p orbitals and is perpendicular to the printed page. The p orbitals are now in neighboring positions (because they are at right angles to the nodal plane) and can coalesce as in Figure 15.6e. Although the molecule is planar (flat) it does zigzag; however, the planar nature of the molecule permits mobility of the π electrons in a π molecular orbital in which four electrons encompass four nuclei. This is a more stable arrangement than that in which the electrons are isolated on individual atoms.

The four electrons in the π molecular orbital are said to be *delocalized*. The increased stability amounting to 3.5 kcal/mole (see Section 16.4) is the delocalization energy of 1,3-butadiene. The σ electrons represented

$$CH_2 \!\cdots\! CH \!-\! CH \!\cdots\! CH_2$$

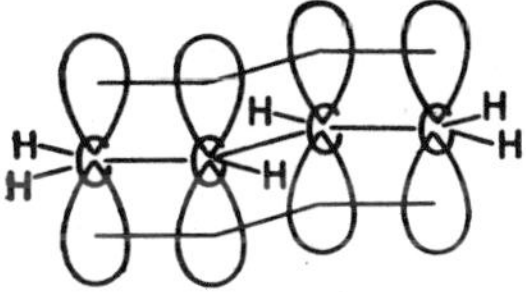

16.1. Adjacent p atomic orbitals in 1,3-butadiene can coalesce to form the molecular orbital containing four electrons.

by the straight line bonds in Figure 16.1 are *localized* electrons. (A different explanation for the behavior of 1,3-butatiene will be found in Section 16.12.)

16.6. The Structure of Carbon Dioxide. In the preceding section we found that when an electron normally restricted to binding a pair of atoms in a molecule has the freedom to encompass three or more nuclei the molecule then has added stability. With butadiene as a model we also found that using the methods of thermochemistry (Section 16.4) this increase in stability can be measured. Another interesting example of enhanced stability is carbon dioxide, $O=C=O$, the heat of formation of which can be calculated to be $\triangle H=-348$ kcal/mole (i.e. twice the value of the $C=O$ bond energy in Table 16.1). The corresponding $\triangle H$ value obtained experimentally, however, is -384 kcal/mole. The compound is therefore 36 kcal/mole more stable than would be expected from its composition. Another way to look at this problem is to compare the actual bond energy of $C=O$ in CO_2 (which is $384/2=192$) with the bond energy in Table 16.1, which is 174. This is a much greater enhanced stability than found for butadiene.

Once again we find that a molecule with unexpected stability has an *indefinite* structure, as compared with compounds like methane, CH_4, which may be said to have a definite structure. Several valence-bond

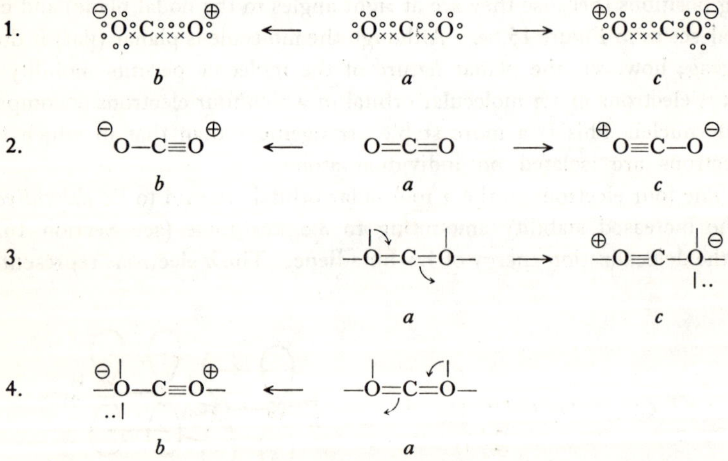

16.2. Electron shifts in the carbon dioxide molecule to form polarized structures. The double bond can be written $C \!\!\div\!\! O$ and the triple bond $C \!\!\vdots\!\! O$ to emphasize the mobility of π electrons and their greater exposure to chemical reagents.

structures can be written for carbon dioxide (see Figure 16.2) by flip-flopping the valence bonds as described in Figures 15.10 and 15.11. In line 1 of Figure 16.2, the normal structure of CO_2 is shown in *a*, where all three atoms are enclosed in the normal octet of electrons. The four electrons of carbon are crosses and the six electrons of each oxygen atom circles. With the relative positions of the atomic nuclei unchanged, an electron shift can be made to form polarized structures *b* or *c*. The oxygen atom at the right in *c* now has one more than its normal quota of six electrons, calculated as explained in Section 8.6, and consequently has a formal charge of minus one. It owns six unshared electrons, and half of the electron pair in the bond to carbon. Observe that superficially the oxygen atom at the right seems to have acquired a pair of electrons, whereas it has acquired an excess of only one (compare Figure 15.10). The oxygen atom on the left in *c* has one less than its quota of six electrons and therefore has a formal charge of plus one. In *b* of line 1 the picture is the same as in *c* but in the reverse direction in the molecule.

In chemical shorthand the polarity of the molecules can be designated as in line 2. Since each electron pair is either an actual or potential valence bond, another shorthand method can be used as shown by lines 3 and 4. This last method is the best one for the student to master because it is extensively used. An electron pair is generally shown at the negative end of the molecule, but it must be remembered that the excess charge is only one electron.

16.7. Resonance. From the remarks on the carbon dioxide molecule in Section 16.6 and the butadiene molecule in Section 15.7 it will be evident they ought to exhibit properties that satisfy the requirements of several plausible valence-bond structures. This is a most important phenomenon, called *resonance*. The term resonance implies the presence in a molecule of *mobile*, or *delocalized*, electrons and also that the structure of the molecule permits the delocalized electrons to *encompass more nuclei than two*. Resonance is possible in a molecule if it has multiple bonds properly placed to react on each other (as in butadiene and in carbon dioxide); a simple test is to determine if the bonds can be flip-flopped as in these illustrations.

The properties of compounds which exhibit resonance will be summarized later in this chapter (Section 16.9). Now we shall picture (Figure 16.3)the molecular orbitals in carbon dioxide, for a final example of the enhanced stability resulting from delocalized electrons.

The carbon-atom valences in CO_2 have the digonal symmetry mentioned in Section 15.2, and when the carbon atom is linked to two other atoms

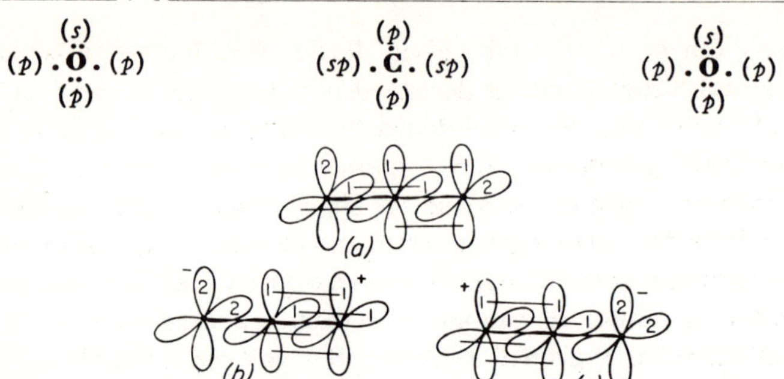

16.3. *Above:* Electron distribution in the carbon dioxide molecule. *Below:* Coalescence of the p atomic orbitals to form the π_b molecular orbitals. Structures a, b, and c correspond to the valence-bond formulas in Fig. 16.2. The numbers indicate how many electrons occupy a given p orbital. The π_b orbitals are similar to those shown for ethyne in Fig. 15.7, parts d and e. Like ethyne, CO_2 is a linear molecule and has cylindrical symmetry about the $O{=}C{=}O$ axis.

the orbitals in the valence shell hybridize (Table 7.1) to sp, sp, p, and p orbitals. The molecule can then be constructed on the basis of the electron distribution shown in the top line of figure 16.3.

Normal bonds of the σ type result when the sp orbitals of the carbon atom overlap with a p orbital of each oxygen; this yields the outline structure O–C–O, where the σ bonds are the two straight lines in the formula. The electrons in these bonds are *localized*. The carbon atom still has two p atomic orbitals, each containing one electron as shown in Figure 16.3a. Each oxygen atom still has two p orbitals; of these one has only one electron, but the other p orbital has two electrons. The single electrons in the p atomic orbitals of carbon and oxygen pair off by overlapping of their orbitals to give molecular π orbitals exactly as illustrated in Figure 15.6e and f. This overlapping is simulated by the tie-lines in Figure 16.3a, the formula of which can be written O⸬C⸬O where the dots are the two pairs of π electrons. The π orbitals in a have a common center line through OCO, but the nodal plane of the orbital on the right is perpendicular to the plane of the paper, whereas the nodal plane of the orbital on the left is the plane of the paper. (Each of the oxygen atoms has a spherical filled s orbital which is omitted from the diagrams.)

Structures b and c of Figure 16.3 correspond to the polarized structures b and c of Figure 16.2. Since the p atomic orbitals on all three atoms in the OCO molecule can align for maximum overlap (coalescence) it is obvious that a path is available for the four π electrons to move around in. They are mobile and delocalized and can reach a lower energy level (Section 15.2) by encompassing three atomic nuclei instead of being restricted to the normal two. This is a situation exactly like that in part d of Figure 15.7, which shows two bonding orbitals (π_b) perpendicular to each other; each orbital contains two electrons. The overall symmetry of the molecule is cylindrical, as in part e of Figure 15.7.

Each oxygen atom in a of Figure 16.3 has a p orbital occupied by a pair of electrons. Electron pairs of this nature are often called "lone-pair" electrons. These do not enter the bonding orbital. They may be nonbonding, or they may occupy antibonding orbitals (π_a) like those shown in Figure 15.8, except that in carbon dioxide they would be perpendicular to each other, at each end of the molecule.

16.8. Bond Length and Atomic Radius.   The distances between the atoms in normal covalent molecules have been quite accurately measured for many valence bonds by a number of methods including X-ray study of crystals and electron-diffraction and spectroscopic study of gaseous molecules. For example, it has been found that the bond distance in C–C is close to 1.54A in a large variety of carbon compounds (A is the Angstrom unit, or 10^{-8} cm, a distance first used in Section 8.3 and Table 14.2). For the corresponding double bond C=C, the distance between the atomic centers is 1.33 A and for the triple bond, C≡C, it is 1.20 A.   Similarly for the oxygen bond O=O, the distance is 1.10 A.

It has been demonstrated that one-half the distance between the atoms in a valence bond may be taken as the effective *atomic radius*,

where r is the radius.   Therefore, if we simply add up the atomic radii of two atoms we can determine the bond length between them. By adding the atomic radius for carbon in C=C and oxygen in O=O, we can find the bond length for C—O. This is $1.33/2 + 1.10/2 = 1.22$ A, using the data just cited.

16.9. Resonance Characteristics. It is now convenient to summarize the statements made regarding the properties of molecules that exhibit resonance. The theory of resonance is a mathematical development from quantum mechanics, but it can be expressed by generalizations that are easily comprehended.

16.9a. Hypothetical Structures. The various structures of carbon dioxide shown in Figure 16.3 do not have an independent existence.

When several such structures can be written, however, the compound has certain characteristics not predictable from the "normal" formula as usually written (in this case, $O=C=O$). No single, fixed structure can be visualized for a compound that exhibits the resonance phenomenon. Although they are hypothetical, the "plausible" formulas we write often offer a clue as to the probable physical and chemical properties of the compound. Hypothetical structures also also called *canonical*.

It is important to observe that the nuclei of the *atoms* in a molecule that exhibits resonance retain their relative positions; only the electrons alter their positions.

16.9b. Bond Length. It has just been observed in section 16.8 that the *calculated* distance between the atoms in the double bond of the CO_2 molecule is 1.22 A. When determined *experimentally*, however, this distance is found to be 1.15 A. In Figure 16.3 it is indicated that the orbitals in structure *a* will overlap. This coalescence of the *p* orbitals readily explains why the bond distances are shorter than those expected from the formula $O=C=O$; the overall length of the molecule is shorter than that corresponding to the structure as usually written. Shortened bond lengths are characterisctic of resonance.

Much information is now available for the distance between carbon atoms in many types of compounds. Some representative data are shown in Figure 16.4. In this figure, a smooth curve has been drawn through the data for ethane ($H_3C–CH_3$), ethene ($H_2C=CH_2$), and ethyne ($HC\equiv CH$), which can be taken as standards—that is, the carbon-carbon distances in these molecules are not affected by the presence of disturbing factors which often affect the distance in more complex molecules.

When data for compounds which have been shown to be in resonance are plotted on this same scale, they fall on the curve as indicated. It enables us to determine quantitatively the extent to which resonance affects a compound. For example, in 1,3-butadiene, $CH_2=CH–CH=CH_2$, the length of the normally single bond between the second and third carbon atoms has been found to be 1.46 A, and from the position of this point on the curve it is apparent that it has some double bond character. The resonance effects in this compound are relatively small.

16.9c. Resonance Energy. This energy is the extent to which a compound is more stable than it would be if it did not exhibit resonance. In Section 16.4, from its heat of hydrogenation it was estimated that the resonance energy of 1,3-butadiene is 3.5 kcal/mole (though some authorities believe this small stabilization energy can be ascribed more to other factors than to resonance as explained in Section 16.12). In Section 16.6

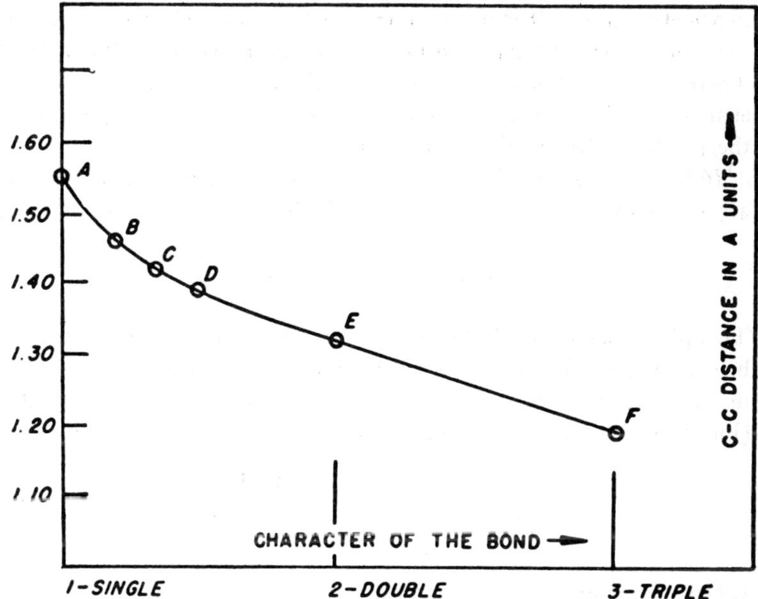

16.4. Relation between character of the bond and distance between carbon atoms. (*A*) Ethane, $H_3C—CH_3$; (*B*) Butadiene (Section 16.4); (*C*) Graphite (Section 19.7); (*D*) Benzene (Section 19.7); (*E*) Ethene, $H_2C=CH_2$; (*F*) Ethyne, $HC\equiv CH$.

the resonance energy of carbon dioxide was calculated to be about 36 kcal/mole. The two major criteria for resonance are this enhanced stability and shorter (tighter) bonds.

1. Resonance energy is observed only in molecules where π electrons can encompass more than two nuclei; there must be at least three atoms in the compound. The enhanced stability (resonance energy) is due to the greater freedom of movement of these electrons.

2. Resonance energy is at a maximum when the plausible valence-bond structures are completely equivalent. This will be found to be the case in benzene (Section 19.3). It is not quite true of carbon dioxide (Section 16.6) where only two of the structures are of equal energy content; however, calculations show that the third structure is not very different from the paired structures in energy, which accounts for the fact that the resonance energy of carbon dioxide is fairly high.

3. The stability of a molecule is greater than that of any of its reso-

nance structures. It is not an intermediate or average value. The structures on paper are a guide to understanding the nature of the molecule. There is no equilibrium state between these structures. The plausible structure with the greatest theoretical stability (lowest energy) will have the greatest effect in determining the properties of the molecule.

16.9d. Symbolism. The plausible valence-bond structures in resonance are generally shown connected with double-headed arrows, thus

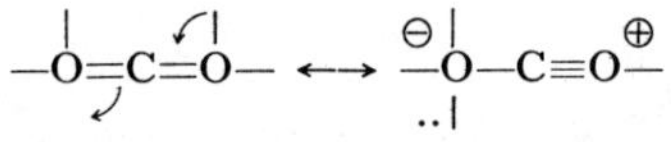

$$-\overset{|}{O}=C=\overset{|}{O}- \longleftrightarrow -\overset{\ominus}{\underset{..}{\overset{|}{O}}}-C\equiv O\overset{\oplus}{-}$$

This phenomenon should be distinguished from that in which structures become polarized during a reaction because of approach of an attacking molecule, or because of some other excitation. An example is ethene (Figure 15.10) where an ordinary arrow sign is used:

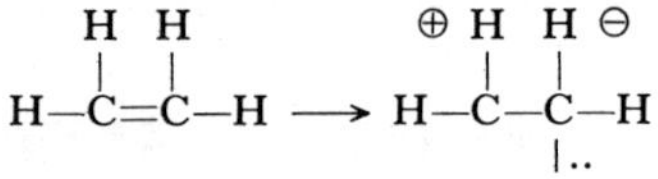

$$H\overset{H}{-}\overset{|}{C}=\overset{H}{\underset{|}{C}}-H \longrightarrow H\overset{\oplus H}{-}\overset{|}{C}-\overset{H \ominus}{\underset{|..}{C}}-H$$

Ethene does not exhibit resonance.

16.9e. Partial Ionic Character. Even though we do not regard resonance structures as forms that can be isolated, if plausible valence-bond structures can be written which show polarization of the molecule it means that the·molecule has what may be considered "partial ionic character." Separation of charge brings about greater reactivity.

16.10. Mesomerism. This is another name for the resonance phenomenon, but is not now widely used in this country. It is from the Greek word *mesos*, meaning intermediate. Resonance structures are sometimes called *mesomers*.

16.11. Hyperconjugation, or No-Bond Resonance. There is still another way of writing a resonance structure, illustrated with propene in Figure 16.5. This scheme is not accepted by all authorities. It was invented to explain the results of dipole-moment measurements on certain compounds. The essential feature of the theory is that a molecule may

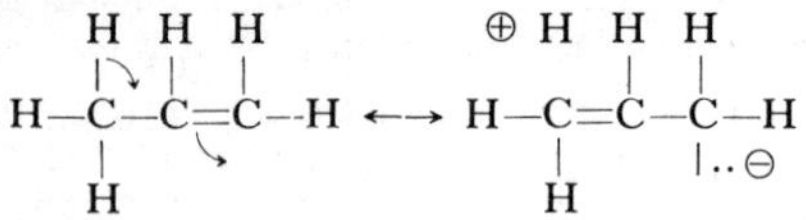

Normal form   A "no-bond" resonance structure

16.5. Illustration of hyperconjugation. The molecule is propene.

lose a hydrogen ion which remains in bonding distance; the electron pair bond that formerly held the proton (H$^+$) to the molecule now makes itself evident elsewhere in the molecule to give the molecule polar characteristics.

16.12. Bond Shortening and Hybridization. Earlier in this chapter emphasis was placed on the fact that a conjugated system of double and single bonds in a molecule will confer on the molecule certain characteristics associated with the *resonance phenomenon*. One of these effects is shortened bond length (which is equivalent to a strengthened bond). As *orbital theory* developed, it was found that in some instances a shortened bond length could also be ascribed at least in part to altered hybridization at the carbon atoms which terminate the bond.

To understand this the student should review the fundamental properties of orbitals shown pictorially in Figure 7.1. The *s* orbital is spherical; an electron in that orbital is *closely* associated with its nucleus, whereas in a *p* orbital the association of an electron with its nucleus is more distant. In Figure 15.5 we see the result when the *s* orbital and the three *p* orbitals of a carbon atom hybridize to form four *sp*^{3} orbitals; the figure illustrates just one of the *sp*^{3} orbitals, and in such a way to indicate it has both *s* and *p* properties. It is apparent, however, that in the *sp*^{3} hybrid the effect of the *p* orbital is to make the bond between an electron and its nucleus a little looser than it was in the original *s* orbital. (Figure 15.5 also shows coalescence of *sp*^{3} orbitals to make the C–C bond, but that does not concern us at this point.)

Now we turn to Figure 15.6, which portrays a flat molecule in which the C atom requires only three valence bonds of the σ type, so that it therefore needs to mix the energy of its *s* orbital with only two of its *p* orbitals, to give *sp*^{2} hybridization. In the *sp*^{2} orbital there is naturally more *s* character than in the *sp*^{3} hybrid, which means closer association of an electron to its nucleus in the *sp*^{2} hybrid than in the *sp*^{3}.

In Section 16.9b of this chapter reference was made to the butadiene molecule, $CH_2{=}CH{-}CH{=}CH_2$, where the shortened bond length between the second and third carbon atoms was ascribed to partial double bond character resulting from resonance (see Figure 15.11). From the point of view of orbital theory this C–C bond is a single bond, but the carbon atoms at each end of the bond are involved in *sp*^{2} hybrid orbitals, indicating the electrons in the bond have more *s* character than in a normal single bond which involves *sp*^{3} hybrid orbitals. Since the electrons in the bond are a little more closely bound to their respective nuclei in *sp*^{2} orbitals, this tends to shorten the bond between the nuclei.

It is claimed that resonance has less effect on the C_2–C_3 bond in butadiene than orbital hybridization of the carbon atoms.

17

How Molecules React

17.1. Reaction Mechanisms. Relatively few reactions were well understood in the first hundred years of organic chemistry. The electron had not yet been discovered. From the subjects discussed in the last few chapters it is obvious that knowledge of the electron structures of molecules is a prerequisite to the understanding of how molecules behave when engaged in chemical reactions. Since about 1900, and more especially since about 1930, there has been a rapidly increasing interest in the intimate details of reaction mechanisms.

Some classes of reactions have already been briefly mentioned in this book. In Section 9.5 it was said that the reaction between CH_3Cl and $AgNO_3$ takes place between ions; in Section 8.5 the reaction between NH_3 and BCl_3 is simple union between molecules; in Section 13.7 the hydrogenation reaction between $CH_2{=}CH_2$ and H_2 was described as an addition reaction; and in Section 9.4 the action of Cl_2 on CH_4 was called substitution. These and other types of reactions will be explored in this chapter, and at its finish the various reactions will be classified with respect to currently recognized mechanisms.

It is the principal purpose of this chapter to explain briefly the language and symbols used in the interpretation of reaction mechanisms, with special reference to the way this language has been affected by the electron theory. We begin by reviewing two phenomena that will be referred to frequently: the *inductive effect* and the *resonance effect*.

17.2. The Inductive Effect. After investigating the fascinating behavior of the movable π electrons in Chapters 15 and 16, it is something of a let-down to return to the more prosaic σ electrons. These are the backbone of the carbon chain; however, it should not be thought that they are firmly anchored. In Section 9.5 it was said that CH_3Cl is more active than CH_4 because the Cl atom causes a displacement of an electron pair in such a direction as to make the Cl atom potentially ionic. A displacement such as that described in Section 9.5 makes itself felt throughout a chain of carbon atoms, although the effect does fall off rap-

153

$$
\begin{array}{ccc}
\text{H H H} & & \\
\text{H :C :C :C :Cl:} & -\overset{|}{\underset{|}{C}}+\to\overset{|}{\underset{|}{C}}+\to\overset{|}{\underset{|}{C}}+\to\text{Cl} & CH_3-CH_2-\overset{\delta+}{CH_2}-\overset{\delta-}{Cl} \\
\text{H H H} & & \\
I & II & III
\end{array}
$$

17.1.   Methods of representing the inductive effect.

idly as it progresses down the chain. The phenomenon is called an inductive effect and can be represented in several ways as illustrated in Figure 17.1. In Formula I it will be seen that the electron pair between any two atoms is unequally shared; this can be shown, as in III, to be equivalent to a dipole (Section 8.3) in which one atom has a minute positive charge $(\delta+)$ whereas the atom toward which the displacement occurs acquires a minute negative charge $(\delta-)$. The inductive effect can also be illustrated as in II, where the arrowhead points to the negative end of the dipole and the plus sign shows the positive end. The molecule is said to be *polarized*.

The inductive effect prevails in the normal state of a molecule. It is essentially a phenomenon associated with the single valence bond. The electron pair (:), which is the normal valence bond, is displaced slightly when there is an atom in the molecule which exerts a greater or lesser attraction for the electron pair than the other atoms do. The direction of the inductive effect in a molecule can often be predicted from the relative electronegativites listed in Table 8.2. If the substituent atom (which is Cl in Figure 17.1) is strongly electron-attracting, the inductive effect is labeled -I; when it is strongly electron-repelling the effect is +I. An inductive effect can usually be detected through measurement of dipole moment (Section 8.3).

17.3. The Resonance Effect. In contrast with the single-bond phenomenon described in the preceding section, the resonance effect is associated with the multiple bond and its mobile π electrons. It was explained in Section 16.9 that in resonance the atoms of a molecule retain their relative positions, but that the π electrons have a greater than normal freedom of motion encompassing three or more atomic nuclei. The nature of this freedom can be pictured in plausible valence-bond structures by flip-flopping the bonds (Section 15.7). The chemical properties of many compounds are explainable by assuming this migration of an electron; when the assumption is found to be applicable it is called the resonance effect.

This principle can be illustrated (Figure 17.2) with allyl chloride, which is a polar compound as a result of the inductive effect. In other words,

$$\text{Allyl chloride} \quad \overset{\text{CH}_2=\text{CH}-\text{CH}_2-\text{Cl}^-}{} \;\rightleftharpoons\; :\ddot{\text{C}}\text{l}:^- \;+\; [\text{allyl ion resonance structures}]^+$$

Allyl chloride
3-Chloro-propene

Resonance structures
of the allyl ion.

17.2.   Stabilization of the allyl ion by resonance.

the C–Cl bond in chemical reactions tends to split off a Cl⁻ ion (compare CH_3Cl in Section 9.4).   If this negative ion does separate during the course of a reaction, as shown in Figure 17.2, the positive fragment is the allyl ion.   Resonance structures can be written for this ion, as indicated, whereas such resonance is not possible for the molecule from which the ion originated.   (Observe that the unshared electron pairs on the Cl atom of allyl chloride are represented by straight lines.)

We have repeatedly stressed the fact that resonance stabilizes *a molecule* (Section 16.9c) in that the molecule will then have a lower energy content; the same thing is true of *an ion* for which resonance forms can be written. In order to cause the chloride and allyl ions to recombine and form allyl chloride, a certain resonance energy must be added, which is considerable because the two resonance structures are equivalent, and this "complete" resonance yields enhanced stability (Section 16.9c (2)).   Once the ions are developed during a reaction, they therefore tend to remain ions.   This is theoretical support for the experimental observation that the C–Cl bond is abnormally active in allyl chloride; it is many times as active as the corresponding bond in propyl chloride, $CH_3CH_2CH_2$–Cl, which is the molecule pictured in Figure 17.1.   There is no enhanced activity of the Cl atom in the four-carbon compound

$$\text{H}-\text{CH}=\text{CH}-\text{CH}_2-\text{CH}_2-\text{Cl}^-$$

4-Chloro-1-butene

because there is no path for moving a π electron through the possible ion of this molecule by flip-flopping the valence bonds as described in Section 15.7.

In Figure 17.2 we have seen that the presence of the $CH_2=CH-$ group in allyl chloride enhances the activity of the C–Cl bond because the ion that can form is stabilized by a resonance effect.   A C–Cl bond further

$$\overset{\text{H \ H}}{:\ddot{C}l:\ddot{C}::\ddot{C}:H} \qquad \overset{\text{H \ H}}{-Cl-C{=}C-H} \longleftrightarrow \overset{\oplus}{-}Cl{=}CH-\overset{\ominus}{CH_2}$$

$$\text{I} \qquad\qquad \text{II} \qquad\qquad \text{III}$$

17.3. Conjugation of an unshared electron pair of the Cl atom with the electrons in the double bond of vinyl chloride (chloroethene).

removed from the $CH_2{=}CH-$ group in the molecule, as just described, behaves normally. When we go to the shorter molecule, where the Cl atom is linked directly to the $CH_2{=}CH-$ group, there is a great *decrease* in chemical activity of the C–Cl bond. The reason for this is shown in Figure 17.3, where each unshared pair of electrons on the Cl atom in Structure I may be considered a potential valence bond. In Structure II they are accordingly replaced by straight line bonds, and it will be seen that they can be regarded as *conjugated* with the double bond in the molecule, similar to the conjugation that exists in butadiene (Figure 15.11). The interaction between these unshared pairs of electrons on the Cl atom with the π electrons in the double bond gives greater stability to the C–Cl bond. It is shorter and quite inactive chemically as compared with the normal C–Cl bond in chloroethane, CH_3-CH_2-Cl, the corresponding saturated compound. This explanation of the reduced activity of the Cl atom in vinyl chloride is based on resonance theory. The student should also consult the reasoning in Section 16.12, which leads to the conclusion that the reduced activity may also be due in part to hybridization effects at the C–Cl bond. This phenomenon will be met with again in Section 24.4.

In Structure II of Figure 17.3 the unshared pairs of electrons on the Cl atom, or the π electrons of the double bond, are sometimes referred to as an electron "source"; an unsaturated atom or group, such as $=CH_2$ at the end of a molecule is an electron "sink."

17.4. Nucleophilic and Electrophilic Reagents. In Section 8.5, which should be reviewed, the formation of a dative bond was illustrated by the following reaction:

$$H_3N: +\ BCl_3 \longrightarrow H_3N \rightarrow BCl_3$$

In this reaction, the BCl_3 molecule is said to be *electrophilic*, which is a classical way of saying *electron-seeking*. The NH_3 molecule is electron-rich, and is willing to share the electron pair which is not actively used for bonding. An *electron donor*, such as the NH_3 molecule, is said to be *nucleophilic*, a term used to signify that such a molecule is ready to react with a molecule deficient in electrons in the sense of the octet theory (Section 6.4)—that is, has a tendency to fill up the quota of eight electrons in the valence shell. Positive ions are electrophilic. Negative ions are nucleophilic.

17.5. Addition of Halogen to the Double Bond. This subject was touched on in Section 13.9, when the general characteristics of compounds containing double bonds were considered. In Figure 17.4 the structure of ethene is written in such a way as to show the exposed nature of the π electron pair which constitutes its second bond. This is a nucleophilic molecule. It extracts a positive bromine ion from the bromine molecule to yield an intermediate stage which probably has a 3-membered ring structure. This is called a bromonium ion, to indicate that the positive charge belongs to the bromine. (Carbonium ions will be described later.) The intermediate ion is an electrophile; it is attacked by a negative bromide ion from a direction opposite to that of the positively charged bromine already in the cyclic structure, to give 1,2-dibromoethane (ethene dibromide).

The dibromoethane molecule has freedom of rotation about the C–C bond, just as described for ethane in Figure 14.5, although the barrier to

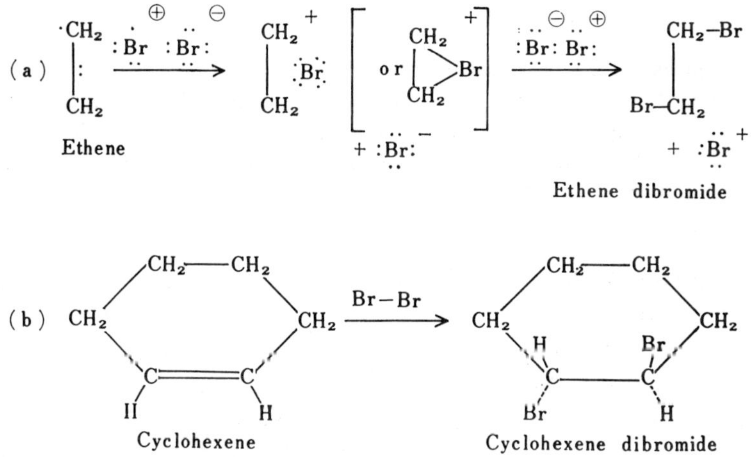

17.4. (a) Addition of bromine to ethene by way of a postulated cyclic intermediate bromonium ion. Note that one mole of ethene consumes one of bromine. Bromine has "partial ionic character" but is not an ionic compound; the ionic division shown here takes place only during a chemical reaction. (b) Addition of bromine to cyclohexene. The mechanism is the same as in *a*.

free rotation is appreciably higher. It is therefore proper to write its structure either as the *anti* conformer as in Figure 17.4, or as the *eclipsed* conformer (compare Figures 12.8 and 12.9). The reason for postulating the cyclic intermediate ion in the bromination of the double bond is that this ring structure would force the two bromine substituent atoms into *trans* positions, and that is the way the reaction apparently takes place with molecules that do not have freedom of rotation. An example is the bromination of cyclohexene. If Figure 15.1 is reviewed, it will be obvious that the C–H bonds of the H–C=C–H group in cyclohexene are in the plane of this page, but the other H atoms are either above or below the page (also see cyclohexane in Figure 12.16). The product of the bromination of cyclohexene is *trans*-1,2-dibromocyclohexane; it is clear the ring must be attacked in two successive steps if the two bromine atoms appear on opposite sides (*trans*). It is currently believed that halogenations of the double bond generally take place this way, although exceptions are known. See last paragraph of Section 17.11.

17.6. 1,4-Addition to Conjugate Systems. When a molecule such as chlorine or bromine splits up into two parts in its reaction with a conjugate system of single and double bonds:

$$CH_2{=}CH{-}CH{=}CH_2 \qquad \text{1,3-butadiene}$$
$$1 \quad\;\; 2 \quad\; 3 \quad\;\; 4$$

these parts will add as usual across the 1,2 and 3,4 double bonds, but will also add 1,4 and a new double bond appears in the 2,3 position. The simplest example is the addition to 1,3-butadiene, a compound described in Sections 15.7, 16.4, and 16.5, and also in Figure 15.11. The products are as follows:

$$
\begin{array}{ccc}
\underset{\text{(A)}}{\overset{\displaystyle \overset{Cl}{|}\;\;\;\overset{Cl}{|}}{CH_2{-}CH{-}CH{=}CH_2}} &
\underset{\text{(B)}}{\overset{\displaystyle \overset{Cl}{|}\;\;\;\overset{Cl}{|}}{CH_2{=}CH{-}CH{-}CH_2}} &
\underset{\text{(C)}}{\overset{\displaystyle \overset{Cl}{|}\qquad\qquad\overset{Cl}{|}}{CH_2{-}CH{=}CH{-}CH_2}}
\end{array}
$$

The compound in (C) is said to be the result of 1,4 addition.

The reaction at the 1,2 double bond probably takes place in the two stages shown in Figure 17.4*a*, the first step being the formation of a complex cyclic positive ion. Now that we understand this mechanism, we can facilitate matters by looking at the 1,3-butadiene molecule as in Figure 17.5*b*. This is an electron shift like that shown in Figure 15.10 for ethene; it may be a slight liberty taken with the facts but it should help the beginner in his working out of this and similar problems with the reactions of double bonds. The positively charged chlorine fragment Cl^+ finds the

$$CH_2{=}CH{-}CH{=}CH_2 \longleftrightarrow \overset{\ominus}{C}H_2{-}\overset{\oplus}{C}H{-}CH{=}CH_2 \xrightarrow{\;:\overset{..}{\overset{\oplus}{Cl}}\;:\overset{..}{\overset{\ominus}{Cl}}:\;}$$

(a)   (b)

$$Cl^- + \underset{\underset{Cl}{|}}{CH_2{-}\overset{+}{C}H}{-}CH{=}CH_2 \longleftrightarrow \underset{\underset{Cl}{|}}{CH_2{-}CH{=}CH}{-}\overset{+}{C}H_2 \xrightarrow{\;:\overset{..}{\overset{\oplus}{Cl}}\;:\overset{..}{\overset{\ominus}{Cl}}:\;}$$

(c)   (d)

$$Cl^+ + \underset{\underset{Cl}{|}}{CH_2}{-}CH{=}CH{-}\underset{\underset{Cl}{|}}{CH_2}$$

(e)

17.5. Addition of one halogen molecule to the conjugate system of 1,3-butadiene. A positive Cl^+ ion is added in the first step, and a negative Cl^- ion is added in the second step. They come from two different molecules of Cl_2.

C_1 position eventually, and the polarized version of butadiene in *b* helps us put it there. After the complex positive ion has formed in *c* we have the possibility of resonance stabilization of the ion, which is a driving force in the reaction. The resonance forms are *c* and *d*. Both are electrophiles, with deficiencies of electrons at the C_2 and C_4 atoms, respectively, and that is where the Cl atom will locate in the second step. Again compare Figure 17.4*a*. Formula *e* in Figure 17.5 shows only the 1,4 addition. The 1,2 product can also form, from structure *c*, and if we had started in *b* with the C_4 atom the products in these equations would be 3,4 and 1,4 addition. The relative amounts of the three possible products depend on the reaction conditions. See last paragraph of Section 17.11.

17.7. The Thiele Theory of Partial Valence. It is of interest to go back into the history of organic chemistry before the advent of electron theory to find out how the 1,4 addition reaction was explained by the Thiele theory—which was quite successful in its day. According to the Thiele theory the valence force of a bond like that designated A in Figure 17.6 is not completely utilized when carbon atoms are linked by a double bond. Valence forces radiate from the atomic center in straight lines, but from elementary physics we know that a force can be resolved into several component forces by the parallelogram method, as illustrated in the figure. Thiele assumed that only the component B takes an active part

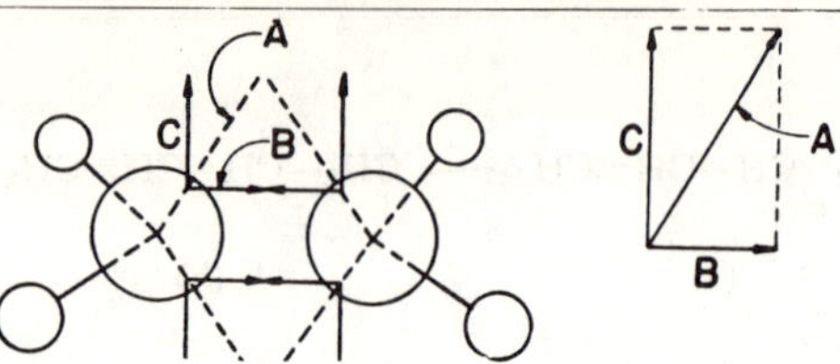

17.6.   Molecule of ethene according to the Thiele theory of partial valence.

in linking the carbon atoms in ethene, with some valence force left over designated by C.

Thiele, accordingly, explained the chemistry of ethene by writing the formula as $CH_2{=}CH_2$, using dotted lines to indicate the fact that the carbon atoms have some residual valence force left over, and the dotted lines were called "partial valence bonds." That was his way of expressing the nature of the reactivity of carbon atoms when they are linked by more than one bond. The addition reaction is then written as follows:

$$CH_2{=}CH_2 + Cl_2 \longrightarrow CH_2{-}CH_2$$

$$\underset{Cl \quad\; Cl}{\qquad\qquad\qquad\qquad\qquad\qquad\;\; |\quad\;\; |}$$

One of the best uses of the partial valence theory has been its application to conjugate systems. This is the way the theory explains the behavior of 1,3-butadiene:

$$CH_2{=}CH{-}CH{=}CH_2 + Cl_2 \longrightarrow CH_2{-}CH{=}CH{-}CH_2$$

$$1 \qquad 2 \qquad 3 \qquad 4 \qquad\qquad\qquad Cl \qquad\qquad\qquad Cl$$

It is assumed that the neighboring valences marked 2 and 3 react on each other to form a loose bond, leaving 1 and 4 open as reactive centers in the molecule. Then, after 1,4 addition has taken place, a normal double bond is formed between 2 and 3.

17.8. Free Radicals. The normal union between two atoms is the covalent bond (:). Up to this point we have referred to only one method by which this link is broken—namely, the formation of ions, as illustrated in Figure 17.5, where a Cl_2 molecule is so strongly polarized in a reaction that it splits up into ions. In that reaction the electron pair remains intact.

$$:\overset{..}{\underset{..}{Cl}}{}^+ + :\overset{..}{\underset{..}{Cl}}:^- \longleftarrow :\overset{..}{\underset{..}{Cl}}:\overset{..}{\underset{..}{Cl}}: \longrightarrow :\overset{..}{\underset{..}{Cl}}\cdot + \cdot\overset{..}{\underset{..}{Cl}}:$$

Formation of ions   Formation of free atoms

A second way to break up this union is for the electron pair to separate.

The result is the pair of free chlorine atoms. The chlorine molecule in the gas state can be broken down into the atoms by light energy equivalent to about 70 kcal/mole (see Table 14.2), whereas the energy required to form the ions would be about four times as great.

The free methyl radical is one-half the molecule of ethane.

$$CH_3:CH_3 \qquad\qquad\qquad \cdot CH_3$$

Ethane   Free methyl radical

It is like a free atom in that it has an unpaired electron. Free radicals, like free atoms, are highly reactive and have a very short life. In Figure 17.7 the free methyl radical is pictured to show how it is to be distinguished from the negative methyl ion present in methylsodium (Section 9.6), and from the positive methyl ion which is present when methyl chloride is in an activated state before it reacts with silver nitrate (Section 9.5). The negative methyl ion is called a *carbanion*. The positive methyl ion is the simplest of the *carbonium* ions; the charge is on the carbon, and this should be contrasted with the bromonium ion in Figure 17.4*a* where the charge is on the bromine.

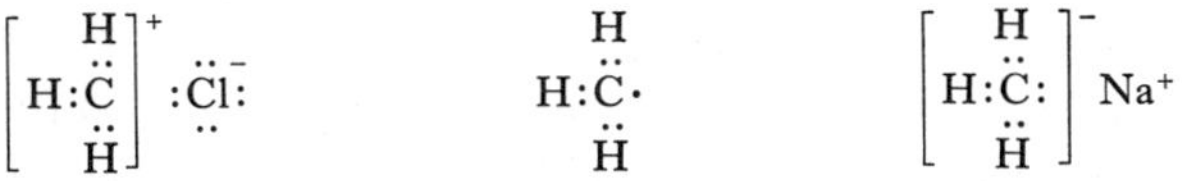

Methyl chloride   Free methyl radical   Methylsodium
                                                  Sodium methide

17.7.   Bond-making possibilities of the methyl group (CH_3).

17.9. Chain Reactions. Free radicals are an important tool for the research chemist and for chemical industry. They can be used to initiate chain reactions, the nature of which will be described in this section using the chlorination of methane as a sample reaction. Although the chemist instinctively associates chain reactions with free radicals, it will be shown in Section 17.12 and in Section 38.7 that chain reactions can also be initiated and propagated by an ionic mechanism. However, the basic principles are similar in both cases, and only the free-radical chain reaction will be discussed at this point.

Methane and chlorine react slowly when mixed in diffused daylight:

$$CH_4 \quad + \quad Cl_2 \quad \longrightarrow \quad CH_3Cl \quad + \quad HCl$$

Methane   Chlorine   Chloromethane   Hydrogen
                                Methyl chloride   chloride

but if this mixture is prepared in a bottle which is then rolled into the sun-

1. Initiation $\quad\quad\quad\quad\quad$ $Cl:Cl \longrightarrow 2Cl\cdot$

2. Propagation $\quad$ (a) $\;H:CH_3 + Cl\cdot \longrightarrow \cdot CH_3 + H:Cl$

$\quad\quad\quad\quad\quad\quad$ (b) $\;Cl:Cl + \cdot CH_3 \longrightarrow Cl:CH_3 + \cdot Cl$

3. Termination $\quad$ (a) $\;Cl\cdot + \cdot CH_3 \longrightarrow Cl:CH_3$

$\quad\quad\quad\quad\quad\quad$ (b) $\;CH_3\cdot + CH_3\cdot \longrightarrow CH_3:CH_3$

$\quad\quad\quad\quad\quad\quad$ (c) $\;Cl\cdot + Cl\cdot \longrightarrow Cl:Cl$

17.8. The three parts of a typical chain reaction. Only those electrons are indicated that are needed in the description of the reaction mechanism.

light, it may explode because of the rapid reaction that takes place. This is an example of a chain reaction, which has the three characteristic stages shown in Figure 17.8.

Chlorine is yellow, which means it absorbs light in the visible spectrum. The chain reaction is initiated when the action of light energy causes chlorine atoms to form. Heat in the absence of light can also be used to initiate the reaction. The chain propagation in Step 2 is self-sustaining: Step 2(a) generates a free methyl radical which is used up in Step 2(b); and Step 2(b) generates a free chlorine atom which is used up in Step 2(a). The products are methyl chloride and hydrogen chloride. In some chain reactions the propagation steps go through a million cycles before they are terminated, but in this chlorination the cycles are numbered in terms of hundreds or thousands.

Chain reactions are stopped when the free radicals make collisions which result in less active molecules. Each collision of this type is very important because it may prevent the formation of thousands of product molecules. The reactions which kill free radicals are listed as termination reactions in Step 3. Substances are sometimes added to control chain reactions; these additives are called chain inhibitors, or free-radical traps. Tetraethyllead is added to gasoline as such an *inhibitor*. The combustion in a gasoline engine is a chain reaction, which causes engine knock (Section 10.7) when the reaction goes too fast, but is controllable by trace amounts of the inhibitor.

The chlorination of methane by a free radical mechanism is used for purposes of illustration. It is not the preferred way to make methyl chloride free from other chlorination products (see Section 9.1).

17.10. Dot-Dot Structures. Free radical reactions and chain reactions were introduced at this point because they will be needed to explain more of the chemistry of the multiple bonds and π electrons which have occupied so much of our attention. Before taking up these new considerations, let us look once more at the ways in which the double bond can be formulated:

$$\overset{\oplus}{\cdots\cdot CH}-\overset{\ominus}{CH}\cdots\cdot \quad \longleftarrow \quad \cdots\cdot CH\dot{-}CH\cdots\cdot \quad \longrightarrow \quad \cdots\cdot CH-CH\cdots\cdot$$

The apparent electron shift when adding dipolar molecules		The apparent disposition of the π electron pair in free radical reactions
(b)	(a)	(c)

In *a* the double bond is shown with the second bond represented by a pair of π electrons. It has been explained (Section 15.6) that ethene itself is highly symmetrical and difficult to polarize to the *b* structure; however, in a larger molecule than ethene much depends on the substituents at other carbon atoms in the chain, which may have a stabilizing effect on this type of structure. When polar reactions occur in which dipoles add to the double bond, it is convenient to overlook some of the intermediate details of mechanism and (on paper at least) work out the generalized aspects of the reaction with the polarized structure written as in *b*.

We have just found that reactions that lead to free radicals tend to "uncouple" an electron pair bond; the free radicals are characterized by an unpaired electron. If it is known that the double bond is going through a free radical reaction, it is convenient to follow the reaction on paper with the formula as written in *c*, which is known as the diradical or dot-dot structure. Diradicals are not necessarily highly reactive. The oxygen molecule $\cdot \ddot{O}:\ddot{O}\cdot$ is a diradical but has an easily controlled activity. Carbon compounds will be described later (Section 24.10) with odd electrons and good stability. Methods used to identify molecules with odd electrons will be mentioned briefly in Section 31.9

17.11. Abnormal Addition to the C=C Bond. The explanation of the addition reaction to a double bond as given in Section 17.5 only covers the simplest possibilities; it was concerned only with addition to a symmetrical olefin like ethene or cyclohexene. To such molecules the addition of hydrogen bromide, H-Br, can be written without regard to position; the H of H-Br can be attached to either of the carbon atoms in the double bond. When H-Br is added to an unsymmetrical olefin there are two ways for addition to take place:

$$CH_3-CH=CH_2 + H-Br \nearrow CH_3-CHBr-CH_3 \text{ (Normal "Markownikoff" addition)}$$

CH₃–CHBr–CH₃ (Normal "Markownikoff" addition)

2-Bromopropane

CH₃—CH=CH₂ + H–Br

Propene

CH₃—CH₂—CH₂Br (Abnormal "peroxide" addition)

1-Bromopropane

The so-called normal addition proceeds by the polar mechanism described in section 17.5, and the name of a principal early investigator (Markownikoff) is usually attached to it.   If we follow the convention in

Section 17.10, the polarized structure of propene would be $CH_3\!-\!\overset{\oplus}{C}H\!-\!\overset{\ominus}{C}H_2$,

and naturally the negative bromine would go to the middle carbon atom and positive hydrogen to the terminal carbon.

Hydrogen bromide and a number of other compounds have also been found to add to the unsymmetrical olefin in a so-called *abnormal* manner (sometimes called Kharasch addition).   This reaction takes the form of a free-radical chain mechanism, initiated by substances which can react with H–Br to generate free bromine atoms:

Initiation $RO\cdot + H{:}Br \longrightarrow RO{:}H + \cdot Br$

Propagation   (a) $CH_3—CH—CH_2 + \cdot Br \longrightarrow CH_3—CH—CH_2Br$

(b) $CH_3—CH—CH_2Br + H{:}Br \longrightarrow CH_3—CH_2—CH_2Br + \cdot Br$

Termination Free radical collisions which form stable molecules like Br_2

Free-radical chain reactions are generally started by adding trace amounts of compounds (usually peroxides) which are easily broken down by gentle heat or by light into their component radicals.   These free radicals are referred to as initiators; an example is the free radical from t-butyl peroxide:

$$(CH_3)_3C—O—O—C(CH_3)_3 \longrightarrow 2(CH_3)_3C—O\cdot$$

This can be symbolized as $RO{:}OR \rightarrow 2RO\cdot$ for a wide variety of such compounds.

The propagation phase described in these reactions for propene is similar to that shown in Figure 17.8; reactions *a* and *b* follow each other in many cycles until terminated when a free radical meets the wrong partner. The structure of propene in *a* is written according to the dot-dot scheme shown in Section 17.10; actually the two electrons are the π bond of the double bond. The unpaired electron on the bromine atom "uncouples" that electron of the pair with opposite spin to form a covalent bond, and the other electron (the same spin) is repelled to another point in the molecule.

The abnormal addition to the olefin bond was recognized long before it was known to be the result of a free-radical mechanism. The reaction occasionally took place in this manner because the reagents and the olefins themselves sometimes had traces of initiators (peroxides) in them. Unlike H–Br, the analogous compounds H–Cl and H–I add only by the normal mechanism.

The reactions described in Sections 17.5 and 17.6 (which should be reviewed) are the generally accepted ionic, or polar, mechanisms that explain the reaction products obtained. However, it is claimed by some that chlorination of an olefin double bond *normally* takes place by a free radical mechanism, and that the mechanism described in Sections 17.5 and 17.6 will predominate only if the proper polar conditions prevail, such as use of a polar solvent to maintain charge separation of the reactants and use of an inhibitor to retard the free radical formation. Some researchers claim that free radical reactions that are not recognized as such often take place because customary conditions for generating free radicals (light and heat on a fairly weak electrovalent bond as in Cl:Cl) are not employed. These workers believe that a pair of reagents like chlorine and cyclohexene (compare Section 17.5) may interact to form free radicals in the dark and at low temperature.

17.12. Carbonium Ion Rearrangements. The literature of organic chemistry is full of rearrangement reactions. For an illustration we shall use hydrocarbon compounds which are already familiar, as depicted in figure 17.9. The rearrangements which will be pointed out follow this simple plan:

$$A:B:C \qquad\qquad\qquad B:C:A$$

$$(a) \qquad\qquad\qquad\qquad\qquad (b)$$

In these structures B and C are atoms, but A may be either a single atom or a large group. If *a* is formed during a reaction it will probably be unstable because it has only six electrons on atom C (note that this is the situation in the carbonium ion V of Figure 17.9). If by doing so it can form a more stable molecule, the group A migrates *with its electron pair* to atom C to give the more stable, but not necessarily completely stable structure, *b*. This is an example of 1,2 shift.

$$\underset{\text{Isobutene (I)}}{CH_3\overset{\oplus}{-}\overset{\overset{\displaystyle CH_3}{|}}{C}-\overset{\overset{\displaystyle H}{|}}{\underset{\underset{\displaystyle H}{|}}{C}}\overset{..}{\underset{\ominus}{}}} + \underset{\text{Isobutane (II)}}{H\!:\!\overset{\overset{\displaystyle CH_3}{|}}{\underset{\underset{\displaystyle CH_3}{|}}{C}}-CH_3} \longrightarrow \underset{\substack{\text{Isooctane (III)}\\ \text{2,2,4-Trimethylpentane}}}{CH_3-\overset{\overset{\displaystyle CH_3}{|}}{\underset{\underset{\displaystyle H}{|}}{C}}-\overset{\overset{\displaystyle H}{|}}{\underset{\underset{\displaystyle H}{|}}{C}}-\overset{\overset{\displaystyle CH_3}{|}}{\underset{\underset{\displaystyle CH_3}{|}}{C}}-CH_3}$$

$$I + \overset{\oplus}{H}-\overset{\ominus}{HSO_4} \longrightarrow CH_3-\overset{\overset{\displaystyle CH_3}{|}}{\underset{\underset{\displaystyle OSO_2OH}{|}}{C}}-CH_3 \rightleftharpoons \left[CH_3-\overset{\overset{\displaystyle CH_3}{|}}{C}-CH_3\right]^{+} + HSO_4^{-}$$

t-Butyl ion (IV)

$$I + IV \longrightarrow \left[CH_3-\overset{\overset{\displaystyle CH_3}{|}}{C}-\overset{\overset{\displaystyle H}{|}}{\underset{\underset{\displaystyle ..H}{|}}{C}}-\overset{\overset{\displaystyle CH_3}{|}}{\underset{\underset{\displaystyle ..CH_3}{|}}{C}}-CH_3\right]^{+}$$

V

$$II + V \longrightarrow III + IV$$

$$V \longrightarrow \underset{VI}{\left[CH_3-\overset{\overset{\displaystyle CH_3}{|}}{\underset{\underset{\displaystyle ..H}{|}}{C}}-\overset{\overset{\displaystyle H}{|}}{C}-\overset{\overset{\displaystyle CH_3}{|}}{\underset{\underset{\displaystyle ..CH_3}{|}}{C}}-CH_3\right]^{+}} \longrightarrow \underset{VII}{\left[CH_3-\overset{\overset{\displaystyle CH_3}{|}}{\underset{\underset{\displaystyle ..H}{|}}{C}}-\overset{\overset{\displaystyle H}{|}}{C}-\overset{\overset{\displaystyle CH_3}{|}}{\underset{\underset{\displaystyle ..CH_3}{|}}{C}}-CH_3\right]^{+}}$$

$$VII + II \longrightarrow \underset{\substack{VIII\\ \text{2,3,4-Trimethylpentane}}}{CH_3-\overset{\overset{\displaystyle CH_3}{|}}{\underset{\underset{\displaystyle H}{|}}{C}}-\overset{\overset{\displaystyle H}{|}}{\underset{\underset{\displaystyle CH_3}{|}}{C}}-\overset{\overset{\displaystyle CH_3}{|}}{\underset{\underset{\displaystyle H}{|}}{C}}-CH_3} + IV$$

17.9. Synthesis of octanes III and VIII. Isobutene (I) is shown as the polarized structure. Carbonium ion rearrangements take place in V→VI and VI→VII. Electron dots are shown only as required by the text.

Figure 17.9 outlines one of the reactions used by the petroleum industry to produce an antiknock gasoline. The product is often called *alkylate*; the reason for the name will be obvious from the material in Section 23.8. The reaction consists in mixing isobutene, $(CH_3)_2C\!=\!CH_2$, with isobutane, $HC(CH_3)_3$, at certain controlled temperatures with concentrated sulfuric acid as catalyst. In I the isobutene structure is written in the polarized form as explained in Section 17.10. Its reaction with II to give III looks like the simple addition across a double bond, but actually it is quite complex as shown in the following reactions.

The sulfuric acid catalyst maintains a constant supply of *t*-butyl ions (IV) by reacting with I. This *t*-butyl ion is a carbonium ion. It becomes twice as big by adding a molecule of isobutene, as shown by the reaction I+IV→V. The ion V decomposes a molecule of isobutane (II) by withdrawing from it an H: unit (the atom and its electron pair is the negative hydride ion, H^-); the product is isooctane (III), but a by-product is the t-butyl ion (IV) which is made available to react with more isobutene (I). It is a self-propagating reaction with an ionic mechanism, analogous to the free radical reactions described earlier in this chapter.

The complex carbonium ion V not only reacts with II as just described, but also suffers a rearrangement to VI as the result of the migration of an :H unit. Carbonium ion VI also rearranges; a :CH_3 unit complete with electron pair (the *methide* ion) migrates as shown to yield the carbonium ion VII. This carbonium ion extracts an :H unit from a molecule of isobutane (II); it yields again the *t*-butyl ion (IV) to continue the reaction with I, and the other reaction product is VIII, which is an isomer of isooctane (III). Other reactions also take place, but the principal products are the two octanes.

The ease with which isobutane (II) loses its :H unit in these reactions is due to the electron-release properties (Section 21.5) of the CH_3 group; when three of these groups are on the same carbon atom as in isobutane the C : H bond in that molecule is much more active. Also involved in these rearrangements are space (steric) considerations of a type that will be discussed later (for example, see Section 22.17 on the hydrolysis of *t*-butyl chloride). When a methide ion migrates from its position in VI to that in VII it indicates that CH_3 is more stable attached to a 2° carbon than to a 3° carbon atom. The meaning of 2° and 3° is defined in connection with Figure 25.1.

17.13. The Kinetics, Order, and Molecularity of a Reaction.
The complex of reactions described in the preceding section was inserted at this point in the chapter to illustrate the problems that may confront a research chemist in working out his reaction mechanisms. These problems are not solved quickly. Some details in the simple addition of bromine to the C=C double bond (Section 17.5) are not yet completely accepted, but those details are beyond the scope of this book.

Reaction mechanisms are generally worked out in the laboratory rather than with pencil and paper. Reactions usually take place in steps, some of which are much faster than others, and the mechanism can be altered by changing the temperature, the concentration of the reactants, the nature of the solvents, etc. All these factors must be studied, usually by varying one at a time while keeping the others fixed. The effect of each of these factors on the reaction rate furnishes the clues as to the reaction mechanism. Such a study is called determination of the *reaction kinetics*. In this elementary introduction we shall look only at the concentration of the reacting substances. For a representative reaction $AB+CD \rightarrow AD +BC$, the rate at which AB and CD react with each other is given by the equation

$$\text{rate} = k\,[AB]^n\,[CD]^{n'}$$

where the brackets symbolize concentration, which is customarily in moles /liter. The effect of [AB] on the reaction rate is found by fixing everything but the concentration of AB; if doubling the concentration of AB doubles the rate—i.e. if rate is proportional to concentration—then the value of n in $[AB]^n$ is 1 and the reaction is said to be first order with respect to AB. Similarly, the rate in the case of CD may vary with the square of the concentration; the reaction is then second order with respect to CD and the value of n' is 2. The overall *order of the reaction* is the sum of the n values; in this case it would be a third-order reaction. The constant k is the *specific rate constant*.

The bromination of ethene and of cyclohexene described in Section 17.5 takes place in two steps. The first step (formation of a bromonium ion) is slow, and has been found to be second order. The second step is fast. The slow step is said to be the rate-determining step for the reaction; these brominations, therefore, are second-order reactions—that is, they have second-order kinetics.

The *order* of a reaction, as just indicated, is an experimental value. It is not necessarily a whole number; for example, the conversion of para-hydrogen to ortho-hydrogen (Section 37.1) is a 1.5-order reaction. The *molecularity* of a reaction is a whole number that is directly apparent from the equation as written, although once more it is based on the rate-determining step. In the bromination of ethene, the reaction is *bimolecular* as well as second order, since in the slow step two molecules are stoichiometrically involved.

17.14. Substitution. The importance of the substitution reaction in organic chemistry was pointed out in Section 9.4, which should be reviewed. Two substitutions were described in that section: the first was replacement of an H atom in CH_4 by a Cl atom; the second was replacement of the Cl atom by the OH group. The reaction is referred to as replacement or displacement as well as substitution, and it will be discussed in greater detail in Section 22.17 when further background material will be available. In this section we shall use the reaction to introduce a few definitions.

Substitution in methyl chloride, when it is hydrolyzed by an alkali, is a nucleophilic attack on an electron-deficient C atom by the :OH$^-$ ion.

$$\overset{\delta^+}{CH_3} \overset{\delta^-}{:Cl} + Na^+ :OH^- \longrightarrow CH_3:OH + Na^+ :Cl^-$$

The English school of chemists who performed a large part of the recorded work on reaction kinetics and mechanism invented the symbolism S_N2 for this reaction. It means substitution, nucleophilic, bimolecular. The number 2 in this symbol, according to its inventors, is the *molecularity* (defined in the preceding section), but many textbooks regard it as the *order*. In this example the reaction is both second order and bimolecular. A substitution viewed as electrophilic is labeled S_E followed by a number that shows its molecularity.

17.15. Elimination. An elimination reaction was described in Section 13.2 when it was shown how a C=C bond can be produced in a molecule that contains H–X on neighboring carbon atoms. The H–X is eliminated by the action of an alkali in alcohol; in the preceding section we found that the same reagent in aqueous solution would result in a replacement reaction. This elimination reaction in which H–X is removed from ethyl chloride is labeled an E2 reaction, meaning elimination, bimolecular.

$$\begin{matrix} H & H & \\ | & | & \\ H-C-C-H & + \text{ NaOH/alcohol} & \longrightarrow CH_2{=}CH_2 + NaCl + H_2O \\ | & | & \text{(Sodium ethoxide,} \\ H & Cl & \text{Section 25.3)} \end{matrix}$$

17.16. Concerted Reaction. From the nature of the interactions between molecules described in this chapter it is obvious that most of them take place in stages. This is due to the intrinsic shapes of molecules as well as to the possible need for the molecules to readjust electrically, either through formation of ions or of radicals. Some reactions have been studied that apparently do not require an intermediate reshuffling of parts or of electrons, but take place by what is called a four-center, or concerted, mechanism. It can be pictured in this way:

$$
\begin{array}{ccc}
A & \cdots & C \\
| & & | \\
B & \cdots & D
\end{array}
$$

Molecules A–B and C–D in this case have four reactive centers which are properly situated and electronically constituted for a practically *simultaneous* interaction. These reactive centers act in concert as illustrated by the HI reaction in Section 18.4. The symmetrical Diels-Alder synthesis in Section 32.10 is believed to be such a four-center reaction.

17.17. Classification of Organic Reactions. Common descriptive words used to describe chemical changes in this chapter are ionic reactions, free radical reactions, nucleophilic reactions, etc. All these can be employed as topic headings when the gamut of organic reactions is summarized. However, the current breakdown of reactions into types recognizes four major classes: addition, elimination, substitution (displacement), and rearrangement.

18

Why Molecules React

18.1. Chemical Reactivity. We have pictured in a qualitative way *how* certain molecules react in accordance with the electron-valence theory. Now we shall develop, in an elementary way, the principles on which the *driving force* in a reaction can be estimated.

Chemical reactivity is not an easy term to define. For example, when discussing the series of paraffin hydrocarbons, CH_4, CH_3-CH_3, $CH_3-CH_2-CH_3$, etc., we made the statement in Section 10.6 that the smaller molecules react more rapidly than the big ones. This is not an exact statement, for by raising the temperature of a large hydrocarbon molecule we may cause it to react faster with chlorine than a smaller molecule does at ordinary temperatures.

When speaking of the chemical reactivity of a certain substance we therefore use relative terms—that is, we compare its behavior with that of some other substance under the same conditions. As in the preceding example, we may compare the rate at which the series of compounds reacts with a given substance (such as chlorine) under specified temperature conditions. Or, conversely, we may measure the temperature to which each member of a series of compounds must be raised in order to react with a certain substance. These are only a few of many experimental methods available for measuring relative reactivities.

In order to prepare a background for a quantitative measurement of reactivity (that is, *why* molecules react), we must digress once more into the field of physical chemistry to review the meaning of thermochemical equations.

18.2. Chemical Reactivity and Free Energy. Our earlier encounter with thermochemical equations was in the opening paragraphs of Chapter 16. At that time we learned how to write the equation for the heat of formation of the simple molecule, methane:

$$C(s) + 2H_2(g) \longrightarrow CH_4(g) \quad [\triangle H_{290} = -18.1 \text{ kcal}]$$

The formation of the molecule of CH_4 is associated with a loss of heat en-

171

ergy to the surroundings (18.1 kcal/mole); since this energy would have to be put back in the molecule to decompose it, it is in a sense a measure of its stability, or lack of reactivity. The heat of formation (or heat content) is also known as the *enthalpy*.

In Section 15.2 and in several other places it was emphasized that all bodies prefer a greater freedom of motion and that this search for greater freedom is really a tendency to attain greater stability and a lower energy level. This principle is the basic idea in the *entropy*, another important word in thermodynamics and symbolized by S. It is sometimes expressed in popular literature as the tendency of a system toward greater randomness.

Tendency toward greater stability can be considered to be the *driving force*, or the *why*. of a chemical reaction. From the preceding paragraphs it is clear there are two such driving forces—namely, $-\triangle H$ and $+\triangle S$; the first, of course, should have a negative value, and the second would be positive. These two energy factors are related to a third quantity, $\triangle G$, as follows:

$$-\triangle G = -\triangle H + T\triangle S$$

In this equation, $\triangle G$ is the change in *free energy* when a reaction is conducted under conditions in which the pressure and the temperature are constant. Free energy measured under these conditions was at one time labeled F, but it was thought advisable recently for several reasons to change it to G, meaning Gibbs free energy. The expression $T\triangle S$ shows the close relation between entropy and temperature.

If we make a slight change in the chemical reaction written in the introduction to this section:

$$C(s) + 2H_2(g) = CH_4(g) \quad [\triangle G_{298} = -12.2 \text{ kcal}]$$

we find we have a new symbol in a familiar background. The symbol $\triangle G_{298} = -12.2$ kcal indicates that the reaction between solid carbon and gaseous hydrogen to form a molecule of methane is accompanied by a decrease in free energy of 12.2 kcal at 25° C (298°K). This heat change is equivalent to the driving force in the reaction. It is said to be a measure of the chemical *affinity* of carbon and hydrogen under these conditions.

Substances will react with each other without outside help (that is, spontaneously) only if the reaction takes place with a decrease in free energy. Some external means, such as a catalyst, may be required to start the reaction, but it then proceeds spontaneously. Methane cannot decompose spontaneously, because the preceding reaction written in reverse is

$$CH_4(g) = C(s) + 2H_2(g) \quad [\triangle G_{298} = +12.2 \text{ kcal}]$$

The free energy change is positive and energy would have to be added to methane in order to break it up.

The *sign* of $\triangle G$ (whether $+$ or $-$) indicates in which direction reaction takes place spontaneously. The *size* of $\triangle G$ gives a rough idea of the extent to which reaction takes place. When $\triangle G = 0$ the system is in equilibrium.

The free energy change ($\triangle G$) of a reaction is important to physical organic chemists, because it is directly related to the equilibrium constant (K), which is the quantitative value of the chemical reactivity of a system. The relation between the free-energy change and the equilibrium constant of a reaction is given by

$$-\triangle G = RT \times 2.303 \log K \tag{18.1}$$

The following example shows how the free-energy change for the methane decomposition (just described) can be calculated from the equilibrium constant. When methane was heated at 400°K at constant atmospheric pressure, the equilibrium constant was found to be

$$K_p = \frac{p_{H_2} \times p_{H_2}}{p_{CH_4}} = 2.86 \times 10^{-6} \tag{18.2}$$

The data show that the decomposition $CH_4 \rightarrow C + 2H_2$ proceeds at this temperature only to a very small extent (about 0.17% by volume of H_2 is present in the gas). When the value of K_p from Equation 18.2 is substituted into Equation 18.1, it is found that $\triangle G_{400}$ is $+10.150$ kcal/mole.

For another useful illustration, the free energy of the reaction between ethene and hydrogen is shown by the equation

$$C_2H_4(g) + H_2(g) = C_2H_6(g) \quad [\triangle G_{298} = -24.1 \text{ kcal}]$$

and the extent of the reaction in the gas phase at constant pressure is given by the equilibrium constant

$$K_p = \frac{p_{C_2H_6}}{p_{H_2} \times p_{C_2H_4}} \tag{18.3}$$

By substituting the value of $\triangle G_{298}$ in Equation 18.1, the value of K_p at 25°C and 1 atmosphere pressure is found to be 5.65×10^{17}, so that the reaction is essentially complete with respect to the formation of ethane under these conditions.

The symbol $\triangle G°$ is often used to represent what is known as the standard change in free energy—in other words, the value calculated for 1 mole at 25°C at 1 atmosphere.

In these examples the equilibrium constants are those for gas reactions, and the gas concentrations are expressed as partial pressures. For reactions in solution, the concentrations are generally expressed in moles/liter, and brackets are used to signify concentration. In the reaction AB +CD→AC+BD the equilibrium constant is given by

$$K = \frac{[AC][BD]}{[AB][CD]} \tag{18.4}$$

18.3. Chemical Kinetics and Activation Energy. The discussion in the preceding section is based on the thermodynamics (Equation 18.1) and the equilibrium constant (Equation 18.2) of a typical chemical reaction. Neither of these factors is concerned with the mechanism by which the reaction takes place; they involve only the initial and final states of the system without regard to the speed or route by which the equilibrium state is reached. The study of reaction mechanisms and reaction speeds is called *chemical kinetics* (Section 17.13).

An obviously important factor in respect to reaction velocities is the frequency with which molecules collide. Under ordinary reaction conditions and concentrations, this is a factor amply provided for. As an illustration of the magnitude of collision frequency, when nitrogen is at a pressure of 1 atmosphere at 25°C, there are 2.46×10^{19} molecules per cubic centimeter and pairs of molecules are colliding at the average rate of 7.8×10^{28} times per second.

Most reactions in organic chemistry are slow. Since frequency of collision of the molecules as indicated here is tremendous, it must be concluded that very few molecular encounters lead to chemical reaction. Reactions in inorganic chemistry (if we restrict the discussion to the chemistry of ions) generally take place so fast that their speed cannot be measured readily. When Na^+Cl^- and $Ag^+NO_3^-$ are mixed in equal molecular quantities in aqueous solution, the reaction yields $Na^+NO_3^-$ and insoluble Ag^+Cl^-. In each gram-mole there are 6.02×10^{23} particles of each species, yet they find partners, and react, practically instantaneously.

The difference between the reaction speeds in organic chemistry (molecules) and inorganic chemistry (ions) can·be attributed to the difference in the degree of activation of the reacting particles. Section 17.2 states that molecules become more active when an electron displacement occurs. Since ions are the product of *complete* displacement of an electron pair, an ionic "molecule" can be regarded as in a state of complete activation when dispersed in a solvent; such a molecule is $Na^+CH_3^-$. The molecule CH_3Cl can be activated, as indicated in Section 9.5, but CH_4, of course, is difficult to activate.

This theory of activated molecules can be placed on a quantitative basis by considering the energy values involved. To break up the ionic crystal lattice of Na^+Cl^- shown in Figure 8.1 and to disperse the ions in aqueous solution requires the expenditure of energy per gram-mole comparable to the energies listed in Table 16.1. This activation energy is supplied by the phenomena accompanying solvation of the crystal and is retained as potential energy by the hydrated ions until called upon for further chemical reaction. To activate organic molecular reactions, the common means in an organic chemistry laboratory is the use of the Bunsen burner, which supplies energy in the form of heat. However, we can show from Equation 14.3 that the kinetic energy of 1 mole of a gas is raised only to the extent of 300 cal if its temperature is raised 100°C. A reaction of the type

$$AB + CD \longrightarrow AC + BD$$

is said to require an activation energy equivalent to 28% of the energy required to dissociate both AB and CD into their constituent atoms. A

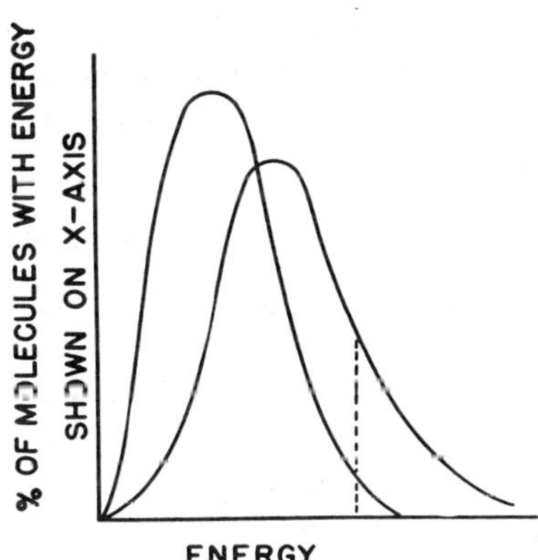

18.1. Typical Maxwellian distribution of energy among molecules of a gas at two temperatures (curve at right is for the higher temperature). To the right of the dotted line are the molecules having abnormally high energy (at least 10 kcal/mole) and capable of chemical reaction.

bimolecular reaction with an appreciable reaction speed at ordinary temperatures, requires an activation energy of about 20,000 cal/mole. The usefulness of a Bunsen burner in the laboratory must therefore have some other explanation than the increase in kinetic energy (calculated previously) which it supplies; this will be discussed further in the following.

The fact that a certain organic reaction is slow means that few molecular encounters lead to chemical reaction; the fact that the reaction takes place at all, however, means that at any one time some of the molecules have the energy required for the reaction to occur. This can be explained by the kinetic theory, according to which a series of collisions in the same direction will give any one molecule an abnormally high velocity (energy). The most probable distribution of velocities among the molecules of a substance is given by the Maxwell distribution law; this can be converted to an analogous expression for distribution of energies. In a fixed number of molecules (see Figure 18.1), at any instant the relative number of molecules with very low or very high energies is extremely small.

A slow reaction is due to lack of a sufficiently large number of molecules in the region to the right of the dotted line. The total area under each of the curves is constant (same total number of molecules), so that at higher temperatures it will be seen there is a considerable increase in the number of high-energy molecules. In fact the increase is so great that at ordinary temperatures a $10°$ rise in temperature doubles the speed of many reactions. This is the reason why using the Bunsen burner increases the rate of reaction (see comment above on the energy supplied by the burner).

According to modern theory, a reaction always takes place by way of the intermediate formation of an "activated complex," which can be illustrated as follows:

$$\text{AB} + \text{CD} \longrightarrow \left\{ \begin{matrix} \text{A B} \\ \text{CD} \end{matrix} \right\} \longrightarrow \text{AC} + \text{BD} \qquad (18.5)$$

$$\text{Reactants} \qquad \text{Activated} \qquad \text{Products}$$
$$\text{complex}$$

The energy change required to form the complex and set the stage for breaking bonds and forming new ones is called the activation energy, $\triangle H_a$. The intermediate complex has a very short life of about 10^{-13} sec. The relation between the heat of activation $\triangle H_a$ and the heat of formation $\triangle H$, which was discussed in Section 18.2, is shown diagrammatically in Figure 18.2 for an exothermic reaction—that is, one in which heat is evolved or $\triangle H$ is negative.

The heat of activation $\triangle H'_a$, shown in Figure 18.2, applies to the reverse reaction; that is, to transform the products back to the reactants

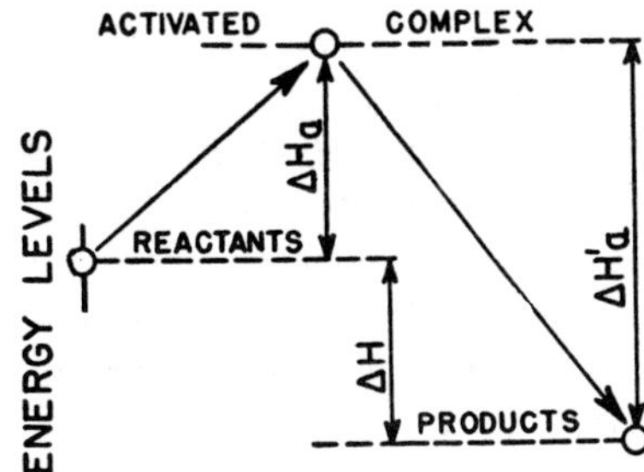

18.2. Energy levels in a chemical reaction. (Adapted by permission from F. Daniels, *Outlines of Physical Chemistry*, John Wiley and Sons, 1948.)

requires absorption of energy equivalent to the height of the line $\Delta H'_a$ before reaction takes place. ΔH is the difference between $\Delta H'_a$ and ΔH_a.

Mention may be made of the methods available for calculating ΔH and ΔH_a. They will be considered together because of the striking resemblance between the formulas used. The heat of formation (or reaction) can be found by determining the equilibrium constants K_1 and K_2 for the reaction (compare Equation 18.2) at two different temperatures T_1 and T_2 and applying Equation 18.6. The heat of activation is calculated similarly from the velocity constants k_1 and k_2 at two temperatures and applying equation 18.7.

$$2.303 \log \frac{K_2}{K_1} = \frac{\Delta H}{R}\left(\frac{1}{T_1} - \frac{1}{T_2}\right) \qquad \textit{Heat of formation} \qquad (18.6)$$

$$2.303 \log \frac{k_2}{k_1} = \frac{\Delta H_a}{R}\left(\frac{1}{T_1} - \frac{1}{T_2}\right) \qquad \textit{Heat of activation} \qquad (18.7)$$

Calculation of the equilibrium constant K has already been described. The velocity constant k refers to the rate of change in the concentration of reactants or products, and is defined in Section 17.13.

18.4. The Arrhenius Equation. Practically all students of elementary chemistry have heard of Svante Arrhenius because of his theory announced in 1887 that acids, bases, and salts ionize in aqueous solution. The same man was also responsible for an equation that has had a major influence in research on chemical kinetics. A modern version is shown in Equation 18.8, and the logarithmic form, better understood by most chemists, in Equation 18.9. Note the resemblance to Equations 18.6 and 18.7.

$$k = pze^{-\frac{\Delta H_a}{RT}} \qquad (18.8)$$

$$\log k = \log pz - \frac{\triangle Ha}{2.3R}\,\frac{1}{T} \tag{18.9}$$

If the rate constant k is determined at several temperatures and is plotted against $1/T$, Equation 18.9 yields a straight line; the slope of this line can be measured, and since slope is equal to $\triangle H_a/2.3R = \triangle H_a/4.6$ we can calculate the heat of activation of the reaction—that is, the $\triangle H_a$ shown in Figure 18.2. In these equations z is a frequency factor (collision frequency) that can be calculated from kinetic theory, and p is the steric or shape factor, the value of which depends on the extent to which the molecules must align themselves in order to react.

The size of $\triangle H_a$ is the principal factor in the speed of a reaction; if it is large, relatively few molecules possess the energy needed for reaction and the rate is slow. If it approaches zero, reaction should be very rapid. Slow reactions can be speeded up by a catalyst which provides intermediate paths with smaller $\triangle H_a$ values. The e term in Equation 18.8 is really a measure of the concentration of activated molecules.

The reader may find $\triangle E$ used in discussions of this nature. The $\triangle H$ values are for constant pressure conditions (most common in the laboratory), and $\triangle H = \triangle E + \triangle (pv)$ if there is a volume change.

The decomposition $2HI \rightarrow H_2 + I_2$ is often cited as a simple reaction in which two molecules take part. The kinetics indicate that it is probably a concerted reaction (Section 17.16). The energy required to bring about the transition state is about 40 kcal/mole, which is much smaller than the 71 kcal/mole listed in Table 16.1, liberated when the H–I bond is broken (see Figure 18.2).

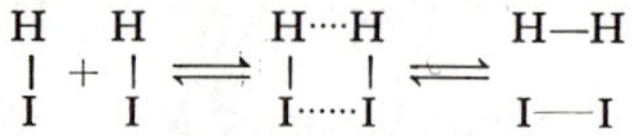

$$
\begin{array}{ccccc}
\text{H} \quad \text{H} & & \text{H}\cdots\text{H} & & \text{H—H} \\
| \ + \ | & \rightleftharpoons & | \quad | & \rightleftharpoons & \\
\text{I} \quad \text{I} & & \text{I}\cdots\cdots\text{I} & & \text{I—I}
\end{array}
$$

Reactants   Activated   Products
complex

18.5. Photochemistry. To introduce the subject of chain reactions in Section 17.9 it was said that a mixture of chlorine and methane explodes when exposed to sunlight because of the very rapid chain reaction initiated by ultraviolet radiation. This is an extreme example of the efficiency of radiation when used to supply energy of activation to a chemical reaction. One of the advantages of this method of activation is that radiation (such as visible radiation or ultraviolet radiation) permeates the reaction mixture and reaches the reacting molecules instantly, whereas raising the temperature of a reaction mixture to speed up the reaction is essentially a slow process.

In order to activate a reaction by radiation, one must select a wavelength which is absorbed by one of the molecular species. If a molecule can absorb the radiation completely, the absorbed energy is converted into heat. The excess energy in the irradiated molecule may be given up by collision with other molecules, thus increasing their kinetic energy, it may be released by collision with the walls of the retaining vessel, or it may be used in a chemical reaction.   The efficacy of the radiation is calculated as follows in terms of the *quantum yield Φ*:

$$\Phi = \frac{\text{number of molecules reacting chemically}}{\text{number of quanta absorbed}}$$

The energy quantum was discussed in Section 14.13, which should be reviewed.   The quantum absorbed by an individual molecule is $\varepsilon = h\nu$. One mole of quanta is equal to $Nh\nu$ and is called an *einstein.*   The efficiency of a source of light used in a photochemical reaction is measured with an *actinometer,* such as that which employs uranyl oxalate.   It has been determined that light in the wavelength range 2540 to 4350 A will decompose 0.57 mole of oxalic acid for each einstein absorbed, when the solution contains uranyl sulphate and oxalic acid at certain concentrations.   The uranyl ion does not react chemically, but simply absorbs energy and transfers it to the oxalic acid.   After a light source has been calibrated with such an actinometer it can be used to determine the Φ value of other photochemical reactions.

Some reactions have a quantum yield close to 1.   When HI is decomposed photochemically (with radiation at about 3,000 A) the quantum yield is about 2.   The primary reaction is $HI \rightarrow H \cdot + I \cdot$, but secondary reactions which can take place, perhaps catalyzed by the walls of the container, are

$$H \cdot + HI \longrightarrow H_2 + I \cdot \qquad \text{and} \qquad I \cdot + I \cdot \longrightarrow I_2$$

When the reaction between hydrogen and chlorine is initiated by light (at about 4,000 A) the quantum yield is obscured by a secondary chain reaction which might lead one to believe the quantum yield is about 100,000.   The chain reaction takes place in three stages, as outlined in Section 17.9.

1. Initiation $Cl:Cl \longrightarrow 2Cl \cdot$

2. Propagation   (a) $Cl \cdot + H:H \longrightarrow H \cdot + H:Cl$

                          (b) $H \cdot + Cl:Cl \longrightarrow Cl \cdot + H:Cl$

3. Termination   Union of free radicals to form molecules

19

The Benzene Ring

19.1. Introduction. At the time when Michael Faraday, the famous English scientist, was preoccupied with the problem of liquefying gases, a new illuminating gas industry had arisen in London. The Portable Gas Company was distilling inflammable gases from whale and cod oil and compressing the gases in portable vessels which could then be employed in private dwellings for lighting purposes.

In 1825, Faraday investigated the liquid in some of these vessels and separated from it a new compound containing hydrogen and carbon. Its composition (in modern terms) he found to be C_6H_6, and it had a melting point of 5°C and boiling point of 80°C. The presence of the compound in coal tar was discovered in 1845. The chemical industry holds this substance in such high esteem that 100 years later a celebration was held, and many essays extolled the virtues of both the compound and its discoverer.

In 1834, a chemist obtained the same compound from a plant substance called *gum benzoin*, and accordingly C_6H_6 was named *benzin*. In 1837, still another chemist proposed the name *phene*, from the Greek word *phainein*, meaning bring to light, because the substance was first found in gas used for lighting purposes. It is now known as *benzol* in most European countries, and as either *benzene* or *benzol* in America. However, many compounds derived from benzene are named as derivatives of *phene*; examples are phenol, phenacetin, thiophene, etc.

It was soon observed that benzene shows a distinct fatherly relationship to an enormous number of compounds that were being isolated in nature or being prepared in the laboratory. However, chemists realized that they were trying to advance in the dark without a guiding star; outside of its composition, C_6H_6, the nature of benzene was still a complete mystery. A satisfactory structural formula for benzene was invented in 1865, and that date marks the beginning of modern progress in organic chemistry.

It required much speculation from 1803 to 1858 before chemists were

180

able to visualize the valence force of an atom in the form of a straight-line bond. That great step forward was made by a Scot named Couper. However, it was the German chemist, August Kekulé, who most clearly saw the carbon atom with its valence of four as we see it today and as we have described it in this book in our structural formulas.

Kekulé showed how carbon atoms can unite with themselves to form chains. This phenomenon was described in Chapter 10 on carbon chains, and we have already seen how fruitful the idea has been in the preparation of large families of compounds. The carbon chain was invented in 1858, but Kekulé could not claim all the credit for the invention because other men had furnished him with many clues. Eight years later, Kekulé not only invented the carbon ring but also the double bond, and for that he is one of the immortals in chemistry.

This crowning achievement of Kekulé came about as a result of his ability as a dreamer as well as laboratory worker. Kekulé put on his night-cap, dreamed about six carbon atoms arranged in a chain like a group of dancing imps,

$$-\text{C}-\text{C}-\text{C}-\text{C}-\text{C}-\text{C}-$$
$$\ \ \ |\ \ \ \ |\ \ \ \ |\ \ \ \ |\ \ \ \ |\ \ \ \ |$$
$$\ \ \ \text{H}\ \ \text{H}\ \ \text{H}\ \ \text{H}\ \ \text{H}\ \ \text{H}$$

and then saw the imps at the ends of the chain join hands in a ring dance, as in formula A:

(A)

(B)

When he awoke, Kekulé realized that this ring might truly be the structure of benzene, except for the valence of carbon, which must necessarily be four to agree with chemical experience. It was a simple step to the assumption that the bonds are alternately double bonds and single bonds as in formula B, which satisfies all the valence requirements for a compound of the composition C_6H_6.

19.2. The Kekulé Benzene Formula. When the Kekulé formula was introduced it raised as many questions as it solved. It is obviously the structure of an olefin, and its name is 1,3,5-hexatriene. It should have the typical, unsaturated properties of the double bond; for example, it should be detectible by two color reactions (Section 13.15) but it is not oxidized by potassium permanganate solution, and it does not decolorize a bromine test solution.

However, the fact that there *are* double bonds in benzene is indicated by a few addition reactions. Under certain catalytic conditions it will add three molecules of hydrogen,

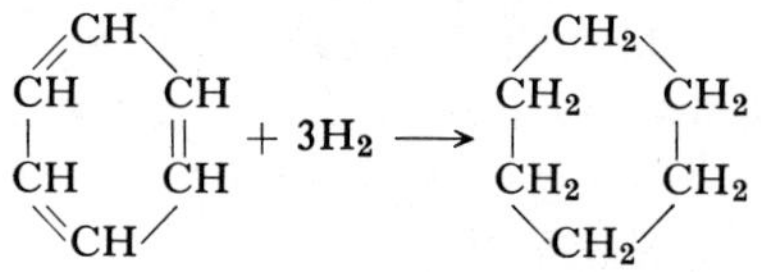

$$
\begin{array}{ccc}
\text{Benzene} & + 3H_2 \longrightarrow & \text{Hexahydrobenzene} \\
& & \text{Cyclohexane}
\end{array}
$$

which suggests the presence of three double bonds. Similarly, by a free-radical chain reaction it will add three molecules of chlorine to yield $C_6H_6Cl_6$, which is chemically benzene hexachloride or hexachlorocyclohexane. Nine isomers of $C_6H_6Cl_6$ are possible, depending on the Cl positions on either side of the ring (compare Figure 17.4*b*); one is the insecticide known as 666.

We see then that the $C=C$ bond in benzene is unusual, because it goes through addition reactions only with great reluctance and is not susceptible to oxidation. In order to explain this state of things, many chemists offered formulas they thought were more satisfactory than that of Kekulé. Most chemists turned to the Thiele variation of the Kekulé formula:

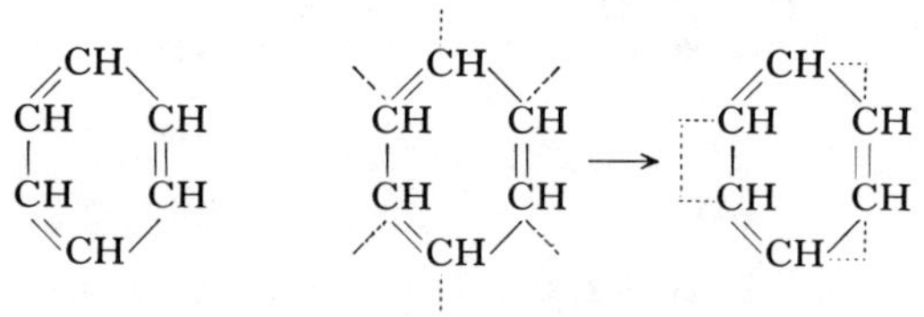

$$
\text{Kekulé formula} \qquad\qquad \text{Thiele formula}
$$

According to the Thiele partial valence theory (Section 17.7), the activity of a double bond is due to residual or partial valences as suggested by the dotted lines. It seemed probable that in benzene all the partial valence bonds unite to form a satisfied ring, with practically no valence force left over for addition reactions. This Thiele formula gave a simple explanation of the reluctance of benzene to act like other compounds containing double bonds. Also, it offered one of the simplest explanations of the following problem, which is the chief drawback of the Kekulé formula.

When two adjacent H atoms of benzene are replaced by chlorine, the Kekulé formula would lead us to expect two different compounds to be formed, in which the Cl atoms are across a single bond in one case and across a double bond in the other:

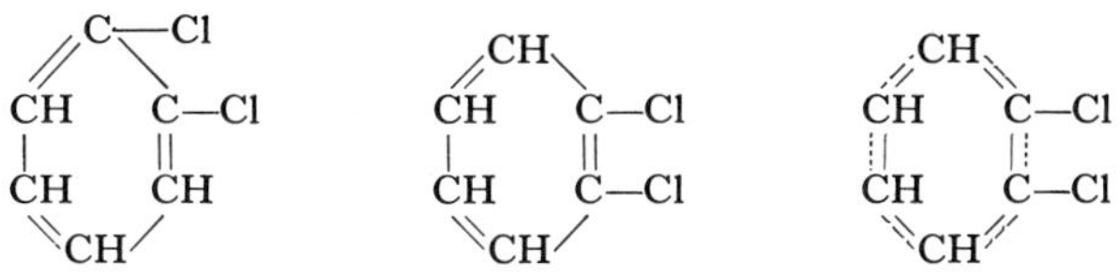

Kekulé   Thiele

Actually, only one such dichlorobenzene has ever been isolated, which agrees with the symmetrical nature of benzene as expressed in the Thiele modification of the Kekulé formula. Supporters of the Kekulé formula had to postulate that the two structures shown here are in dynamic equilibrium.

Although benzene has to be coaxed to go through *addition* reactions, this is true only of the first C=C bond. After the first double bond has been added to, the others behave just as actively as they do in the olefin series of hydrocarbons. In the *replacement* of its H atoms by halogen atoms, the benzene ring behaves like the paraffin hydrocarbons. In Section 20.2 this substitution reaction will be described in detail and it will be shown that the procedure used for introducing a halogen atom into the ring is quite different from that employed for chain compounds. Under the proper conditions, all six hydrogen atoms can be replaced by halogen.

19.3. Resonance of Benzene. The carbon-carbon bond distance in benzene is the same between any two carbon atoms, and the observed length is 1.39 A, which is exactly half-way between the single-bond and double-bond character in Figure 16.4. Benzene therefore satisfies one of the two principal requirements for resonance mentioned in Section 16.9—namely, a shortened bond length. The second criterion is greater stability than is apparent from the structure as normally written.

An accurate determination of the resonance energy of benzene can be obtained from heat-of-hydrogenation measurements (compare Section 16.4) on benzene and on cyclohexene:

$$3H_2 \longrightarrow \qquad \qquad \overset{H_2}{\longleftarrow}$$

Benzene   Cyclohexane   Cyclohexene
Heat of hydrogenation   *Heat of hydrogenation*
= 49.8 kcal   *= 28.6 kcal*

One would expect the heat of hydrogenation of benzene to be 3×28.6 or 85.8 kcal/mole, instead of the observed value of 49.8 kcal. The difference $85.8 - 49.8 = 36$ kcal is the measure of the increased stability of the benzene structure due to its resonance possibilities.

Half a dozen valence-bond structures can be written for benzene, but only two are important—the two Kekulé structures:

These have identical energy contents and therefore enter into *complete* resonance, with the corresponding increase in stability. Since the actual state of the molecule of benzene is intermediate between these structures, it explains why only one dichlorobenzene can exist of the type shown in the preceding section. This will be even more apparent when the structure of benzene is described according to the molecular orbital theory.

19.4.   Molecular-Orbital Representation of Benzene.   The flat benzene ring is pictured in figure 19.1*a*, where the valence angles are 120°, just as in the graphite rings shown in Figure 19.3. The reason for the flatness was discussed in connection with Figure 15.6 (which should be reviewed), where it was shown that σ-type molecular orbitals are formed in a plane surface when the carbon atom is linked with three other atoms.

The residual p electron on each carbon atom is shown in Figure 19.1*b* in atomic orbitals perpendicular to the ring. The coalescence of adjacent p atomic orbitals to form molecular π orbitals follows the reasoning applied to butadiene in Section 16.4 and Figure 16.1. Instead of the four π electrons on four carbon atoms, we now have six π electrons on six carbon atoms and these mobile electrons can travel a continuous circular path. The closed orbital should be visualized as two concentric doughnuts.

The wave motions of the π electrons around the ring have been calculated by the methods of wave mechanics. Under the influence of a magnetic field perpendicular to the plane of the benzene ring, the π electrons flow preferentially in one direction. This results in diamagnetism (Sections 24.10 and 32.3).

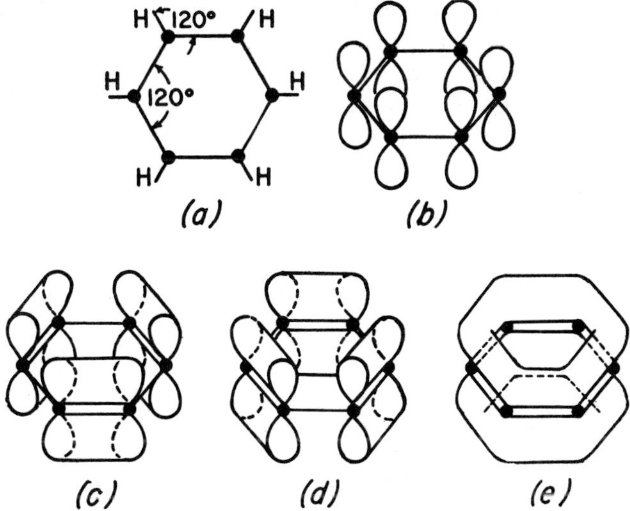

19.1. Structure of benzene on the basis of molecular-orbital theory.

(*a*)   Three of the four valence bonds on each carbon atom are σ type and are all in the same plane. The valence angles are 120°. The molecular-orbital nature of the C—C and C—H bonds was described in Fig. 15.6.

(*b*)   One p electron is left on each carbon atom. The H atoms are not shown.

(*c*) and (*d*)   Overlapping of adjacent p atomic orbitals to form three π molecular orbitals. The two arrangements correspond to the two Kekulé structures in Section 19.3.

(*e*)   The resultant π molecular orbital in benzene offers a continuous path for the six π electrons. It should be remembered that an electron occupies both portions of an orbital.

(See comments under Fig. 15.8.)

It is obvious that all the carbon-carbon bonds in benzene are alike and that its high degree of stabilization is due to the great freedom enjoyed by its six π electrons. However, for textbook purposes it is still the practice to write the Kekulé structure for benzene, using only one of the two resonance forms shown in Section 19.3. Occasionally, the formula will be found with a dotted circle inscribed in the hexagon (which is usually referred to as the benzene ring); the dotted circle represents the presence of *delocalized* electrons. It is seldom found necessary to show the H atoms in the benzene formula; in the normal molecule it is understood that there is one at each corner. These conventions are illustrated by Formulas IV, V, and VI in Section 20.2.

19.5. Hydrocarbons Containing the Benzene Ring. A large number of hydrocarbons derived from benzene are known. Many of these are found in coal tar and in resins, balsams, or other plant substances. Because many of the benzene derivatives first prepared were found to have a pleasant odor, they came to be known as the "aromatic" compounds. The term is still used to differentiate these compounds from the compounds described earlier in this book.

The nature of the benzene hydrocarbons is shown by the formulas of some representative members of the series given in Table 19.1. The student should examine these formulas rapidly at first, just to get acquainted with the new type of architecture being described. Then the names should be studied from the following points of view: Many compounds have a *common* name, often acquired from the nature of their source; for example, toluene is so-called because it was discovered among the distillation products of balsam of Tolu. It can also be named *systematically* as derived from benzene (i. e. methylbenzene), or as derived from a chain hydrocarbon (i.e. phenylmethane). The radical C_6H_5-, which is a benzene ring — with one H atom missing, is called *phenyl*, from the obsolete name for benzene, which is phene. This radical has a valence of one, just like the methyl radical, CH_3-, and the ethyl radical, CH_3-CH_2-.

19.6. Nucleus and Side Chain. Benzene is such a stable unit that chemists like to consider it as the principal part of a compound when it is found combined with other radicals. For example, ethylbenzene $-CH_2-CH_3$ is looked on more as a derivative of benzene than of ethane, so it is more often called ethylbenzene than phenylethane.

In a compound like this one, the benzene ring is called the *nucleus*, while the chain of carbon atoms attached to the nucleus is called the *side chain*. The side chain can have any of the various forms we have already studied, and to illustrate this fact, we have included examples of these various types in Table 19.1. The side chain can be of the *open* type, in which the carbon atoms are joined by single bonds ($-CH_2-CH_3$), by double bonds ($-CH=CH_2$), or by triple bonds ($-C\equiv CH$). This side chain can also be of the *closed* type, as shown by the last compound in the table, phenylcyclohexane.

In every case, the benzene hydrocarbons have two sets of properties —the properties of the nucleus as given in this chapter and the properties of the side chain as given in preceding chapters. Each set of properties,

TABLE 19.1.   Hydrocarbons Containing the Benzene Ring

With the exception of the last three compounds, these hydrocarbons belong to the *benzene series* of hydrocarbons. The general formula of the family is C_nH_{2n-6}.

 Benzene (Phene) C_6H_6 b.p. 80°C	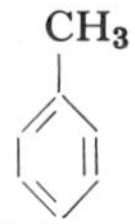 Toluene Methlbenzene Phenylmethane C_7H_8 C_6H_5—CH_3 b.p. 110°C	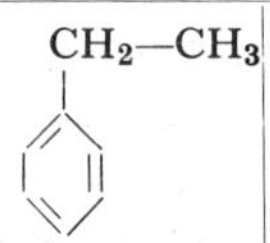 Ethylbenzene Phenylethane C_8H_{10} C_6H_5—C_2H_5 b.p. 134°C	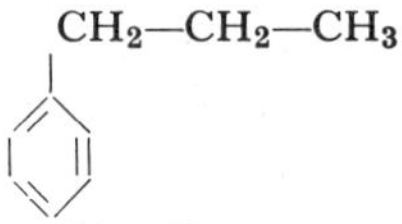 Propylbenzene Phenylpropane C_9H_{12} C_6H_5—C_3H_7 b.p. 157°C
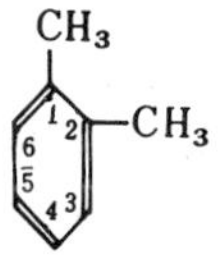 *ortho* Xylene 1,2-Dimethyl-benzene C_8H_{10} o-$C_6H_4(CH_3)_2$ b.p. 144°C	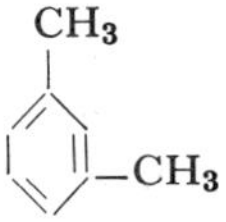 *meta*-Xylene 1,3-Dimethyl-benzene C_8H_{10} m-$C_6H_4(CH_3)_2$ b.p. 139°C	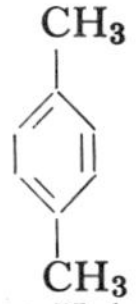 *para*-Xylene 1,4-Dimethyl-benzene C_8H_{10} p-$C_6H_4(CH_3)_2$ b.p. 138°C	**Note.** The three xylenes are isomers of ethylbenzene.
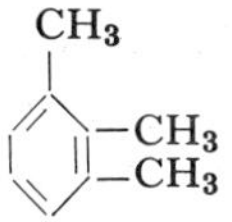 Hemimellitene 1,2,3-Trimethyl-benzene *vicinal*-Trimethyl-benzene C_9H_{12} vic-$C_6H_3(CH_3)_3$ b.p. 175°C	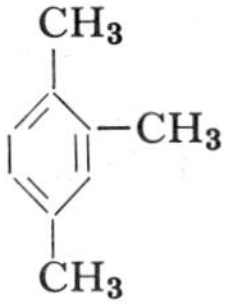 pseudo-Cumene 1,2,4-Trimethyl-benzene *unsymmetrical*-Trimethylbenzene C_9H_{12} *unsym*-$C_6H_3(CH_3)_3$ b.p. 169°C	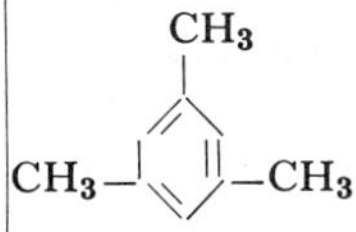 Mesitylene 1,3,5-Trimethyl-benzene *symmetrical*-Trimethylbenzene C_9H_{12} *sym*-$C_6H_3(CH_3)_3$ b.p. 165°C	**Note.** The three trimethylbenzenes are isomers of propylbenzene.
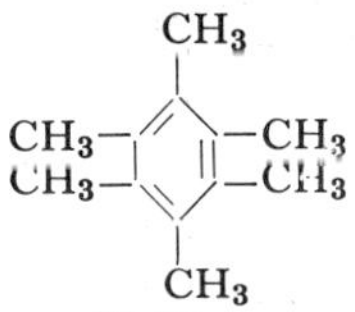 Mellitene Hexamethyl benzene $C_{12}H_{18}$ $C_6(CH_3)_6$ m.p. 164°C	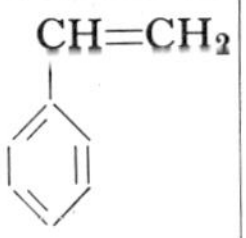 Styrene Phenylethene C_8H_8 C_6H_5—CH=CH_2 b.p. 144°C	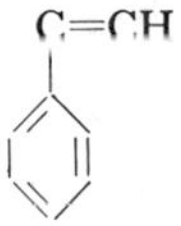 Phenylacetylene Phenylethyne C_8H_6 C_6H_5—C≡CH b.p. 142°C	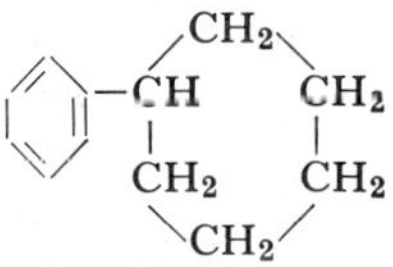 Phenylcyclohexane $C_{12}H_{16}$ b.p. 237°C

of course, will be affected by the presence of the opposing type of structure in the molecule. The nucleus is generally more stable (less reactive) than the side chain, as we shall learn more definitely in later chapters.

19.7. Graphite and Diamond. We have just stated that the nucleus and the side chain differ considerably in chemical properties. We also find that there is a marked difference when we examine their probable physical structures. This difference is best shown by contrasting the structures of graphite and diamond, which are apparently related to each other in much the same way as the benzene compounds are related to the chain compounds.

Carbon occurs in nature in two forms. One is the cheap variety known as graphite and used as a lubricant. The other is the expensive variety known as diamond and worn as jewelry. These two kinds of carbon have been thoroughly studied by X-ray methods and their structures are shown in Figures 19.2 and 19.3.

A diamond crystal is a giant molecule. It consists of an interlacing network of the tetrahedral carbon atoms described in detail in Chapter 12. Each carbon atom is linked to four others, and the angle between valence bonds is the theoretical 110°, which generally means great stability. The distance between any two atoms is 1.54 A units. The chain (and closed-chain) compounds are built on the diamond principle. In the model for diamond, the reader will recognize the zigzag chains mentioned in Section 12.3 and the crumpled rings described in Section 12.5.

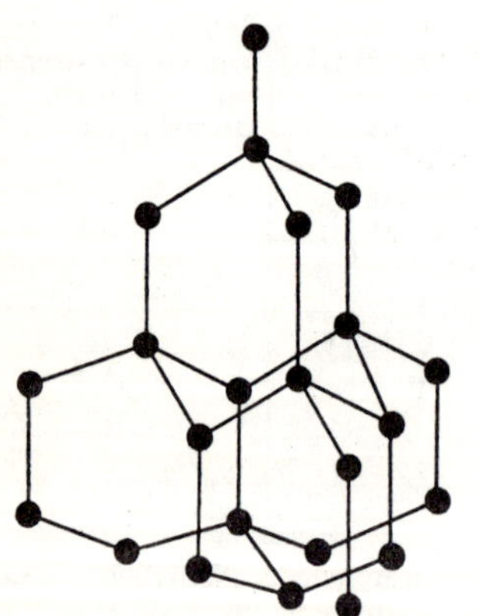

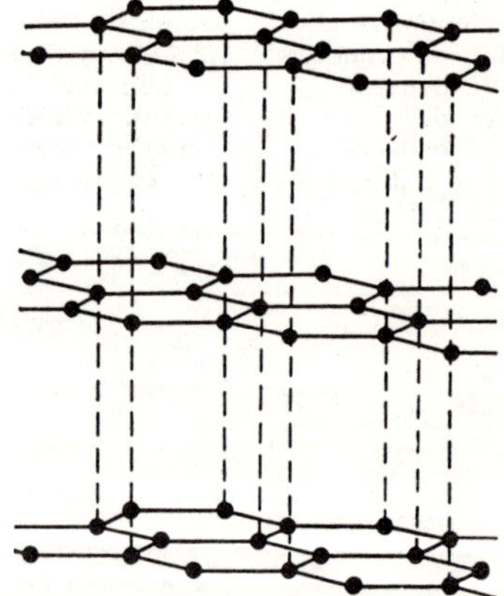

19.2. Arrangement of carbon atoms in diamond.   **19.3.** Arrangement of carbon atoms in graphite.

Benzene derivatives are apparently built on the graphite principle. The benzene ring is flat, and atoms attached to the ring are in the same plane as the ring. In the graphite molecule, there is a network of plane rings to form a continuous sheet. Each sheet is a giant molecule, whereas in diamond the whole crystal is a giant molecule. The angle between valence bonds in graphite is 120°, and each carbon atom is linked to three others. The sheet molecules are 3.4 A units apart, and they are therefore too far apart to be bound by covalent bonds, but are held together by weak molecular forces. The experts who made these measurements ascribed the lubrication properties of graphite to the "slippage" of these plane, sheet molecules. However, it has been found that graphite is not a lubricant in a vacuum. The lubrication is due to adsorbed gases packed between the layers.

In Figure 19.3 only three valence bonds are shown on each carbon atom. The position of the fourth is evident from Figure 19.1; the p atomic orbitals coalesce to a molecular orbital in which the π electrons have freedom of motion, so much freedom in fact that graphite is an electrical conductor. If a positive pole is connected to one end of a piece of graphite and a negative pole to the other, there is a flow of electrons (current) to the positive pole. On the basis of resonance theory, several reasonable structures can be written for graphite, of which one is the following:

$$>\!C\!=\!C<_{\displaystyle >C=C<}^{\displaystyle >C=C<}>\!C\!=\!C<$$

Each ring consists of six linkages, of which two are $C\!=\!C$ bonds. The ring, therefore, has $33\frac{1}{3}\%$ double-bond character, instead of the 50% double-bond character found for benzene with its three double bonds in six linkages. The distance between carbon atoms in the graphite ring is 1.42 A. The relation between extent of double-bond character and C–C distance is shown graphically in Figure 16.4.

19.8. Benzyne. The smallest stable ring compound with a triple bond is cyclooctyne, as stated in Section 13.3. However, an unstable, 6-carbon ring structure is known which has a triple bond and a very brief life in a number of synthetic reactions. This is benzyne, or dehydrobenzene.

Benzyne

Biphenylene
(Diphenylene)

It has been identified spectroscopically. Several reactions have been developed in which benzyne can be made in the presence of reagents with which it reacts instantly. If such reactions do not take place, benzyne adds to itself to yield a dimer called biphenylene, which is structurally analogous to biphenyl (Section 24.4). The *phenylene* group is benzene minus two hydrogen atoms, analogous to *phenyl* which is benzene minus one H atom (Section 19.5).

19.9. Chemists Have Fun. After Kekulé published his tri-ene formula for benzene in 1864 several other ring formulas were also given consideration. Two of these are the Ladenburg structure I, without double bonds and with the shape of a prism, and the Dewar structure II with two double bonds and a bond between the *para* positions. On the one-hundredth anniversary of the Kekulé publication, several groups of chemists announced they had made derivatives of these *valence isomers* of benzene. Two of these are illustrated in III and IV. Compound III

I   II   III   IV

Ladenburg prism   Dewar formula   A Ladenburg   Hexamethyl Dewar
formula for   for benzene   "prismane"   benzene
benzene   derivative

is a trimer of t–C≡C–F, where t is the *tertiary*-butyl group and F is a fluorine atom. Compound IV is the trimer of 2-butyne, which is CH_3–C≡C–CH_3.

Many cyclic compounds are known with somewhat bizarre shapes and have been given appropriate names. An example is *cubane*, which has the shape of a cube. Also, a *catena* compound has been made in which a ring consisting of 26 carbon atoms is intertwined with a ring of 28 carbon atoms, like two links in a chain (*catena* is Latin for chain). In 1963, at a Congress of the IUPAC, it was suggested that organic chemists might try to make a new compound the structure of which was published and was to be called *congressane*; two years later its synthesis was accomplished. Congressane is related to *adamantane*, which has been isolated from petroleum and has been synthesized. Adamantane is a saturated compound

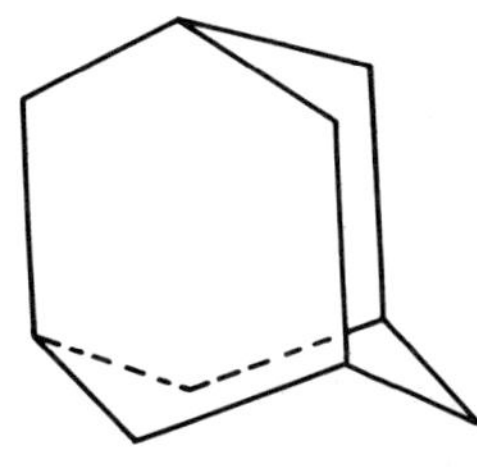

Adamantane, $C_{10}H_{16}$

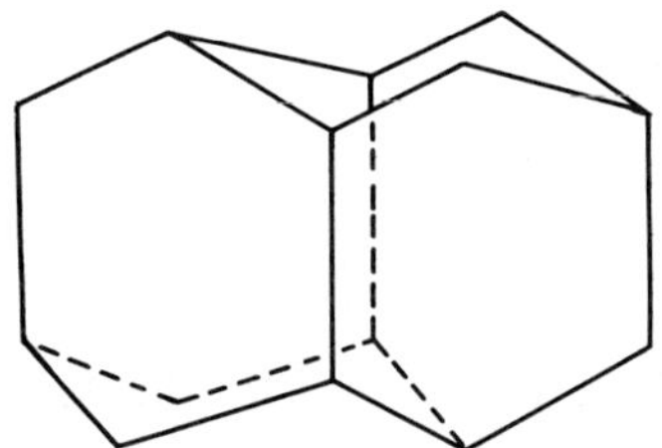

Congressane, $C_{14}H_{20}$

with a C atom at each corner and enough H atoms to fill up four valence bonds at each C atom. The structure is analogous to that of diamond (Figure 19.2) and the compound is very stable; the diamond structure is free from strain.

20

Nuclear Reactions

20.1. Introduction. The nucleus of an atom is described briefly in Chapter 4 of this book; the study of the atomic nucleus is called *nucleonics*. The inventor of the benzene ring called that a nucleus also, to distinguish it from the side chains that may be found attached to the ring. The benzene nucleus has unique properties. In the study of the double bond in such molecules as ethene, $CH_2=CH_2$, we found that its chemistry could be made more apparent by writing the formula $CH_2 \cdots CH_2$ to emphasize the presence of a pair of π electrons which are more exposed to attack than the σ electrons of the C–C bond. There is a similar situation in the benzene molecule (see Formula V in Section 20.2) where six π electrons are easily accessible from either side of the ring. The nucleus is therefore readily attacked by electrophilic reagents.

Through their π electrons, compounds like benzene and toluene form so-called π *complexes* with other molecules; these are generally unstable but some have been identified and isolated. Benzene couples with the halogens in a 1:1 ratio; it is probably the result of a π electron pair from the ring filling up an empty orbital in the halogen molecule. The π complex is often referred to as a charge-transfer complex because of the resemblance to the dative-bond formation described in section 8.5, benzene being the donor, i.e. the molecule that is transferring charge in the form of a pair of electrons to an acceptor. This behavior has much to do with the chemistry that will now be described.

20.2. Halogenation. An H atom in the benzene ring is easily replaced by chlorine or bromine, but the assistance of a catalyst is necessary. A commonly used catalyst is anhydrous aluminum chloride, in the presence of which a complex (I) is formed consisting of a molecule of benzene, a molecule of chlorine, and a molecule of catalyst. The probable function of the $AlCl_3$ catalyst is to polarize the Cl:Cl molecule, to promote an attack by its positive pole on the negative π electrons of the benzene ring (Figure 19.1). The $AlCl_3$ molecule is like the BCl_3 molecule in Section 8.5 in that it has room for an electron pair; this makes possible its catalytic

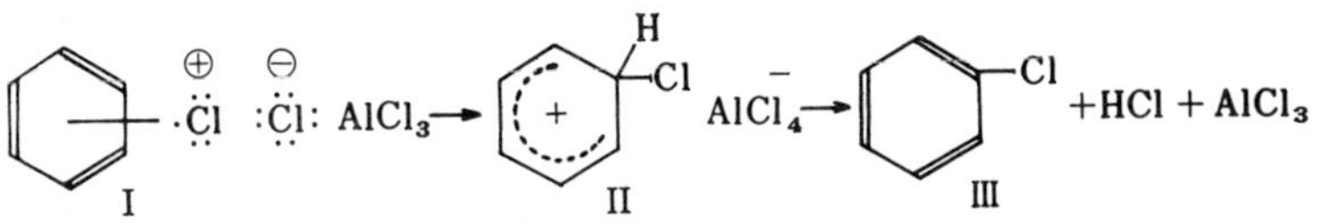

I   II   III

| The π complex plus catalyst | A benzenonium ion | Chloroalumi- nate ion | Chlorobenzene |

action toward the π complex as shown in the equation (through the Cl:Cl molecule). In this charge-transfer process, benzene is donor of an electron pair and Cl^+ is the acceptor. The result is an unstable benzenonium ion (II), which breaks down by elimination of a proton (H^+). This proton is the H atom displaced from the ring by Cl; the proton now reacts with the $AlCl_4^-$ ion to give HCl and regenerate the catalyst.

This attack by a polarized Cl:Cl molecule on the benzene ring is reminiscent of the addition of bromine to ethene as described in Figure 17.4, where the polarized bromine molecule ($Br^{\oplus}:Br^{\ominus}$) forms an intermediate bromonium ion.

This is a convenient place at which to elaborate further on representations of the benzene ring (Section 19.4).

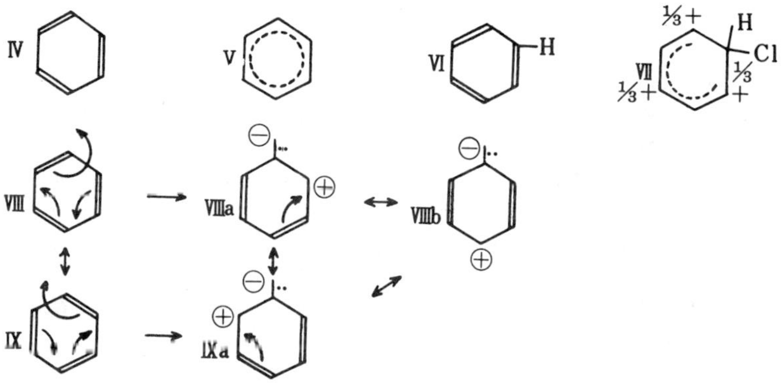

IV   V   VI   VII

VIII $\longrightarrow$ VIIIa $\longleftrightarrow$ VIIIb

IX $\longrightarrow$ IXa

In IV, the reader must understand that each angle has a C–H bond (the total is C_6H_6, the formula for benzene). The dotted inner circle in V is equivalent to the use of electron dots in the formula $CH_2\cdots CH_2$; it is a pictorial way to show the presence of 6 *delocalized* electrons distributed over the 6 carbon atoms. To call attention to a particular H atom in the ring it is shown as in VI. The benzenonium ion (II), which is an intermediate stage in the chlorination reaction just described, can be found written as in VII to show the distribution of charge in that ion, which should add up to one + charge as indicated in II. The partial dotted circle in VII shows that there are now 4 delocalized electrons, distributed over 5 carbon atoms. The H atom shown in VII is part of the original C_6H_6 molecule; the Cl was introduced in the chlorination reaction; the C–H and C–Cl bonds shown are σ bonds at a saturated carbon atom that is tetrahedral (Figure 12.1), whereas the rest of the ring is flat.

The use of partial charges in VII is related to the way we handled similar problems in the chain compounds which have multiple bonds. In the simple case of ethene (Figure 15.10) it was said to be convenient to regard its reactions with ions as taking place through a polarized structure in which there is an electron shift that looks like transfer of two negative charges, but actually only one:

$$CH_2\overset{..}{}CH_2 \longrightarrow \overset{\oplus}{CH_2}-\overset{\ominus}{\underset{|..}{CH_2}}$$

Similarly, we used 1,3-butadiene to illustrate movement of electrons through a conjugate system, and introduced a flip-flop procedure (Figure 15.11) for manipulating the valence bonds. The benzene ring can be treated in an analogous fashion since it is essentially a closed conjugate system.

Structures VIII and IX are the equivalent Kekulé formulas for benzene, in resonance as indicated by the double-headed arrows. In excitation, an electron pair is presented to an attacking molecule, as shown by VIIIa and IXa. Electron movements to yield VIIIb are also possible, which provides a third position for the $\oplus$ charge. This explains the charge distribution in VII, which is a hybrid structure resulting from charge distributions as shown in formulas VIIIa, VIIIb, and IXa.

It should be understood that although the $Cl^{\oplus}$ in formula I lacks an electron pair it has only one positive charge due to the fact it has only one electron less than its normal seven. The benzene ring in VIIIa presents an electron pair to an attacking molecule, but the negative charge at that point is only one, exactly as explained for ethene in Figure 15.10. These positive and negative charges are neutralized when Cl is attached to the ring, but that still leaves positive charge elsewhere in the ring as shown by VIIIa, etc. This explains the + charge on the benzenonium ion in II.

In the terminology of Section 17.14, the replacement of an H atom in the benzene ring by an atom of Cl is typically S_E2—that is, substitution, electrophilic, bimolecular.

It has already been shown that the tendency of benzene to undergo ionic reactions is similar to that of the olefin hydrocarbons. The attack by a polarized halogen molecule on the π electrons of benzene and of ethene (Figure 17.4) is analogous in the initial stage; with benzene the final result is replacement of an H atom, whereas with ethene it is addition across the double bond.

Halogenation (replacing of H atoms) of the saturated paraffin hydrocarbons is typically a chain reaction with a free-radical mechanism (Section 17.9). These reactions are photochemical (induced by light energy), and high temperature is generally helpful, whereas displacement reactions in the nucleus are better conducted in the absence of strong light. The reader should refer to Section 24.5 where the chlorination of toluene is discussed. This compound consists of a nucleus (ionic behavior) and a side chain (free radical behavior). It was mentioned in Section 19.2 that benzene does not easily go through addition reactions, despite the three double bonds usually shown in its formula, but that with the proper inducement addition of halogen can be made to take place. The conditions for *addition* to the nucleus are those mentioned at the start of this paragraph for *displacement* of H atoms in open-chain compounds (Section 24.6).

20.3. Nitration. Concentrated nitric acid mixed with concentrated sulphuric acid acts on benzene hydrocarbons to give products in which the OH group of the nitric acid molecule is replaced:

$$C_6H_6 + HO{-}NO_2 \longrightarrow C_6H_5{-}NO_2 + H_2O$$

Benzene   Nitric acid   Nitrobenzene

The process is called nitration (or nitronation) and the nitro compounds are stable products in which there is a C–N link. Later, we shall describe compounds called esters (Section 27.11) in which only the H atom of nitric acid is replaced; since there is no C–N union the properties are much different.

The mechanism of nitration of the ring is similar to that described for halogenation. The active reagent is the highly electrophilic NO_2^+ ion which results from the interaction of the mixed acids.

$$HO{-}NO_2 + 2\ HO{-}SO_2{-}OH \longrightarrow NO_2^+ + H_3O^+ + 2\ O{-}SO_2{-}OH^-$$

Nitric acid   Sulfuric acid   Nitronium   Hydronium   Bisulphate
ion   ion   ion

The NO_2^+ ion attacks the π electrons of the ring to form the usual π complex, which rearranges to a benzenonium ion that splits off a proton to yield the nitro compound. The benzenonium ion, like that described in Formula VII of Section 20.2 is a hybrid, with the charge characteristics represented by formulas VIIIa, VIIIb, and IXa.

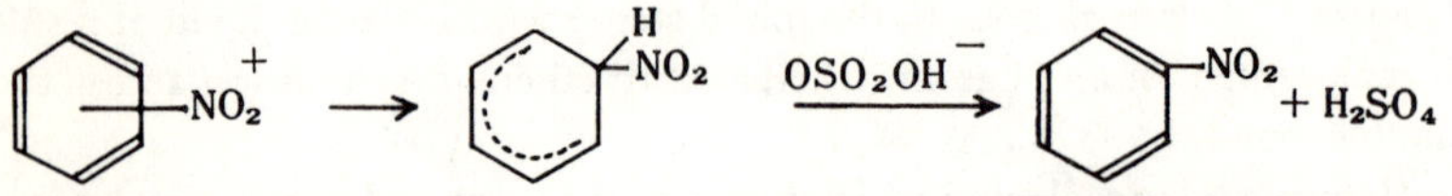

$$\text{π Complex} \quad\longrightarrow\quad \text{A benzenonium ion} \quad\xrightarrow{\;\overset{-}{O}SO_2OH\;}\quad \text{Nitrobenzene} + H_2SO_4$$

π Complex   A benzenonium ion   Nitrobenzene

Nitration of open-chain hydrocarbons is more difficult and must be carried out with more dilute acid at temperatures above 100°C, the temperature being a more important factor than the concentration. In some cases, as in the manufacture of nitroethane, $C_2H_5NO_2$, better results are obtained by nitrating in the vapor phase. The chain hydrocarbons most readily nitrated are those of tertiary structure (Table 10.2). In the olefin series, the action of nitric acid generally causes decomposition of the compound.

The nitro compounds derived from the benzene hydrocarbons are yellow liquids or solids, with a great many practical uses. They are especially useful in the dye and explosives industries. The nitro compounds of the paraffin series are colorless liquids and are useful as solvents. This is discussed further in Section 29.15.

20.4. Sulphonation. When the benzene hydrocarbons are treated with concentrated sulphuric acid (preferably fuming sulphuric acid, which has sulphur trioxide dissolved in it), characteristic compounds are formed that are known as sulphonic acids:

$$C_6H_6 + HO{-}SO_2{-}OH \longrightarrow C_6H_5{-}SO_2{-}OH + HO_2$$

Benzene   Sulphuric acid   Benzenesulphonic   acid

The process is called sulphonation, and the sulphonic acids are stable compounds in which there is a C–S link. There is no such bond in the sulphates (review the first paragraph on nitration in Section 20.3). A typical organic sulphate is dimethyl sulphate, $CH_3O{-}SO_2{-}OCH_3$, or $(CH_3)_2SO_4$.

Most of the sulphonic acids are water-soluble solids and are strong acids in solution. Sulphonic acids of the paraffin hydrocarbons, like methanesulphonic acid, $CH_3{-}SO_2{-}OH$, can be made (Section 30.4) but usually by indirect methods; the straight-chain hydrocarbons are not easily sulphonated. When olefins are treated with concentrated sulphuric acid, addition takes place as described in Section 13.16.

The mechanism of sulphonation of benzene is similar to that of nitration.

The active reagent is sulphur trioxide, SO_3, which is a *neutral molecule*, but highly polar and electrophilic due to electron deficiency at the sulphur atom.   It will act as an acceptor of an electron pair the way BCl_3 does in Section 8.5.

The π complex   Benzenesulphonic acid

The intermediate complex is a highly polarized structure, but not an ion (see Section 8.1 for the difference between $+$ and $\oplus$ signs).   The complex is unstable and as usual loses a proton.   This, of course, is the displacement of an H atom from the benzene molecule as recorded in the first paragraph of this section.   Loss of the proton converts the complex into a negative ion.

Sulphonation differs from the other displacements described in this chapter in that the reaction can be reversed under the proper operating conditions.

20.5. π and σ Complexes.   All the substitution reactions in the benzene ring that have been described in this chapter are based on the attack on the π electrons in the ring by a positively charged ion or by the positive pole of a polarized molecule. In every case there is subsequent formation of an intermediate complex in which there is a C atom that is tetrahedral (Figure 12.1) and has attached to it four normal σ bonds.   This is the important intermediate in these reactions; it is called a σ complex.   All these substitution reactions can be written without the π complex, but the presence of the π complex in the equation helps to understand what brings the molecules together in the first place.

In the nitration reaction, the slow (rate-determining) step is the formation of the σ complex, and the break-up that benzenonium ion to yield the final products is fast.   In the sulphonation reaction, the formation of the σ complex is fast, and the removal of a proton from that intermediate complex to yield final product is the slow step.   One of the methods used to find the rate-determining step will be mentioned in the chapter on isotopes (Section 37.5).   After the intermediate σ complex has broken up, the substitution product once again has the aromatic ring structure. See Section 17.13 for the importance of knowing which step in a reaction is rate-determining   The difference between the π and σ complexes is illustrated by the compounds shown in formulas X and XI.

X ⬡ — HCl

π Complex

XI ⬡ BF$_4^-$

σ Complex

The solution of HCl in benzene forms a 1:1 complex which is a nonconductor; the HCl has retained control of its proton while in contact with the π electrons of the benzene ring. A solution of HF and BF$_3$ in benzene is colored and is a conductor; a salt is formed in which the $+$ ion is the σ complex (see Formulas VIIIa and b and IXa in Section 20.2 for the charge distribution in this ion). That part of the σ complex in which the formula shows a dotted inner line is flat; only one of the C atoms in the ring is tetrahedral, as already stated. The aromatic benzene ring in the π complex, of course, is all flat.

20.6. Friedel-Crafts Synthesis. With the assistance of a catalyst of the AlCl$_3$ type, many structural groups can be attached to the benzene ring by displacement of a hydrogen atom. A simple example is the synthesis of ethylbenzene:

$$C_6H_6 + CH_3CH_2{-}Br \xrightarrow{AlCl_3} C_6H_5{-}C_2H_5 + HBr$$

Benzene Ethyl bromide   Ethylbenzene

This is representative of a reaction that has wide use in organic chemistry under the name of the Friedel-Crafts synthesis. The basic principle is the one already described in this chapter; the reagent must be one that can generate a polarized or a positively charged entity, and the catalyst assists attack by this entity on the π electrons in the ring. This introduction of an ethyl group into the ring is an example of alkylation (Section 23.8).

The π complex   Benzenonium ion   Ethylbenzene

With this type of synthesis it is possible to introduce six CH$_3$ groups into the ring, to form hexamethylbenzene (see Table 19.1).

Ethylbenzene is made commercially in enormous quantities. It is the source of styrene, from which polystyrene plastics are manufactured (Section 38.7). It is made economically without recourse to the Friedel-Crafts catalyst, from benzene and ethene, using phosphoric acid as the catalyst. The reaction between ethene and the catalyst supplies the ethyl carbonium ion ($C_2H_5^+$) with which to attack the ring:

$$CH_2{=}CH_2 + H_3PO_4 \longrightarrow CH_3{-}CH_2^+ \ H_2PO_4^-$$

Ethene   Phosphoric acid   Ethyl phosphate

20.7. Preparation of Benzene Hydrocarbons. Soon after the nature of benzene compounds was discovered, it was recognized that the chemistry of benzene is quite different from that of the chain compounds. Its properties are so different it was even thought that the benzene ring cannot be synthesized from a chain of carbon atoms. Several methods of preparation of benzene ring compounds from straight-chain compounds have been developed, however, and one such method will be found mentioned in Section 26.9. Another and much more direct method is to pass a straight-chain compound of six or more carbon atoms at 450 to 700°C (this is an example of *pyrolysis*) over a special catalyst:

$$CH_3{-}(CH_2)_5{-}CH_3 \longrightarrow C_6H_5{-}CH_3 + 4H_2$$

Heptane (Table 10.2)   Toluene (Table 19.1)

Before closing this chapter it may be mentioned that the Wurtz synthesis for hydrocarbons described in Section 9.4 is also useful in the benzene series. When the method is used for the preparation of benzene hydrocarbons it is called the Fittig reaction:

$$C_6H_5{-}Cl + 2Na + Cl{-}CH_3 \longrightarrow C_6H_5{-}CH_3 + 2NaCl$$

Chlorobenzene   Chloromethane   Toluene

21

The Geography of the Benzene Ring

21.1. How to Name Compounds Derived from Benzene. The benzene ring is a hexagon, and at each of its six corners there is an H atom that can be replaced by other elements or radicals. Suppose we were to replace two of these hydrogen atoms at the corners of the hexagon, forming dichlorobenzene, $C_6H_4Cl_2$. The problem before us is this—which of the six hydrogen atoms are replaced? The enterprising reader will take out paper and pencil and figure out three possibilities:

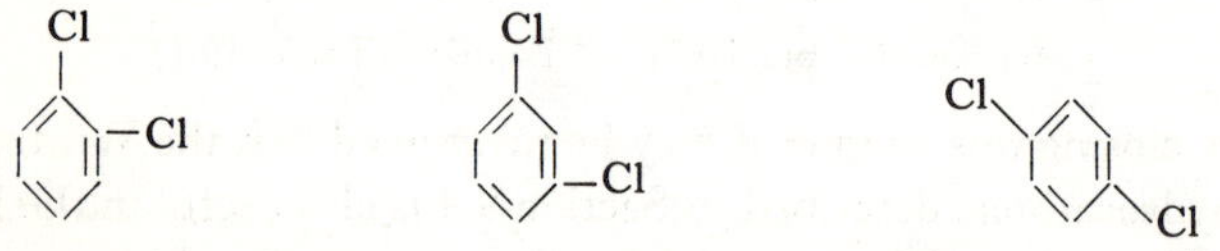

Either one, two, or all three of these substances are possible, and it is typical of our tasks in this chapter to find out how the organic chemist knows which one of them he is dealing with and how he can tell them apart.

Before going into this little question in geography, let us first discuss the nomenclature of benzene derivatives. When the compound is a substitution product, obtained by replacing the H atoms in the ring by other elements or groups, the number system is always a safe one to use:

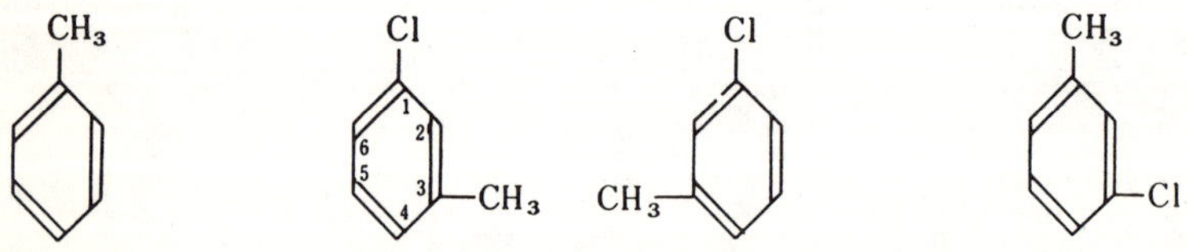

Methylbenzene
(Toluene)

1-Chloro-3-methylbenzene
(The same compound written three ways)

If there is only one substituent in the ring, as in methylbenzene, no number is necessary because all six positions in the ring are equivalent. When there are two or more substituents in the ring, the number system

200

tells us definitely where they are. The 1 position is assigned on an alphabetical basis, as explained in Section 11.3.

In the case of the *di*substitution compounds, such as those in the preceding illustration, it is a more general practice to locate the radicals or side chains by means of the prefixes *ortho* (*o-*), *meta* (*m-*), and *para* (*p-*):

o-Chloromethylbenzene   m-Chloromethylbenzene   p-Chloromethylbenzene

The prefix *para* is common in the English language, meaning opposite, as in the word paradox. The para positions in the benzene ring are those opposite each other. The prefix *ortho* is also common, and is usually employed to indicate a regular, direct (or orthodox) course. In the benzene ring the ortho positions are those at the ends of a direct valence bond. The prefix *meta* means between and the meta position in the ring is that between the ortho and the para. This system of nomenclature using prefixes derived from corresponding Greek words is important because it is used extensively.

When there are three substituent elements or radicals in the ring, there is still another system for locating their positions without using numbers. This is illustrated by the formulas of the three trimethylbenzenes given in Table 19.1. This method of nomenclature, however, is not widely used. Finally, the reader should review the method of nomenclature in Section 19.5 where it was shown that benzene derivatives can often be named as substitution products of chain compounds, using the name *phenyl* for the radical C_6H_5-.

The *addition* products of benzene are named in the customary manner by simply adding together the names of the reacting substances, as in the following example:

$$C_6H_6 \ + \ 3Cl_2 \ \longrightarrow \ C_6H_6Cl_6$$

Benzene Chlorine   Benzene
hexachloride

21.2. How to Locate the Positions of Radicals in the Ring. When only one element or only one radical replaces an H atom in the benzene ring, the resulting compound exists in only one form. No isomers have ever been prepared of a monosubstitution product like chlorobenzene, C_6H_5 Cl. For this reason we say that all six positions in the ring are equivalent, and therefore it does not matter where we write the Cl atom in the ring.

In the case of disubstitution products, we have already found there are three possibilities.

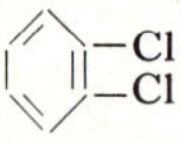

o-Dichlorobenzene
1,2-Dichlorobenzene
m.p. −17.6°C

m-Dichlorobenzene
1,3-Dichlorobenzene
m.p. −24.8°C

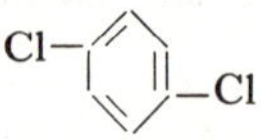

p-Dichlorobenzene
1,4-Dichlorobenzene
m.p. +52.C

The chemists who state that the ortho compound is the one with a melting point of –17.6°C must have had some proof that the two Cl atoms in that particular compound are really in the ortho position. The process of finding out where in the ring the substituting elements or radicals are situated is called *orientation*. A simple orientation method for the disubstituted compounds, developed by Körner, is often spoken of as Körner's 2-3-1 method.

The Körner orientation method consists in the introduction of a *third* element or radical into the ring. For instance, if all three of the *dichloro*benzenes are put through a series of reactions by which trichlorobenzenes are made we obtain the following compounds. The third Cl atom in each ring is represented by an X:

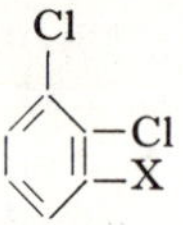

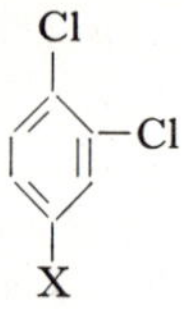

From the
dichlorobenzene
which melts at
−17.6°C

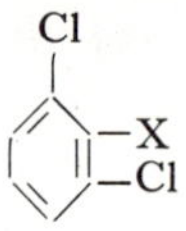

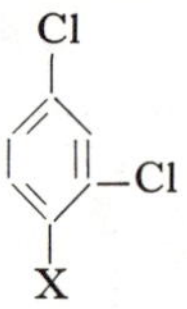

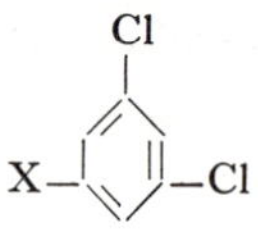

From the dichlorobenzene which
melts at −24.8°C

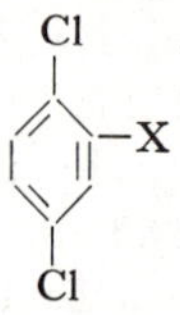

From the di-
chlorobenzene
which melts
at +52.°C

The reader should copy these formulas and satisfy himself that no other products are possible. It will then be clear that ortho, meta, and para *di*substitution compounds respectively result in 2,3, and 1 *tri*substitution compounds.

If we start, then, with a dichlorobenzene from which we can prepare two different trichlorobenzenes we know definitely that the original dichlorobenzene is the ortho compound. If we should determine the melting point of this dichlorobenzene we would also find that it melts at –17.6°C. In this way it can be found that the *m*-compound is the one melting at −24.8°C and that the *p*- compound melts at +52°C.

21.3. Reference Compounds. Naturally, when somebody has already gone through the Körner procedure to definitely locate the two Cl atoms in the dichlorobenzenes, it is not necessary for anyone else to repeat it. All we need to do, when we have a sample of Cl—C_6H_4—Cl and when we do not know whether the Cl atoms are in *o-*, *m-*, or *p-* position, is to find the melting point of the compound. Then, by looking in the tables of the textbooks of organic chemistry, we can find out which one it is from its melting point.

It is also possible to use the three dichlorobenzenes as reference compounds for determining the structure of other disubstitution compounds. For example, suppose we are the first persons to have made a sample of dinitrobenzene, NO_2–C_6H_4–NO_2 and we do not know the positions of the two NO_2 groups in the ring. By a roundabout process it is possible to convert NO_2–C_6H_4–NO_2 into Cl–C_6H_4–Cl and from the melting point of this compound, we can tell where the Cl atoms are situated. If the product happens to be the *p*-dichlorobenzene, we feel certain that the original compound was *p*-dinitrobenzene:

$$NO_2-\langle\;\rangle-NO_2 \longrightarrow Cl-\langle\;\rangle-Cl$$

Data on reference compounds of many types are part of the stock equipment of the synthetic chemist. In order to find out the structure of an unknown substance he tries to convert it by as direct a route as possible to a reference compound of known structure and known constants. Sometimes an element or radical may wander from one position in the compound to another position during the transformation. That makes the game more interesting. This action is beyond our scope and will not be discussed here.

21.4. Rules for Substitution in the Benzene Ring. In Section 13.12 (which should be referred to) it was indicated that a structural group such as the C=C double bond may have a dramatic effect on other elements or groups in the molecule and that a great deal depends on the positions of these groups relative to each other. This is an aspect of organic chemistry that supplies most of the material in this chapter. Before going into details, we shall first summarize what will be learned in the rest of this chapter by reciting a few simple *rules for substitution* in the ring:

a) When the benzene ring contains a substituent which is an element

like Cl, or a group like H $\overset{\text{H}}{\underset{\text{O}}{\overset{|}{\text{C}}}}$ —H in which the atom linked to the ring is united to other elements by single bonds, the second substituent is directed mainly to the *ortho* and *para* positions.

b) If the benzene ring already has a substituent like the $O=N=O$ group or the $-C\equiv N$ group, in which the atom linked to the ring is united to other elements by multiple bonds, the second substituent will be directed mainly to the *meta* position.

These rules are illustrated very briefly by the following compounds, in which the substituent at the top of the ring was the first occupant and the other substituent was introduced second:

o-Chlorotoluene   p-Chlorotoluene   m-Chloronitrobenzene

The rules are not at all quantitative and there are exceptions, but they are useful generalizations. To explain them, we employ resonance theory (Section 16.9), inductive effect (Section 17.2), and dipole moment data (Section 8.3).

It should be observed that if the first tenant in the ring exerted no directive effect whatever on the position of the second, then every disubstitution product would be 40% *o-*, 40% *m-*, and 20% *p-*, because there are twice as many *o-* and *m-* positions as there are *p-*, and it would not be necessary to remember any rules.

21.5. Inductive Effect and Dipole Moment. The inductive effect, as explained in Section 17.2, is essentially a phenomenon associated with a single valence bond; the electron pair (:) which is the normal valence bond is displaced in the direction of a more electronegative element. This displacement is not only one of *direction*, but also of *size*, and its magnitude is found by a determination of the dipole moment (Section 8.3). The dipole moments (in Debye units) of some substituents in the benzene ring are as follows:

	CN	NO$_2$	Cl	H	CH$_3$	OH
Dipole moment	−4	−4	−1.7	0	+0.4	+1.6

The plus or minus sign indicates whether the inductive effect of the group is positive or negative with respect to the benzene nucleus. The CH$_3$ group releases electrons to the ring, whereas the tendency of Cl is to withdraw electrons.

When *two* substituents are in the ring the dipole moment will depend on the symmetry of the molecule. A *p-* compound like Cl—⟨benzene ring⟩—Cl has a zero moment, for the dipoles Cl←C and C→Cl cancel each other and the benzene ring is flat. The *o-* and *m-* compounds, however, have high dipole moments.

Throughout Chapter 20 it was emphasized that the reagents that attack the benzene ring are those which are electrophilic; they seek out the exposed π electrons. It is obvious, then, that the inductive effect of a substituent already in the ring will have a marked bearing on the ease with which a second group can be introduced. The CH_3 group with its electron-release properties should make the electrons in the ring even more accessible to a second reagent; it is said to be *activating*. The NO_2 group, with its electron-withdrawal properties, makes electrons in the ring less accessible and its effect on subsequent substitution is said to be *deactivating*.

21.6. Ortho and Para Substitution. The mechanism by which a Cl atom is inserted into the benzene ring was described in detail in Section 20.2. The reaction is essentially an attack on the π electrons of the ring, and the structures VIIIa and b and IXa in Section 20.2 show how the polarization of the benzene ring by the attacking molecule results in withdrawal of electron charge from the ring. The positive charge left *in* the ring is spread over two *ortho* positions and the *para* position. This situation in chlorobenzene, C_6H_5–Cl, is summed up in the dipole moment of -1.7 listed in Section 21.5. All groups below H in that list withdraw electron charge from the ring and thus tend to deactivate it—that is, make it a little more difficult for a second group to enter.

Now according to rule *a* in Section 21.4, a Cl in the ring will direct a second group into *o-* and *p-* positions. However, we have just seen that due to the inductive effect of Cl in the ring, the *o-* and *p-* positions carry partial positive charge (see VIIIa and b and IXa in Section 20.2) rather than the negative charge we associate with substitution in the benzene ring. The answer to this seeming contradiction is that the inductive effect must compete with a resonance effect described in Figure 21.1. The inductive effect is the more important one in this molecule, for we know that although the Cl in the ring does indeed orient *ortho, para* in accordance with the resonance effect it does so with deactivation. The effect of resonance to charge the *o-, p-* positions negatively as indicated in Figure 21.1 must be rather weak.

Structure A of Figure 21.1 is written in such a way as to emphasize the

21.1. Resonance structures of chlorobenzene.

fact that the atom attached to the ring contains *unshared* pairs of electrons. In Structures I and II the same molecule is shown in the two electronic configurations corresponding to the two resonance structures for benzene (Section 19.3). The Cl atom is shown with the unshared electron pairs (potential valence bonds) as lines.

From Structures I and II, we can construct three other resonance structures, marked III, IV, and V, by flip-flopping the valence bonds starting at the Cl atom, as suggested in Figure 15.11. Structure III is arrived at by shifting the bonds in a nearly complete cycle, the terminus being indicated by the electron pair at an *ortho* carbon atom which now has a negative charge. The atom from which the shift started acquires a positive charge. Similarly, from Structure II, we can arrive at V.

From both I and II we can construct Structure IV simply by stopping the shift at the *para* carbon atom in the ring and conferring a negative charge or potential valence bond on that atom. It is not possible by this system to write any structure in which the *meta* position acquires a negative charge. The structures in Figure 21.1 are resonance forms which could be connected by the usual double-headed arrow ($\leftrightarrow$), the symbol for resonance.

An OH group in the ring has a dipole moment on the positive side of hydrogen (see Section 21.5). It promotes o- and p- substitution *with activation* because in this case the resonance effect (similar to that of Cl) reinforces the inductive effect.

21.7. Meta Substitution. Although the $-NO_2$ group is often written $O=N=O$ in order to show the nitrogen atom with its maximum

valence of five, the electronic conception of this group presents quite a different picture. The oxygen atom has six valence electrons ($:\ddot{O}:$) and the nitrogen atom has five such electrons ($\cdot\ddot{N}\cdot$); the pairing of these electrons in the formation of valence bonds in a molecule is as follows, where R is any radical to which the group is attached.

$$:\ddot{O}:N::\ddot{O}: \leftrightarrow :\ddot{O}::N:\ddot{O}: \qquad\qquad \overset{\ominus}{O}\leftarrow\overset{\oplus}{N}=O \leftrightarrow O=\overset{\oplus}{N}\rightarrow\overset{\ominus}{O}$$

The first two formulas show all the electrons written out; they are equal and opposite resonance forms. The formulas show that the NO_2 group can have only one conventional double bond. The other link must be a dative bond (Section 8.5), as indicated more explicitly in the last two formulas. The nitro group is a polar, hybrid structure. Formation of one of the resonance structures is shown pictorially in Figure 21.3.

The important thing about the nitro group in nitrobenzene is not the fact that it is a resonance hybrid, but that the double bond in the nitro group is *conjugated* with the double bonds in the benzene ring. Throughout Figure 21.2 we shall use only one of the resonance forms of the NO_2 group, shown electronically in B and with straight line bonds for unused electron pairs in VI and VII. Structures VI and VII are the two Kekulé resonance forms of benzene (Section 19.3). By flip-flopping the bonds as indicated by the curved arrows we can work out a series of polarized structures for nitrobenzene. The structures in Figure 21.2 are resonance forms and may be connected by the standard double-headed arrow signs ($\leftrightarrow$). Nitrobenzene is a hybrid of all these structures, of which the important ones in this discussion are VIII, IX, and X, in which an electron charge is transferred out of the ring, leaving partial + charge distributed

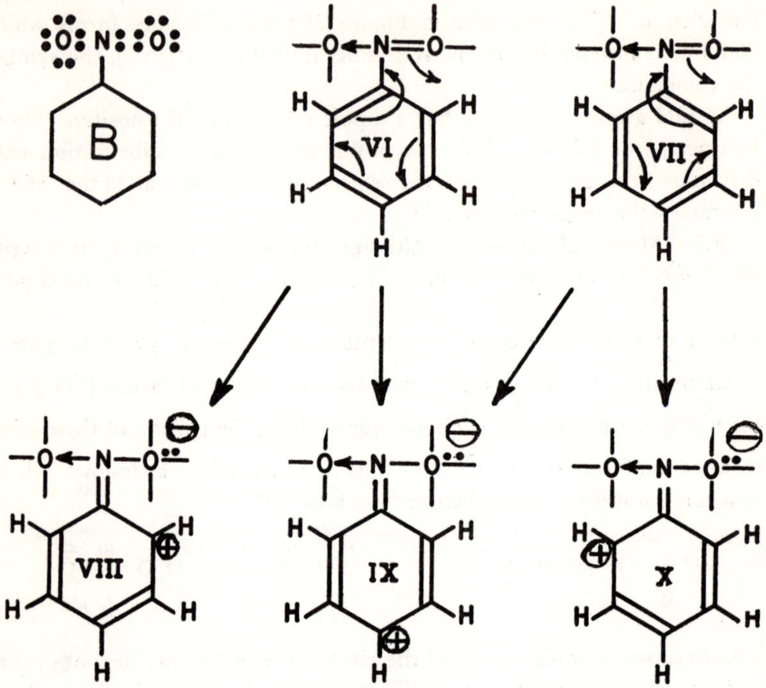

21.2. Resonance structures of nitrobenzene.

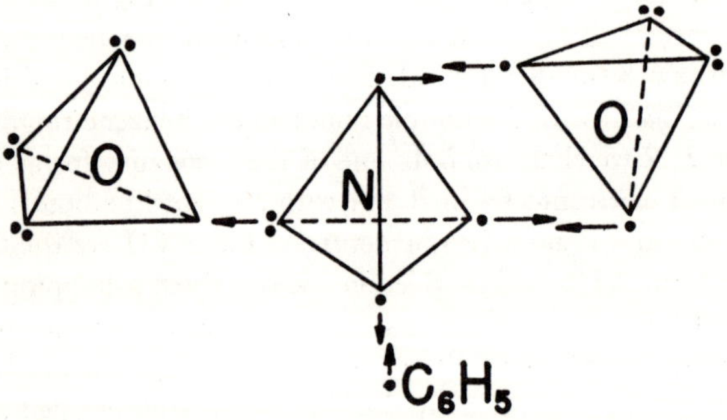

21.3. Electron-sharing in the nitro group.

on the *o*- and *p*- carbon atoms. In other words, the *m*- positions are more negative than the others, and that is where the next substitution will most likely take place.

A nitro group already in a benzene ring will therefore orient a second substituent into a *m*- position because of a resonance effect that works in the same direction as the inductive effect mentioned in Section 21.5 (resonance effect is predominant). Subsequent substitution into the ring will be more difficult, because negative charge removed from the ring by these two effects is highly deactivating. In fact, all substituents that orient *m*- are deactivating, whereas we found that those that orient *o*- and *p*- may either activate or deactivate. Many groups with high electron-attracting properties are *m*- directing, though this may violate the rules in Section 21.4; one example is $-CCl_3$.

22

Stereochemistry and Isomerism

22.1. Introduction. In Chapter 12 we showed how a large part of our chemistry would be confusing if we persisted in thinking of molecules simply as flat figures no thicker than the ink on this page. We found that to understand the molecule we must examine it as something with a definite shape. As usual, "the Greeks had a word for it," and in this book we have called the study of the shapes of molecules *morphology*, as in the title for Chapter 12. This term is familiar to biologists, who use it when dealing with the classification of plants and animals according to form and structure.

The scholars who put organic chemistry on its feet in 1860 to 1890 knew so much Greek they could not help using it in chemical literature. They introduced the term *stereo*chemistry to designate the branch of chemistry devoted to the shapes of molecules. The word *stereos* in Greek means solid, but to the modern mind it connotes space. For example, our unit of space is the *stere*, or one cubic meter. According to a standard dictionary, stereochemistry is the "chemistry dealing with the arrangements of the parts of a molecule in space."

A stereoscope, it will be recalled, is an instrument for viewing special kinds of picture post cards; it is capable of transforming a flat landscape into a scene in which the cows and the farmer's daughter are substantial objects rather than flat wallpaper sketches. In stereochemistry, likewise, we put on our stereoptical spectacles, and instead of looking at the flat

carbon atom $-\overset{\displaystyle |}{\underset{\displaystyle |}{C}}-$ in which the valence bonds point north, east, south,

and west, we see the bonds occupy space in the shape of a regular tetrahedron, as shown in Section 12.2.

22.2. Structural Isomers. In Section 10.4 we defined isomers as compounds with the same composition and same molecular weight, but different properties. To illustrate the meaning of isomerism, we described two different compounds of the same composition, C_3H_7Cl, and showed

210

that the difference between them is due to the *position* of the Cl atom in the molecule. These isomers are known as position isomers as well as structural isomers:

$$CH_3{-}CH{-}CH_3 \qquad\qquad\qquad CH_3{-}CH_2{-}CH_2Cl$$
$$\underset{Cl}{|}$$

As we have already pointed out, the word *isomer* is another term descended from the Greeks.

In the case of simple isomers, like those in the preceding paragraph, it is not necessary to examine the shapes of the two molecules to see that they are different.

22.3. Geometric Isomers. The isomers we shall consider in this section cannot be explained as easily as the isomers that were just described. For our illustration we shall go back to the chapter on olefins, the compounds containing $C{=}C$ bonds, and pick out from Table 13.1 the compound 2-butene, $CH_3{-}CH{=}CH{-}CH_3$. This looks like the formula for a single substance of the composition C_4H_8. There are, however, two chain compounds of this composition, and their structures are written as follows:

$$
\begin{array}{cc}
CH_3 \quad CH_3 & \qquad\qquad CH_2 \quad H \\
| \qquad | & \qquad\qquad | \qquad | \\
C{=}C & \qquad\qquad C{=}C \\
| \qquad | & \qquad\qquad | \qquad | \\
H \qquad H & \qquad\qquad H \qquad CH_3
\end{array}
$$

cis-2-Butene   *trans-2-Butene*
Z-2-Butene   *E-2-Butene*

In Section 14.11 it was explained that there is a barrier to free rotation about a single bond, which has been measured for certain molecules. The energy to overcome for free rotation is so small, however, that isomeric forms have never been isolated. The chemist imagines a single bond as normally allowing free rotation. In the last paragraph of Section 15.4 it was calculated that the energy required to break the π bond in a molecule like butene is about 60 kcal/mole. Breaking down this rather high barrier to rotation would quite likely be accompanied by other changes in the molecule.

In Section 12.3 it was explained that the different shapes a molecule can assume by rotation about a single bond are called its *conformations*; they cannot be isolated. The different shapes a molecule assumes because of a double bond are called *configurations*; they can be isolated. Although we do not have space in this book to discuss the methods of distinguishing between *cis* and *trans* isomers, a brief note on the problem will be found in Section 31.4.

This phenomenon due to a double bond in the molecule is referred to as geometric isomerism, and since it depends on space arrangements it is also called stereoisomerism. Throughout the history of organic chemistry these stereo isomers have been known as *cis* and *trans* isomers; *cis* is the Latin word meaning on the same side and *trans* is the corresponding prefix for on the other side. Although the simple compound used here for illustration does not offer any difficulty in this method of nomenclature, more complex molecules have presented difficulties not overcome by modifications occasionally introduced. A new system using the descriptors *E* and *Z* has been adopted by *Chemical Abstracts* beginning with the 1967 volume, and will be employed by succeeding generations of organic chemists. The *E, Z* method will be described in Section 22.16a, since it is based on a set of rules explained in Section 22.16.

22.4. Mirror-Image Isomers. This is the third, and last, of the various kinds of isomerism to be described in this chapter. It all depends on a simple bit of solid geometry, as we can easily discover by examining a tetrahedron. If we make a tetrahedral model, as described under Figure 22.1 and place four different things at the corners, it will be found that two different arrangements are possible. These arrangements are I and II of Figure 22.2. A few minutes spent in trying to superpose one of these

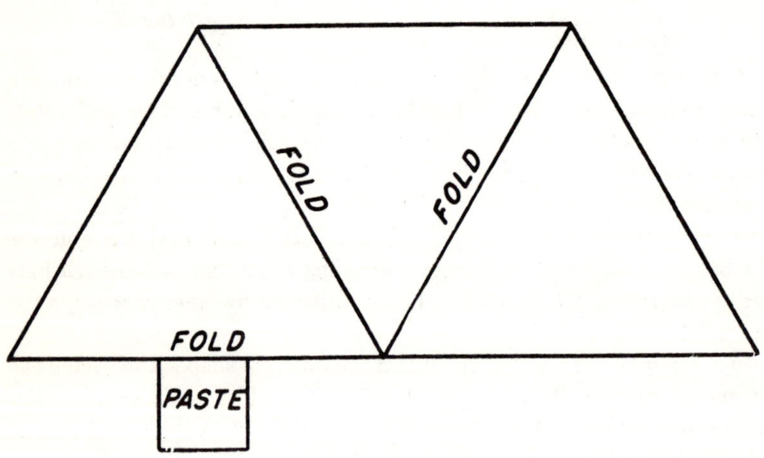

22.1. Design for making tetrahedral models. Cut a sheet of stiff paper and fold as indicated. A little paste on the flap will hold the model together.

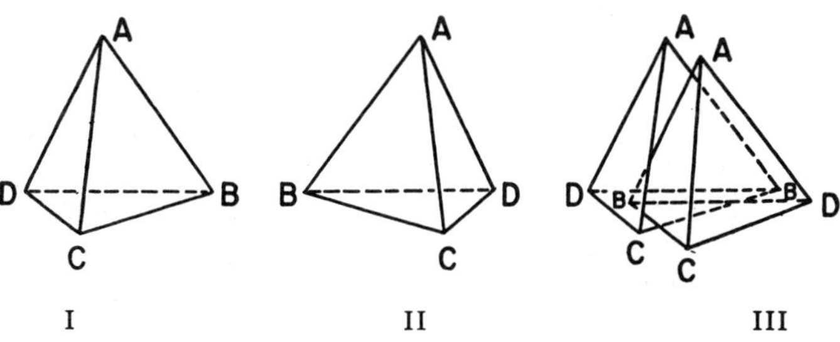

22.2. Mirror-image isomers.

models on the other (as illustrated by Model III) will convince any skeptic that they are decidedly different.

The interesting thing about these models is that they are mirror images of each other. This relationship is more clearly expressed by the models shown in Figure 22.3, which may be compared to a man standing in front of a mirror. If the man adjusts his necktie with the right hand his reflection in the mirror does the same thing with the left hand. Figure 22.3 shows an opposite arrangement of the same groups around the central carbon atoms, but we shall explain presently that the two molecules have a certain opposite physical effect, just like the right-handed man and his left-handed image. That is the reason for the arrows in Figure 22.3.

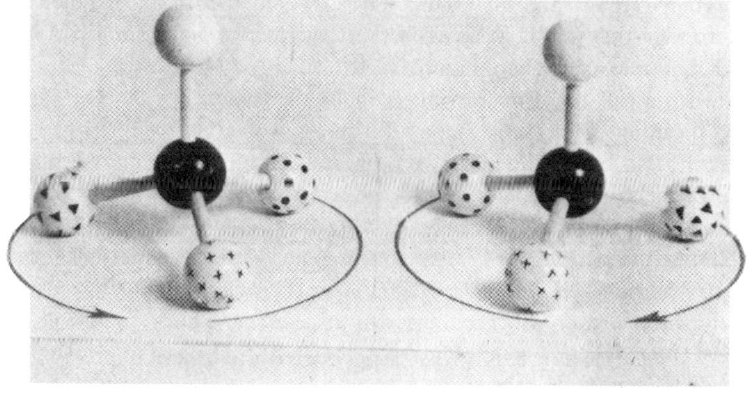

22.3. Models of mirror-image isomers.

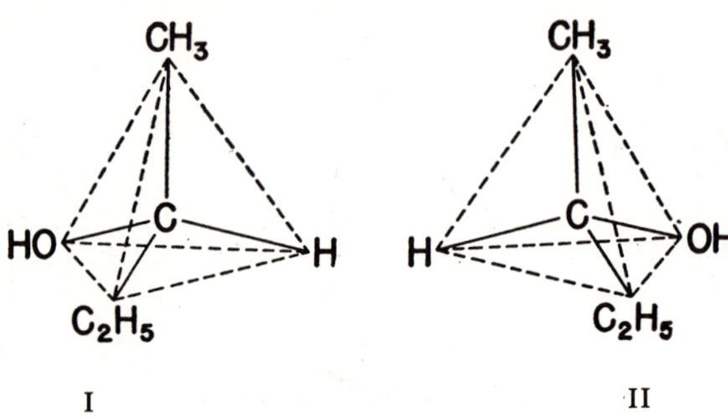

22.4.   Mirror-image isomers of 2-butanol.

Suppose we examine the structure of 2-butanol, the formula of which is $CH_3-CH_2-{}^*CH-CH_3$. The carbon atom in this compound indicated
$$\underset{\displaystyle OH}{|}$$
by an asterisk (*) has four different elements or groups attached to it—namely, –H, –OH, $-CH_3$, and $-CH_2-CH_3$. Writing the formula in the

$$HO-{}^*\overset{\displaystyle H}{\underset{\displaystyle CH_3}{\overset{|}{\underset{|}{C}}}}-C_2H_5$$

usual flat way gives no indication at all that there are two ways of arranging the groups around the *C atom. But if we go to the third dimension and arrange the groups in space, we get the pair of right-hand and left-hand molecules shown in Figure 22.4.

Chemists call these mirror-image molecules *enantiomorphs*, the Greek word meaning opposite in shape. Since the isomerism of these pairs of molecules can be explained only on the basis of a different space arrangement of the same groups, we also call them *stereo*isomers, but it is not as satisfactory a name as enantiomorphs.

22.5. Optical Isomers.   Up to this point we have simply called to the attention of the reader the fact that certain substances can exist in two forms which are mirror images of each other. This state of affairs comes about when a compound contains a carbon atom to which *four different* elements or groups are attached.

In preceding chapters, when simpler isomers were described, it was usually sufficient to indicate that they had different melting points or

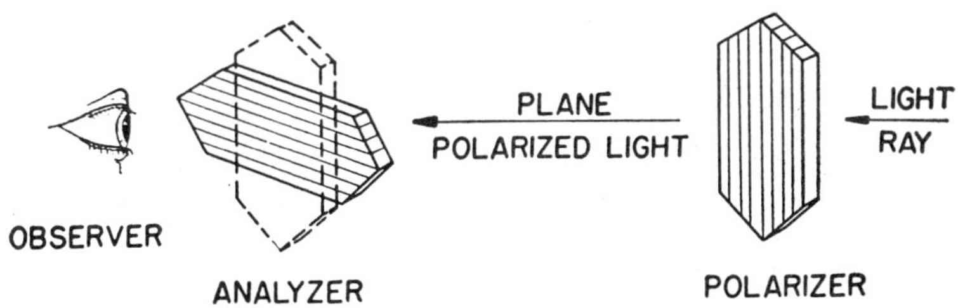

22.5. An elementary illusration of the formation and detection of plane polarized light.

boiling points and were therefore different substances. The mirror-image isomers, however, have the same boiling point, same melting point, same density—in fact, all their properties are alike except one. The only physical difference we can discover between mirror-image isomers is that they have either a right-hand or left-hand effect on a certain property of light. Because of this action on light, mirror-image isomers are generally called *optical* isomers; optics is the science dealing with the properties of light.

Before we explain more definitely why mirror-image isomers are also called optical isomers, we must first recall the phenomenon in optics called polarized light. Certain crystalline substances, such as tourmaline, when placed in the path of a ray of light will cut out all vibrations except those in one plane, as indicated in Figure 22.5. The light that issues past this substance is called plane-polarized light, and the crystal causing the polarization is known as the polarizer.

If a second piece of tourmaline, called the analyzer, is placed so that its axis is at right angles to that of the polarizer as indicated in the diagram, this plane polarized light is intercepted and the light is completely blocked out. If the analyzer is rotated into the position indicated by the dotted lines, so that its axis is parallel to that of the polarizer, the light is able to pass through it and the observer sees the crystal illuminated.

Assume that the analyzer is in the dotted position in Figure 22.5 and is illuminated. If now we place a quartz plate between the polarizer and the analyzer the light will be blotted out. Then, on rotating the analyzer, at some point the light will go through again.

This shows that the quartz plate rotates the plane-polarized light, and that by rotating the analyzer until it is illuminated we can measure the rotating effect of the quartz plate. The rotation is usually expressed in

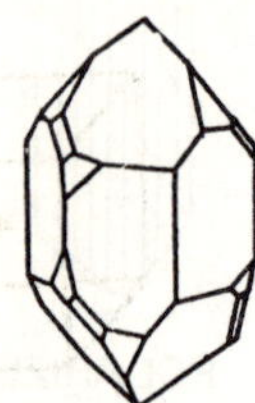 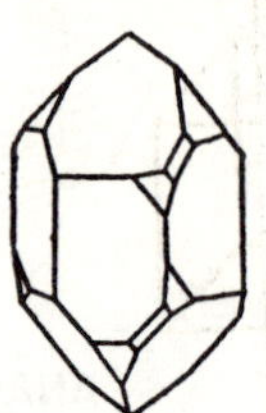

22.6. Quartz crystals. These are mirror images (*enantiomorphs*, in Greek). A section of quartz 1 mm thick, cut perpendicular to the principal axis of a quartz crystal, rotates the plane of yellow light (D line of the spectrum) through an angle of 22°.

degrees—for example, a certain thickness of quartz may have a rotating effect of 30°, a slightly thicker sample may rotate the light 45°, and so on. The laboratory instrument known as the polariscope or polarimeter is based on this principle and is used to measure the light-rotating activity of various substances (Section 34.10).

All this was known shortly after the year 1800, and it was also common knowledge at that time that certain samples of quartz rotate plane polarized light to the left and others rotate it to the right. This, the crystallographers showed, is because some quartz crystals are right-handed and others are left-handed. In other words, there are two kinds of quartz crystals, identical in every respect except that they are not superposable but are mirror images (Figure 22.6). They also knew that certain organic compounds like sugar, tartaric acid, camphor, etc., when in solution, are able to rotate polarized light.

All these substances affecting polarized light are said to be optically active, and since some substances even in solution are optically active, it is clear that the phenomenon can be caused by *molecules* as well as by *crystals*.

In 1848, Pasteur's work on tartaric acids showed that the structure of the molecule itself, just like the structure of quartz crystals, may give rise to right-hand and left-hand effects. Then the world had to wait another 25 years for the invention of the tetrahedral carbon atom, which explained how it is possible to obtain two mirror-image isomers if there are four different groups attached to it, as we demonstrated earlier in the chapter. The tartaric acids, which played such an important role in the explanation of optical activity, are described briefly in Section 28.7.

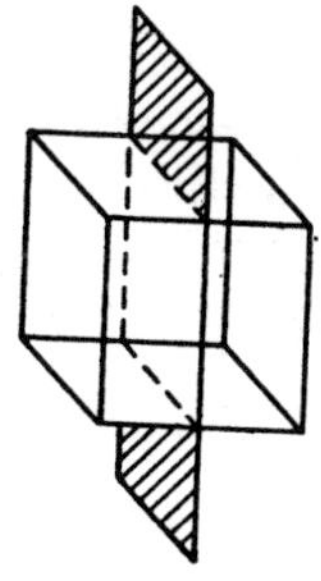 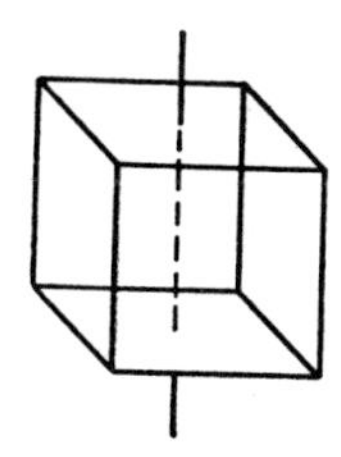 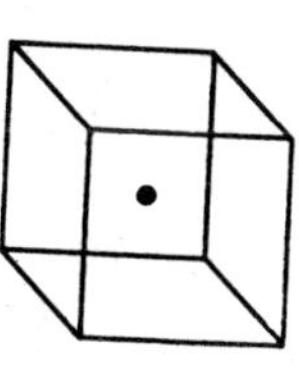

22.7.   Illustrations of the meaning of symmetry.

22.6. Symmetry of Crystals and Molecules. The phenomenon of optical activity is associated with the symmetry of substances. A cube has a high order of symmetry and is a good model for an elementary explanation as shown in Figure 22.7.

When a plane can be drawn through a crystal in such a way as to cut it into two halves that are mirror images of each other, the crystal is said to have a plane of symmetry. If a line, or axis, can be drawn through a crystal, so that when the crystal is rotated about this axis it presents the same face to an observer several times in a revolution, it is said to have an axis of symmetry. The cube presents the same face, or the same edge, four times in one revolution about its axis of symmetry. A crystal has a center of symmetry when its faces are arranged in pairs, as they are in the cube.

22.7. Asymmetry and Dissymmetry. If we apply these principles of symmetry to the quartz crystals pictured on one of the preceding pages, it will be found that they have a very low order of symmetry. They have no principal plane of symmetry and no center of symmetry; such a substance is said to be *asymmetric*—not symmetrical. It should not be inferred that the crystal of quartz has no symmetry whatever, for it may have several minor planes (or axes) of symmetry. What is meant is that the crystal lacks symmetry in that two forms are possible that are not superposable, but are mirror images.

Recently, the term *dissymmetric* has been introduced for describing mirror-image isomers that are not superposable, and *Chemical Abstracts* uses *asymmetric* and *dissymmetric* interchangeably in its indexing. Some molecules that are dissymmetric are not asymmetric, but the explanation is beyond the scope of this book.

Mirror-image isomers are due to lack of symmetry of the molecule as a whole. Chemists, however, often state that mirror-image isomers are produced when ı molecule contains an asymmetric carbon atom. By an *asymmetric carbon atom* we mean one that has four different elements or radicals attached to it. In the following examples the asymmetric carbon atoms are indicated by (*). Each of these molecules can exist in two mirror-image forms.

$$CH_2\underset{\underset{\displaystyle OH}{|}}{\overset{\overset{\displaystyle H}{|}}{\underset{\underset{\displaystyle CH_3}{|}}{C^*}}}Cl \qquad CH_3\text{—}CH_2\text{—}\underset{\underset{\displaystyle CH_3}{|}}{{}^*CH}\text{—}CH_2OH \qquad CH_3\text{—}CH_2\text{—}\underset{\underset{\displaystyle Cl}{|}}{{}^*CH}\text{—}\underset{\underset{\displaystyle Cl}{|}}{CH_2}$$

Molecules can be asymmetric without containing such carbon atoms. A relatively simple example is found among certain derivatives of allene (propadiene), a hydrocarbon with the formula $CH_2{=}C{=}CH_2$. If the H atoms are replaced by X and Y groups, mirror-image isomers are possible, as shown in Figure 22.8, and have been prepared in the laboratory. The student should refer back to the structure of ethene shown in Figure 15.1.

The only requirement for a compound to exist in the form of right-hand and left-hand isomers is that it must lack certain forms of symmetry. Whether or not the molecule also contains an asymmetric carbon is of less importance. Thus in Figure 22.8 it will be seen that there are no carbon atoms in the molecule that conform to the strict definition of "an atom to which four different atoms or groups are attached."

Later we shall describe compounds in which asymmetry is due to nitrogen atoms (Section 29.6), to sulphur atoms (Section 30.6), and to the phenomenon of steric hindrance (Section 28.10). The student may at this time also refer to the Walden inversion (Section 22.17).

A term being revived for use in stereochemistry is *chirality*, which is

22.8.   Enantiomorphic forms of certain allene derivatives.

from the Greek for "handedness." It identifies the right-hand or left-hand effects caused by asymmetric molecules. Mirror-image isomers have opposite chiralities. An asymmetric carbon atom is a chiral center.

22.8. Dextro and Levo Compounds. Pairs of optical isomers have identical melting points, boiling points, surface tensions, etc. They differ in but one physical property, and that is in their *directional* effect on light. Since their only physical difference is in their action on polarized light, we may logically name them on that basis. The form producing a right-hand rotation is called a *dextro* compound; the opposite form is then a *levo* compound. Although these isomers have *opposite* effects on light, the *amount* of rotation is the same, i.e., if one rotates the light 30° to the right, under the same conditions the other rotates it 30° to the left. The compounds pictured in Figure 22.4 are *dextro*-2-butanol and *levo*-2-butanol, but we make no attempt as yet to indicate whether the *dextro* molecule is Model I or Model II.

Chemistry is a science in which much ingenuity is expended in the effort to transform one compound into another. Often this can be done with full confidence that the structure of the new compound will be obvious from that of the original compound. This principle was touched on in the last chapter (Section 21.3) in a discussion of *reference compounds*. In the case of optically active compounds the chemist has had a particularly difficult time proving his transformations. In fact, before 1951 no one knew for certain which of the two 2-butanols was *dextro* and which *levo*. Now we do know, but it is still not easy to make compounds related to the 2-butanols and prove that they are derivatives of the *dextro* or the *levo* compound. The relationship cannot safely be made on the basis of the direction of optical rotation. For example, the *dextro* isomer of glyceraldehyde (Figure 22.9) can be converted readily to lactic acid (Section 28.7) but the product is *levo*.

22.9. Family Configurations. Since the direction of optical rotation cannot be depended on to establish identity of the many compounds in this series, a system was developed in which they were placed in configurational families. This was started about 1900 by Emil Fischer, who laid the foundations for carbohydrate chemistry (Section 34.2) and needed a scheme for systematizing that branch of chemistry. The nature of this system can be learned from a homely illustration. Let us say there is a right-handed Smith and a left-handed Smith, and that each has sons, grandsons, great-grandsons, etc. A descendant of R. H. Smith may be left-handed, but we find that that is a minor consideration compared with the fact that he belongs to the family of R. H. Smith rather than to that of

$$
\begin{array}{cc}
\text{D(}dextro\text{)-Glyceraldehyde} & \text{L(}levo\text{)-Glyceraldehyde} \\
\text{Head of the D family} & \text{Head of the L family}
\end{array}
$$

CHO
H——OH
CH$_2$OH

CHO
H—C—OH
CH$_2$OH

CHO
OH——H
CH$_2$OH

CHO
HO—C—H
CH$_2$OH

D(*dextro*)-Glyceraldehyde   L(*levo*)-Glyceraldehyde
Head of the D family   Head of the L family

22.9. The reference compounds for determining D or L configuration of optically active compounds.

L. H. Smith. However, we can designate him in such a way as to bring out both factors; thus, RH(*levo*)-Smith shows he is left-handed although descended from a right-handed patriarch.

In a similar fashion, the experts in this field of chemistry have designated a few simple asymmetric compounds as progenitors of a D series and an L series of compounds. The structures of a pair of these compounds are shown in Figure 22.9. These are called *projection* formulas and are discussed in detail in Sections 39.3 and 39.5. The asymmetric carbon atom is drawn as a regular tetrahedron, with a *vertical edge* resting *in the plane* of the paper. The chemical nature of the compound pictured need not concern us in this chapter, but it should be emphasized that the CHO group (the principal functional group) in this molecule is always at the top. The ruling was made that the molecule with OH on the right is a D compound. It is conventional to print D and L in small capitals.

In 1951 it was demonstrated that Emil Fischer had made a very fortunate guess, and that the *dextro* compound formulated in Figure 22.9 is the one that really is *dextro* in aqueous solution. The proof was made with a salt of *dextro*-tartaric acid (Section 28.7), in which the actual space arrangement of the atoms was determined by a complex X-ray method. The configuration of the tartaric acid had already been correlated with that of the *dextro*-glyceraldehyde in Figure 22.9. Before 1951 all decisions on whether a compound was a member of the D or L family

C$_2$H$_5$
H——OH
CH$_3$

CH$_3$
CH$_2$
H—C—OH
CH$_3$

C$_2$H$_5$
OH——H
CH$_3$

CH$_3$
CH$_2$
HO—C—H
CH$_3$

D(*levo*)-2-Butanol   L(*dextro*)-2-Butanol
(Model I of figure 22.4)   (Model II of figure 22.4)

22.10. Assignment of the 2-butanols to D and L families.

were *relative*; in other words, a compound could only be related to the postulated formulas in Figure 22.9. Now the configurational family is on an *absolute* basis. For example, the lactic acid in Section 28.7 which is *levo* has the configuration as shown and is definitely a D compound.

A recently developed scheme for naming optical isomers so as to show the chirality of the individual asymmetric carbon atoms will be described in Section 22.16.

We return now to compounds already mentioned in this chapter. To determine the configuration of compounds I and II pictured in figure 22.4, we must obey the conventions and employ projection formulas. This is done in figure 22.10, where the D and L family assignments are based on proofs in the chemical literature as to their relations to the reference standards in Figure 22.9. The D compound is *levo* with respect to light, although it belongs to the right-handed family.

Among optical isomers, we do not necessarily convert all compounds to the family heads in Figure 22.9 in order to find out to which family they belong. In the course of time, the D and L configurations of many compounds have been determined, and these in turn can serve as reference standards. As an example, the compounds in Figure 22.10 can now serve as standards; a compound which can be directly prepared from the D-2-butanol, or converted to it, is a D compound regardless of its rotational direction. An illustration of the use of such a reference compound is the following reaction:

$$CH_3—CHOH—CH{=}CH_2 + H_2 \longrightarrow CH_3—CHOH—CH_2—CH_3$$

3-Buten-2-ol 2-Butanol

The butenol has an asymmetric C atom and can be prepared in D and L forms. The compound that hydrogenates to L-2-butanol is necessarily an L compound, regardless of its rotational direction (which happens to be *dextro*).

In performing an experiment of this kind, the experimenter must be careful that the groups in a series of reactions retain their relative positions in the molecule (see the Walden inversion, discussed in Section 22.17). Reference compounds used extensively for determination of configuration are malic acid, lactic acid, glyceric acid, and tartaric acid (Section 28.7), all of which are believed to be thoroughly established as regards their relationships to the family heads in figure 22.9.

22.10. Compounds with More than One Asymmetric C Atom.
The first asymmetric carbon atom (in most families of compounds)

controls the nomenclature of all of its descendants. In Figure 22.9, the head of the D family has one asymmetric C atom (with —OH to the right of the observer and —CHO at the top). The chain can be lengthened by reactions which introduce more asymmetric C atoms, thus,

$$\overset{\displaystyle H}{\underset{\displaystyle OH}{\underset{5}{CH_2OH}\overset{*}{-}\underset{5}{C}\overset{*}{-}\underset{4}{CHOH}\overset{*}{-}\underset{3}{CHOH}\overset{*}{-}\underset{2}{CHOH}\overset{*}{-}\underset{1}{CHO}}}$$

(* represents asymmetric C atom)

but no matter what the effect of the succeeding asymmetric C atoms may be with respect to direction of rotation, these longer molecules are all D compounds if the original asymmetric C atom (in this illustration, C atom 5) is a D asymmetric atom. This is discussed fully in the study of sugars (Section 34.3), the big family of compounds for which this convention holds. However, among the α-amino acids (Section 33.8) the carbon atom controlling the configuration of the molecule is the α carbon atom, and that is the asymmetric carbon atom with the lowest number.

22.11. Racemic Mixture. When an asymmetric compound is made in the laboratory, the product is usually optically inactive. In the course of its preparation, by the laws of probability, equal amounts of both isomers are formed and these neutralize each other. The product may crystallize either as a mechanical mixture or as a loose double compound. It is known as the DL-compound, which indicates that it is a 50-50 combination of the D- and L- forms. The mixture of the D- and L-forms is also known as a racemic mixture or racemic compound. The name racemic is derived from the Latin word *racemus*, meaning a bunch of grapes. The first racemic substance (D- and L-tartaric acid) was isolated from grapes, and was known originally by the trivial name racemic acid.

The possible varieties of a compound with an asymmetric carbon atom, therefore, are the D and L modifications and the DL mixture. If we refer back to our example of 2-butanol in Figure 22.10, there are these three possibilities: D-2-butanol, L-2-butanol, and DL-2-butanol.

22.12. Meso Compounds. The butanediol in Figure 22.11 is different from the butanol of Figure 22.10 in that it can exist not only as the inactive racemic mixture but also as a simple compound which is optically inactive, known as the *meso* compound.

The new factor to explain is the *meso* modification, which does not affect plane-polarized light. This type of isomer is possible because of the

D(*levo*)2-3-butanediol   L(*dextro*)-2,3-butanediol   *meso*-2,3-butanediol

$$\begin{array}{c} CH_3 \\ | \\ HO-C-H \\ | \\ H-C-OH \\ | \\ CH_3 \end{array} \qquad \begin{array}{c} CH_3 \\ | \\ H-C-OH \\ | \\ HO-C-H \\ | \\ CH_3 \end{array} \qquad \begin{array}{c} CH_3 \\ | \\ H-C-OH \\ | \\ H-C-OH \\ | \\ CH_3 \end{array}$$

22.11. Assignment of 2,3-butanediols to D and L families.

potentially greater symmetry of this molecule than of the others we have studied. The molecule in figure 22.9 shows a lower degree of symmetry because one end is –CHO and the other is –CH₂OH. In Figure 22.11, however, both ends are the same.

If a horizontal plane is passed through the *meso* structure in Figure 22.11, between the second and third carbon atoms, the two halves will be seen to be mirror images. One of these two carbon atoms is *dextro* and the other is *levo*; they effectively neutralize each other, rendering the molecule inert to polarized light. We should stress again the importance of the way in which we observe the molecules. In Figure 22.9 this is simplified, because we always put the CHO at the top when the tetrahedra

22.12. A meso molecule (optically inactive).

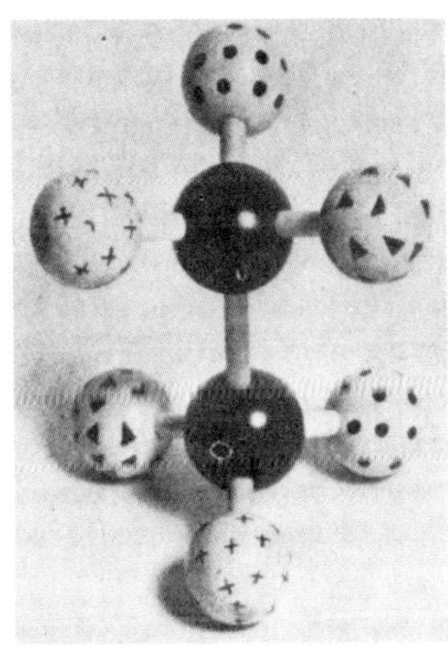

22.13. A D or L molecule (optically active).

are drawn according to rule. In Figure 22.11, however, both ends are the same, so that in order to prove that the D compound contains two asymmetric C atoms that are similar (*levo*) we must look at the molecule from each end. This is explained further in the discussion which follows.

The model shown in Figure 22.12 was built by combining the two carbon atoms of figure 22.3. These two carbon atoms are of opposite rotational effect; therefore, if we link them into the same molecule through their vertical valence bonds (after removing the unmarked balls shown in Figure 22.3) they neutralize each other to yield an optically inactive compound. The fact that the two carbon atoms in Figure 22.12 are of opposite rotational effect can be seen by viewing the molecule from both ends. From the top, the direction is counterclockwise when reading in the order *crosses*, *polka dots*, and *triangles*. From the bottom, however, the direction is clockwise for the same order of markings. It is obvious that the carbon atoms can be rotated into such a position that one-half of the molecule will be the mirror image of the other half.

The reader will be helped in visualizing this phenomenon if he places his hands together, palms touching, and then imagines light passing through his hands by entering through the back of the left hand or through the back of the right hand. It is apparent that after the light has passed through one hand it meets the other hand with opposite directional effects.

The model in Figure 22.12, as we stated before, was built by linking each of the carbon atoms in Figure 22.3; these carbon atoms have *opposite* rotational effects. We can also construct a model using two carbon atoms of the *same* kind, as shown by the model in Figure 22.13. Each of these carbon atoms represents a clockwise directional effect when reading the attached balls in the order *crosses*, *polka dots*, and *triangles* (looking at the molecule from each end). The model, therefore, represents an optically active modification—that is, it is either a D- or L-form.

Tartaric acid (Section 28.7) has a structure exactly equivalent to that in Figure 22.11, with an acid group (—COOH) replacing the —CH$_3$ groups. Some data on the possible isomers of tartaric acid will be given in Section 22.14.

22.13. Externally and Internally Compensated Compounds.

In a *meso* compound, there is no optical activity, and the molecule is said to be *internally* compensated because the neutralizing effect arises from the presence of two opposite kinds of atoms *in the same molecule*. In a *racemic* compound there is likewise no optical activity, but the substance is called *externally* compensated because the neutralization is brought about *by two different molecules*. It is easy to tell the difference between racemic compounds and meso compounds. Since the racemic substance is a mixture of the D- and L-forms it can be separated into its two components by methods explained in Section 22.15. The meso compound, being one single substance, cannot be separated.

22.14 Diastereoisomers. We have now seen there are compounds that are isomeric in that they have the same composition, and also contain asymmetric carbon atoms, but are not mirror images. The example of such isomerism in this chapter is a D-compound and its isomeric but optically inactive *meso* compound. In Section 34.3, when discussing more complex substances, we shall find that compounds may be optically active as well as isomeric, yet not be mirror images. Since they are not mirror images, they do not have the same properties. Such compounds are called *diastereoisomers*, which may be briefly defined as compounds that are isomeric and contain asymmetric carbon atoms, but are not mirror images and therefore have different properties. They are mentioned at this point in order to clarify the subject to be discussed in Section 22.15.

The tartaric acids (Section 28.7) have the same general configuration as the models shown in Figure 22.11, with respect to their asymmetric carbon atoms. The difference in properties between D- or L-tartaric acid and the diastereoisomeric *meso*-tartaric acid is shown by the data in table 22.1.

TABLE 22.1.   A Few Properties of the Tartaric Acids

	Melting Point, °C	*Specific Rotation at 20°C*
L(*dextro*)-Tartaric acid	170	+12°
D((*levo*)-Tartaric acid	170	−12°
DL-Tartaric acid	206	0°
meso-Tartaric acid	140	0°

22.15.   Resolution of Mirror-Image Isomers. This refers to separation of the iscmers. Mirror-image isomers, as we have often remarked, have the same solubility, same melting point, same boiling point, etc. One of the usual means of separating two substances (like sugar and salt) by making use of their different solubilities in water cannot, therefore, be used to separate a D- compound from its L-isomer. There are, however, three well-known methods for separating the D- and L-varieties in a racemic compound, and all of them were originally developed by Pasteur in his famous work on the tartaric acids.

In the first place, if we crystallize the racemic substance, the D- and L-isomers will sometimes separate and crystallize in their respective mirror-image forms.   One shape corresponds to the D-compound; the other is the L-compound (compare the diagrams of the quartz crystals in Figure 22.6).   These crystals can be separated mechanically by picking them out one by one with tweezers, but of course the process is a tedious one.

Secondly, certain molds and bacteria show a preference for either the D- or the L- isomers in a racemic mixture for metabolic use.   The other isomer will be left practically untouched.   The opposite effect is also known; right-hand and left-hand molecules have different activities on certain organisms.   For example, adrenaline (Section 35.5) is used in medicine as a powerful stimulant.   It is optically active, and the *levo* compound is more than ten times as effective as the *dextro* isomer.

Lastly, Pasteur found that D- and L-isomers in a racemic mixture can sometimes be separated by reacting them with another optically active substance.   If we can combine them, for example, with a substance like L-adrenaline, they are no longer mirror images and will have different properties.   The D-form combined with L-adrenaline may have a different solubility from the L-form combined with L-adrenaline, and in this way the two can be separated.

This principle is difficult to make clear to a beginner in the study of optical isomers.   The following outline may help;

D-compound + inactive substance L-compound + inactive substance	Internal nature of the products is not different from that of the original substances. They are still mirror images.
D-compound + L-compound L-compound + L-compound	These products cannot possibly be mirror images. They are diastereoisomers, which can be separated because they have different physical properties.

A popular illustration of this last principle is to imagine two golfers of equal ability, one of them right-handed (D-) and the other left-handed (L-).   If they are both given left-handed (L-) golf clubs they instantly become unequal.

22.16. The Sequence Rule.   In the opening paragraphs of Section 22.9 it was stated that the system used for assigning most optically active compounds to D or L configurational families originated with carbohydrate chemists.   The family assignment in carbohydrates is controlled by the asymmetric carbon atom with the highest number; however, among the α-amino acids of the protein chemists family assignments are governed by the asymmetric carbon atom with the lowest number (Sections 33.8 and 34.3).   To minimize the confusion resulting from this situation, the D, L symbols are sometimes written D_g and L_g or D_s and L_s to show whether the carbohydrate system (based on glyceraldehyde in Figure

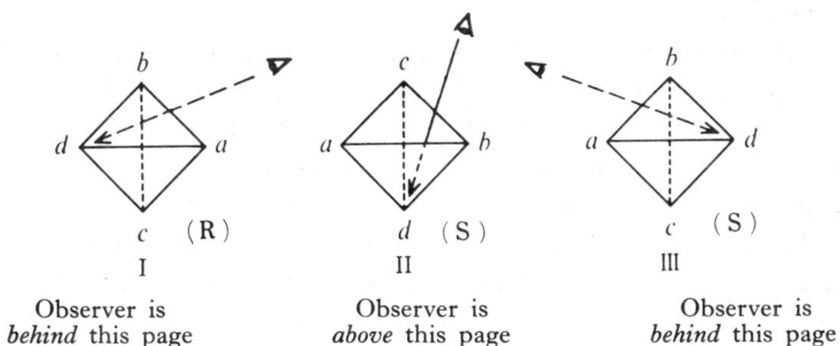

22.14. A convention for viewing an asymmetric carbon atom (a chiral center) when determining its chirality according to the Sequence Rule (see text).

22.9) or the α-amino acid system (based on serine in Figure 33.4) is being followed. In this book all the family assignments are based on the carbohydrate convention.

In the period 1950 to 1960 a new procedure was introduced which does not have as its principal object the assignment of a compound to a right-hand or left-hand family, but instead determines the right-hand or left-hand configuration of *each asymmetric atom* in the compound. This is known as the Sequence Rule, and its basic principles are available in a readily accessible article in the literature.*

To apply this system, the tetrahedral carbon atom, as usual, must be viewed in a standardized way. In the models of Figure 22.14, the letters *a, b, c, d* represent four different groups arranged in descending order a>b>c>d according to a sequence rule which will be explained in a following paragraph. The observer looks at *d* ("low man on the totem pole") through the triangle *a, b, c*. The observer of I is behind the printed page, because the dotted line in our conventional model of an asymmetric carbon atom (Section 22.9) is in the plane of the printed page and the rest of the model is *above* it. The observer now determines if the sequence *a, b, c* in the triangle is to the right or to the left. If the sequence is right-handed the asymmetric carbon atom has the configuration R (Latin, *rectus*), as it is in Model I. Model II is an example where the observer has to look at *d* from above the page in order to look through the triangle *a, b, c*; in this case, the sequence *a, b, c* is to

*R. S. Cahn, "An Introduction to the Sequence Rule," *Journal of Chemical Education* **41**, 116–125 (1964).

the left and the configuration is S (Latin, *sinister*). In III, once more, the observer is back of the page, and the chiral center happens to have an S configuration.

For a practical example, refer to the 2-butanol formulas in Figure 22.10. The Sequence Rule (see below) gives this descending order of precedence for the groups in the *levo* compound: $a=OH$, $b=CH_2CH_3$, $c=CH_3$, $d=H$. We observe the molecule by looking at the H atom through a triangle like that in Figure 22.14, Model I, and note that the sequence OH, CH_2CH_3, CH_3 is clockwise. This asymmetric carbon atom has an R configuration. The compound is (2R)-*levo*-2-butanol, which indicates that the second atom is asymmetric and has an R configuration; the (R) symbol is not a family designation, but simply defines the status at the second carbon atom.

The sequence in which the groups are to be read is based on the atomic number of the atoms *directly linked* to the asymmetric atom. The atomic number is the element number listed on the inside cover of this book. That group has precedence (is observed first) in which the atom linked to the asymmetric carbon atom has the higher atomic number. In the preceding paragraph, the OH group is first (that is, *a*) because the atomic number of oxygen is eight, and the H atom is last with an atomic number of one. The other two groups each have a C atom (atomic number six) linked to the asymmetric carbon atom; in this case, we look further to the next atoms in each group to determine precedence. In the CH_2CH_3 group we pass from C to another C atom, but in the CH_3 group we pass from C only to H atoms; the larger group therefore has the precedence. The preceding paragraph should now be read once more. This is the sequence priority in descending order for all the groups in the compounds used for illustration in this chapter:

$$-OH \quad -\overset{\displaystyle \|}{\underset{\displaystyle O}{C}}-OH \quad -\overset{\displaystyle \|}{\underset{\displaystyle O}{C}}-H \quad H-\overset{\displaystyle |}{\underset{\displaystyle CH_3}{C}}-OH \quad H-\overset{\displaystyle |}{\underset{\displaystyle OH}{C}}-H \quad H-\overset{\displaystyle |}{\underset{\displaystyle CH_3}{C}}-H \quad H-\overset{\displaystyle |}{\underset{\displaystyle H}{C}}-H \quad -H$$

If we return now to the 2-butanols in Figure 22.10, we find that *dextro*-2-butanol when observed by the procedure indicated in Figure 22.14 Model III, has an S configuration, and is (2S)-*dextro*-2-butanol. For another practical example, consider *dextro*-glyceraldehyde in Figure 22.9. The observer looks at H through the opposing triangle and reads the other groups in the priority sequence OH, CHO, CH_2OH and finds it to be clockwise; this molecule has an R asymmetric carbon atom. The priority of CHO over CH_2OH is due to the fact that CHO has a double bond,

—C=O, which means that it can be regarded as —C—O in which the (with H below the C on the left, and O above / H below the C on the right)

oxygen atom (atomic number 8) can be counted twice.

The Sequence Rule was so designed that the D, L and R, S systems would agree for the compounds in stereochemistry that have been guide-posts in the past, such as glyceraldehyde and serine (Section 39.5). For small molecules like those studied in this chapter the chirality of the asymmetric carbon atom (which is all the Sequence Rule sets out to find) is essentially a family assignment for the molecule. For all three compounds used as illustrations in this chapter (Figures 22.9, 22.10, and 22.11) the D *family* configuration under the old D, L system is identical with the R configuration in the new R, S system. However, the tartaric acids are reversed by this new system; the D, L configurations in Section 28.7 are based on glyceraldehyde (the reference in the carbohydrate system) and are the reverse of the configurations which are based on serine (the reference in the amino acid system). The Sequence Rule in this case gives configurations agreeing with those based on serine.

In an earlier paragraph we gave the name (2R)-*levo*-2-butanol to a compound in Figure 22.10. The number in (2R) is hardly necessary in this small molecule. The numbers are needed for larger molecules (see carbohydrates in Section 34.3) to identify the asymmetric carbon atoms (the chiral centers).

The Sequence Rule is regarded as a *general* system for establishing the configuration of a chiral center. It is so planned that it will work for all types of optically active compounds, even those for which other systems are known to fail. However, it has not been proposed to replace with the Sequence Rule the so-called "local" systems that are useful in certain areas, such as the D, L system in carbohydrate chemistry.

22.16a. The E, Z System for Naming Geometric Isomers. The fundamental idea behind the Sequence Rule just described is the basis for a system of nomenclature in stereoisomerism due to a double bond (see Section 22.3). In the new method the *cis, trans* descriptors

$$\underset{\substack{| \quad | \\ H \quad I}}{\overset{\substack{Cl \quad Br \\ | \quad |}}{C=C}} \quad (I) \qquad\qquad \underset{\substack{| \quad | \\ H \quad Br}}{\overset{\substack{Cl \quad I \\ | \quad |}}{C=C}} \quad (II)$$

E-1-Bromo-2-chloro-1-iodoethene   *Z*-1 Bromo-2-chloro-1-iodoethene

22.15. Bimolecular reaction between CH_3Cl and OH^- ion.

are replaced by *Z* from the German *zusammen* meaning together, and *E* from the German *entgegen* which means opposite.

Two rules are applied to determine which descriptor is appropriate. (1) For each of the carbon atoms in the double bond, find out which of the attached groups has a higher priority according to the Sequence Rule explained in Section 22.16. (2) Label the compound *Z* if the two groups with the higher priority are on the same side of the double bond, and *E* if they are on opposite sides.

In examples (I) and (II) the Cl atom, based on atomic number, has the higher priority on its carbon atom, and the I atom has the higher priority on the other carbon atom. In (I) they are on opposite sides of the C=C bond, and it is therefore an *E* molecule, whereas in (II) they are on the same side and therefore *Z*. Each C=C bond in a larger molecule is named by the same procedure. The system leaves no ambiguity. The reader should observe how difficult it would be to name the compounds described in this section when the *cis, trans* system in Section 22.3 is employed; it is an easy system when both groups on the same side are the same.

The substituents in examples (I) and (II) are given alphabetically in the names of the molecules in accordance with *Chemical Abstracts* practice.

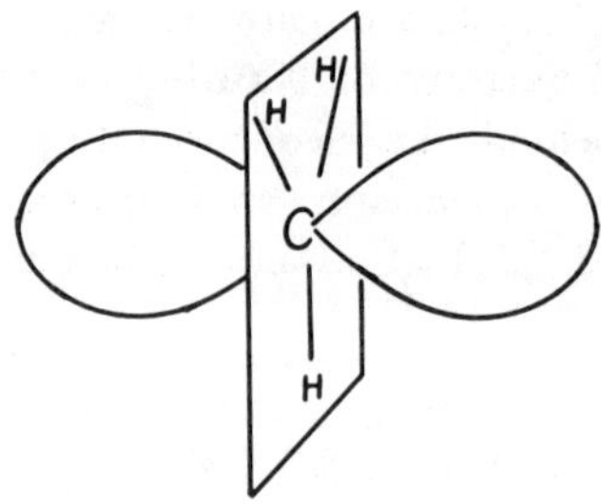

22.15a. The *p* atomic orbital of the carbon atom in the transition state.

22.17.   Stereochemistry, Reaction Mechanisms, and Walden Inversion.   In Section 18.3 it was explained that molecules go through chemical reactions only after they have become activated, and Equation 18.5 shows how reactions in which two molecules take part (bimolecular reactions) pass through a transition phase in which an activated complex is formed.   The subject matter presented in Section 17.14 can be made more understandable by study of Figure 22.15, which postulates a mechanism for the S_N2 reaction between CH_3Cl and the OH^- ion.

The negative OH^- ion will approach the polarized molecule of methyl chloride from the rear.   In the transition state the three atoms (or groups) attached to the carbon atom are in a *plane* perpendicular to the line HO . . . .C. . . .Cl.   The *flat* structure of the CH_3 group in the transition complex is similar to that in figure 15.6*d*, which shows two carbon atoms with *p* atomic orbitals resulting from sp^2 hybridization.   In Figure 22.15 all the carbon valence bonds of the original CH_3Cl molecule are sp^3 hybrids but, to form the intermediate complex, sp^2 hybridization takes place, characteristic of the flat (plane) structure in Figure 15.6.   Each of the two lobes of the *p* orbital in this transition state (see Figure 22.15a) is attacked (one by OH and the other by Cl) and they are on opposite sides of the molecule.   The group that is displaced in the reaction (in this case, the Cl ion) is called the *leaving group*.

In the final stable state there is an *inversion of configuration* of the groups and atoms around the carbon atom;   the phenomenon is exactly like re-

22.16.   Unimolecular reaction between $(CH_3)_3$ CCl and OH^- ion. R represents CH_3.

versal of an umbrella in a wind. In the final state there is once more a tetrahedral arrangement around the central carbon atom, but the product (CH_3OH) has the mirror-image arrangement of that in the reactant (CH_3Cl). If the four groups distributed around the carbon atom were different—that is, if the carbon atom were asymmetric—there would be an inversion of optical activity; a member of a D family would change over to one of an L family, or vice versa. This is known as *Walden inversion.* Inversion of optical activity of this nature was first recognized by Walden about 1895. The observation created wide interest, coupled with mystification, until the explanation was found about 40 years later.

All bimolecular nucleophilic reactions (S_N2) take place with inversion, whether or not a change in optical activity is associated with the reaction. Unimolecular reactions (S_N1) generally lead to racemization; in other words, even if the reactant is optically active, the product will be a mixture (generally in nearly equal amounts) of the mirror-image isomers. To explain this, we shall refer to the hydrolysis of *t*-butyl chloride described in Section 24.8 and draw the structures as shown in Figure 22.16.

The OH^- ion displaces Cl^- in Figure 22.15 because it is more basic (more nucleophilic). The displacement occurred from the rear because room was available for OH^- back there among the small H atoms. In Figure 22.16, however, backside attack does not take place because there is so little room when three bulky groups are attached to the carbon atom. The OH^- thus must wait for the carbon compound to ionize (relatively slow); this is the rate-determining step and the reaction is classed as S_N1, *substitution, nucleophilic, unimolecular,* because the rate depends only on the ionization step and not on concentration of OH^- ion. As soon as the *t*-butyl chloride ionizes to give a positive carbonium ion (the intermediate complex) it becomes planar, with three groups in the same plane as the central carbon atom to which they are attached. The OH^- can attack from both front and rear (refer to Figure 22.15a which shows a *p* atomic orbital on a trivalent carbon atom in a flat structure).

In this transition ion each lobe of the *p* orbital can be engaged by an OH^- ion. If the three groups on the C atom are all different from one another and from the OH group, the product of this reaction will be a racemic mixture of asymmetric isomers. An optically active compound hydrolyzed by this mechanism, in which the first step is ionization, will become optically inactive. Racemization is not necessarily complete; there may be a shielding effect, for example, by the leaving group, which will produce more molecules by the backside attack and that means some excess of an *inversion* product.

From the nature of this discussion it should be obvious why the research chemists engaged in determination of configuration of asymmetric molecules are careful to convert one molecule to another by a reaction in which it is definitely known whether inversion or *retention* of configuration takes place.   In the type of reaction illustrated by Figure 22.16, much depends on the stability of the intermediate carbonium ion, which depends in turn on the nature of the solvent and of other experimental conditions. A few reactions are known in which the entering group can attack the C atom from the front before it reaches the flat shape of the carbonium ion and before the leaving group has completed its departure.   Under these conditions the reaction takes place with retention of configuration.

We may now turn to the consideration of stereochemistry as it applies to certain of the reactions at the olefin double bond.   If the ethene molecule is regarded in the old conventional manner (Figure 15.1), addition of a molecule of bromine would be expected to take place in a single step by opening one of the bonds to yield *cis*-ethene dibromide.   However, kinetics of the reaction show that it takes place in several steps, as explained in Figure 17.4(a).   A cyclic bromonium ion is formed, which then suffers a backside attack by a second molecule of bromine.   The product is *trans* rather than *cis*, but rotation can occur freely about the single C–C bond so that the product has no isomers that can be isolated.

Part 3

THE CLASSIFICATION OF CARBON COMPOUNDS

There is an almost endless number of compounds consisting of carbon and hydrogen, combined with just a few other elements. Fortunately, as explained in Part 2 of this book, the carbon compounds can be grouped into a relatively few distinct families, depending on their structures, and this simplifies our study to an enormous degree. When an organic chemist glances at the following two compounds

he classifies them immediately among the *chain* and *ring* compounds, respectively. This helps him to compare the properties of these compounds with the properties of other compounds of similar structure.

We shall find, however, that the organic chemist in some cases is not as much interested in the shapes of these molecules as he is in the fact that they both contain the halogen atom, chlorine. He may therefore classify both these compounds in a certain family which will be called the "halogen compounds." Other methods of classification sometimes employed are outlined in the next chapter, and constitute the principal problem in Part 3 of this text.

23

The Common Methods of Classification in Organic Chemistry

23.1. Classification According to General Structure. In our study of the architecture of carbon compounds we have emphasized two broad types of structures, one of which we call chain compounds and the other, ring compounds. Before proceeding to other methods of classification we should recall that the cyclic compounds may be of two types:

Type I   Type II
OPEN-CHAIN   CYCLIC

Aliphatic Compounds   (a) *Aromatic Compounds*   (b) *Alicyclic Compounds*
(Section 10.3)   (Section 19.1)   (Section 11.1)

$$CH_3-CH_2-CH_2-CH_3$$

These examples illustrate the possible arrangements of the carbon atoms in organic compounds, as far as the general architecture is concerned. For a classification of this kind it is only necessary to consider general outlines.

Aliphatic. The word aliphatic is derived from *aleiphatos*, which means fat in Greek. The name has been applied to the chain compounds because some of them were originally obtained from fats such as butter and lard. To the modern chemist the word aliphatic simply means a string of carbon atoms in straight-chain or branched-chain formation. It does not mean that the compound has any "fatty" characteristics.

Aromatic. This term is applied to all cyclic compounds containing the benzene ring, or rings with similar properties (Section 32.3). Most of

237

the benzene derivatives isolated in the early days of organic chemistry had a pleasant odor and for a while it was thought that an aromatic odor was a characteristic of all benzene compounds. However, not all benzene derivatives have an aromatic odor. Some have practically no odor, and others have very bad odors. Aromatic compounds are also known as *arenes*.

Alicyclic. In Chapter 11 we described certain ring compounds that strongly resemble the open-chain compounds in properties. Because of their similarity to chain compounds they are called closed-chain derivatives. They are also known as alicyclic (aliphatic-cyclic).

23.2. Classification According to Homologous Series. Up to this point, most of our time has been devoted to the study of chains and rings that can be made from carbon atoms and hydrogen atoms. The families of compounds that we have studied in detail are the following, and it will be recalled that they are all hydrocarbons:

Name of Family	*General Formula*	*Typical Member*
Paraffin hydrocarbons	C_nH_{2n+2}	CH_3—CH_2—CH_3
Cycloparaffin hydrocarbons	C_nH_{2n}	CH_2——CH_2 with CH_2
Olefin hydrocarbons	C_nH_{2n}	CH_3—CH=CH_2
Acetylene hydrocarbons	C_nH_{2n-2}	CH_3—C≡CH
Benzene hydrocarbons	C_nH_{2n-6}	(benzene ring)—CH_3

Each of these families of compounds is known as a homologous series, for reasons explained in Section 10.6. In every homologous series, each member differs from the next succeeding member by CH_2, as in the paraffin family, CH_4, CH_3–CH_3, CH_3–CH_2–CH_3, CH_3–CH_2–CH_2–CH_3, etc.

All the members of any one homologous series have the same general characteristics, the differences being only that of degree. As the molecules in a series grow larger the solubility may increase or decrease, and the reactivity may increase or decrease, but the essential nature of all the compounds in the series is the same. The principle of homology has proved useful as a method of classification in organic chemistry, as we have already seen. It allows us to arrange a multitude of compounds into just a few families or series.

23.3. Classification According to Composition. From now on we must use still another common method of classifying organic compounds. We are now to begin the study of interesting classes of compounds that owe their characteristic properties to the fact that in addition to carbon and hydrogen they also contain elements such as oxygen, sulphur, nitrogen, and others.

These compounds will be classified in general according to the most important elements or groups present in the molecule—that is, the classification is based on the *composition* of the molecule. For example, compounds containing only C and H atoms are known as *hydrocarbons*. If the compound also contains a Cl atom this element may stand out as the most important one in the molecule, and we may therefore refer to it as a *halogen* compound. As we have already found (Section 9.4), chain compounds containing an –OH group are called *alcohols*, and we shall learn presently that when the –OH group is attached to an aromatic ring structure the compound is called a *phenol*.

Hydrocabons	*Halogen Compounds*	*Alcohols and Phenols*
CH_4	$CH_3—Cl$	$CH_3—OH$
$CH_3—CH_3$	$CH_3—CH_2—Br$	$CH_3—CH_2—OH$
$CH_2=CH_2$	$CH_2=CH—Cl$	
⬡$—CH_3$	⬡$—Cl$	⬡$—OH$

The names of many of the important classes of compounds in organic chemistry can be found by looking at the chapter headings in the rest of this book. When a compound contains two or more characteristic elements or groups it is called a mixed compound. The substance $CH_2=CH–CH_2Cl$ is a mixed compound, for it is both an olefin and a halogen compound.

23.4. Alkyl Radicals. Two different ways of looking at the same molecule are illustrated by the formulas that can be used for chloroethane:

$$\text{(A)} \quad H-\overset{\overset{\displaystyle H}{|}}{\underset{\underset{\displaystyle H}{|}}{C}}-\overset{\overset{\displaystyle H}{|}}{\underset{\underset{\displaystyle H}{|}}{C}}-Cl \qquad\qquad \text{(B)} \quad C_2H_5—Cl$$

Chloroethane, or ethyl chloride

Formula (A) focuses attention on the structure of the molecule; the Cl atom is shown as replacing the H atom in a member of the *paraffin family of hydrocarbons*.

Formula (B) directs our attention to the fact that a reactive Cl atom is in the molecule; the molecule is more clearly designated as a member of the *halogen class of compounds*. Accordingly, we give the name *ethyl* to the radical, $–C_2H_5$; and the substance C_2H_5 Cl, ethyl chloride, is then regarded as consisting of two parts just as sodium chloride, Na–Cl, is always thought of as composed of two units. It is evident that the *ethyl* radical is nothing more than the paraffin hydrocarbon $CH_3–CH_3$ with one

H atom removed.

The radicals like $-C_2H_5$ that are paraffin hydrocarbons with one H atom missing are known as alkyl radicals.

	Paraffin Hydrocarbons		*Alkyl Radicals*	
CH_4		Methane	CH_3-	Methyl
CH_3-CH_3		Ethane	C_2H_5-	Ethyl
$CH_3-CH_2-CH_3$		Propane	C_3H_7-	Propyl
$CH_3-CH_2-CH_2-CH_3$		Butane	C_4H_9-	Butyl

The alkyl radicals are named by replacing the ending -ane by -yl in the name of the hydrocarbon. They have a valence of one, being found combined always with one atom of a univalent element like chlorine, or with one univalent radical like $-OH$.

To represent *any* alkyl radical the chemist uses the symbol R. Thus, $R-Cl$ is an alkyl chloride, and stands for any of the compounds CH_3-Cl, methyl chloride, C_2H_5-Cl, ethyl chloride, etc. In the same way, $R-OH$ stands for all alcohols such as CH_3-OH, methyl alcohol, and $R-CN$ is any alkyl cyanide such as CH_3-CN, methyl cyanide.

The use of R to represent an alkyl radical is quite common in chemical literature. The student should be warned, however, that it is also used as a symbol for other radicals. For example, if the author of an article wishes to designate a series of compounds such as CH_3-CH_2-Cl, $CH_3-CH_2-CH_2-Cl$, and $CH_2=CH-CH_2-Cl$ by $R-Cl$, that is his privilege. (The unsaturated compound in this list is not an alkyl chloride.)

23.5. Alkanes, Alkenes, and Alkynes. If we call R an alkyl radical, and define it as a paraffin hydrocarbon minus one H atom, then a paraffin hydrocarbon itself is $R-H$. On this basis, the *paraffin* hydrocarbons are also called *alkanes*.

Whenever someone develops a new family name, like alkanes, for the paraffin hydrocarbons, we can always work out corresponding names for the olefins and acetylenes. The following table explains this plan.

Hydrocarbon Families

Common Name	*Systematic Name*	*General Formula*	*Typical Member*	
Paraffin	Alkane	C_nH_{2n+2}	CH_3-CH_3	Ethane
Olefin	Alkene	C_nH_{2n}	$CH_2=CH_2$	Ethene
Acetylene	Alkyne	C_nH_{2n-2}	$CH\equiv CH$	Ethyne

23.6. Aryl Radicals. The aryl radicals bear the same relation to the aromatic cyclic compounds as the alkyl radicals do to the aliphatic chain compounds.

Aromatic Hydrocarbons
C_6H_6
Benzene
(Phene)

$-CH_3$ C_6H_5—CH_3
Toluene

Aryl Radicals
C_6H_5—
Phenyl

$-CH_2$— C_6H_5—CH_2—
Benzyl

$-CH_3$ $-C_6H_4$—CH_3
Tolyl

The aryl radicals are the aromatic hydrocarbons with one H atom missing. They have a valence of one, just like the alkyl radicals.

To represent *any* aryl radical, the chemist may use the symbol Ar. Thus, Ar-Cl is any aryl chloride and stands for such compounds as

$-Cl$ $-CH_2Cl$ $-CH_3$ $-Cl$

Phenyl chloride   Benzyl chloride   Tolyl chloride

The student should notice particularly that the radical corresponding to benzene itself is not called benzyl, but is called phenyl, for the reason given in Section 19.1. Just like the chain compounds, aromatic hydrocarbons can be regarded as Ar–H. When a compound consists of both a nucleus and a side chain, like toluene, it can be written Ar–R:

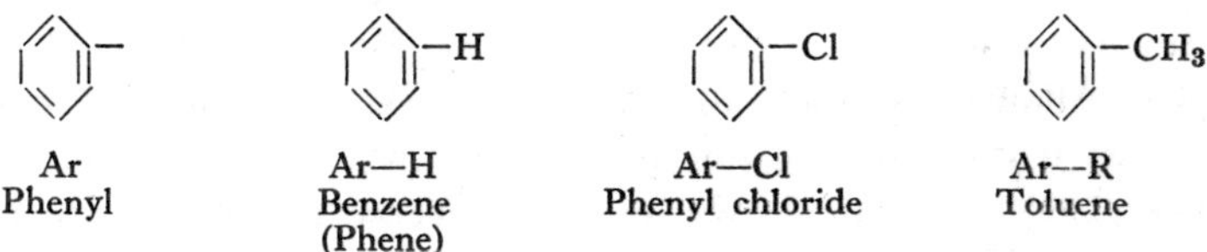

Ar   Ar—H   Ar—Cl   Ar--R
Phenyl   Benzene   Phenyl chloride   Toluene
          (Phene)

23.7.   Free Radicals and Ions.   To this summary of the radicals appearing frequently in carbon compounds we should add a brief review of the *free radicals* and *ions* that were introduced earlier in this book.   It is often said that the chemistry of chain compounds (aliphatic chemistry) consists mostly of the free-radical reactions described in Sections 17.8 and 17.9, whereas the chemistry of benzene compounds (aromatic chemistry) is generally ionic as described in Chapter 20.   Free-radical reactions are really common to both fields of chemistry, and ionic mechanisms are likewise common to both; however, the generalization is worth keeping in mind even though it is not at all quantitative.

The student should review Figure 17.7, which shows the structures of the free methyl radical, the methyl carbonium ion, and the methyl carbanion. To these we add below the corresponding structures derived from benzene:

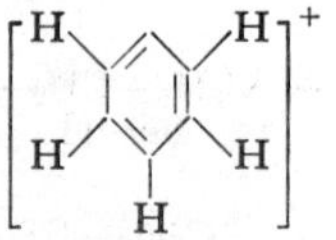

Phenyl cation
Phenyl carbonium ion

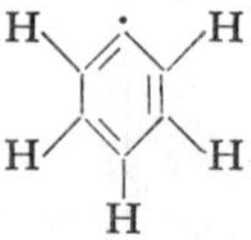

Free phenyl radical

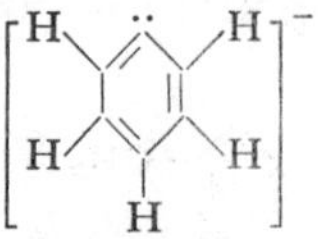

Phenyl anion
Phenyl carbanion

It should be observed that the phenyl cation is not the same thing as the benzenonium ion, which is an intermediate in the substitution reactions described in Chapter 20. The *onium* ion nomenclature is sometimes confusing. The classical example with a long history, and probably the original, is ammonium ion. Some onium ions are illustrated here, and

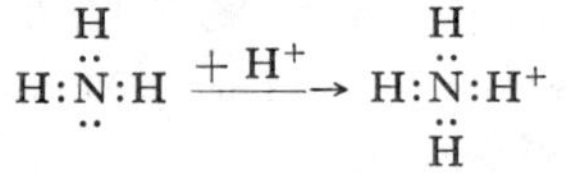

Ammonium ion

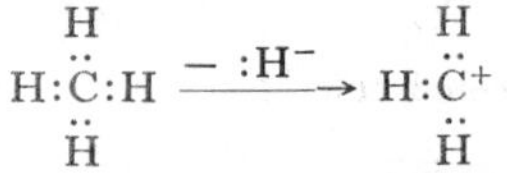

Methyl carbonium ion

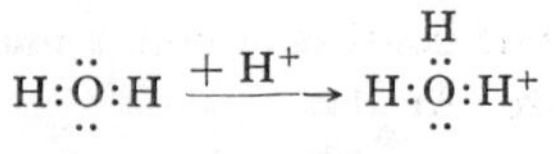

Hydronium ion

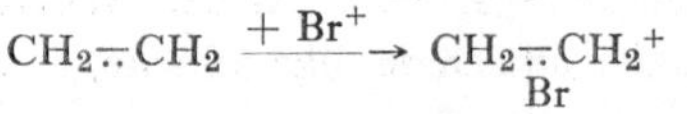

A bromonium ion

others will be found in the index. Structurally, the only thing these onium ions have in common is the positive charge.

The two phenyl ions do not appear frequently in chemical reactions. The free phenyl radical takes part in the addition reaction of chlorine to the benzene ring mentioned in Sections 19.2 and 24.6.

23.8. Alkylation. This is one of the most frequently used terms in the synthesis of organic compounds. It is understandable, because synthesis is a building-up process and introduction of an alkyl group into another structure is a direct and quick way to make a bigger compound with more carbon atoms per molecule. The Friedel-Crafts reaction in Section 20.6 is an alkylation, since it is introduction of an alkyl radical into a ring structure; it can equally be called an *arylation*, since it is introduction of a ring radical into a chain compound. The petrochemicals industry makes extensive use of alkylation processes; in Section 17.12 we described a highly interesting, though complex, synthesis involving rearrangement of carbonium ions.

24

Halogen Compounds and Free Radicals

24.1. Introduction. The general nature of the compounds to be described in this chapter is shown by the following examples, in which X stands for any of the halogens–fluorine, chlorine, bromine, and iodine. We shall consider the chain compounds first.

$$CH_3-CH_2-CH_2-X \qquad \langle\!\!\!\bigcirc\!\!\!\rangle-CH_2-X \qquad \langle\!\!\!\bigcirc\!\!\!\rangle{-CH_3 \atop -X}$$

An aliphatic halogen compound. Aromatic halogen compounds. Halogen in side chain. Halogen in nucleus.

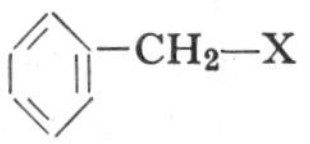

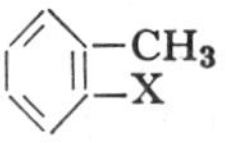

24.2. Alkyl Halides. The simplest halogen compounds are those in which one X atom replaces one H atom in a chain compound. Examples we have met quite often are methyl chloride, CH_3–Cl, and ethyl chloride, CH_3–CH_2–Cl. These compounds are known as alkyl halides; they may be represented by the general formula R–X, in which R is the symbol of the alkyl radical and X is the symbol of the halogen atom.

Because of the reactivity of the halogen atom in the alkyl halides, these compounds are of great importance in synthetic organic chemistry. They react very easily with metals and metallic compounds, as may be seen in the following:

1. $\quad CH_3-CH_2-Cl + 2Na + Cl-CH_3 \longrightarrow CH_3-CH_2-CH_3 + 2NaCl$

Propane

This, it will be recalled, is the Wurtz synthesis (Section 10.3) for building up larger molecules.

243

2. CH_3—CH_2—Cl + KOH/water $\longrightarrow$ CH_3—CH_2—OH + KCl

Ethyl alcohol

This is a method for preparing alcohols (Section 9.4).

3. CH_3—CH_2—CH_2—Cl + KOH/alcohol $\longrightarrow$ CH_3—CH=CH_2 + KCl + H_2O

Chloropropane   Propene

This is a method for preparing compounds in the olefin series (Section 13.2).

4.   ⟨ ⟩ + C_2H_5Br $\xrightarrow{\text{AlCl}_3}$ ⟨ ⟩—C_2H_5 + HBr

Benzene   Ethylbenzene

This is the Friedel-Crafts synthesis by which many of the benzene hydrocarbons can be made (Section 20.6).

In addition to these, a highly important reaction will be described in Section 30.15 in connection with the preparation of the Grignard reagents.

There is a considerable variation in the rate at which alkyl halides will react with other substances, depending on the chain length and on the halogen atom in the molecule. Thus, methyl halides (CH_3X) react much faster than the homologs with longer chains of carbon atoms and iodides react faster than other halides. Most organic chemists prefer to work in the laboratory with bromides or iodides because of their greater reactivity, if their cost is not prohibitive. Fluorine compounds will be considered separately in Section 24.7.

In Section 10.5 we defined primary, secondary, and tertiary structures among the hydrocarbons, and stated that ease of chlorination to alkyl halides is in the order $3° > 2° > 1°$. Similarly, the reactivity of alkyl halides depends on what is attached to the C atom of the C-X group. In CH_3-Cl, which is 1°, only H atoms are attached, and hydrolysis is mainly by a bimolecular mechanism; the 3° structure, $(CH_3)_3C$-Cl, with three CH_3 groups attached, hydrolyzes mainly by a unimolecular process (Section 22.17). The order of reactivity for the **bimolecular** hydrolysis is generally $1° > 2° > 3°$, but for the unimolecular **hydrolysis** it is generally $3° > 2° > 1°$.

An elimination reaction involving **an alkyl** halide was illustrated in Section 17.15. That was an E2 type of **reaction**, and is illustrated once again in Reaction 3 above. The same **alkyl halide**, under the conditions of Reaction 2 above, would go through the S_N2 displacement as described in Section 22.17. Tertiary and secondary alkyl halides hydrolyze by an

S_N1 mechanism as shown in Section 22.17, but at the same time will decompose by an E1 elimination reaction to give an olefin by-product:

$$CH_3-\underset{\underset{Cl}{|}}{\overset{\overset{CH_3}{|}}{C}}-CH_3 \xrightarrow{\text{alcohol/water}} \left[CH_3-\overset{\overset{CH_3}{|}}{C}-CH_3\right]^+ + Cl^-$$

t-Butyl chloride

$$CH_3-\overset{\overset{CH_3}{|}}{C}=CH_2 + H_2O\ H^+$$

Isobutene
(See I in Figure 17.9)

The ionization of *t*-butyl chloride to a carbonium ion is depicted in Figure 22.16; this is the rate-controlling step and makes it a first order reaction. The subsequent reaction with solvent as shown in Figure 22.16 is the major event, but some elimination takes place as shown here to give as products isobutene and hydrogen chloride.

The student who is making his first acquaintance with organic chemistry by reading this book will soon become aware of an intricate, closely woven relationship existing among the various classes of carbon compounds. This has just been indicated by the equations which show how the halogen compounds can be used to prepare both alcohols and olefins, as well as higher paraffin hydrocarbons.

In order to explain the relationships among the various classes of compounds we are not going to be practical in the sense that the equations and reactions we shall write can always be translated into processes useful in chemical industries. Our chief object is to show how the classes of compounds are related to each other, even though the relations described may represent difficult procedures in the laboratory.

24.3. Preparation of Alkyl Halides. Now that we have seen how useful alkyl halides can be, we should also find out how the alkyl halides themselves can be made. Instead of listing a series of reactions showing how to make the halogen compounds, we shall refer the student to the other places in this book where their preparation is described. This will make the relation of the halogen compounds to the other classes of compounds more evident.

In Section 13.11 we found that H–X (in other words HCl, HBr, etc.) can be added to the double bond, C=C, in an olefin hydrocarbon. Some halogen compounds can be made on a large scale by this method because the olefins are a by-product of gasoline refining, and are therefore very cheap.

In Section 25.4 the preparation of halogen compounds from alcohols by replacement of the –OH group by a halogen atom will be described. This is also a favorite method of preparation, because many alcohols can be obtained in large quantities from natural substances.

Free radical chlorination of methane was described in detail in Section 17.9, where it was said not to be a preferred way to make pure chloromethane. However, this type of reaction is used commercially for making many halogenated hydrocarbons; modern distillation procedures yield products pure enough for most purposes.

24.4. Aryl Halides. Among the aromatic compounds, as we have already indicated in the introduction to this chapter, there are two general types of halogen derivatives—those in which the halogen atom is in the nucleus, and those in which the halogen atom is in the side chain.

ortho-Chlorotoluene
1-Chloro-2-methylbenzene

Benzyl chloride
Chloromethylbenzene

Both of these compounds can be represented by the general formula Ar–X; they consist of the aryl radical and the halide radical.

The position of the halogen atom in the compound is of great importance. The chlorine atom in the nucleus is very stable, so stable in fact that chlorobenzene, ⟨ ⟩–Cl, does not react with KOH except under pressure and at high temperature, whereas we have observed that a boiling solution of KOH converts C_2H_5–Cl into C_2H_5–OH quite easily.

The side-chain chlorine atom is even more reactive than when the benzene ring is not present. Water alone will hydrolyze benzyl chloride. The student should review what was said in Section 17.3 about the *enhanced* activity of Cl in allyl chloride. In the structure of benzyl chloride there is the same arrangement as in allyl chloride—namely, a–CH_2–group between a double bond and a halogen atom. The benzyl ion therefore has the same resonance possibilities as the allyl ion, as shown in Figure 24.1, and the benzyl chloride molecule has a reactive Cl atom for the same reason allyl chloride does.

At the same time the enhanced activity of Cl in allyl chloride was discussed in Section 17.3, we also presented an explanation for the reduced activity of Cl in vinyl chloride. Chlorobenzene and vinyl chloride are alike

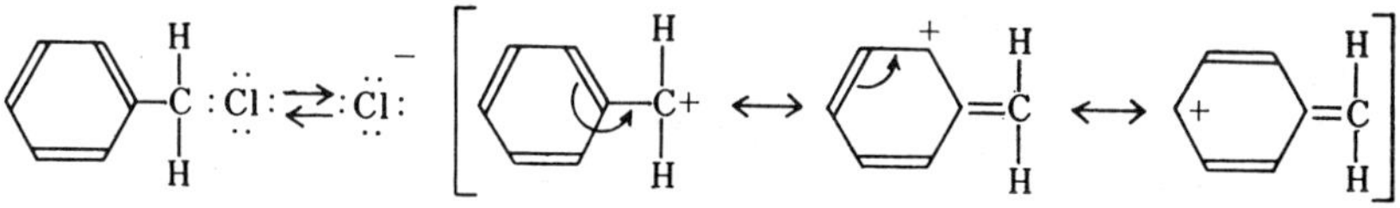

24.1.   Stabilization of the benzyl ion by resonance (compare Fig. 17.2).

in that in both molecules the unshared electron pairs on the Cl atom are conjugated with a double bond.   Resonance structures are therefore possible, as in Figure 24.2, which should render the compounds more stable than one would expect from their normal formulas.   These resonance forms are obtained by flip-flopping the real and potential valence bonds as described in section 15.7.   The effect of resonance is to give the C–Cl bond some double bond character—that is, shorten and stabilize it. In both instances, however, it is also believed that most of the stabilization may be due to sp^2 hybridization at the $>$C—Cl bond, as described in Section 16.12.

The Cl atom in the ring, although relatively inactive, retains some activity toward metals.   Attention has already been called to the Fittig synthesis (Section 20.7) by which the benzene ring can be linked to other radicals, and by the same method two rings can be joined:

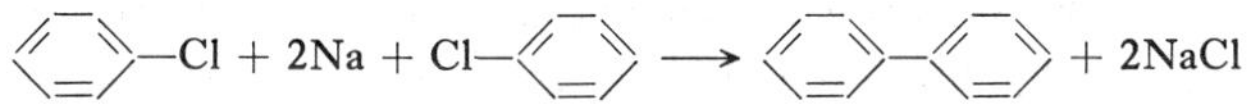

Phenyl chloride
Chlorobenzene

Biphenyl (Diphenyl)
Xenene

Biphenyl is a crystalline solid (it melts at 71°C and boils at 254°C) and is more properly known as *xenene*, but this name is seldom used.   It has many properties of considerable interest, of which we shall mention

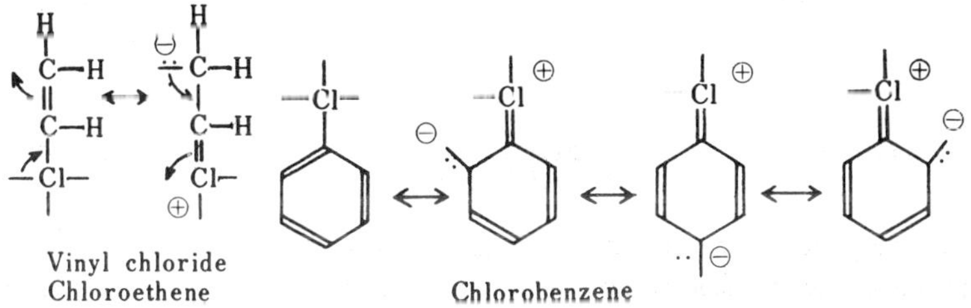

Vinyl chloride
Chloroethene

Chlorobenzene

24.2.   Resonance in vinyl chloride and chlorobenzene.

an important one in Section 28.10. Its thermal properties are favorable for its use in certain types of "steam" engines in which it replaces the steam.

24.5. Preparation of Aryl Halides. The experimental conditions for substitution in a chain are different from the conditions necessary for substitution in the ring. The mechanism of halogenation of an open-chain compound has already been described (Section 17.9) and need not be repeated. It will be recalled that it is a photochemical reaction with a *free-radical* mechanism, that takes place best at high temperature in strong light. Halogenation of the benzene ring (Section 20.2) is an *ionic* reaction which requires catalysts known as halogen carriers. The carrier described in the earlier chapter was $AlCl_3$, but iron filings is more generally used (the catalyst in that case is an iron chloride). Sometimes it is best to halogenate ring structures in the absence of light. Substitution of H

$$\text{C}_6\text{H}_5\text{-H} + \text{Cl-Cl} \xrightarrow[\text{halogen carrier}]{\text{low temperature}} \text{C}_6\text{H}_5\text{-Cl} + \text{HCl}$$

Phenyl chloride
Chlorobenzene

atoms in the ring by iodine is generally more difficult than with the other halogens, but when the ring contains certain radicals, substitution even by iodine takes place rapidly, as will be shown in Section 25.12.

When we wish to introduce a halogen atom into an aromatic compound that contains both a nucleus and a side chain, we must use one of the two sets of experimental conditions which have just been outlined in order to direct the halogen atom into the correct position. If toluene is chlorinated in the dark and at low temperature, the Cl atom goes into the nucleus, but if chlorine is passed into toluene in sunlight, the following compounds are formed:

CH_2Cl $CHCl_2$ CCl_3

Benzyl chloride	Benzal chloride Benzylidene chloride	Benzo trichloride

These products are mentioned here because they illustrate the use of the names *benzyl*, *benzal*, and *benzo* for three radicals that occur often in other classes of compounds (see *benzyl* alcohol, *benzal*dehyde, and *benzo*ic acid in the index).

24.6. Addition of Halogen to Benzene. The benzene ring will reluctantly undergo certain addition reactions. Chlorine can be *added* under conditions that facilitate side-chain *substitution* if side chains happen to be present in the ring. The reaction takes place by a free-radical mechanism, as already mentioned in Section 19.2, and the ultimate product is benzene hexachloride.

24.7. Fluorine Compounds. Fluorine is the most electronegative of the elements and chemically the most active (see Table 8.2). Its activity is so great that comparatively little was done with it in the laboratory until the urgency of the development of fluorine compounds required for the "atomic" bomb in World War II brought about a thorough study of how to tame it.

An important characteristic is the high heat of reaction of fluorine with hydrocarbons. In a substitution, for example, the heat (104 kcal) developed per mole of product is higher than that required to break the C–C bond (see Table 16.1).

$$-\overset{|}{\underset{|}{C}}-\overset{|}{\underset{|}{C}}-H + F-F \longrightarrow -\overset{|}{\underset{|}{C}}-\overset{|}{\underset{|}{C}}-F + HF$$

Fluorinations with F_2 are therefore accompanied by decomposition of the product unless the reaction is carefully controlled by such means as dilution with an inert gas to dissipate the heat of reaction. A common practice is to fluorinate with salts such as CoF_3 or AgF_2. The reaction takes place as follows:

$$-\overset{|}{\underset{|}{C}}-\overset{|}{\underset{|}{C}}-H + 2CoF_3 \longrightarrow -\overset{|}{\underset{|}{C}}-\overset{|}{\underset{|}{C}}-F + 2CoF_2 + HF$$

The salt can be regenerated by treatment with excess fluorine.

Fluorinations such as those described generally tend to go to completion—that is, substitution of all the H atoms usually takes place. Compounds containing a restricted number of F atoms are often made by special methods; one such method is an *exchange reaction* with a corresponding chlorine compound. Thus, to make fluoroethane, CH_3CH_2F, it is convenient to treat chloroethane, CH_3CH_2Cl, with potassium fluoride in a suitable solvent. Another example is the following:

$$3Cl-\overset{\overset{\textstyle Cl}{|}}{\underset{\underset{\textstyle Cl}{|}}{C}}-Cl + 2SbF_3 \longrightarrow 3Cl-\overset{\overset{\textstyle F}{|}}{\underset{\underset{\textstyle F}{|}}{C}}-Cl + 2SbCl_3$$

Carbon tetrachloride   Freon-12

Tetrachloromethane   Dichlorodifluoromethane

In this reaction, a small amount of antimony pentafluoride, SbF_5, is used as a catalyst. The product is one of the Freons which is used as a refrigerant (Section 9.1).

Fluorine compounds are the most stable of the alkyl halides, the order of stability of which is $RF > RCl > RBr > RI$. For example, methyl chloride (CH_3Cl) can be converted slowly to the corresponding alcohol by a concentrated alkali solution, but the reaction does not take place with methyl fluoride (CH_3F).

A remarkably stable compound is tetrafluoroethene, $CF_2{=}CF_2$, which can be polymerized to a plastic (Section 38.9) marketed under the name of Teflon. This is so stable that it is used to make gaskets for engines, bottle-cap linings inert to nearly all chemical reagents, and linings for pots and pans in the kitchen.

24.8. Free Radicals that are Really Free. The student will find it of interest to compare the *t*-butyl chloride discussed in Section 22.17 with triphenylmethyl chloride, which is described here.

$$\underset{\underset{\displaystyle CH_3}{|}}{\overset{\overset{\displaystyle CH_3}{|}}{CH_3{-}C{-}Cl}} \qquad\qquad \underset{\underset{\displaystyle C_6H_5}{|}}{\overset{\overset{\displaystyle C_6H_5}{|}}{C_6H_5{-}C{-}Cl}}$$

t-Butyl chloride   Trityl chloride
Trimethylmethyl chloride   Triphenylmethyl chloride

It was explained that *t*-butyl chloride is readily hydrolyzed in basic solution (by OH^- ions) but that there is not room enough at the back of the molecule for the OH^- ion to displace the Cl^- from the rear; it waits for the molecule to divest itself of the Cl^- by the ionization process before attaching itself to carbon.

The Triphenylmethyl Ion. With triphenylmethyl chloride we have another steric (space) effect; this molecule is so crowded it hydrolyzes to the corresponding alcohol simply when standing in water solution. This high reactivity of the Cl atom is due not only to a steric effect but also to resonance possibilities in the positive (carbonium) ion that is formed when it ionizes (see Figure 24.3). This action is like that described for allyl chlo-

24.3. Resonance structures of the triphenylmethyl carbonium ion.

ride (Section 17.3), but the resonance effects that stabilize the ion are much greater in this benzene compound.

The Triphenylmethyl Radical. Triphenylmethyl chloride can be readily converted in a benzene solution containing powdered zinc to a compound the development of which was a milestone in the history of chemistry. This is hexaphenylethane, which splits in solution to a free radical, triphenylmethyl, the first *stable* free radical discovered (the year, 1900). At equilibrium the benzene solution contains a few per cent of the radical. The split in hexaphenylethane is due partly to steric strain effects and partly to resonance stabilization of the free radical. The odd electron in the free radical can distribute itself over many positions in the molecule, as illustrated in figure 24.4.

Hexaphenylethane

Triphenylmethyl free radical
(note the odd electron)

24.9. Resonance with an Odd Electron.

In order to obtain a picture of the extent of resonance possible in triphenylmethyl, the reader should observe that there are two resonance possibilities for each ring—the Kekulé stuctures A and B (Section 19.3).   The odd electron can enter into resonance possibilities with each of these, because of conjugation (Section 15.7) with the double bonds, to give Structures I, II, and III. Since there are three rings in each structure, the possibilities can be multiplied by three, the odd electron being found in nine positions *in addition* to that on the central carbon atom. The odd electron in these structures appears at the *o-* and *p-* positions. The student should work out the structures

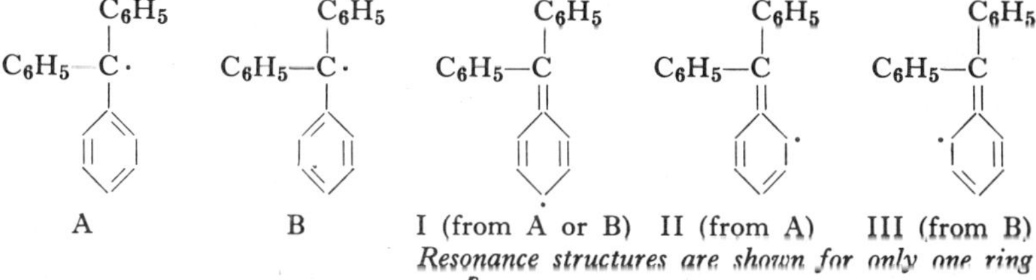

A   B   I (from A or B)   II (from A)   III (from B)
Resonance structures are shown for only one ring

24.4. Some resonance structures of triphenylmethyl free radical.

on paper, using a system illustrated by the formulation of Structure I, as follows, keeping in mind that each straight-line bond represents a pair of electrons:

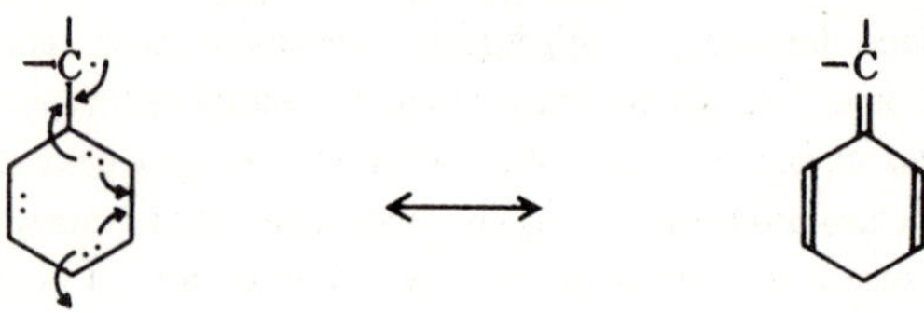

It will be seen that a conjugate system of double and single bonds is necessary for such an interaction of electrons. The electron pairs shown in this example are the π electrons described in Section 19.4.

A study of the possible positions in the molecule of the odd electron, together with the various possible combinations of Kekulé resonance forms for the benzene ring, leads to the conclusion that a total of 44 resonance structures can be written for the triphenylmethyl radical. This represents a considerable resonance energy, and accounts for the radical's fair stability. Even more stable radicals of this type have been made. If instead of the *phenyl* group in hexaphenylethane one uses the *xenyl* group (see the xenene formula in Section 24.4), the resulting hexaxenylethane in dilute solution in benzene dissociates *completely* to the trixenylmethyl radical, whereas under the same condition the triphenylmethyl radical is present to the extent of only a few per cent. (The free methyl radical, $\cdot CH_3$, ordinarily has a life of the order of 0.01 second. It unites with another methyl radical to form ethane, $CH_3 : CH_3$, if there is nothing else with which to react.)

24.10. Other Odd Electron Structures. In Chapter 9, which introduced Part 2 of this book, much emphasis was placed on the constancy of valence of carbon, which was said to have a valence of four in nearly all of its compounds. We have discussed some of the exceptions, but it is obvious that the exceptions indicate even more strongly the normal tendency of the carbon atom toward quadrivalence.

It is of interest now to mention briefly a group of compounds with a *diradical* type of structure. These are structurally related to oxygen, which has the electron configuration $:\ddot{O}:\ddot{O}:$ although it may be found more generally written $:\ddot{O}::\ddot{O}:$ or $O{=}O$. A fundamental assumption about the duet of electrons making a chemical valence bond is that the two electrons have opposite spins. Such electrons are said to be *paired*. A molecule in which all the electrons are paired is *diamagnetic*. The oxygen molecule contains two electrons that are not paired; they spin in the same direction and the molecule is *paramagnetic*. These terms refer to the *magnetic susceptibility* of the molecule. If the magnetization induced in a body has a value I when placed in a field in which the magnetizing force is H, the magnetic susceptibility is the ratio I/H. If the ratio is greater than one the molecule is said to be paramagnetic, whereas if the ratio is less than one it is diamagnetic. (A paramagnetic substance is one we commonly refer to as magnetic.)

Paramagnetism is due to an excess of electrons spinning in one direction. Most molecules have paired electrons, but are feebly diamagnetic because of the characteristic behavior of electrons, rotating in their orbits, called precession. Since paramagnetism is a property characteristic of an unpaired electron, it is a useful test for odd electron structures and consequently for free radicals. The existence of triphenylmethyl was postulated originally on the basis of its chemical properties, before the development of electron valence theory, but there is now this physical property which proves its structure (see Section 31.9).

Diradical structures were discussed in Section 17.10. It was indicated that the C=C bond in a compound is often better written C÷C in order to explain its chemistry, and that if the bond enters into reactions with free radicals it is also convenient to write C—C for the double bond. This was called a *dot-dot* structure, as well as diradical.

It is possible to prepare molecules in which two such unpaired electrons are known to be present. In Compounds I and II to save space, certain of the phenyl rings are represented by Ph.

Ph   Ph   Ph   Ph
 | | | |
Ph—C—⟨   ⟩—⟨   ⟩—C—Ph ↔ Ph—C=⟨   ⟩=⟨   ⟩=C—Ph
 •   •

I   Ia

This compound is not paramagnetic

Ph—C—Ph   Ph—C—Ph
    •   •

II

This compound is paramagnetic

Formula I is that of a compound that can be written with unpaired electrons, but as shown in Ia these electrons can interact through a conjugate system of double and single bonds. In Formula II, where the end groups are *meta*, such resonance interaction is not possible (the student should prove this to himself on paper). The odd electrons in II are therefore fixed and the compound has been found to be paramagnetic. Like free $\cdot CH_3$ radicals, these molecules will tend to react with each other, but measurements show that there is an appreciable dissociation to free diradicals at ordinary temperatures.

24.11. Carbene and the Singlet State. The unpaired electrons of the diradical molecules described in I and II of the preceding paragraph are so far apart they behave as isolated bodies. Of greater interest is the small *methylene* molecule (H–Ċ–H) in which the unpaired electrons are close enough to react on each other; we shall study this substance in Section 24.12 after examining its sister compound *carbene* which usually appears for a brief instant before the formation of methylene takes place in a reaction. Carbene is an extremely reactive molecule that can be made in several ways; we shall use here the reaction presented in Section 29.8 under the subject of diazomethane:

$$CH_2N_2 \xrightarrow{-N_2} \underset{H}{\overset{H}{>}}C: \longrightarrow H-\overset{\cdot}{\underset{\cdot}{C}}-H$$

Diazomethane   Carbene   Methylene

When diazomethane is decomposed by light (or heat) the first product is carbene, unless it is intentionally prevented from forming by introducing certain reaction conditions; this initial product is converted spontaneously to methylene, which is more stable. The reaction must be conducted in the presence of the planned reagent because the life of carbene is only about 10^{-8} second; that of methylene is "much longer," perhaps as long as one second. These free radicals have been observed spectroscopically.

The C atom in carbene is an sp^2 hybrid. The electronic nature of this compound can be visualized with the help of Figure 15.6, by fixing our attention on one C atom of the ethene molecule: in *b* there is an s-sp^2 molecular orbital, which represents a C–H bond; two such bonds on the same C atom can be picked out from structure *c*; the *c* structure also shows a third sp^2 orbital, which connects to a second C atom, but this is not used in carbene; instead it has a "lone pair" of electrons in that orbital, and the *p* orbital at right angles to it as shown in *d* is empty. The sp^2 orbital containing the unshared pair of electrons in carbene has more s character than the other two. The electrons in this lone pair have the normal opposing spins ↓↑, and from the effect that this pair produces spectroscopically it is called a singlet; in carbene the pair is in the singlet state.

If carbene from the decomposition of CH_2N_2 is generated in the presence of an olefin, it adds to the olefin double bond in a single step before it has time to change into methylene. Because of its free electron pair carbene looks like a nucleus-seeking molecule, but the molecule actually lacks two electrons and is in fact strongly electrophilic; it reacts instantly with the C=C bond and its exposed π electrons. The *cis*-butene shown in the equation is the one described in Section 22.3 as an example of geomet-

$$
\underset{\substack{\text{cis-2-Butene or Z-2-Butene}\\ \text{(see Sec. 22.16a)}}}{\overset{\text{CH}_3}{\underset{\text{H}}{>}}\text{C}\!\cdots\!\text{C}\overset{\text{CH}_3}{\underset{\text{H}}{<}}}
\;+\;
\underset{\text{Carbene}}{\overset{\,\cdot\cdot}{\underset{\text{H}}{}}\text{C}\underset{\text{H}}{}}
\;\longrightarrow\;
\underset{\text{cis-Dimethylcyclopropane}}{\overset{\text{CH}_3}{\underset{\text{H}}{>}}\text{C}\!-\!\text{C}\overset{\text{CH}_3}{\underset{\text{H}}{<}}\;\;\underset{\text{H}_2}{\text{C}}}
$$

ric isomerism. It should be pictured here with the CH_3 groups above the page, and the H atoms below the page. The reaction product is a derivative of cyclopropane, with the ring in the plane of the page. Carbene is said to be stereospecific (or stereoselective); if it adds to a *cis* compound the product is *cis*; the reason for this will be made apparent after the discussion of methylene.

24.12. Methylene and the Triplet State. The two unshared electrons in methylene are different from those in carbene in that they are not paired in opposing spins. They have parallel spins ↑↑ . These electrons are also different from the carbene pair in that they are in two separate orbitals, which are the p type; they cannot occupy the same orbital because they are not paired. The electronic structure of methylene can be worked out by reference to Figures 7.1 and 7.2. The four atomic orbitals of the carbon atom are those shown in Figure 7.1. The s and p_x orbitals hybridize to two sp orbitals (see Table 7.1), and each C—H bond would have the appearance of c in Figure 7.2 but along the x axis. In methylene, which has an H–C–H structure, the H–C and C–H molecular orbitals are in a straight line along the x axis. Perpendicular to this line are the p_y and p_z orbitals, each containing one electron. The unpaired electrons in methylene are in the triplet state; a pair of electrons with parallel spins produces in spectroscopy an effect called a triplet (Section 24.13).

We return now to the *cis*-butene reaction described in the preceding section. The CH_2N_2 molecule can be decomposed with the aid of a photosensitizer that will produce methylene without intermediate carbene. The addition of methylene to *cis*-butene follows a diradical mechanism. One of the unpaired electrons uncouples the π bond in butene and pairs with

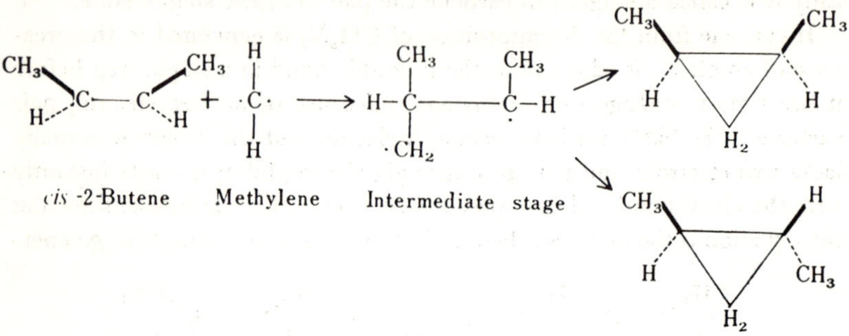

$$
\underset{\text{cis -2-Butene}}{\text{CH}_3\text{C}\text{=}\text{C}\text{CH}_3\;(\text{H, H})}
\;+\;
\underset{\text{Methylene}}{\cdot\text{C}\text{H}_2\;(\text{H, H})}
\;\longrightarrow\;
\underset{\text{Intermediate stage}}{\text{H--}\overset{\text{CH}_3}{\underset{\cdot\text{CH}_2}{\text{C}}}\text{----}\overset{\text{CH}_3}{\text{C}}\text{--H}}
$$

that electron with opposite spin. This is the intermediate stage shown in the equation. To form a second electron pair bond, one of the two remaining electrons must reverse its spin. This takes time as well as energy. Until that reversal takes place the butene molecule is not restricted by a double bond. It can rotate about a single bond in the intermediate step, and when the reaction is completed the CH_3 groups may be on opposite sides of the cyclopropane ring (*trans*) as well as on the same side (*cis*). Addition of methylene to *cis*-butene is not stereospecific; the product may be a mixture of *cis* and *trans* isomers because of the two steps required in the reaction. The carbene reaction takes place in only one step and leaves the isomerism unchanged. The carbene addition to *cis*-butene is an example of a concerted reaction (Section 17.16).

Nomenclature of these compounds is not consistent. Carbene is referred to as singlet carbene or triplet carbene. Likewise, methylene is called singlet methylene and triplet methylene. In this introduction to "carbene chemistry" we have mentioned only the parent hydrocarbons, carbene and methylene. Derivatives are easily made, such as dichlorocarbene, $Cl_2C{:}$, from chloroform, Cl_3CH, and this makes possible many important syntheses.

24.13. Phosphorescence and the Triplet State. The two electrons in a normal bond are spinning and therefore are magnetized, but since they spin in opposite directions their magnetic effects cancel. In a strong external magnetic field this electron pair can have only one energy level, and is accordingly a singlet.

The unpaired electrons in methylene have parallel spins. In a magnetic field their spin directions can be (1) parallel to each other in the direction of the field, (2) parallel but opposite to the field direction, or (3) parallel and perpendicular to the field direction. These constitute three energy levels, or three lines in the spectrum, and therefore the triplet state.

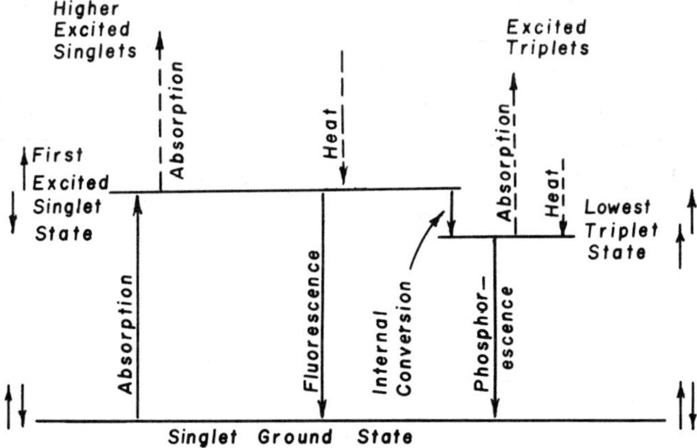

24.5. Excitation of electrons by light. (Adapted principally from *Flash Photolysis* by Leonard Grossweiner. Copyright © May 1960 by Scientific American, Inc. All rights reserved.)

In certain molecules the energy of light (visible or ultraviolet) will uncouple an electron pair bond and create free radicals (Section 17.8). If the bond is not broken, the absorbed energy will raise an electron to an excited singlet state (Figure 24.5). If the substance fluoresces readily it is because the electron falling back to its ground state emits light; the emitted light is often at a longer wavelength than the incident radiation and the time lag is about 10^{-8} second (longer wavelength corresponds to a lower energy as shown in Table 14.3). Instead of falling back to its original energy level, an electron of a singlet pair may have its spin direction reversed by the momentary excess energy; this is the "internal conversion" in Figure 24.5. The pair with parallel spins is now in the triplet state. This state absorbs visible light and undergoes transitions to higher triplet levels, but eventually the electrons return to the ground state. Another way to return to the ground state is by phosphorescing; an appreciable time is needed to lose triplet state energy, generally about 10^{-4} to 10^{-3} second, but lifetimes of several seconds may be reached. In rigid media, phosphorescence may last a few minutes. Large polycyclic ring systems exemplified by naphthalene and anthracene (Section 32.6) are popular subjects for the study of phosphorescence in organic chemistry. The oxygen molecule (Section 24.10) is a triplet in its stable, ground state.

Phosphorescence is different from fluorescence in that it lasts longer. Fluorescence occurs only while a substance is being irradiated; phosphorescence persists after the irradiation is stopped. From this discussion it is obvious that a methylene free radical, with its unpaired electrons, may live a million times as long as the carbene free radical. The eminent biochemist, Dr. Albert Szent-Gyorgyi, proposes the theory that compounds that are energy transmitters in physiological processes and need a time delay for their reactivity are those that can exist in the triplet state. The pioneering work on phosphorescence and the triplet state was done by G. N. Lewis, who is also referred to on pp. 26, 55, and 296 of this book.

25

Alcohols, Phenols, and Ethers

25.1. The Hydroxyl Group. In this chapter we take up the study of oxygen compounds, beginning with the alcohols. There are two ways of looking at alcohols, as shown by the two ways ethyl alcohol is written:

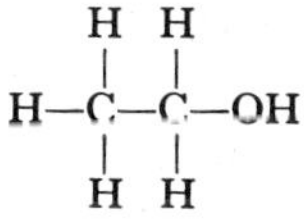

$$H-\overset{\overset{\displaystyle H}{|}}{\underset{\underset{\displaystyle H}{|}}{C}}-\overset{\overset{\displaystyle H}{|}}{\underset{\underset{\displaystyle H}{|}}{C}}-OH \qquad\qquad C_2H_5-OH$$

Ethanol
Hydroxyethane

Ethyl alcohol

According to one system, ethyl alcohol is a compound derived from the hydrocarbon, ethane, by replacing an $-H$ atom with the $-OH$ group. It is named ethanol to indicate it is an alcohol derived from ethane, the ending *-ol* being used as a designation for alcohols.

In the other system, ethyl alcohol is considered as derived from water, $H-O-H$, by replacing an $-H$ atom with the ethyl radical, $-C_2H_5$. Any such alcohol can be represented by R–OH or R–O–H. The $-OH$ group is called *hydroxyl*, but the combining form in the naming of compounds is *hydroxy*, as indicated under the formula of ethanol.

The student should turn back to the first page in Chapter 24 and compare the halogen compounds listed at that time with the following compounds containing $-OH$ groups:

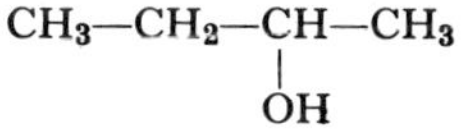

$$CH_3-CH_2-\overset{}{\underset{\underset{\displaystyle OH}{|}}{CH}}-CH_3$$

An aliphatic alcohol
($-OH$ in a chain
compound)

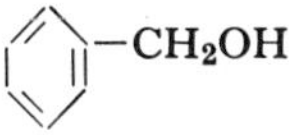

An aromatic alcohol
($-OH$ in a side chain of
an aromatic compound)

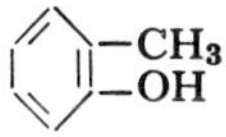

A phenol
($-OH$ in nucleus of an
aromatic compound)

When the $-OH$ group replaces an H atom in the *nucleus* of an aromatic compound, the product is called a phenol. The phenols will be discussed separately (Section 25.11), because the presence of the benzene ring structure introduces properties different from those of alcohols. In Chapter 24, we found that the Cl atom also has different properties, depending on whether it is in the side chain or nucleus of a compound. There is a direct relation between the —OH compounds to be described now and the halogen compounds we have already studied, for they can often be converted into each other under the proper conditions:

$$\cdots -\overset{|}{\underset{|}{C}}-\overset{|}{\underset{|}{C}}-Cl + H-OH \overset{hydrolysis}{\rightleftharpoons} \cdots -\overset{|}{\underset{|}{C}}-\overset{|}{\underset{|}{C}}-OH + H-Cl$$

25.2. How to Name Alcohols. We have already indicated (Sections 10.9 and 13.3) that when the —OH group is the principal "functional" group in a molecule, the ending *-ol* is used in the main part of the name of the compound. For this reason we write for CH_3CH_2OH ethanol instead of hydroxyethane. Similarly, we assign to $CH_3-CH_2-CHOH-CH_3$ the name 2-butanol instead of 2-hydroxybutane; in this example *butan* shows there are four carbon atoms in the chain and *ol* shows it is an alcohol. The —OH group is on the second carbon atom.

By this systematic method it is a comparatively simple matter to name fairly complex compounds such as this:

$$\overset{\overset{\displaystyle H\ \ H\ \ H}{|\ \ \ |\ \ \ |}}{Cl-C=C-\underset{\underset{\displaystyle OH}{|}}{C}-H} \qquad\qquad \text{3-Chloro-2-propen-1-ol}$$

This is a halogen compound (*chloro*), and an olefin (*en*), as well as an alcohol (*ol*). It is also a three-carbon chain (*prop*) and the numbering is started from the right-hand end in order to give the functional group (—OH) the lowest number.

One of the old methods for naming alcohols is based on an old name for wood alcohol, which used to be called carbinol. Other alcohols are named as derived from carbinol, and the carbon atom in the C–OH bond is called the carbinol carbon atom.

Carbinol		Methylcarbinol	
Methanol	$\overset{\overset{\displaystyle H}{|}}{\underset{\underset{\displaystyle H}{|}}{H-C-OH}}$	Ethanol	$\overset{\overset{\displaystyle H}{|}}{\underset{\underset{\displaystyle H}{|}}{CH_3-C-OH}}$
Wood alcohol		Grain alcohol	
Methyl alcohol		Ethyl alcohol	

$$CH_3{-}CH_2{-}\overset{\displaystyle H}{\underset{\displaystyle OH}{C}}{-}CH_3 \qquad\qquad CH_3{-}CH_2{-}\overset{\displaystyle H}{\underset{\displaystyle OH}{C}}{-}CH_2{-}CH_3$$

Methylethylcarbinol
2-Butanol (a butyl alcohol)
        Diethylcarbinol
3-Pentanol (an amyl alcohol)

These examples also indicate a third common way to name alcohols—that is, according to their source. At one time, the only commercial source of methanol was the distillation of wood; consequently, it is known as wood alcohol. Ethanol is still obtained in large quantities from the fermentation of cereals and grains, such as rye, barley, and corn, and is therefore called grain alcohol. The five-carbon alcohols, corresponding to pentane and called pentanols, were originally obtained from starch fermentation and named amyl alcohols, because the Latin word for starch is amylum.

Lastly, the alcohols are often named in such a way as to indicate their two-component structure. Thus, methyl alcohol consists of the methyl radical, $CH_3{-}$, and the hydroxyl radical, $-OH$, which is the "alcohol" group.

The aromatic alcohols should give no difficulty in regard to nomenclature. As an illustration, ⬡$-CH_2CH_2OH$ is phenylethyl alcohol, or 2-phenyl-1-ethanol, or hydroxyethylbenzene.

25.3. Alcohols as Acids. The alcohols have polar characteristics, and the lower alcohols in particular behave very much like water toward certain reagents. The lower alcohols will dissolve many water-soluble inorganic substances. Like water, the alcohols are amphoteric—they can react either as acids or as bases. The chemical activity of alcohols is due more to the R–OH bond than to the RO–H bond. However, the properties are to be summed up in the type formula R–O–H, analogous to H–O–H for water. When an alcohol acts as an acid the active bond in the molecule is RO–H:

$$R{:}\overset{..}{\underset{..}{O}}{:}H + {:}\overset{..}{\underset{..}{O}}{:}H \rightleftharpoons [R{:}\overset{..}{\underset{..}{O}}{:}]^- + [H{:}\overset{..}{O}{:}H]^+$$
$$\qquad\qquad\ \ \overset{\textstyle H}{} \qquad\qquad\qquad\qquad\qquad H$$

An alcohol reacts as an acid to yield hydronium ion (hydrated H^+ ion).

The hydronium ion was referred to in Section 23.7 as a member of a large family of *onium* ions. The negative ion in this equation is an *alkoxide* ion. The acidity of alcohols is very weak; ethanol has an ionization constant as an acid a little lower than that of water, so there are relatively few

$$CH_3-CH_2-CH_2-OH \qquad CH_3-\overset{\displaystyle H}{\underset{\displaystyle OH}{C}}-CH_3 \qquad CH_3-\overset{\displaystyle CH_3}{\underset{\displaystyle OH}{C}}-CH_3$$

Propanol   2-Propanol   2-Methyl-2-propanol
A 1° alcohol   Isopropyl alcohol   *t*-Butyl alcohol
      A 2° alcohol   A 3° alcohol

25.1. A structural classification of alcohols.

hydronium ions in solution. Acids and bases will be discussed more fully in Section 27.6.

Alcohols are often classified on the basis of the location of the OH group in the molecule. This is illustrated in Figure 25.1. The C atom to which OH is attached serves as the basis for the classification (see Section 10.5); if it is attached to only one other C atom the alcohol is primary, if attached to two other C atoms the alcohol is secondary, and if attached to three other C atoms the alcohol is tertiary (*t*).

The acidity of alcohols is greatest in the primary alcohols and least in the tertiary. The CH_3 group has electron-donating properties (Section 21.5). The lower acidity with increase in number of CH_3 groups on the alcoholic C atom is probably due to increase in electron-donor activity, which raises electron density at the O atom and increases its hold on the proton. This reasoning is similar to that in the last paragraph of Section 27.5 concerning the acidity of acetic acid. When an active metal such as sodium is added to an alcohol, the hydroxylic H atom is replaced since it is acidic:

$$2CH_3OH + 2Na \longrightarrow 2CH_3ONa + H_2$$
Sodium methylate

The reaction takes place most rapidly with 1° alcohols, less rapidly with the 2°, and least rapidly with the 3° type. The product is a salt that belongs to the class of *alcoholates*. It is preferable to call them *alkoxides*; the sodium compound is then *sodium methoxide*.

The organic chemist makes wide use of the fact that he can insert a metal so easily into a carbon compound. Such compounds can easily be made to react with compounds containing halogen atoms, to produce larger molecules. An example of such a preparation will be given in the study of ethers (Section 25.18). The alcoholates resemble the hydroxides of inorganic chemistry, such as NaOH, in their method of preparation. Since alcoholates are salts of extremely weak acids they hydrolyze in water to give basic solutions.

25.4. Alcohols as Bases. The availability for reaction of the free pairs of electrons on the oxygen atom of an alcohol is made evident by the ease of formation of valence bonds with electron-seeking reagents. An alcohol

$$R\!:\!\overset{..}{\underset{..}{O}}\!:\!H + H^+\!:\!\overset{..}{\underset{..}{Cl}}\!:^- \longrightarrow \left[R\!:\!\overset{\overset{\textstyle H}{..}}{\underset{..}{O}}\!:\!H\right]^+ :\!\overset{..}{\underset{..}{Cl}}\!:^-$$

An alcohol reacts as a base to yield a salt.

reacts with acids to form salts that can be isolated at low temperatures (about $-50°C.$). These are *oxonium* salts; the structure of the positive ion should be compared with that of the hydrated H^+ ion. It is obvious that in this reaction an alcohol is acting as a base to form a salt, whereas in the reaction with sodium it acts as an acid to form a salt.

Alcohols are converted to alkyl halides by means of halogen acids such as HBr. Essentially this is the reverse of the reactions in Figures 22.15 and 22.16, where alkyl halides are hydrolyzed to alcohols by displacement reactions that depend on an obliging *leaving* group—namely, the Cl^- ion. However, the OH^- ion is so strongly basic it is not a willing leaving group. The reaction is assisted by using excess of the halogen acid or a strong acid like sulphuric acid to furnish H^+ ions:

$$CH_3\!-\!CH_2\!:\!\overset{..}{\underset{..}{O}}\!:\!H \xrightarrow{H^+} \left[CH_3\!-\!CH_2\!:\!\overset{\overset{\textstyle H}{..}}{\underset{..}{O}}\!:\!H\right]^+ \xrightarrow{:Br^-} CH_3\!-\!CH_2\!:\!Br + H\!:\!\overset{..}{\underset{..}{O}}\!:\!H$$

In strong acid solution the action of HBr yields ethoxonium bromide; the complex ion in this salt now has a more willing leaving group, which is the water molecule (HOH), and displacement reactions can occur similar to those shown in Figures 22.15 and 22.16. The type of reaction that occurs will depend on many factors, but mainly on the structure of the alcohol and the nature of the halogen acid. In an S_N2 mechanism the order of the reactivity of the alcohols is primary>secondary>tertiary. In S_N1 mechanisms this order is reversed.

25.5. Reaction with PX$_3$. When alcohols are treated with one of the chlorides of phosphorus the entire $-OH$ group is replaced:

$$3C_2H_5\!-\!OH + P\overset{\textstyle Cl}{\underset{\textstyle Cl}{\overset{\displaystyle |}{-}Cl}} \longrightarrow 3C_2H_5\!-\!Cl + H_3PO_3$$

Phosphorus   Ethyl Phosphorous
trichloride   chloride   acid

This is one of the standard methods by which $-OH$ groups are replaced by halogen atoms in many kinds of compounds.

25.6. Dehydration of Alcohols. Most alcohols can be dehydrated by the following scheme:

$$
\begin{array}{c}
\text{H}\ \ \text{H} \\
|\ \ \ | \\
\text{H—C—C—H} + \text{dehydrating agent} \longrightarrow \text{H—C=C—H} + \text{H}_2\text{O} \\
|\ \ \ | \qquad (\text{H}_2\text{SO}_4\ \text{or}\ \text{Al}_2\text{O}_3 \\
\text{H}\ \ \text{OH} \qquad \text{at high temp.})
\end{array}
$$

The requirement here is that the C–OH is next to a carbon atom holding an H atom. This is a convenient way to prepare olefins, as pointed out in Section 13.2. The dehydration reaction takes place through intermediate formation of an oxonium ion, due to the proton (H^+) from the acid catalyst "sitting down" on one of the electron pairs of the oxygen atom. This oxonium ion then splits off water as it usually does (see Section 25.4),

$$
\left[\begin{array}{c}\text{H H H}\\ \text{H:C:C:O:H}\\ \text{H H}\end{array}\right]^{+} \xrightarrow{\ -\ \text{HOH}\ } \left[\begin{array}{c}\text{H H}\\ \text{H:C:C}\\ \text{H H}\end{array}\right]^{+} \longrightarrow \begin{array}{c}\text{H H}\\ \text{C::C} + \text{H}^+\\ \text{H H}\end{array}
$$

An oxonium ion   A carbonium ion   Ethene

yielding a carbonium ion that is even less stable and in turn proceeds to eliminate H^+ ion. The H^+ ion that is eliminated reacts with an acid radical to regenerate a molecule of the acid catalyst, or it can attack another molecule of the alcohol to decompose it to an olefin.

On a commercial scale, alcohols are often dehydrated by passing their vapor over an aluminum oxide catalyst at about 350°C. This catalyst is similar in its action to the H^+ ion. The student is advised to read ahead in Section 25.18 for still another mechanism by which alcohols may be dehydrated—that is, to yield *ethers*.

25.7. Oxidation. Most alcohols are easily oxidized, but the course of the reaction depends on the position of the –OH group in the molecule—that is, whether it is primary, secondary, or tertiary. A primary alcohol loses a pair of H atoms and yields an *aldehyde*; a secondary alcohol also loses a pair of H atoms, but the product is a *ketone* (Section 26.1). In both cases there is no change in the carbon content of the molecule or its general structure; these reactions should be consulted but will not be discussed further at this point. A tertiary alcohol when oxidized will break down into molecules with a smaller number of carbon atoms; strong oxidation may ultimately yield carbon dioxide and water. Alcohols of unknown structure are sometimes identified by oxidizing them with an oxidizer such as chromic acid and determining the nature of the end products.

25.8. Polyhydric Alcohols. The alcohols with only one OH group in the molecule are called monohydric to distinguish them from the large family of polyhydric alcohols with more than one. Glycols are the dihydric alcohols. Very few stable compounds are known in which two OH groups reside on the same carbon atom; an example is chloral hydrate (Section 26.11). Methylene glycol, $HO-CH_2-OH$, is known in aqueous solution but has not been obtained in the free state; it dehydrates to formaldehyde (Section 26.11) when isolation is attempted. (Carbonic acid,

$$HO-\overset{\overset{\displaystyle O}{\|}}{C}-OH,\ \text{also, exists only in solution.})$$

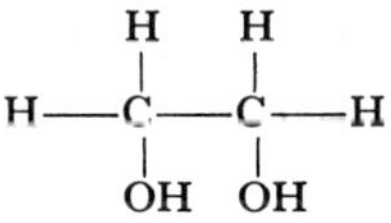

1,2-Ethanediol
Ethylene glycol

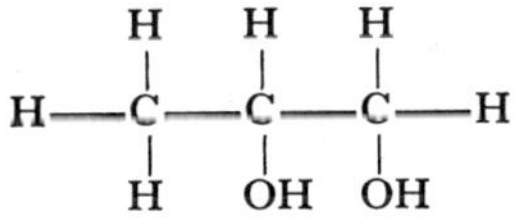

1,2-Propanediol
Propylene glycol

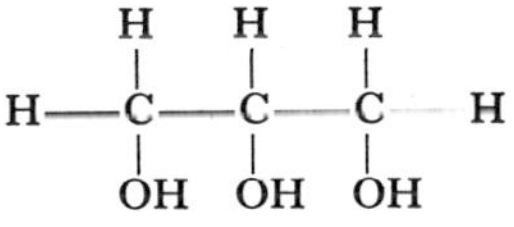

1,2,3-Propanetriol
Glycerol
Glycerin

Ethylene glycol is used as an antifreeze ingredient in autos and is rather toxic. Glycerol and other polyhydric alcohols are used extensively as sweetening agents, in cosmetics, and as humectants. The trinitrate of glycerol is an explosive (dynamite).

25.9. Aromatic Alcohols. The alcohols containing benzene rings behave in general like the ordinary aliphatic alcohols. The simplest compound in this class is ⟨◯⟩$-CH_2OH$, which is the phenyl derivative of methanol. It can be called phenylmethyl alcohol, or phenylmethanol, or phenylcarbinol, but is generally known as *benzyl* alcohol. The benzyl radical is defined in Section 24.5.

Benzyl alcohol gives most of the characteristic reactions of the –OH group which have just been described. The ring part of the compound, however, does not behave normally—that is, it is not possible to sulphonate and nitrate it as described in Chapter 20, because the alcohol group is sensitive to oxidation and these reagents may cause oxidation and decomposition of the molecule.

The aromatic alcohols usually have a pleasant, ethereal odor.

25.10. The Hydrogen Bond. It has been known for many years that water tends to combine with itself to form polymers, or giant molecules; this is the phenomenon referred to as association (Section 8.13). A satisfactory mechanism for the polymerization of water was not available until the electron theory of valence was proposed. It is now understood that H:Ö:H molecules are linked through hydrogen bridges; thus

$$\text{H:Ö:H:Ö:H:Ö:} \qquad \text{or} \qquad \text{H—O} \ldots \text{H—O} \ldots \text{H—O} \ldots \text{etc.}$$

The hydrogen bond is currently believed to be due to the ability of pairs of electrons on small elements (such as oxygen and nitrogen) to share a proton between them. This gives rise to dipole forces, measurements of which are in substantial agreement with the conventional picture of the hydrogen bridge. It is an electrostatic effect.

Alcohols polymerize in the same way; if we represent any alcohol by R:Ö:H the continuous chain in the liquid phase is

$$\text{H:Ö:H:Ö:H:Ö: etc.}$$

PHENOLS

25.11. Examples of Phenols. These compounds were defined in the introduction to this chapter as hydroxyl compounds in which the OH group is in the benzene ring. The simplest phenol is phenol itself. It is the compound derived from benzene, one of the original names of which was *phene* (Section 19.1). All four compounds listed here are found in coal tar. They all have penetrating odors, and are all used as antiseptics as well as intermediates in many industrial processes. The toluene deri-

Phenol	o-Cresol	m-Cresol	p-Cresol
Carbolic acid	o-Hydroxytoluene	m-Hydroxytoluene	p-Hydroxytoluene
Hydroxybenzene	o-Methylphenol	m-Methylphenol	p-Methylphenol
m.p. 43°	m.p. 30.8°	m.p. 10.9°	m.p. 35°

vatives are known as the *cresols* because they are found in wood creosote. These examples of phenols containing but one –OH group are known as monohydric phenols to distinguish them from the polyhydric phenols to be described later.

25.12. Properties of Phenols. Phenols are more acidic than alcohols. The chemically active bond in phenols is ArO–H, whereas we found the properties of alcohols represented better by R–OH than by RO–H. The acidity of phenol can be explained with the aid of the resonance theory, the usual Kekulé structures A and B being in resonance with Structures I, II, and III in Figure 25.2. These structures are worked out by the scheme described in Section 21.6. They contribute resonance energy which increases the stability of the phenol molecule, but the point to be emphasized is the fact that the O atom acquires a partial + charge because of shift of negative charge (electrons) into the ring. The electropositive H atom is therefore held less firmly, which accounts for the stress laid on the ArO-H bond.

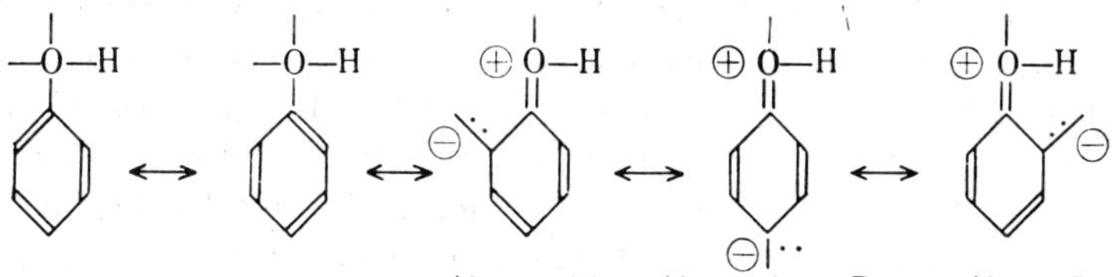

A   B   I (from A)   II (from A or B)   III (from B)

25.2.   Resonance structures of phenol.

In an alkaline solution of phenol the H^+ ion in Figure 25.2 is removed, liberating the phenoxide ion, $C_6H_5O^-$, which has greater stability through resonance than the original molecule. Resonance energy in dipolar structures like those in Figure 25.2 is less than it would be in corresponding nonpolar structures. (The resonance energy decreases with greater separation of the two charges.) Removal of the H atom in the form of an H^+ ion from the resonance forms in Figure 25.2 in alkaline solution, to free the phenoxide ion as shown in Figure 25.3, eliminates the *charge-separation* effect and permits stronger interaction of the electron pairs on the O atom with the π electrons of the ring.

Because of its acid properties, phenol reacts with inorganic bases to form salts:

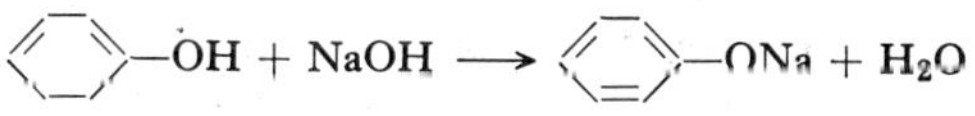

$$\text{C}_6\text{H}_5\text{—OH} + \text{NaOH} \longrightarrow \text{C}_6\text{H}_5\text{—ONa} + \text{H}_2\text{O}$$

Phenol   Sodium phenoxide
                          Sodium phenolate

25.3. Resonance in the phenoxide ion (compare Fig. 25.2).

Because phenol is a stronger acid than alcohol, the phenolates are more saltlike and more stable than the alcoholates of Section 25.3.   Although phenol is definitely an acid, it is one of the weakest of acids.   Its salts are decomposed even by carbonic acid, which is very weak indeed (in general, an acid is liberated from its salts by stronger acids).

Now that we have seen how the nature of the –OH group is modified by the simple fact that it is attached to the nucleus, let us consider the other side of the question and find out how the presence of –OH in the ring modifies the properties of the ring itself.   The presence of –OH in the nucleus has a great activating effect on the o- and p- positions for the reason shown in the resonance Structures I, II, and III of Figure 25.2. Normally, iodine does not directly replace hydrogen atoms in the nucleus, but phenol reacts with an alkaline solution of iodine in water almost instantly and quantitatively.   The mechanism is probably similar to that already described for reactions resulting in substitution in the ring (Section 20.2). This reaction may be used for estimating the quantity of phenol in a solution.

$$C_6H_5\text{—}OH + 3I_2 \longrightarrow C_6H_2I_3\text{—}OH + 3HI$$

Phenol	Iodine	2,4,6- Triodophenol	Hydrogen iodide

Another example of this activating effect of –OH on other positions in the ring is that shown by the preparation of the high explosive, *picric acid*.   This compound is made by nitrating phenol (see Section 20.3 for the meaning of nitration).   Three –NO$_2$ groups are inserted into the ring in positions o- and p- to the OH group.   Phenol is much easier to nitrate than benzene itself.

25.13.   Preparation of Phenols.   One of the favorite laboratory methods for the preparation of alcohols is the hydrolysis of halogen compounds.   This is not easily accomplished when the halogen atom is in the ring, $\langle\!\!\!\!\!\!-\!\!\!\!\!\rangle$–Cl → $\langle\!\!\!\!\!\!-\!\!\!\!\!\rangle$–OH, because the Cl atom in the ring is firmly attached and does not react readily (Section 24.4). Industrial methods have been developed for introducing the OH group into the ring. One of these is the use of strong alkali at high temperature and pressure. Phenol must be made synthetically because the supply from coal tar can hardly meet the demand for the enormous quantities needed in the manufacture of resins, plastics, and other products.

25.14. Dihydric Phenols. There are three dihydroxybenzenes, with the formula $C_6H_4(OH)_2$, in which the two OH groups are *ortho*, *meta*, and *para* to each other. Their names are

o-Dihydroxybenzene	*m*-Dihydroxybenzene	*p*-Dihydroxybenzene
Pyrocatechol	Resorcinol	Hydroquinone
		Quinol

These are important substances in the dye industry and in photography. Hydroquinone particularly is employed as a photographic developer. They are all easily oxidized.

25.15. Trihydric Phenols. There are also three trihydroxybenzenes, and like the dihydric phenols they are known generally by their common names rather than by their systematic names, are easily oxidized, and are useful as reducing agents:

1,2,3-Trihydroxybenzene	1,2,4-Trihydroxybenzene	1,3,5-Trihydroxybenzene
Pyrogallic acid	Hydroxyhydroquinone	Phloroglucinol
Pyrogallol		

Pyrogallic acid is so readily oxidized that it is used in quantitative analysis for estimating the amount of oxygen in a gas mixture. Phloroglucinol is an excellent example of a compound that exhibits *tautomerism* (Section 28.11); this is indicated in these formulas by a dynamic equilibrium between *keto* and *enol* forms. The compound takes part in reactions as if it had either structure.

(enol form)   (keto form)

Phloroglucinol

ETHERS

25.16. Examples of Ethers. Structurally, alcohols and phenols may be thought of as H–O–H derivatives in which one of the H atoms is replaced by a radical. The ethers complete the possibilities; they are compounds in which both H atoms are replaced by radicals. There are aliphatic ethers, aromatic ethers, and aliphatic-aromatic ethers, as shown by the examples listed here. Dimethyl ether is referred to as a *simple* ether, in contrast

Dimethyl ether CH_3-O-CH_3

Methyl ethyl ether $CH_3-O-C_2H_5$

Diethyl ether $CH_3-CH_2-O-CH_2-CH_3$

Diphenyl ether ⟨ ⟩–O–⟨ ⟩

Methyl phenyl ether ⟨ ⟩–O–CH_3

to methyl ethyl ether, which is called a *mixed* ether because there are two different radicals in it. The best known ether is diethyl ether, called ether, as ethyl alcohol is called alcohol.

The rules permit use of the name ethyl ether for diethyl ether, just as sodium sulfate is used instead of disodium sulfate for Na_2SO_4. Naming the ethers is generally an easy problem. In the form R–O–R or Ar–O–Ar all we need do is name the radicals on each side of the oxygen bridge, as in the examples already given. In more complicated cases, the –OR

$$\text{(benzene ring)} \begin{matrix} -OCH_3 \\ -Cl \end{matrix}$$

group is called the *alkoxy* radical. In this particular instance, the compound is *o*-chloro-methoxybenzene. Similarly, $-OC_2H_5$ is *ethoxy* and $-OC_6H_5$ is *phenoxy*.

Brief mention may also be made of a system of nomenclature, employed principally for certain rather complicated molecules, in which the oxygen atom is regarded as a substituent for a $-CH_2-$group of a chain of carbon atoms (either open-chain or closed-chain). According to this system, diethyl ether is 3-oxapentane, indicating that the oxygen atom is the third member in a chain of five members. Another example will be found in Section 25.19 in the alternative name for 1, 4-dioxane.

25.17. Properties of Ethers. As a class the ethers are stable compounds. In fact, the ethers are often used as solvents in which other, very reactive substances are dissolved—that is, they often serve as a reaction medium, just as water is a common medium for reactions in inorganic chemistry. The oxygen bridge does not introduce the profound changes in the properties of the carbon chain that we generally associate with other foreign elements. For example, the boiling point of pentane (from Table 10.1) is 36° and that of diethyl ether 35°. The C–O–C link, however, is

$$CH_3—CH_2(—CH_2—)CH_2—CH_3 \qquad\qquad CH_3—CH_2(—O—)CH_2—CH_3$$

Pentane   3-Oxapentane
Diethyl ether

weaker than the C–C link because of a very important factor. The oxygen atom still has two unshared pairs of electrons, which offer points of attack to electron-seeking reagents. The ethers accordingly form salts similar to those described in Section 25.4. These salts are stable only at quite

$$CH_3\!:\!\overset{..}{\underset{..}{O}}\!:\!CH_3 + HCl \longrightarrow \left[CH_3\!:\!\overset{\displaystyle H}{\underset{..}{O}}\!:\!CH_3 \right]^+ Cl^-$$

Dimethyl ether   Dimethyl ether hydrochloride
Dimethylhydrogenoxonium chloride

An ether acts as a weak base to form a salt.

low temperatures (about $-100°C$) and are practically completely dissociated at ordinary temperatures. The ether molecule can also add a second molecule of HCl to form a dihydrochloride in which the oxonium ion has a positive charge of two.

Although the ether salts are unstable, they offer an explanation for at least part of the usefulness of ethers as a reaction medium. The intermediate complexes formed with dissolved reagents generally serve to activate them. It will be evident that an ether cannot combine with itself through hydrogen bridges (Section 25.10), but can associate with other compounds containing loosely held hydrogen. The Grignard reaction (Section 30.15) makes extensive use of the tendency of ethers to form molecular complexes through the unshared electron pairs on its oxygen atom.

When acted on by strong acids an ether molecule can be broken at the oxygen bridge.

$$CH_3—CH_2—O \;\vdots\; CH_2—CH_0 + H—I \quad\longrightarrow\quad CH_3—CH_2—OH + CH_3—CH_2—I$$

Diethyl ether   Ethyl alcohol   Ethyl iodide

An intermediate oxonium compound is first formed, as described for hydrogen chloride, and this breaks down to give the products indicated here. The best halogen acid for this cleavage is HI.

On exposure to air, ethers slowly form explosive peroxides of the structure $R:\ddot{O}:\ddot{O}:R$. Ordinary diethyl ether is usually sold in bottles containing iron wire, which is oxidized preferentially and protects the ether. Distillation of an ether containing peroxides is dangerous because the peroxides are concentrated by distillation and explode when most of the ether has been distilled off. This is true of many compounds containing the ether linkage.

25.18. Preparation of Ethers. The Williamson method for preparing ethers is a general one and depends on alcohols for the raw materials. The ethoxide and iodide are both made from alcohols.

$$C_2H_5-ONa + I-CH_3 \longrightarrow C_2H_5-O-CH_3 + NaI$$

Sodium ethoxide   Methyl iodide   Methyl ethyl ether

This method is especially useful for preparing *mixed* ethers.

Ethers can be prepared easily from some alcohols by dehydration of two molecules of the alcohol. Several catalysts are available for this

$$C_2H_5-OH + HO-C_2H_5 \xrightarrow{\text{Catalyst}} C_2H_5-O-C_2H_5 + H_2O$$

Ethyl alcohol   Diethyl ether

process, such as Al_2O_3 at an optimum temperature. At a certain temperature, the product of dehydration of an alcohol is an olefin, as explained in Section 25.6 where it was shown that an acid catalyst causes formation of an unstable *carbonium* ion. This section should be reviewed. At 180°C, the carbonium ion expels a proton to yield an olefin. At 140°C, however, the carbonium ion reacts with a second molecule of the alcohol

$$\left[\begin{array}{c} H\ H \\ H:\ddot{C}:\ddot{C} \\ H\ H \end{array}\right]^{+} + :\ddot{O}:C_2H_5 \longrightarrow \left[C_2H_5:\ddot{O}:C_2H_5\right]^{+} \longrightarrow C_2H_5OC_2H_5 + H^+$$

to give an intermediate *oxonium* ion which is unstable and expels a proton to yield an ether. The positively charged fragments during these reactions are paired with the negative ions of the catalyst, such as the sulfate ions of sulfuric acid.

25.19. Other Representative Ethers. An industrially important ether is diethylene glycol, a straight-chain condensation product of

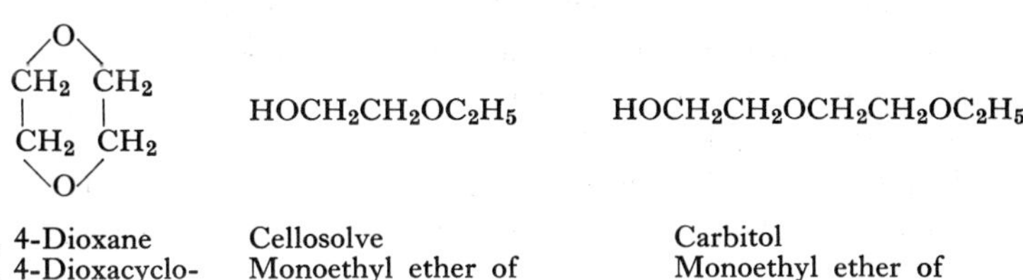

$$HOCH_2CH_2OCH_2CH_2OH \xleftarrow[-H_2O]{\text{2 molecules}} \underset{\underset{OH}{|} \quad \underset{OH}{|}}{CH_2-CH_2} \xrightarrow{-H_2O} \underset{O}{CH_2-CH_2}$$

Diethylene glycol　　　　　　Ethylene glycol　　　Ethylene oxide

ethylene glycol (Section 25.14).　An inner anhydride of ethylene glycol is ethylene oxide, which is a cyclic ether widely used as an intermediate in chemical processes and as a fumigant.　These reactions illustrate the interrelations of the compounds rather than their commercial methods of preparation.

The cyclic ether known as 1, 4-*dioxane* may be regarded as derived from two molecules of ethylene glycol by elimination of two molecules of water, or from diethylene glycol by removal of a molecule of water from the terminal OH groups.　Among the important straight-chain ethers used in large quantity as solvents for lacquers are *Cellosolve* and *Carbitol*.

$$\underset{CH_2 \quad CH_2}{\overset{O}{\underset{|\qquad|}{\overset{/\ \backslash}{CH_2 \quad CH_2}}}}\overset{}{\underset{O}{}}\qquad\qquad HOCH_2CH_2OC_2H_5 \qquad\qquad HOCH_2CH_2OCH_2CH_2OC_2H_5$$

1, 4-Dioxane　　　　Cellosolve　　　　　　Carbitol
1, 4-Dioxacyclo-　　Monoethyl ether of　　Monoethyl ether of
　hexane　　　　　　ethylene glycol　　　　diethylene glycol

26

Aldehydes and Ketones

26.1. Formation from Alcohols. The alcohols described in the last chapter, the aldehydes and ketones to be presented in this chapter, and the acids to be studied in the next chapter are all very closely related. If we keep this in mind while reading about the properties of aldehydes and ketones their chemistry will be much easier to understand. The student should review what was said about primary (1°) and secondary (2°) alcohols in Sections 25.3 and 25.7.

When the vapors of alcohols are led over a suitable catalyst at an elevated temperature there is a direct removal of two atoms of hydrogen:

$$
\begin{array}{ccc}
\underset{\substack{|\\ \text{H}\ \ \text{OH}}}{\overset{\substack{\text{H}\ \ \text{H}\\ |\ \ \ |}}{\text{H}-\text{C}-\text{C}-\text{H}}}
& \underset{+\text{H}_2}{\overset{-\text{H}_2}{\rightleftharpoons}}
& \underset{\substack{|\ \ \ ||\\ \text{H}\ \ \text{O}}}{\overset{\substack{\text{H}\\ |}}{\text{H}-\text{C}-\text{C}-\text{H}}}
\end{array}
$$

Ethanol	Ethanal	2-Propanol	Propanone
Ethyl alcohol	Acetaldehyde	Isopropyl alcohol	Acetone

Formation of an aldehyde, R—C—H   *Formation of a ketone,* R—C—R

Since the —OH group is at the end of the chain in a primary alcohol, the product of dehydrogenation has a –C–H group. The resulting compound is called an aldehyde, and –CHO is the *aldehyde* group. In a secondary alcohol the —OH group is inside the chain; dehydrogenation results in a ketone, where the –C– group is inside the chain and called the *keto* group.

We shall find very shortly that these H atoms can be put back again by the process of *hydrogenation,* and the alcohols regenerated. The dehydrogenation process explains how the aldehydes as a class obtained their family name. The word aldehyde is a contraction of *al*cohol *dehyd*rogenated.

The same results are obtainable by oxidizing the alcohols in a dilute oxidizing acid solution (a chromate dissolved in sulphuric acid):

$$
\begin{array}{ccccccc}
& \text{H} & \text{H} & & & \text{H} & \\
& | & | & & & | & \\
\text{H}- & \text{C} & -\text{C}-\text{H} & \xrightarrow{\text{O}_2} & \text{H}-\text{C}-\text{C}-\text{H} & & \\
& | & | & & | \quad \| & & \\
& \text{H} & \text{OH} & & \text{H} \quad \text{O} & &
\end{array}
$$

Ethyl alcohol　　　　Acetaldehyde　　　　Isopropyl alcohol　　　　Acetone

It is an equally simple matter to pass from an aldehyde or ketone to an acid

(see Section 27.2). The $-\overset{\textstyle |}{\text{C}}=\text{O}$ group is one of the most important in organic chemistry and is given the name *carbonyl*. Since it will be met in so many different surroundings it is important at this point to make sure its position in aldehydes and ketones is well understood.

26.2. Oxidation and Reduction. Many organic chemists still use these two terms in the old-fashioned sense. For example, we just showed how dehydrogenation of ethyl alcohol (CH_3-CH_2OH) yields acetaldehyde (CH_3-CHO). This is usually called an oxidation because the compound formed in this reaction has a higher oxygen content (at the terminal carbon atom) than the original molecule, which has been at the same time deprived of hydrogen. Similarly, hydrogenation of acetaldehyde is often called a reduction reaction, because hydrogen is added and the oxygen content thus decreased. Reduction, to many organic chemists, is a reaction involving hydrogen in which the hydrogen content is increased.

In modern terminology, oxidation and reduction refer, respectively, to the loss or gain of electrons by an element or its compounds. As an illustration, when a sodium atom combines with a chlorine atom as shown in Figure 4.3, the sodium is oxidized (loss of electron) while the chlorine is reduced (gain of electron):

$$\text{Na}^0 + \text{Cl}^0 \longrightarrow \text{Na}^+\text{Cl}^-$$

This is an oxidation, although it does not involve oxygen, the oxidation state of the sodium atom being raised from 0 to $+1$. The chlorine atom gains an electron and is said to pass to a lower state of oxidation, to a charge of -1. Oxidation and reduction take place simultaneously.

Applying these principles to compounds of carbon is not quite so simple because of the tendency of carbon to form covalent rather than ionic compounds. Electrons in carbon compounds are usually shared rather than transferred, but a calculation of the *oxidation number* of the carbon atom may arbitrarily be made by use of the oxidation numbers of the elements attached to carbon, as in the following examples:

$$\text{H:C:O:H} \rightleftharpoons \text{H:C::O:} + \text{H}_2$$

Methyl alcohol　Formaldehyde

The H atom, in most of its compounds, is referred to as having an oxidation state of $+1$, as in hydrochloric acid, $\text{H}^+ \text{Cl}^-$. The carbon atom in methyl alcohol may therefore be said to have an oxidation state of -2; this charge is the sum of -3 due to the electrons gained from the three hydrogen atoms and $+1$ due to an electron donated to the oxygen atom. There is no actual transfer of electrons, but carbon is looked upon as slightly more electron-attractive than hydrogen and oxygen more electron-attractive than carbon (see Table 8.2).

Conversion of the alcohol to the aldehyde is therefore an oxidation, the carbon atom being raised to an oxidation state of O; this is the sum of -2 due to the H atoms and $+2$ due to the oxygen atom. In the reverse reaction—formation of the alcohol from the aldehyde—the process is a reduction.

This conception of oxidation states of the elements has wide use in inorganic chemistry, with ionic compounds, but, as already stated, its application in organic chemistry is quite limited. As carbon tends to share electrons, it is often meaningless to attempt to assign a definite oxidation state or oxidation number to carbon in its compounds. However, oxidation-reduction systems are an important factor in modern biochemistry (see cystine-cysteine in Section 30.3 and in Section 33.6).

26.3.　Nomenclature. Typical aldehydes and ketones occurring in the straight-chain and cyclic families of compounds are shown in Tables 26.1 and 26.2. Aldehydes, like most organic compounds, can be named in two different ways, one method giving the common names and the other method resulting in the systematic names. In order to understand how the aldehydes received their common names, it is necessary to know something about the relationship between aldehydes and organic acids. This will be discussed later in this chapter in connection with the chemical properties of the aldehydes (Section 26.6).

The systematic names are obtained as usual from the longest chain of carbon atoms containing the aldehyde group, with the 1 position assigned

Table 26.1. Aldehydes: *Aliphatic*

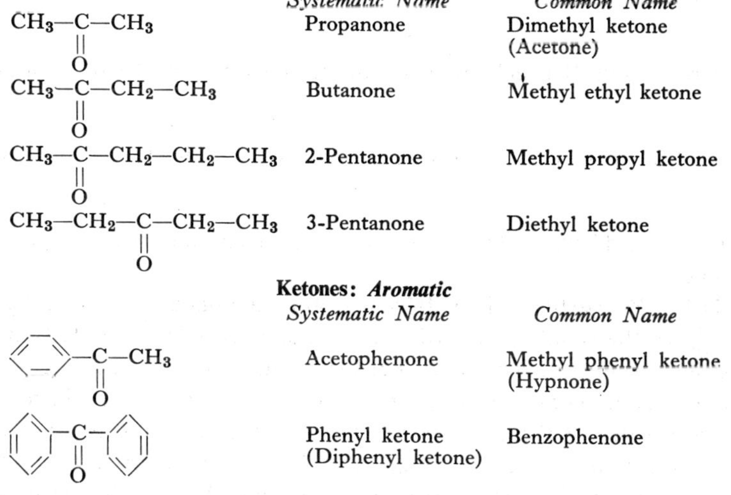

	Systematic Name	Common Name
H—C—H (||O)	Methanal	Formaldehyde
CH_3—C—H (||O)	Ethanal	Acetaldehyde
CH_3—CH_2—C—H (||O)	Propanal	Propionaldehyde

Aldehydes: *Aromatic*

	Systematic Name	Common Name
C_6H_5—C—H (||O)	Benzaldehyde	Benzaldehyde Oil of bitter almonds

Table 26.2. Ketones: *Aliphatic*

	Systematic Name	Common Name
CH_3—C—CH_3 (||O)	Propanone	Dimethyl ketone (Acetone)
CH_3—C—CH_2—CH_3 (||O)	Butanone	Methyl ethyl ketone
CH_3—C—CH_2—CH_2—CH_3 (||O)	2-Pentanone	Methyl propyl ketone
CH_3—CH_2—C—CH_2—CH_3 (||O)	3-Pentanone	Diethyl ketone

Ketones: *Aromatic*

	Systematic Name	Common Name
C_6H_5—C—CH_3 (||O)	Acetophenone	Methyl phenyl ketone (Hypnone)
C_6H_5—C—C_6H_5 (||O)	Phenyl ketone (Diphenyl ketone)	Benzophenone

to the aldehyde group. The aldehyde group is regarded as a "functional" one (Section 10.9); its presence in the molecule is designated by the ending *-al* which is used in the main part of the name as follows:

$$CH_3—CH—CH_2—CHO$$
$$|$$
$$CH_3 \qquad \text{3-Methylbutanal}$$

In relatively simple compounds like this one, it is not necessary to use a numeral with *-al* because it is at the beginning of the chain.

In the nomenclature of aldehydes, frequent use is made of the Greek alphabet (compare the names of acids in Section 27.3), carbon atom 2 being marked α. The scheme can be represented by this skeleton structure.

$$C-C-C-C-CHO$$
$$\delta \quad \gamma \quad \beta \quad \alpha$$

On this basis, 3-methylbutanal would be called β-methylbutanal.

In the case of ketones, the name is obtained by counting the number of C atoms in the chain, writing down the name of the corresponding hydrocarbon, and adding the suffix *-one* in place of *-e*. A number is then added to show the position of the $C=O$ bond in the chain. As illustrated in Table 26.2 by the two formulas of pentanone, this number is not necessary until we reach the compound with a chain of five carbon atoms. Ketones with the structure Ar–C–R are *phenones*, and the rest of the name

is derived from the acid corresponding to the group –C–R; thus, aceto-

phenone is a phenone of acetic acid (Section 27.15). The Ar–C–Ar ketones

have the word ketone added to the names of the Ar radicals as in diphenyl ketone.

26.4. Properties of Aldehydes and Ketones. Formaldehyde (HCHO), the simplest aldehyde, is a gas and acetaldehyde (CH_3CHO), the next member of the family, boils at about 21°C. Higher aldehydes are liquids or solids. Although the ketones differ from the aldehydes in the position of the $C=O$ bond, this difference does not bring about many important variations in properties. Most of the common aldehydes have a disagreeable odor, whereas ketones are liquids or solids with pleasant, aromatic odors; in other respects, they are similar physically.

As we have already stated, the $C=O$ bond is present in many types of compounds, but it is most active in an aldehyde molecule. From the following illustrations, it will be apparent that the reactivity of aldehydes is due almost entirely to the nature of the $C=O$ bond. The ketones are not quite as reactive as the aldehydes, but the general nature of their chemistry is the same.

The reactivity of aldehydes and ketones is easy to understand if we relate their properties to their electron structures. For the double bond in the $\overset{|}{\underset{|}{C}}$=O group we can write either a normal covalent structure or a dipolar structure. Formulas I and II are two different ways of showing

$$
\begin{array}{cccc}
CH_3{-}\overset{||}{\underset{|}{C}}{-}H & CH_3{:}\overset{..}{\underset{..}{C}}{:}H & \overset{\oplus}{CH_3}{-}\overset{|}{\underset{|}{C}}{-}H & \overset{\oplus}{CH_3}{:}C{:}H \\
O{-} & \overset{..}{O}{:} & {::}\overset{|}{O}{-} & {:}\overset{..}{\underset{..}{O}}{:} \\
 & & \ominus & \ominus \\
(I) & (II) & (III) & (IV)
\end{array}
$$

Acetaldehyde

$$
\begin{array}{cccc}
CH_3{-}\overset{||}{\underset{|}{C}}{-}CH_3 & CH_3{:}\overset{..}{\underset{..}{C}}{:}CH_3 & \overset{\oplus}{CH_3}{-}\overset{|}{\underset{|}{C}}{-}CH_3 & \overset{\oplus}{CH_3}{:}C{:}CH_3 \\
O{-} & \overset{..}{O}{:} & {::}\overset{|}{O}{-} & {:}\overset{..}{\underset{..}{O}}{:} \\
 & & \ominus & \ominus \\
(I) & (II) & (III) & (IV)
\end{array}
$$

Acetone

that there are unshared electron pairs on the oxygen atom. Formulas III and IV show the polarized structures. Aldehydes and ketones can be considered as resonance hybrids of II and IV, and from dipole-moment measurements they appear to have about 50% ionic character. Accordingly, either the normal covalent or the polarized structure can be used for the formula of an aldehyde or ketone, but it is better to keep in mind the polarized structures when studying the chemistry of these compounds. The student may wish to refresh his memory as to why the apparent shift of a *pair* of electrons leads to separation of only *unit* charge in the dipolar structures (see text accompanying Figure 15.10).

26.5. Reactions as a Hydrocarbon. Aldehydes and ketones show the principal characteristic property of hydrocarbons—namely, substitution of H atoms by the halogens (Section 9.1). The important observation in this chapter is that substitution takes place most readily with the H atoms on α-carbon atoms. Chlorination of butanal and of diethyl ketone, for example, yields the following products:

$$
\begin{array}{cc}
CH_3{-}CH_2{-}\underset{\underset{Cl}{|}}{CH}{-}\overset{\overset{||}{O}}{CH} & CH_3{-}\underset{\underset{Cl}{|}}{CH}{-}\overset{\overset{||}{O}}{C}{-}CH_2{-}CH_3
\end{array}
$$

α-Chlorobutanal   Ethyl α-chloroethyl ketone
2-Chlorobutanal   2-Chloro 3 pentanone

This activity of a C–H bond when it is adjacent to the C=O group will be discussed in Section 28.11 under *tautomerism*, and in Section 37.4 a proof of the mobility of the α hydrogen atom will be presented. In this section we shall briefly examine the phenomenon on the basis of the principles of resonance which have already been used for analogous interpretations.

As explained in Section 28.11, compounds in which C–H is adjacent (α) to the C=O group often react in an *enol* form:

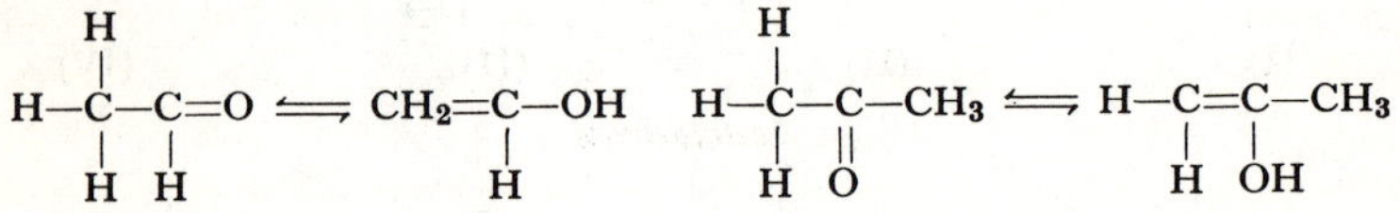

$$
\begin{array}{cc}
\underset{\substack{| \\ H}}{\overset{\substack{H \\ |}}{H-C}}-C=O \;\rightleftharpoons\; CH_2=\underset{\substack{| \\ H}}{C}-OH \qquad\qquad
\underset{\substack{| \\ H}}{\overset{\substack{H \\ |}}{H-C}}-\underset{\substack{|| \\ O}}{C}-CH_3 \;\rightleftharpoons\; H-\underset{\substack{| \\ H}}{C}=\underset{\substack{| \\ OH}}{C}-CH_3
\end{array}
$$

     Acetaldehyde   Enol form   Acetone   Enol form

In the enol modification, the H atom in the OH group has acidic properties; this property is so pronounced that acetone will react with an active metal like sodium to evolve H_2 gas and yield a salt known as an *enolate:*

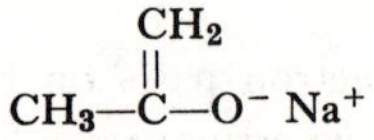

$$
\begin{array}{c}
CH_2 \\
|| \\
CH_3-C-O^-\;Na^+
\end{array}
$$

Although sodium metal is frequently used in the laboratory to dry organic liquids (by reacting with the water present), it cannot be used to dry acetone—a fact that many chemistry students learn from unfortunate experience.

In the chlorination of aldehydes and ketones it is an H atom on carbon adjacent to the C=O group that is similarly involved. The reaction is catalyzed by both acids and bases, but the base catalysis is far more effective and will be discussed here with acetone as an example. In dilute wa-

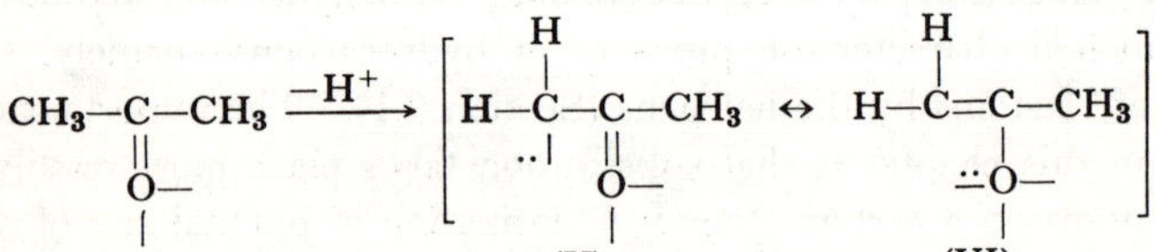

$$
CH_3-\underset{\substack{|| \\ O-}}{C}-CH_3 \xrightarrow{-H^+} \left[\; \underset{\substack{| \\ O- \\ (V)}}{\overset{\substack{H \\ |}}{H-C}}-\underset{\substack{|| \\ O-}}{C}-CH_3 \;\leftrightarrow\; \underset{\substack{| \\ O- \\ (VI)}}{\overset{\substack{H \\ |}}{H-C}}=C-CH_3 \;\right]^-
$$

     Acetone (see I)   Resonance structures of the enolate ion

ter solution of a base, a proton is removed from an α carbon atom. This is a relatively slow process; the product is an ion for which resonance structures can be written. Stabilization of the ion by this resonance effect is the principal reason for acidity of the α hydrogen atoms (compare acidity of phenol in Section 25.12). Halogenation of acetone has been exhaustively studied; the rate-determining (Sections 17.13 and 18.3) step, or steps, is enolization and ionization. As soon as an ion appears in solution it is instantly attacked by a reagent molecule, and the rate is therefore the same no matter which of the halogens is used.

In Structure V, it will be seen that the unshared electron pair on the α carbon atom offers a point of attack for an electron-seeking reagent. The reaction with chlorine probably occurs through an *ionic* mechanism like that described in Section 20.2, the positive fragment of the chlorine molecule, Cl^+ :Cl^-, adding to the α carbon atom in Structure V and the negative chloride ion pairing with an H^+ ion to give HCl as by-product.

Halogenation products of acetone are lacrimatory. Examples are bromoacetone, $CH_2Br–CO–CH_3$, and *sym*-dichloroacetone, $CH_2Cl–CO–CH_2Cl$. The prefix *sym* means symmetrical. In chemical warfare terminology, these compounds are tear "gases," but the first is a liquid and the second a solid. One of the most powerful of lacrimators is known in chemical warfare as CN; it is also used for riot control and "burglar-proofing." It is a crystalline solid (m.p. 54°C) of the following structure:

$$\text{C}_6\text{H}_5-\text{C}-\text{CH}_2\text{Cl} \qquad \alpha\text{-Chloroacetophenone}$$
$$\|$$
$$\text{O}$$

This compound is also known as ω-chloroacetophenone, because the prefix *omega* is often used to designate the last carbon atom in a chain (ω is the last character in the Greek alphabet).

26.6. Oxidation. Aldehydes, in general, can be oxidized without much trouble:

$$\text{CH}_3-\text{C}=\text{O} \xrightarrow{\ \text{O}_2\ } \text{CH}_3-\text{C}=\text{O}$$
$$\quad\ \ | \qquad\qquad\qquad\ \ |$$
$$\quad\ \ \text{H} \qquad\qquad\qquad\ \ \text{OH}$$
$$\text{Acetaldehyde} \qquad\quad \text{Acetic acid}$$

It is from this property of easy oxidation to acids that aldehydes derive their common names. An aldehyde is named from the acid to which it can be oxidized. Acetaldehyde (or acetic aldehyde) is convertible to acetic acid, formaldehyde to formic acid, and so on. It will be seen that the aldehyde and its corresponding acid have very similar structures; the two molecules differ only by one oxygen atom.

A substance that is easy to oxidize is a correspondingly strong reducing agent. This strong reducing action of aldehydes is used in one of the methods for detecting aldehydes in unknown substances. Aldehydes reduce the copper in "Fehling's solution" to cuprous oxide and the silver in ammoniacal silver nitrate to metallic silver. Silver mirrors can be made by the interaction of formaldehyde and certain silver solutions on glass.

Whereas the aldehydes are easily oxidized, the ketones are oxidized only with some difficulty. They cannot, therefore, be detected by their action on solutions of silver or copper salts. The changes that take place in the oxidation of a ketone are much more complex than those taking place in the oxidation of an aldehyde, and we. shall therefore reserve its discussion for the next chapter.

26.7. Addition to the C=O Group. By far the most important characteristic of the C=O bond in aldehydes and ketones is its tendency toward addition reactions. The following reaction is a *reduction:*

$$
\underset{\text{Ethanal}}{\begin{array}{c} H \\ | \\ CH_3{-}C \\ || \\ O \end{array} \begin{array}{c} \leftarrow\cdots H \\ + | \\ \leftarrow\cdots H \end{array}} \longrightarrow \underset{\text{Ethanol}}{\begin{array}{c} CH_3{-}CH_2 \\ | \\ OH \end{array}}
$$

Thus, when an aldehyde is treated with active hydrogen, it easily adds on one molecule of hydrogen to form the corresponding alcohol. In this illustration, acetaldehyde is reduced to ethyl alcohol. Notice that the *hydrogenation,* or reduction of aldehydes to alcohols, is exactly the reverse of the *dehydrogenation,* or oxidation of alcohols to aldehydes which was mentioned at the beginning of the chapter. In the same way, ketones are also reduced to alcohols. As we showed earlier in this chapter, aldehydes give 1° alcohols on reduction whereas ketones give 2° alcohols.

The reason for the facility with which the C=O group undergoes addition reactions is easily found in the electronic structures III and IV of Section 26.4. It is obvious that the C=O bond has two centers of attraction; at the C end it will accept electrons, and at the O end it can donate them. It therefore can accommodate *both* nucleophilic and electrophilic reagents, whereas it will be recalled that the C=C bond is often written C⊸C to indicate its preference for electrophilic reagents. Since the C=O bond is an active dipole, compounds add to it that are themselves dipolar, or can furnish + and − fragments when acted on by the reaction medium. In a basic solution, H:CN yields :CN⁻ ions through the reaction HCN +OH⁻⇋HOH+:CN⁻. The :CN⁻ adds to the C atom; the complex then has a negative charge and extracts a proton (H⁺) from a molecule of water. This frees an OH⁻ ion, which can then continue the reaction.

Similar *double compounds* can be made with ammonia (H-NH$_2$), with

$$\underset{\substack{\text{An aldehyde}\\ \text{RCHO}}}{\overset{\text{H}}{\underset{-\text{O}-}{\overset{\oplus}{R-C}}}\ \ \ \ +\ \text{H-CN}} \ \ \xrightarrow[\text{(HCN in basic solution)}]{:\text{CN}^-}\ \ \left[\overset{\text{H}}{\underset{-\text{O}-}{R-C-CN}}\right]^{-}\ \ \xrightarrow{\text{HOH}}\ \ \underset{\substack{\text{An aldehyde cyanohydrin}\\ \text{RCHO}\cdot\text{HCN}}}{\overset{\text{H}}{\underset{\text{OH}}{R-C-CN}}\ +\ \text{OH}^-}$$

sodium bisulfite (Na^+ SO_3H^-), and with many other substances. These addition compounds are often simply the first step in more complex rearrangements and eliminations. Some ketones (R–C–R) may find it impossible to add certain reagents unless one of the R groups is small enough, like CH_3, to provide space for the attachment to carbon (this is steric hindrance). The addition reactions mentioned here played a significant part in the development of organic chemistry. The cyanohydrin is an intermediate in the synthesis of acids, and the other two compounds have been used in isolating and purifying aldehydes and ketones.

26.8. Replacement of O in the Carbonyl Group. The O atom in the C=O group of aldehydes and ketones can be replaced in reactions with many compounds. The mechanism is one in which an intermediate addition compound is formed, as in the cyanohydrin synthesis just described, which then either breaks down internally or reacts further in some other way to eliminate the oxygen atom. Reactions of this type are usually written in this short form:

$$\underset{\text{Acetone}}{\overset{CH_3}{CH_3-C=O}}\ +\ \underset{\text{Methylamine}}{H_2N-CH_3}\ \longrightarrow\ \underset{\text{Acetone methylimine}}{\overset{CH_3}{CH_3-C=N-CH_3}}\ +\ H_2O$$

Methylamine will be described in Section 29.2. Its reaction with aldehydes and ketones has been found to be catalyzed by acids—in other words, by H^+. The complex initially formed is a carbonium ion; its addition com-

$$\overset{\text{R}}{\underset{-\text{O}-}{\overset{\oplus}{R-C}}}\ \ \xrightarrow[\text{Acid solution}]{H^+}\ \ \left[\overset{\text{R}}{\underset{\text{OH}}{R-C}}\right]^{}\ \ \xrightarrow{:NH_2-R}$$

$$\left[\underset{\substack{\text{OH H}}}{R\ \overset{\text{R H}}{C}\ N\ R}\right]^{+}\ \ \xrightarrow[-H^+]{-H_2O}\ \ \overset{\text{R}}{R\ C=N\ R}$$

plex with the amine breaks down in several stages to eliminate first a molecule of water and then the proton (regeneration of the catalyst). Important products obtainable by reactions analogous to this are listed in Figure 26.1.

Reagent	*Acetaldehyde*	*Acetone*
H_2N-NH_2 Hydrazine	$CH_3-\overset{\overset{\textstyle H}{\vert}}{C}=N-NH_2$ Acetaldehyde hydrazone	$CH_3-\overset{\overset{\textstyle CH_3}{\vert}}{C}=N-NH_2$ Acetone hydrazone
H_2N-OH Hydroxylamine	$CH_3-\overset{\overset{\textstyle H}{\vert}}{C}=N-OH$ Acetaldehyde oxime Acetaldoxime	$CH_3-\overset{\overset{\textstyle CH_3}{\vert}}{C}=N-OH$ Acetone oxime Acetoxime
CH_3CH_2-OH Ethanol (2 molecules)	$CH_3-\overset{\overset{\textstyle H}{\vert}}{\underset{\underset{\textstyle OC_2H_5}{\vert}}{C}}-OC_2H_5$ Acetal	
PCl_5 Phosphorus pentachloride	$CH_3-\overset{\overset{\textstyle H}{\vert}}{\underset{\underset{\textstyle Cl}{\vert}}{C}}-Cl$ 1,1-Dichloroethane	$CH_3-\overset{\overset{\textstyle CH_3}{\vert}}{\underset{\underset{\textstyle Cl}{\vert}}{C}}-Cl$ 2,2-Dichloropropane

26.1. Compounds formed by replacement of the O atom of the $C=O$ group in aldehydes and ketones.

The oximes and hydrazones are generally solids, easy to crystallize, with definite melting points and crystal shapes. It is often a simple matter to identify an aldehyde or ketone from these characteristics of its derivatives. The chemistry of the acetal reaction will be considered in more detail in the study of the sugars (Section 34.12). Ketals corresponding to the acetals can be made, but not from the listed reagent. The reaction with PCl_5 is of interest because it affords a method of placing two chlorine atoms on the same carbon.

26.9. The Aldol Reaction. The aldehydes and ketones are so susceptible to the formation of addition products that they even show a marked tendency to combine with themselves, thus:

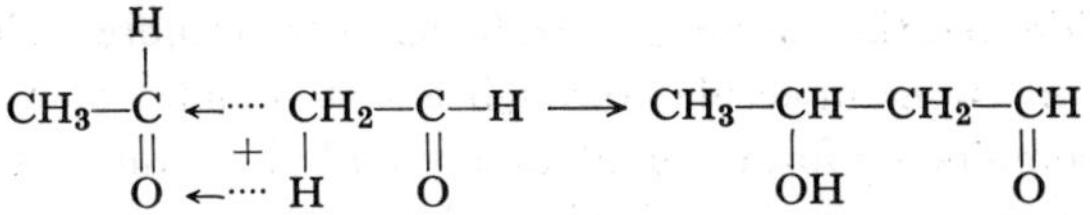

$$\text{CH}_3\text{-}\overset{\text{H}}{\underset{\overset{\|}{\text{O}}}{\text{C}}} \overset{}{\underset{+}{\longleftarrow}} \overset{}{\underset{\text{H}}{\text{CH}_2\text{-}\overset{}{\underset{\|}{\text{C}}}\text{-H}}} \overset{}{\underset{\text{O}}{}} \longrightarrow \text{CH}_3\text{-}\overset{}{\underset{\text{OH}}{\text{CH}}}\text{-CH}_2\text{-}\overset{}{\underset{\text{O}}{\text{CH}}}$$

Two molecules of acetaldehyde   Aldol
(An aldehyde and an alcohol)

This building up to the double molecule takes place readily in dilute alkalies. It is a good example of synthesis in organic chemistry (Section 10.2).

When aldehydes containing more than two carbon atoms are combined, it is the C atom next to the –CHO group that participates in the synthesis, because of the relatively great activity of hydrogen on an α carbon atom:

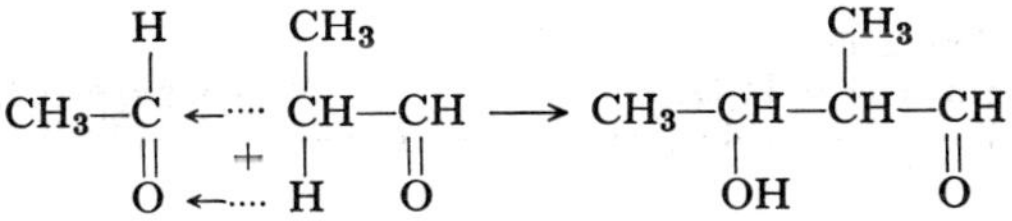

$$\text{CH}_3\text{-}\overset{\text{H}}{\underset{\overset{\|}{\text{O}}}{\text{C}}} \overset{}{\underset{+}{\longleftarrow}} \overset{\text{CH}_3}{\underset{\text{H}}{\text{CH}\text{-}\overset{}{\underset{\|}{\text{CH}}}}} \overset{}{\underset{\text{O}}{}} \longrightarrow \text{CH}_3\text{-}\overset{}{\underset{\text{OH}}{\text{CH}}}\text{-}\overset{\text{CH}_3}{\underset{}{\text{CH}}}\text{-}\overset{}{\underset{\text{O}}{\text{CH}}}$$

Ethanal   Propanal   2-Methyl-3-hydroxybutanal

The mechanism of the reaction is not so simple as that pictured, but takes place in steps as described for other addition reactions of the $\text{C}=\text{O}$ group. Much depends on the reaction medium (acid or base) and on the concentrations used. Ketones go through reactions of this type, and one example is the formation of mesitylene from acetone in strong acid. This reaction

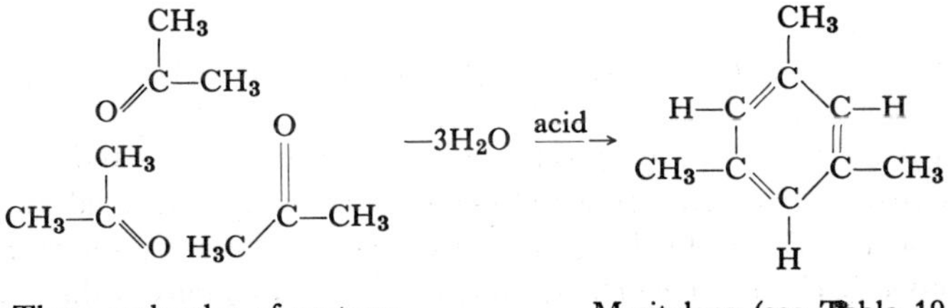

Three molecules of acetone   Mesitylene (see Table 19.1)

is of interest because it illustrates the ease with which we can pass from the straight-chain (aliphatic) series of compounds to a member of the cyclic (aromatic) series.

The union of aldehyde molecules to form an aldol is now called aldol *addition*, rather than aldol *condensation*, the original name. The condensation type of reaction is now said to include only those in which small molecules such as H_2O are eliminated as a by-product. Mesitylene is a product of condensation.

In the aldol reaction, as already explained, two or more molecules combine to form larger molecules through a C–C bond. Some aldehydes, however, combine by formation of an oxygen link between carbon atoms, as shown in general by the following scheme:

$$\ldots {}^+C{-}O^- \ldots {}^+C{-}O^- \ldots {}^+C{-}O^- \ldots {}^+C{-}O^- \ldots$$

In this way, cyclic compounds of the following types can be easily made:

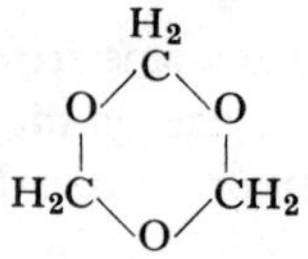

Trioxane
Trioxymethylene $(CH_2O)_3$
(Three molecules of formaldehyde)

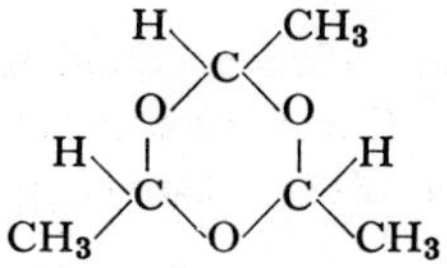

Paraldehyde $(CH_3CHO)_3$
(Three molecules of acetaldehyde)

These compounds are not so stable as the compounds in which C–C links are formed; for example, paraldehyde quickly reverts to acetaldehyde when it is boiled, and under the proper experimental conditions trioxane can be used as a convenient source of formaldehyde for chemical reactions.

Formaldehyde (and possibly other aldehydes) combines with itself to form long-chain molecules as well as the cyclic compounds just described. These are called polyoxymethylenes and have the following general structure,

$$\ldots -CH_2-O-CH_2-O-CH_2-O-CH_2-O-CH_2- \ldots$$

the principal variations being in the nature of the end groups and the length of the chain. The chains contain from forty to perhaps one hundred CH_2O units, with –OH or –OCH$_3$ groups at the ends. They are solids, readily converted back to formaldehyde on heating. Compounds of this type are called *polymers*, from the Greek meaning many parts of the same thing (compare *isomer* in Section 10.4). The general subject of polymerization will be discussed in Chapter 38.

26.10. The Cannizzaro Reaction. In Sections 26.6 and 26.7 it was observed that an aldehyde is easily *oxidized* to an acid or *reduced* to an alcohol. An aldehyde, therefore, occupies an intermediate state between oxidation and reduction. With the proper assistance, two molecules of an aldehyde can interact in a way known as mutual oxidation and reduction. One molecule oxidizes the other to an acid while it is reduced to an alcohol:

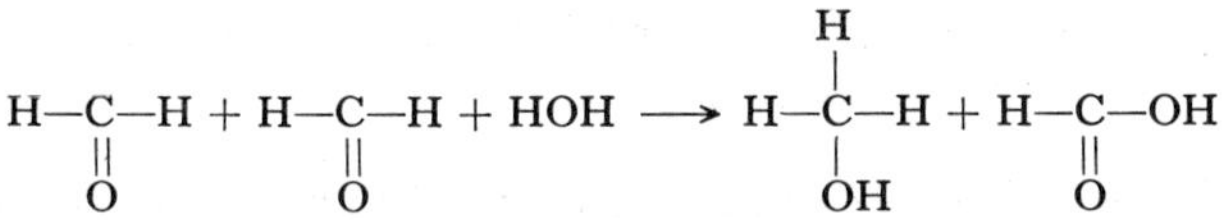

$$H-C-H + H-C-H + HOH \longrightarrow H-\overset{\overset{\displaystyle H}{|}}{\underset{\underset{\displaystyle OH}{|}}{C}}-H + H-C-OH$$

Formaldehyde   Methyl   Formic acid
                  alcohol

The reaction takes place when the aldehyde is warmed with aqueous sodium hydroxide. Since they are in a basic solution, the products contain the salt of the acid rather than the free acid.

The interaction of two similar molecules to give other dissimilar products, as in the Cannizzaro reaction, is referred to as *dismutation* or *disproportionation*.

It should be observed that the experimental conditions for the Cannizzaro reaction are those that promote the aldol addition described in the preceding section. The aldol reaction, however, takes place only with aldehydes in which there is hydrogen on an α carbon atom. Formaldehyde has no α carbon atom; nor has benzaldehyde, which therefore dismutates when heated with alkali:

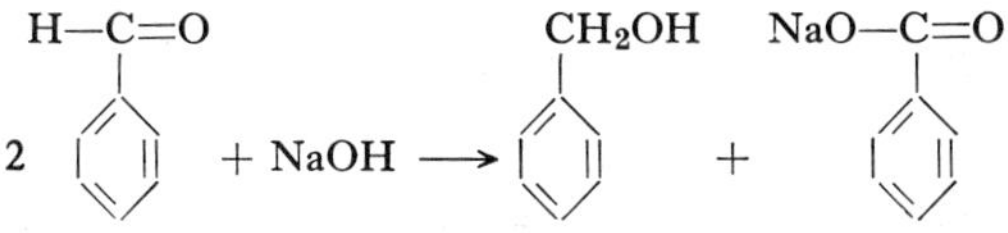

Benzaldehyde   Benzyl alcohol   Sodium benzoate

26.11. Representative Aldehydes and Ketones.

Formaldehyde (see Table 26.1) is the simplest aldehyde. Since it contains only one carbon atom it does not have some of the characteristics of other aldehydes which contain a chain of carbon atoms. For example, it cannot react with halogens in the same way as the other aldehydes do, because of the absence of the carbon chain. It reacts with itself in the Cannizzaro reaction (Section 26.10) and, unlike other aldehydes, it combines with phenols to form hard condensation products of which Bakelite (Section 38.11) is an example. In aqueous solution containing a mild base, formaldehyde (CH_2O) can be made to combine with itself to give a mixture of sugars $(CH_2O)_6$ called *formose*, similar to the natural sugars. The water solution of formaldehyde is completely hydrated to $CH_2(OH)_2$, which is unusual in that there are two OH groups on one carbon atom (see Section 25.8).

Formaldehyde is a gas. It is available in a 40% solution in water called *formalin*. Formaldehyde candles, used for disinfecting, are polymerized formaldehyde (Section 38.5); the solid polymer is known as *paraformaldehyde*.

Acetone, $CH_3-\underset{\underset{O}{\|}}{C}-CH_3$, is the simplest ketone. It is obtained by the

dry distillation of wood, from the products of fermentation of starch and molasses induced by certain bacteria, and also by treatment of the propylene (Section 13.16) in cracked petroleum oils. It is one of the most widely used commercial solvents.

Glyoxal, $H-\underset{\underset{O}{\|}}{C}-\underset{\underset{O}{\|}}{C}-H$, is the simplest of the di-aldehydes. It is interesting in that it is a simple combination of two aldehyde groups. It can be obtained by oxidizing the simplest di-alcohol *glycol* (Section 25.8), and gets its name from the fact that it can be further oxidized to oxalic acid (Section 27.16). Glyoxal is the simplest organic compound with color (Section 36.14). It melts at 15°C and boils at 50°C. The crystals are yellow and the vapor is green.

Chloral can be obtained by chlorination of acetaldehyde. It is a liquid

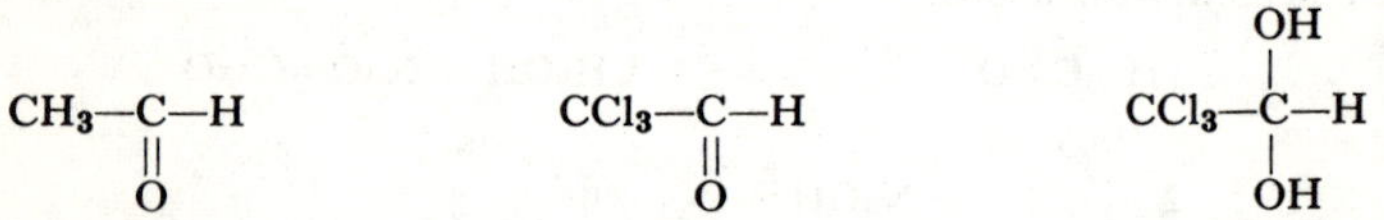

$$CH_3-\underset{\underset{O}{\|}}{C}-H \qquad\qquad CCl_3-\underset{\underset{O}{\|}}{C}-H \qquad\qquad CCl_3-\underset{\underset{OH}{|}}{\overset{\overset{OH}{|}}{C}}-H$$

Acetaldehyde   Chloral   Chloral hydrate

Trichloroacetaldehyde

that forms a stable hydrate with water when in aqueous solution. The hydrate is so stable that it can be obtained from the solution as a crystalline solid. This is an exception to the generalization (Section 25.8) that two OH groups on the same carbon atom form an unstable system.

It was pointed out above that formaldehyde is completely hydrated in water. Acetaldehyde is partially hydrated in solution, but the hydration of acetone is negligible.

Chloral hydrate is used in medicine as a soporific.

Benzaldehyde is known as oil of bitter almonds. It is present in bitter almonds, combined with glucose and hydrogen cyanide. It can be prepared from benzal chloride (Section 24.5) by hydrolysis in weak alkalies:

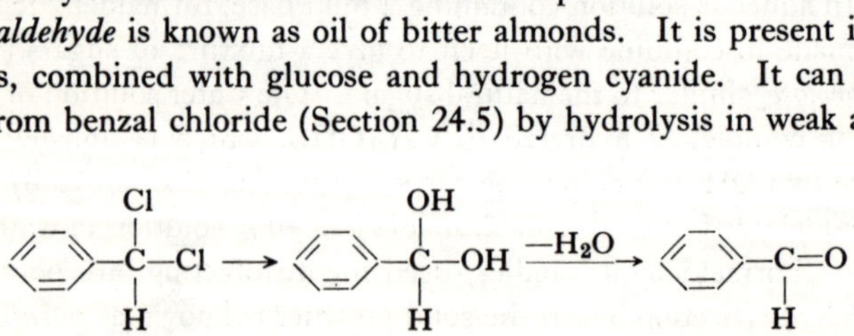

Benzal chloride   (Unstable)   Benzaldehyde

Quinones. In Section 25.14 we described several derivatives of benzene containing two –OH groups. When these are oxidized, they are converted to ketones of the following type:

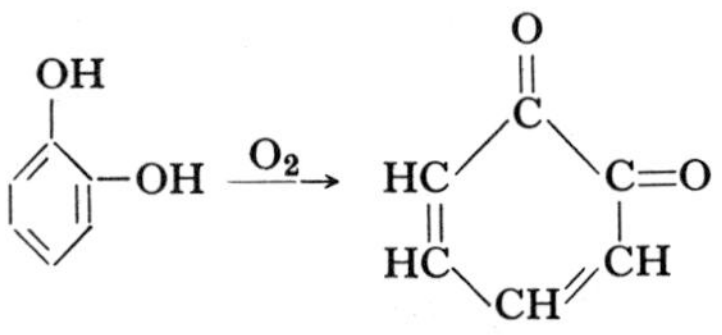

o-Quinol o-Quinone 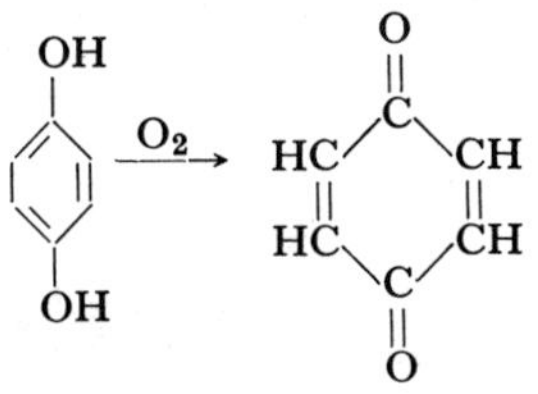 p-Quinol p-Quinone

These quinones do not belong to the benzene series of compounds. The quinones are cyclic compounds, but they have the properties of olefin hydrocarbons in the straight-chain series and also the properties of ketones. They are unusual in that they are highly colored, and when we take up the question of dyes (Section 36.2) we will find that the theory of color in dyes is closely associated with the "quinoid" structure.

27

Carboxylic Acids

27.1. Introduction. Organic acids differ from inorganic acids in that they are compounds of carbon, in addition to being acids. The following compounds are typical organic acids:

$$H_2CO_3 \qquad\qquad HCN \qquad\qquad C_6H_5OH \qquad\qquad C_2H_4O_2$$

Carbonic acid   Hydrocyanic acid   Carbolic acid   Acetic acid
(Phenol)

The composition of acetic acid is $C_2H_4O_2$, and it is correctly listed as an organic acid. It is used quite often, however, in the routine laboratory work of inorganic chemistry, and in the textbooks of inorganic chemistry it is written $HC_2H_3O_2$. The inorganic chemist, when he writes the formula of an acid, likes to single out that particular H atom with acid properties, and experience has shown that only one of the four H atoms in acetic acid is acidic. A typical salt obtained from acetic acid is sodium acetate, $NaC_2H_3O_2$.

Once again we find that the organic chemist is not content with the simpler formulas of inorganic chemistry. If *one* of the H atoms in acetic acid is different from the others, this must be due to a unique position of that one H atom in the molecule, and we ought to show its position when we write the formula of the compound. From a study of the properties of acetic acid which we shall give in this chapter, it is clear that the acetic acid molecule consists of a methyl group, $-CH_3$, linked to what is called a carboxyl group, $-C{=}O$. The H atom in the carboxyl group has acid

$$\qquad\qquad\qquad\qquad\qquad \underset{\displaystyle OH}{\vert}$$

properties when the compound is dissolved in water, and this H atom is the one that is replaceable by metals:

$$CH_3{-}COOH + NaOH \longrightarrow CH_3{-}COONa + H_2O$$

290

27.2. Formation of Carboxylic Acids. In Section 26.1, we showed that alcohols can be oxidized to aldehydes; similarly, aldehydes are easily oxidized one step further to acids:

$$CH_3-CH_2-OH \longrightarrow CH_3-\underset{H}{\overset{\displaystyle =O}{C}} \longrightarrow CH_3-\underset{OH}{\overset{\displaystyle =O}{C}}$$

| Ethyl alcohol | Actaldehyde | Acetic acid |
| Ethanol | Ethanal | Ethanoic acid |

The aldehyde originates from a primary alcohol, and the number of carbon atoms does not change when these reactions take place. However, secondary alcohols on oxidation yield ketones. A ketone is more difficult to oxidize, but when it is accomplished, a mixture of acids results, because the chain of carbon atoms will break at the $C=O$ group to give acids of smaller carbon content:

$$R-CH_2-\underset{O}{\overset{\|}{C}} \mid CH_2-R' \xrightarrow{O} R-CH_2-\underset{O}{\overset{\|}{C}}-OH + HO-\underset{O}{\overset{\|}{C}}-R'$$

$$R-CH_2 \mid \underset{O}{\overset{\|}{C}}-CH_2-R' \xrightarrow{O} R-\underset{O}{\overset{\|}{C}}-OH + HO-\underset{O}{\overset{\|}{C}}-CH_2-R'$$

As indicated, the chain can be broken at either side of the $C=O$ group. The two carbon atoms at the point where the chain is broken are converted into carboxyl groups, so that no matter where the break occurs two acid molecules are formed. It is probable that in the course of oxidation the *enol* form of the ketone adds two OH groups across the double bond $C=C$ and then breaks (Sections 26.5 and 13.10).

The relationship between the carboxylic acids and the other series of compounds we have studied can be shown as follows:

$$CH_3-CH_3 \rightarrow CH_3-CH_2Cl \rightarrow CH_3-CH_2OH \rightarrow CH_3-CHO \rightarrow CH_3-COOH$$

| A paraffin hydrocarbon | A halogen compound | An alcohol | An aldehyde | An acid |

The student should note here the course of events in the transformation of one $-CH_3$ group in CH_3-CH_3 to the carboxyl group, $-COOH$, in CH_3-COOH.

27.3. How to Name Carboxylic Acids. The nomenclature of acids related to the principal type hydrocarbons is apparent from Tables 27.1 and 27.2. The common names of the acids, in most cases, indicate their original sources in nature. Formic acid is obtainable by distillation of ied ants, the Latin name of which is *formica*; acetic acid is present in vinegar, the Latin name of which is *acetum*; and butyric acid gets name from the Latin *butyrum* butter.

TABLE 27.1. Aliphatic Acids

Corresponding Hydrocarbon	Carboxylic Acid	Common Name	Systematic Name
CH_4 Methane	$H-C=O$ $\quad\quad\vert$ $\quad\quad OH$	Formic acid	Methanoic acid
CH_3-CH_3 Ethane	$CH_3-C=O$ $\quad\quad\quad\vert$ $\quad\quad\quad OH$	Acetic acid	Ethanoic acid
$CH_3-CH_2-CH_3$ Propane	$CH_3-CH_2-C=O$ $\quad\quad\quad\quad\quad\vert$ $\quad\quad\quad\quad\quad OH$	Propionic acid	Propanoic acid
$CH_3-CH_2-CH_2-CH_3$ Butane	$CH_3-CH_2-CH_2-C=O$ $\quad\quad\quad\quad\quad\quad\quad\vert$ $\quad\quad\quad\quad\quad\quad\quad OH$	Butyric acid	Butanoic acid
$CH_3-(CH_2)_{16}-CH_3$ Octadecane	$CH_3-(CH_2)_{16}-C=O$ $\quad\quad\quad\quad\quad\quad\quad\vert$ $\quad\quad\quad\quad\quad\quad\quad OH$	Stearic acid	Octadecanoic acid
$CH_3-CH=CH-CH_3$ 2-Butene	$CH_3-CH=CH-C=O$ $\quad\quad\quad\quad\quad\quad\quad\vert$ $\quad\quad\quad\quad\quad\quad\quad OH$	Crotonic acid	2-Butenoic acid
$CH\equiv C-CH_3$ Propyne	$CH\equiv C-C=O$ $\quad\quad\quad\quad\vert$ $\quad\quad\quad\quad OH$	Propiolic acid	Propynoic acid

In systematic nomenclature, the main part of the name is derived from the name of the longest chain of carbon atoms containing the –COOH group, and this group is always written at the end of the chain. The carboxyl group is regarded as the "chief function" (Section 10.9) in the molecule and its presence is indicated in the main part of the name by calling the compound an -oic acid, thus: CH_2Cl-CH_2-COOH, 3-chloropropanoic acid. The carboxyl group does not require a numeral because the carbon atom in the carboxyl group is assigned the 1 position in the chain.

The Greek alphabet is used instead of numbers in an older systematic method for naming acids containing substituent elements or groups. The lettering starts at the first carbon atom after the COOH group, thus:

$$C-C-C-C-COOH$$
$$\delta \quad \gamma \quad \beta \quad \alpha$$

According to this method, CH_2Cl-CH_2-COOH is called β-chloropropionic acid, the older common names being used as last names. As already

TABLE 27.2. Aromatic Acids

Corresponding Hydrocarbon	*COOH Group in the Nucleus*

Benzene
Phene

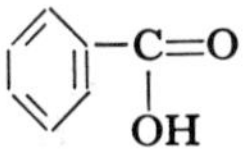

Benzoic acid

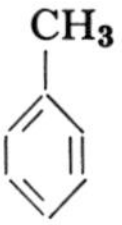

Toluene
Methylbenzene
Phenylmethane

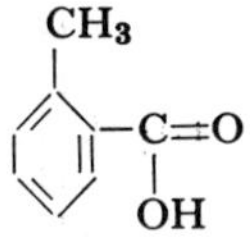

o-Toluic acid
o-Methylbenzoic acid

COOH Group in the Side Chain

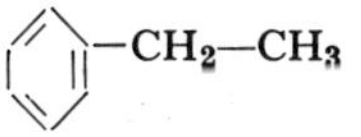

Ethylbenzene
Phenylethane

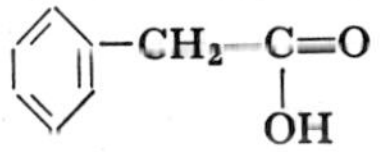

Phenylacetic acid

explained, according to the new systematic procedure this compound is 3-chloropropanoic acid. The Greek-alphabet method of designating the carbon atoms in the chain is used more often for acids than for other types of compounds.

27.4. Properties of Carboxylic Acids. The straight-chain acids in which the chain of C atoms is small are liquids of pungent odor. Formic acid and acetic acid have sharp, irritating odors and the next few acids in the series have terribly rancid odors. The acids with a chain of ten carbon atoms or more are waxy solids at room temperature, and the odor diminishes with the increasing chain length until it is practically absent in stearic acid, $C_{17}H_{35}$ COOH, the substance used in candles.

The melting points of the first ten straight-chain acids are given in Table 27.3 and it will be seen that the molecule with an even number of carbon atoms has a higher melting point than the next member of the series with an odd number of carbon atoms (see Figure 12.12). There is no such irregularity in the boiling points, for the reason given in Section 12.4.

The lower acids are water-soluble, but as the chain of C atoms is lengthened, the solubility in water decreases. The characteristics of the compounds will, of course, approach the properties of the corresponding hydrocarbons as the chain grows larger and the –COOH group at the end

TABLE 27.3.   Straight-chain Acids (R—COOH)

Acid	Formula	Melting point, °C	Boiling point, °C
Formic	H—COOH	8.4	100.7
Acetic	CH_3—COOH	16.6	118.1
Propionic	CH_3—CH_2—COOH	−22	141.1
Butyric	CH_3—$(CH_2)_2$—COOH	− 7.9	163.5
Valeric	CH_3—$(CH_2)_3$—COOH	−58.5	187
Caproic	CH_3—$(CH_2)_4$—COOH	− 1.5	205
Heptylic	CH_3—$(CH_2)_5$—COOH	−10	223.5
Caprylic	CH_3—$(CH_2)_6$—COOH	16	237.5
Nonylic	CH_3—$(CH_2)_7$—COOH	12	254
Capric	CH_3—$(CH_2)_8$—COOH	31.5	269

of the chain becomes a comparatively insignificant part of the molecule.

The aromatic acids—that is, compounds derived from benzene hydro-carbons—are generally colorless, odorless, crystalline, and only slightly soluble in water. In the table of representative acids in this series (Table 27.2), the student should note that we have listed examples in which the –COOH is in the *nucleus* or in the *side chain*, as in benzoic acid and in phenylacetic acid, respectively. There is little difference in the chemical nature of the –COOH group in these two positions.

27.5. Nature of the Carboxyl Group. This group is an intimate union of two other groups already described and with readily recogniz-able characteristics. These are the carbonyl group, $-\overset{|}{C}{=}O$, and the hydrox-yl group, –OH. Their effect on each other in this close association is a a prelude to the subject matter of the next chapter (Mixed Compounds) where we discuss the interaction of several functional groups when they are contained in the same molecule.

In Section 26.4 it was stated that for the $-\overset{|}{C}{=}O$ group in aldehydes and ketones we can write resonance structures $-\overset{|}{C}{=}\overset{|}{O}-$ and $-\overset{|}{\underset{\oplus}{C}}-\overset{|}{\underset{\ominus}{O}}{\cdot\cdot}$ which give it a dipolar nature (about 50% ionic character). These properties carry over into the carboxyl group, and possible resonance structures for the acetic acid molecule can be formulated as shown in I, II, and III. One obvious inference from these resonance structures is that the O atom in the OH group (see III) has a lower electron density than would be expec-ted from the so-called "normal" structure in I. The H atom is therefore held more loosely, and has more freedom to react with water (i.e. ionize, as explained in Section 27.6). This is a condition that does not exist in the OH group of alcohols, where the acidity is almost negligibly small; in the carboxyl group, acidity is brought about by the presence of the $C=O$ group.

The ionization tendency of acetic acid is enhanced by the much greater resonance effect in the acetate *ion* (see IV and V) as compared with that in the acetic acid *molecule*. In the ion, the two structures are completely equivalent, and as stated in Section 16.9c(2) the resonance effect is then at a maximum, with its accompanying increase in stability.

Another factor that controls the acidity of the carboxylic acids is the inductive effect, described in Section 17.2 and Section 21.5. In Section 21.5 it was shown that the CH_3 group has electron-release properties; in $CH_3-C=O$ this effect will increase the electron density of the carboxyl group and in that way increase the ability of the group to hold on to its H atom. Acetic acid accordingly has a lower acidity than formic acid, as seen in Table 27.4. On the other hand, the inductive effect of Cl is $-I$; it is electron-attracting as illustrated in Figure 17.1. In the molecule of chloroacetic acid, listed in Table 27.4, the effect of the Cl substituent is to pull electrons away from the carboxyl group, thus lowering its ability to hold the H atom. The acidity of chloroacetic acid is many times as great as that of acetic acid.

27.6a. The Proton Theory of Acids and Bases. An ionic substance like Na^+Cl^- (Figure 8.1) is simply pulled apart when dissolved in water, $Na^+Cl^- \rightarrow Na^+ + Cl^-$. This is ionic dissociation or *ionization*. Hydrogen chloride, however, is a covalent molecule. When HCl is dissolved in water, a molecule of water extracts a proton from it to form the hydronium ion referred to in Section 23.7:

$$H_2O + HCl \rightleftharpoons H_2O \cdot H^+ + Cl^-$$

or

$$\begin{matrix} H & & H \\ \ddot{\cdot O \cdot} + H \colon \ddot{Cl} \colon & \rightleftharpoons & \colon \ddot{O} \colon H^+ + \colon \ddot{Cl} \colon^- \\ H & & H \end{matrix}$$

The solution contains hydronium chloride, and the ionization is practically complete in the direction of the arrow to the right. Water is not a mass of single molecules of $H \colon \ddot{O} \colon H$. These are hydrogen-bonded and form clusters of molecules. The Na^+ and Cl^- ions are not bare, but clothed with molecules of water (that is, hydrated). The hydration of H^+ is probably more complex than shown in these equations, but it is conventional to represent the hydration by hydronium $(OH_3)^+$. The oxonium ions described in Section 25.4 are substitution products of the hydronium ion.

According to the proton theory of acids and bases, an acid is a substance capable of donating protons, and a base is a substance that accepts protons. Water, from this point of view, can be regarded as a base because it takes up the protons from the HCl molecules dissolved in it. The OH^- ion obtainable from an alkali in solution is also a base, because it tends to react with protons to form water $(H_2O.H^+ + OH^- \rightarrow 2H_2O)$.

27.6b. The Electron Theory of Acids and Bases. The proton theory just described is included in a theory of much wider scope, according to which any molecule is an acid if it can accept an electron pair and any molecule is a base if it can donate an electron pair. According to this *electron* theory, the hydrogen ion, or proton, is an acid because it is an acceptor of an electron pair, as in the reaction $H^+ + \colon \ddot{O} \colon H \rightarrow H \colon \ddot{O} \colon H^+$, where water is the base. For a better illustration we may refer to the discussion of the dative bond in Section 8.5, where it was shown that BCl_3 and $\colon NH_3$ form a simple molecular compound, $Cl_3B \colon NH_3$, in which an electron pair is donated by ammonia. The BCl_3 molecule here is an acid and ammonia is the base. The BCl_3 molecule is representative of compounds referred to as *Lewis acids*, after G. N. Lewis, one of the founders of electron-valence theory. The NH_3 molecule, of course, is a Lewis base.

27.7. Acid Strength. When HCl is dissolved in water $(H_2O + HCl \rightleftharpoons H_2O.H^+ + Cl^-)$ there is practically no tendency for the reverse reaction to take place to form HCl molecules. The right-hand reaction is nearly complete, and for this reason we state that hydrochloric acid is a strong acid. The Cl^- ion in this solution is not as basic as water.

With acetic acid the conditions are quite different:

$$CH_3-COOH + H_2O \rightleftharpoons H_2O \cdot H^+ + CH_3-COO^-$$

Acetic acid   Acetate ion

The negative ion in this case tends to react with H^+ ions to form molecular acetic acid. The reaction to the left predominates, and since there are relatively few hydrated H^+ ions in solution, compared with the status in hydrochloric acid solution, the solution of acetic acid is said to be weak. Moreover, the acetate ion is more basic than water.

The relative strengths of acids are usually stated in terms of the *dissociation constant*. This constant is obtained from the law of mass action with reasonable accuracy for weak acids in dilute solution. If the dissociation of a weak acid, HA, is represented by $H_2O + HA \rightleftharpoons H_2O.H^+ + A^-$, then

$$K = \frac{[H_2O \cdot H^+][A^-]}{[HA]}$$

K is the dissociation constant and the quantities in brackets are the concentrations of the reacting substances. The effect of the solvent H_2O molecules can be neglected in the case of dilute solutions of weak electrolytes.

The dissociation constants for a number of representative acids are shown in Table 27.4. The open-chain acids all have about the same de-

TABLE 27.4. **Dissociation Constants of Acids (at 25°C.)**

Substance	*Formula*	*Dissociation Constant*
Hydrochloric acid	H—Cl	Complete dissociation
Formic acid	H—COOH	0.000 17
Acetic acid	CH_3—COOH	0.000 018
Propionic acid	CH_3—CH_2—COOH	0.000 014
Butyric acid	CH_3—CH_2—CH_2—COOH	0.000 015
Chloroacetic acid	CH_2Cl COOH	0.001 5
Dichloroacetic acid	$CHCl_2$—COOH	0.050
Trichloroacetic acid	CCl_3—COOH	Complete dissociation
Benzoic acid	⬡—COOH	0.000 063
Carbonic acid	HO—C—OH (‖O)	0.000 000 35
Phenol (carbolic acid)	⬡—OH	0.000 000 000 13
Hydrocyanic acid	H—CN	0.000 000 000 72
Water (for comparison)	H—OH	0.000 000 000 000 01

gree of acidity, as illustrated by the data for acetic acid and butyric acid. Acidity of the open-chain acids increases rapidly, however, when *electronegative* elements are present in the chain, as shown by the chloroacetic acids. Trichloroacetic acid, CCl_3–COOH, is practically as strong as HCl, but no accurate data are available for strong acids, because their behavior in solution does not follow the simple form of the law of mass action. (For information concerning the equilibrium of strong acids, the student is referred to recent texts on physical chemistry in which the "activity coefficient" is defined.) The benzene ring is electronegative in character, like the Cl atom, and consequently benzoic acid, C_6H_5–COOH, also, is somewhat stronger than acetic acid.

27.8. Hydrogen Bonding. The electronic arrangement in carboxylic acids is highly conducive to the formation of hydrogen bonds (Section 25.10). The H atom in the COOH group is rather loosely held, as we have just pointed out, and the space availability is also good for this H atom to attach itself to the =O of a second molecule of the acid. Two molecules of acetic acid unite to form a *dimer* through two hydrogen bridges. This double molecule is present in the crystalline compound, and it is so stable that it holds together even when the acid is vaporized. Some of the physical properties of carboxylic acids were more readily understood after the discovery of hydrogen bonding.

$$CH_3-C\underset{OH\,\cdots\,O}{\overset{O\,\cdots\,HO}{\lessgtr}}C-CH_3$$

27.9. Acyl Halides. The OH group in a carboxylic acid is easily replaced. It is convenient to name the radical obtained by removing OH from a carboxylic acid the *acyl* group, to create an analogy with the alkyl

$CH_3-C=O$ |	$CH_3-C=O$ | Cl	CH_3-	CH_3-Cl
Acetyl group	Acetyl chloride	Methyl group	Methyl chloride

An acyl group is R—C=O   An alkyl group is R—
|

group. The acyl chlorides are among the very important compounds used in the synthetic work of organic chemistry. In Section 22.17 it was shown how the :Cl⁻ ion can be displaced from an alkyl radical by an entering group that is more nucleophilic (can attack the C atom with a more active electron pair). Now if we inspect Formula II in Section 27.5

of this chapter, we see that the C atom in the $C=O$ group has considerable $+$ character; it is more readily attacked than the alkyl C atom by a nucleophilic reagent. The Cl atom in an acyl chloride is therefore easily displaced; it is also an excellent *leaving* group. Acyl chlorides hydrolyze rapidly in water to regenerate the original acids.

To make an acyl chloride, the carboxylic acid is chlorinated with reagents like PCl_3, PCl_5, and $SOCl_2$ (thionyl chloride). The latter is often preferred because the by-products are easily removed:

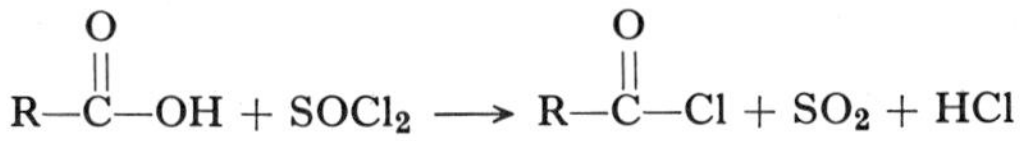

$$R-\overset{\overset{\displaystyle O}{\|}}{C}-OH + SOCl_2 \longrightarrow R-\overset{\overset{\displaystyle O}{\|}}{C}-Cl + SO_2 + HCl$$

A base, such as pyridine (Section 32.4), is usually added to neutralize the acid by-product. It is also of interest to point out that if PCl_5 is used in the chlorination of acetic acid, it replaces not only the –OH but also the oxygen in the carboxyl group:

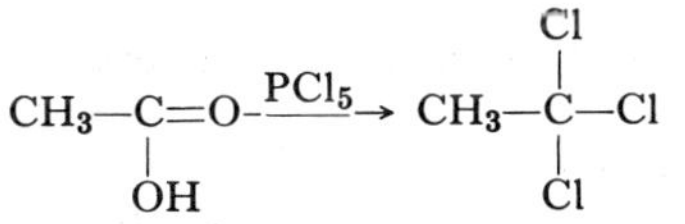

$$CH_3-\underset{\underset{\displaystyle OH}{|}}{C}=O-\xrightarrow{PCl_5} CH_3-\overset{\overset{\displaystyle Cl}{|}}{\underset{\underset{\displaystyle Cl}{|}}{C}}-Cl$$

Ethanoic acid   1,1,1-Trichloroethane
Acetic acid

This yields a compound with three Cl atoms on the same C atom. This should be compared with the action of the same chlorinating agent on aldehydes and ketones (Figure 26.1).

The acyl chlorides are often used to detect an –OH group in organic compounds because of the ease with which they react with this group:

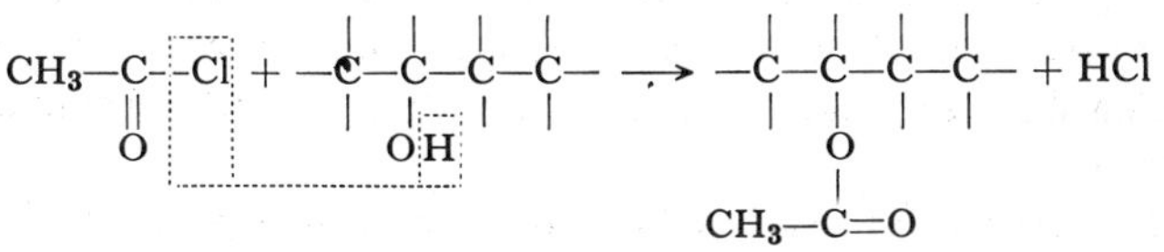

$$CH_3-\underset{\underset{\displaystyle O}{\|}}{C}-Cl + -\overset{|}{C}-\underset{\underset{\displaystyle OH}{|}}{\overset{|}{C}}-\overset{|}{C}-\overset{|}{C}- \longrightarrow -\overset{|}{C}-\underset{\underset{\displaystyle O}{|}}{\overset{|}{C}}-\overset{|}{C}-\overset{|}{C}- + HCl$$
$$CH_3-C=O$$

The product is an ester, a type of compound to be described in Section 27.11. If several –OH groups are present in the molecule, they can usually all be replaced by the acetate radical in this way. The process is called *acetylation.*

Another interesting application is shown by the following reaction, which is an instance of the Friedel-Crafts synthesis (Section 20.6). The product is an aromatic ketone:

$$\bigcirc\!-\!\underset{\underset{O}{\|}}{C}\!-\!\boxed{Cl + H}\!-\!\bigcirc \xrightarrow{AlCl_3} \bigcirc\!-\!\underset{\underset{O}{\|}}{C}\!-\!\bigcirc + HCl$$

Benzoyl chloride   Benzene   Benzophenone

27.10. Salts. The acid nature of the carboxyl group is shown by the ease with which salts are formed through the interaction of these organic acids with bases:

$$CH_3—COOH + Na^+OH^- \longrightarrow CH_3—COO^-Na^+ + H_2O$$

Acetic acid   Sodium acetate

This is a *neutralization* as defined in inorganic chemistry—that is, a reaction between an acid and a base to form a salt and water. The salts of the carboxylic acids are generally crystalline substances. The salts of the open-chain acids with a large number of C atoms in the chain, like sodium stearate, $CH_3–(CH_2)_{16}–COONa$, are called *soaps*, and will be discussed further in Section 27.17. The sodium salt of benzoic acid, $\bigcirc$–COONa, is used as a food preservative.

The salts of the open-chain acids, when fused with a strong alkali, lose the carboxyl group; they are said to be *decarboxylated*:

$$CH_3—CH_2—\boxed{COONa + NaO}H \longrightarrow CH_3—CH_2—H \text{ (or } CH_3—CH_3) + Na_2CO_3$$

Sodium propionate   Ethane

This gives the chemist a method of *reducing* the length of the carbon chain by one carbon atom. To pass from propane to ethane, we can convert $CH_3–CH_2–CH_3$ to $CH_3–CH_2–COOH$ by the reactions outlined in Section 27.2, and then take off the –COOH group as just indicated. The decarboxylation discussed under malonic acid in Section 27.16 should also be referred to.

27.11. Esters. The carboxylic acids react with alcohols, in the presence of strong acids like H_2SO_4, to form esters, the formulas of which resemble those of salts:

$$CH_3—\underset{\underset{O}{\|}}{C}—OH + HO—C_2H_5 \xrightarrow{H^+} CH_3—\underset{\underset{O}{\|}}{C}—OC_2H_5 + H_2O$$

Acetic acid   Ethyl alcohol   Ethyl acetate

It should be understood that esters can be made from inorganic acids as well as organic. This was mentioned in Sections 20.3 and 20.4 in the discussions of nitro compounds and sulphonic acids. The structure of dimethyl sulphate, an ester, was given in Section 20.4. An important nitrate is the trinitrate of glycerol (see Section 25.7), the formula of which is CH_2—CH—CH_2. This is the explosive popularly known as *nitrogly-*

$$ONO_2 \quad ONO_2 \quad ONO_2$$

cerin; it is also used to lower blood pressure, a purpose for which ethyl nitrite, C_2H_5–ONO, another inorganic ester, is also used.

It is instructive to examine the mechanism of esterification of carboxylic acids because it illustrates the way in which the carboxyl group behaves in many of its acid-catalyzed reactions. The resonance structures I, II, and III of the carboxyl group shown in Section 27.5 indicate that the $\overset{\oplus}{C}{=}\overset{\ominus}{O}$ group has a polarized structure. This is intensified by the strong acid catalyst used in the esterification; it adds a proton to the negative end as shown here in II and III and in that way makes the other (positive) end even more reactive toward reagents with free electron pairs. The formulas in this sequence are written with straight lines to represent electron pairs not shared with other atoms (compare Figure 17.2).

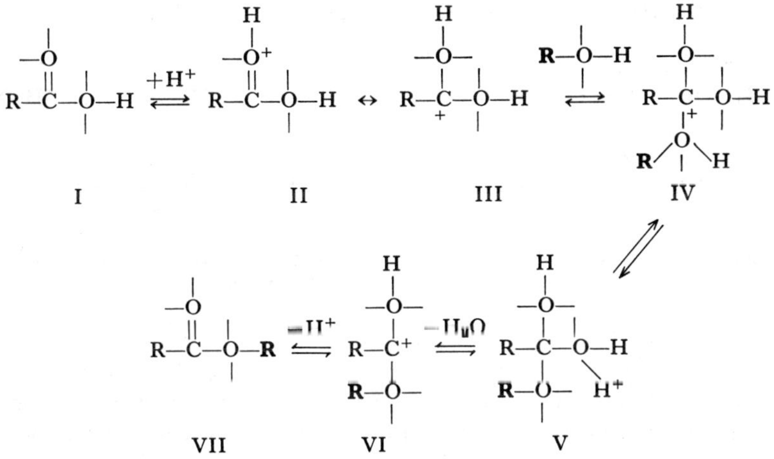

The electron pairs on the alcohol molecule ($R{:}\overset{..}{O}{:}H$) are only weakly nucleophilic; the addition reaction from III to IV takes place because of the strong | charge shown in III. Structure IV is an unstable complex ion; a proton shift takes place to yield V, which then eliminates a molecule of water, VI, and finally a proton is transferred to a molecule of the reaction medium (water) to give OH_3^+ and final product, VII. Formulas II and III are resonance structures, which differ only in arrangement of electrons. The OH group on carbon (as in IV) is not a good "leaving" group because of its strongly basic nature, but the protonation in V creates a water molecule which leaves readily.

Esters formed from the lower open-chain acids and the lower alcohols are sweet-smelling liquids, used for flavoring and as solvents for lacquers. The *waxes* are essentially esters of the long-chain acids and the long-chain alcohols. For example, spermaceti is a crystalline wax obtained from sperm oil of the whale; one of its ingredients is cetyl palmitate:

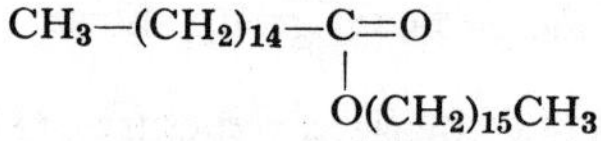

$$CH_3-(CH_2)_{14}-C{=}O$$
$$|$$
$$O(CH_2)_{15}CH_3$$

the ester of a C_{16} acid, known as palmitic acid, and of a C_{16} alcohol, called cetyl alcohol.

Fats are the esters of the open-chain acids and glycerol. This alcohol contains three —OH groups and can therefore react with three molecules of a carboxylic acid. These substances occur plentifully

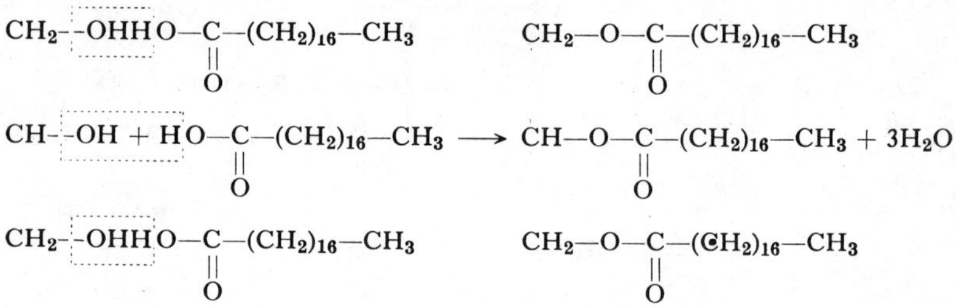

$$CH_2{-}OH \quad HO{-}\underset{\underset{O}{\|}}{C}{-}(CH_2)_{16}{-}CH_3 \qquad CH_2{-}O{-}\underset{\underset{O}{\|}}{C}{-}(CH_2)_{16}{-}CH_3$$

$$CH{-}OH + HO{-}\underset{\underset{O}{\|}}{C}{-}(CH_2)_{16}{-}CH_3 \longrightarrow CH{-}O{-}\underset{\underset{O}{\|}}{C}{-}(CH_2)_{16}{-}CH_3 + 3H_2O$$

$$CH_2{-}OH \quad HO{-}\underset{\underset{O}{\|}}{C}{-}(CH_2)_{16}{-}CH_3 \qquad CH_2{-}O{-}\underset{\underset{O}{\|}}{C}{-}(CH_2)_{16}{-}CH_3$$

Glycerol   Stearic acid   Glyceryl tristearate (a typical fat) "Tristearin"

in nature in fats and oils, such as tallow, lard, olive oil, butter, etc. In most cases, the acid part of the ester has a long chain, like stearic acid, or palmitic acid $CH_3-(CH_2)_{14}-COOH$. Butter fat, however, contains a high percentage of the ester of butyric acid, which is only a four-carbon acid, $CH_3-(CH_2)_2-COOH$. It is curious that acids built up of odd numbers of carbon atoms have not been found in fats and oils that occur in nature.

The fats were among the first compounds thoroughly studied in the early days of organic chemistry and substances isolated from them were known as "fatty" or aliphatic, in contrast with the "aromatic" compounds obtained from coal tar. It was a long time before it was discovered that the "fatty" materials and "aromatic" materials differ only in that the first are open-chain compounds whereas the second are mostly cyclic compounds of the benzene series.

Since the esters can be hydrolyzed without much difficulty, the reaction for the formation of an ester is really reversible:

$$CH_3-\underset{\underset{\displaystyle OH}{|}}{C}=O + C_2H_5-OH \quad \underset{\text{hydrolysis}}{\overset{\text{esterification}}{\rightleftarrows}} \quad CH_3-\underset{\underset{\displaystyle OC_2H_5}{|}}{C}=O + H_2O$$

     Acetic acid   Ethyl alcohol   Ethyl acetate

Whether or not the reaction proceeds to the right or to the left depends on the concentrations of the reacting substances on either side of the equation. Both right and left reactions are catalyzed by acids and bases. The hydrolysis of fats is one of our major industries. When glyceryl tristearate is heated with a large amount of water and alkali, it is hydrolyzed to glycerol and stearic acid. If the alkali used in the hydrolysis of the fat is sodium hydroxide, the sodium salt of stearic acid is formed. We have already explained that this salt is known as a *soap*. In fact, it is one of our common soaps. The hydrolysis of fats is technically termed *saponification* and the same term is now used for all types of hydrolysis of esters, even though no soap is formed.

The first stage of the digestion of fats in the animal body consists in the hydrolysis of the fat to glycerol and the free carboxylic acid. It is brought about by the action of enzymes called *lipases*.

In animal and vegetable life, oily substances that are not esters are often found. Since they are not esters, they cannot be broken down by enzymes or hydrolysis into an acid and an alcohol. They are said to be nonsaponifiable. The carotenes (Section 38.3) are present in the non-saponifiable fraction of many oils and fats, such as cod-liver oil.

27.12. Dehydration to Acid Anhydrides. Carboxylic acids can be condensed by the action of strong dehydrating agents. A typical reaction is the following:

$$CH_3-\overset{\displaystyle O}{\overset{\displaystyle \|}{C}}-OH + HO-\overset{\displaystyle O}{\overset{\displaystyle \|}{C}}-CH_3 \quad \xrightarrow[\text{dehydration}]{P_2O_5} \quad CH_3-\overset{\displaystyle O}{\overset{\displaystyle \|}{C}}-O-\overset{\displaystyle O}{\overset{\displaystyle \|}{C}}-CH_3 + H_2O$$

        Acetic acid   Acetic anhydride

The acid anhydrides are of importance in synthesis. They are often useful where the original acids cannot be used because of certain experimental requirements. When warmed with water, the original acid is regenerated.

Mixed anhydrides can be obtained in which two different acyl radicals are joined by the oxygen bridge, such as the anhydride of acetic acid and propionic acid, CH_3–CH_2–$\overset{\displaystyle \|}{\underset{\displaystyle O}{C}}$–O–$\overset{\displaystyle \|}{\underset{\displaystyle O}{C}}$–$CH_3$.

The student should compare the structures of ethers, esters, and anhydrides:

$$R\!-\!O\!-\!R \qquad\qquad R\!-\!\overset{\|}{\underset{O}{C}}\!-\!O\!-\!R \qquad\qquad R\!-\!\overset{\|}{\underset{O}{C}}\!-\!O\!-\!\overset{\|}{\underset{O}{C}}\!-\!R$$

$$\text{An ether} \qquad\qquad \text{An ester} \qquad\qquad \text{An acid anhydride}$$

$$\Big\Updownarrow \qquad\qquad\qquad \Big\Updownarrow$$

$$R\!-\!\overset{\|}{\underset{O}{C}}\!-\!OH + HO\!-\!R \qquad\qquad R\!-\!\overset{\|}{\underset{O}{C}}\!-\!OH + HO\!-\!\overset{\|}{\underset{O}{C}}\!-\!R$$

$$\text{An acid} \quad \text{An alcohol} \qquad\qquad \text{Two acids}$$

Ethers, it will be recalled, are very stable, but esters and anhydrides can be hydrolyzed by water to their original constituents.

27.13. Reduction to Primary Alcohols. In Section 26.2 it was shown that the C=O group in an aldehyde is easily reduced, by addition of hydrogen across the double bond. The product is an alcohol. The C=O group in a carboxylic acid, however, is difficult to reduce by methods that work well for most compounds with double bonds. The problem has been solved by the discovery that certain reducing agents —for example, borane (BH_3) and lithium aluminum hydride ($LiAlH_4$) —can be used with great efficiency on these acids. The overall reaction is

$$R\!-\!\overset{\displaystyle O}{\overset{\displaystyle \|}{C}}\!-\!OH + 2H_2 \longrightarrow R\!-\!\overset{\displaystyle H}{\underset{\displaystyle H}{\overset{\displaystyle |}{\underset{\displaystyle |}{C}}}}\!-\!OH + H_2O$$

$$\text{A carboxylic acid} \qquad\qquad \text{A primary alcohol}$$

27.14. Hydrocarbon Properties. The outstanding property of a hydrocarbon chain is its susceptibility to halogenation, and this is still true when the chain is part of a carboxylic acid. Halogens are introduced most readily at the α carbon atom if a hydrogen atom is available at that place. In fact, a reaction has been found in which bromine will replace only the H atoms on the α carbon atom of these acids. The reason why H atoms on an α carbon atom are so reactive was given for aldehydes and ketones in Section 26.5; the same general principles prevail among the carboxylic acids. Chlorine, because it likes to react by a free radical mechanism, has a greater tendency to replace H atoms elsewhere in the chain as well as at the α position.

Carboxylic acids containing a benzene nucleus (for example, benzoic acid in Table 27.2) can be halogenated, nitrated, and sulfonated as described in Chapter 20. The presence of the negative –COOH group, however, makes the introduction of other negative groups a little more difficult.

27.15. Representative Carboxylic Acids. *Formic acid*, H–C=O,
                                                                                           |
                                                                                         OH

is the simplest member of the series of carboxylic acids. It is related to methane in that they both contain but one carbon atom; it is also known as methanoic acid. This colorless liquid, corrosive to the skin, occurs in nature in various plants, in bee stings, and in ants. Formic acid merits a paragraph to itself because of its unusual properties. If the formula of formic acid is examined, it will be seen that it is not only an acid but also an aldehyde, for it contains not only the carboxyl group, –C=O, but also
                                                                                                                              |
                                                                                                                            OH

the aldehyde group, –C=O. It therefore has the properties of both
                                        |
                                        H

types of compounds. It differs from all the other carboxylic acids in that it is a reducing agent (Section 26.2)—that is, it can be easily oxidized:

$$\text{H–C=O} \xrightarrow{\text{O}} \text{HO–C=O} \longrightarrow CO_2 + H_2O$$
$$\quad\ |\qquad\qquad\quad\ |$$
$$\text{OH}\qquad\qquad\text{OH}$$

Formic acid   Carbonic acid

Oxidation converts it to carbonic acid, which decomposes to carbon dioxide and water. Formic acid is also a stronger acid than the succeeding members of its family of compounds.

Acetic acid, CH_3–COOH, is the only edible free acid in this series of compounds, about 4% being present in vinegar. Pure, anhydrous acetic acid is called *glacial acetic acid*, because it crystallizes easily in cold weather to form icelike crystals at 16.7°C. The dry acid has a corrosive effect on the skin and is a good solvent for many organic compounds.

Oleic acid is probably the most important of the acids which contain a double bond in the chain. It is derived from the hydrocarbon octadecene; in other words, there are eighteen carbon atoms in the chain. Experiment has shown that the $C=C$ group is in the middle of the chain, so its formula is $CH_3-(CH_2)_7-CH=CH-(CH_2)_7-COOH$.

Oleic acid occurs combined with glycerol as the fat "triolein," similar in structure to the "tristearin" pictured in Section 27.11. Triolein (or glyceryl trioleate) is found, together with tristearin, in fats, like lard, and in many vegetable oils. Triolein is liquid, whereas tristearin is solid. The vegetable oils are often converted into fats by hydrogenation (Section 13.8). Hydrogen is bubbled through the oil in the presence of a catalyst, and as the H–H molecules add on across the double bonds, the triolein changes into tristearin which is solid. Many oils, like cottonseed oil and peanut oil, are converted to fats by this process and sold at a higher price. In the early 1960's the oils (popularly called *polyunsaturates*) were given a big boost commercially when some medical authorities claimed they were more beneficial than solid fats in human metabolism.

27.16. Polycarboxylic acids. In this section we shall consider a few acids that contain more than one carboxyl group. They are not usually classified among the "fatty" acids defined in Section 27.11.

Oxalic acid, $HO-\underset{\underset{O}{\|}}{C}-\underset{\underset{O}{\|}}{C}-OH$, is the simplest acid in this group. Its

systematic name is ethanedioic acid, which shows its relationship to ethane, CH_3-CH_3. Oxalic acid can be made by the oxidation of glycol (Section 25.7) or of glyoxal (Section 26.11) and can be obtained from many other substances, such as sugars and cellulose, by oxidizing them with nitric acid. Oxalic acid occurs in many plants, especially of the *oxalis* variety, in the form of its salts. It is a crystalline solid and is poisonous.

Malonic acid, or propanedioic acid, $HO-\underset{\underset{O}{\|}}{C}-CH_2-\underset{\underset{O}{\|}}{C}-OH$, is of im-

portance because of its extensive use in organic synthesis, since it has properties similar to those of ethyl acetoacetate, as described in Section 28.12. When malonic acid is heated above its melting point one of the

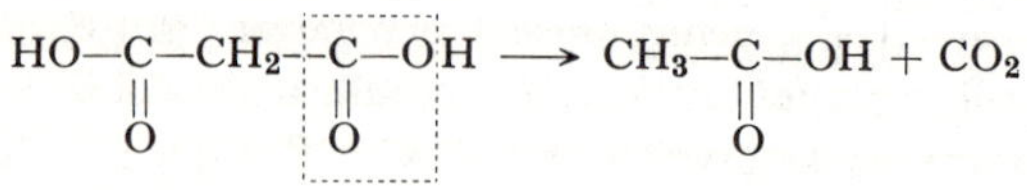

$$HO-\underset{\underset{O}{\|}}{C}-CH_2-\underset{\underset{O}{\|}}{C}-OH \longrightarrow CH_3-\underset{\underset{O}{\|}}{C}-OH + CO_2$$

Malonic acid   Acetic acid

carboxyl groups is decomposed with elimination of CO_2. This is called *decarboxylation*. Acids are most easily decarboxylated by heating them if the α carbon atom carries a group with a strong attraction for an electron pair (high -*I* effect as discussed in Section 17.2). Malonic acid has an α C atom (the CH_2 group) and the COOH group attached to it is electronegative. A similar situation is present in trichloroacetic acid, CCl_3–COOH, which has three electronegative Cl atoms on the α carbon (see Section 27.10) and is easily decarboxylated.

Succinic acid, or butanedioic acid, $HOOCCH_2$–CH_2COOH occurs in amber, the Latin name of which is *succinum*. An interesting property of succinic acid (and of other dibasic acids that contain a chain of at least four carbon atoms) is its ability to form an inner anhydride:

$$\text{Succinic acid} \xrightarrow[\text{dehydrating agent}]{\text{heat and}} \text{Succinic anhydride} + H_2O$$

This behavior is similar to the formation of anhydrides from two molecules of an acid, as described in Section 27.12. The anhydride is a ring compound, whereas the original acid is a chain compound. They are both crystalline solids.

Phthalic acid, $HOOC$–C_6H_4–$COOH$, is a dicarboxylic acid in which the two carboxyl groups are attached to the benzene ring. There are three acids corresponding to this formula. The three acids differ in the positions of the –COOH groups in the ring; they can be *o*-, *m*-, and *p*- to each other.

Phthalic acid (see Section 32.6)   Isophthalic acid   Terephthalic acid (see Section 38.10)

When benzene hydrocarbons are oxidized either in solution or by hot air over a catalyst, the side chains are oxidized down to COOH groups. For example, *o*-xylene in Table 19.1 gives phthalic acid when oxidized; so does naphthalene (Section 32.6). Phthalic acid when heated gives an anhydride (see Section 32.6) which can be decarboxylated to benzoic acid, ⬡–COOH. And benzoic acid can be further decarboxylated to benzene; in fact, benzene got its name from this reaction (the benzoic acid came from gum benzoin).

27.17. Soaps. In Section 27.10 soaps were defined as salts of long-chain carboxylic acids. The common soap with which everybody is familiar is sodium stearate, the principal ingredient of ordinary bar soap. Its manufacture from fats was briefly described under esters. Many other soaps are used commercially in enormous quantities and in a great many ways. They are often made from the sodium soap by precipitation with compounds·of other metals; thus, the aluminum stearate soap is generally made from the sodium soap and aluminum sulphate.

Many of the soaps have lubricating' properties; thus, zinc stearate is used in face powders to lubricate the human skin. It is also a water repellent. Calcium, zinc, and magnesium soaps are used as lubricants on metal and leather surfaces. Some soaps can be dispersed in suitable liquids to give "gels." The well-advertised solid alcohol, or *canned heat*, is such a gel in which the semisolid structure is due to a soap. A related gel made by adding aluminum soaps to gasoline was used in tremendous quantities in World War II as a filling in incendiary bombs and was highly effective as a destructive agent. More concentrated mixtures of such soaps in oils yield thick lubricants or "greases."

27.18. Detergents. The cleansing action of ordinary soap (sodium stearate) has been explained as due to its surface activity. Surface-active compounds tend to concentrate at the surface of a solution and the molecules are then forced to align themselves in a characteristic manner because of their polar nature. Representative surface-active compounds are shown in Table 27.5. Figure 27.1 shows how such molecules are arranged in a water surface. The hydrocarbon chain is not water-soluble and tends to escape from the surface, leaving the polar end groups anchored in the water. If there are not too many molecules present, the hydrocarbon chains can move like tall grass in a wind. When the concentration is high, they are packed like the molecules in a crystal. It is apparent that a water solution, with a monomolecular layer of such molecules on its surface, actually presents a hydrocarbon surface to the air. The surface tension is lowered from the normal value of about 72 dynes/cm to about 25.

If we place a typically "organic" liquid layer, such as benzene, on a soap solution in water, the polar carboxyl group remains anchored to the water layer, but the hydrocarbon chain extends into the benzene layer and anchors that layer. The interfacial tension is greatly reduced. When the mixture is shaken, an emulsion forms, which consists of benzene droplets dispersed in water, the droplets being stabilized by the soap film between the two liquids.

TABLE 27.5. Representative Detergents

(The effective surface-active, long-chain structure can be either a negative ion, a posititive ion, or a nonionic compound)

$$CH_3-CH_2-CH_2-CH_2-(CH_2)_{10}-CH_2-CH_2-CH_2-\overset{\overset{O}{\|}}{C}-O^-Na^+$$
Sodium stearate (anionic)

$$CH_3-CH_2-CH_2-(CH_2)_{10}-CH_2-CH_2-CH_2-\overset{\overset{CH_3}{|}}{\underset{\underset{CH_3}{|}}{N}}-CH_3{}^+Cl^-$$
Cetyltrimethylammonium chloride (cationic)

$$CH_3-CH_2-CH_2-CH_2-(CH_2)_{10}-CH_2-CH_2-CH_2-\overset{\overset{O}{\|}}{C}-O-CH_2-\overset{\overset{OH}{|}}{CH}-\overset{\overset{OH}{|}}{CH_2}$$
Glyceryl monostearate (nonionic)

The "soil" of soiled clothing is understood to consist of solid particles coated with a thin oil film. Removal of such soil makes use of the phenomena just described. A solution of a surface-active agent is added, which necessarily must wet the object. A compound which makes it easy to wet the soiled object, emulsify the oil film, and keep the particles in suspension until washed away is called a *detergent*. Some surface-active agents merely wet a surface, but have little cleansing action and are not regarded as detergents. These are called wetting agents.

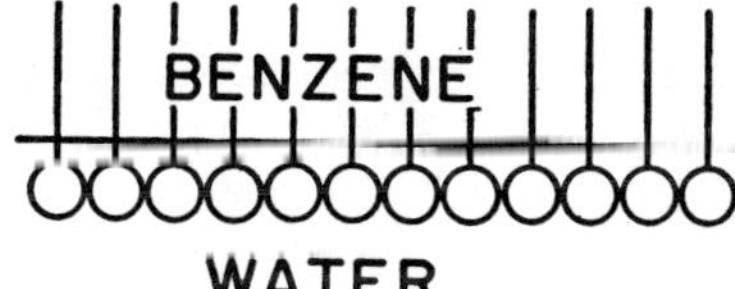

27.1.   Distribution of a monomolecular soap film between water and benzene. The small circles are polar groups.

A great many detergents have been made and marketed. They are frequently called soapless soaps. Although ordinary soap is still the best cleanser when used in pure water, it is inefficient if the water is *hard*— that is, contains calcium and magnesium salts. These salts precipitate the corresponding soaps, which are insoluble in water. Of all the detergents, only soap is affected by hard water. This is a very important reason why soap is being replaced by other detergents in many industrial applications.

The cationic type of surface-active compound (Table 27.5) is generally highly bactericidal. Since they are relatively nontoxic, they have important applications as cleansers for surgical instruments, dishwashing machines, etc. Some of the nonionic surface-active compounds have such weak polar characteristics that they are used more for wetting than for cleaning in nonaqueous liquids. They are often useful when mixed with other detergents or with soap. (See also ethanolamine and morpholine in Section 29.10, and the sulphur compounds in the concluding paragraph of Section 30.4)

The common carboxylic acid soaps are essentially *straight-chain* compounds, because of their manufacture from natural fats and oils (Section 27.11). The common detergents are mainly salts of the sulphonic acids (section 30.4) made synthetically by economical processes which in the past yielded chains that were extensively branched. The *branch-chain* detergents were found to pollute surface waters and sewer effluents because they were not decomposed by bacteria; in other words, they were not biodegradable. Synthetic processes had to be altered to give straight-chain detergents that are biodegradable (BDG).

28

Mixed Oxygen Compounds

28.1. Introduction. We have now described the principal simple families of compounds in organic chemistry. We have concentrated our attention on the fact that one particular group in a molecule gives the compound its family properties. In this chapter, we shall discover what happens to the molecule when it contains two or more characteristic groups. Such substances are called *mixed compounds*. Several examples of these compounds have already been mentioned in this book. For example, formic acid, $H\text{–}C\text{=}O$, has the usual acid properties of the car-

$$\underset{\text{OH}}{|}$$

boxyl group, –COOH, but it differs from other carboxylic acids in that it has reducing properties due to the aldehyde group, –CHO, present in its structure.

28.2. Review of Simple Family Types. Before proceeding to a study of the mixed types, let us pause for a moment to review the family relationships of the simple compounds as shown in Figure 28.1. As indicated in the chart, the various families of compounds are regarded as derived from *chain* or *ring* structures. Each family is obtained by replacing an H atom in the chain or in the cycle by *one* of the characteristic groups studied in the past chapters.

The arrow pointing from one class of compounds to another is used to illustrate how more complex compounds can be developed from simpler ones. The actual manufacture of these various substances seldom follows the outline indicated on the chart. An outline of this kind, however, is of assistance to the beginner in visualizing the relationships in organic chemistry.

28.3. Importance of Mixed Compounds. It would be very pleas ant indeed if we could say that the properties of a molecule are the simple sum of the properties of its two or more characteristic groups. However,

311

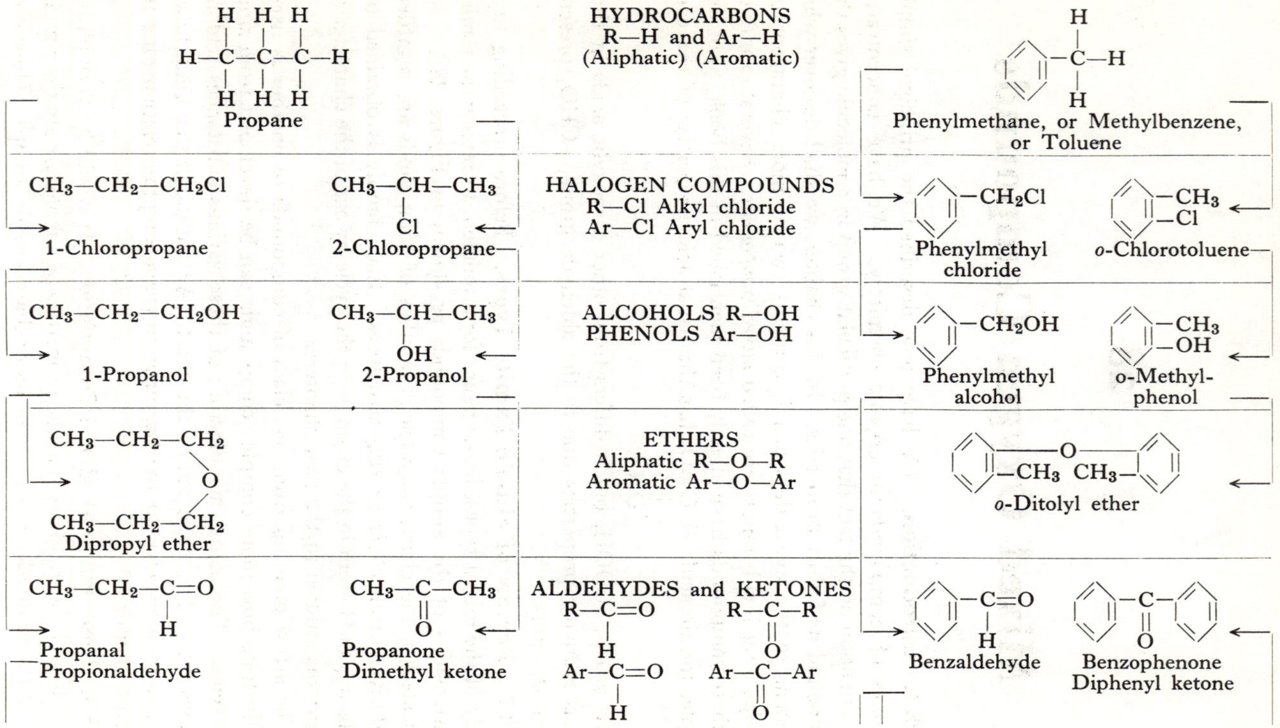

HYDROCARBONS
R—H and Ar—H
(Aliphatic) (Aromatic)

Propane

Phenylmethane, or Methylbenzene,
or Toluene

HALOGEN COMPOUNDS
R—Cl Alkyl chloride
Ar—Cl Aryl chloride

$CH_3—CH_2—CH_2Cl$
1-Chloropropane

$CH_3—CH—CH_3$
|
Cl
2-Chloropropane

—CH_2Cl
Phenylmethyl chloride

—CH_3
—Cl
o-Chlorotoluene

ALCOHOLS R—OH
PHENOLS Ar—OH

$CH_3—CH_2—CH_2OH$
1-Propanol

$CH_3—CH—CH_3$
|
OH
2-Propanol

—CH_2OH
Phenylmethyl alcohol

—CH_3
—OH
o-Methyl-phenol

ETHERS
Aliphatic R—O—R
Aromatic Ar—O—Ar

$CH_3—CH_2—CH_2$
O
$CH_3—CH_2—CH_2$
Dipropyl ether

—O—
—CH_3 CH_3—
o-Ditolyl ether

ALDEHYDES and KETONES
R—C=O   R—C—R
| ‖
H   O
Ar—C=O   Ar—C—Ar
| ‖
H   O

$CH_3—CH_2—C=O$
|
H
Propanal
Propionaldehyde

$CH_3—C—CH_3$
‖
O
Propanone
Dimethyl ketone

—C=O
|
H
Benzaldehyde

—C—
‖
O
Benzophenone
Diphenyl ketone

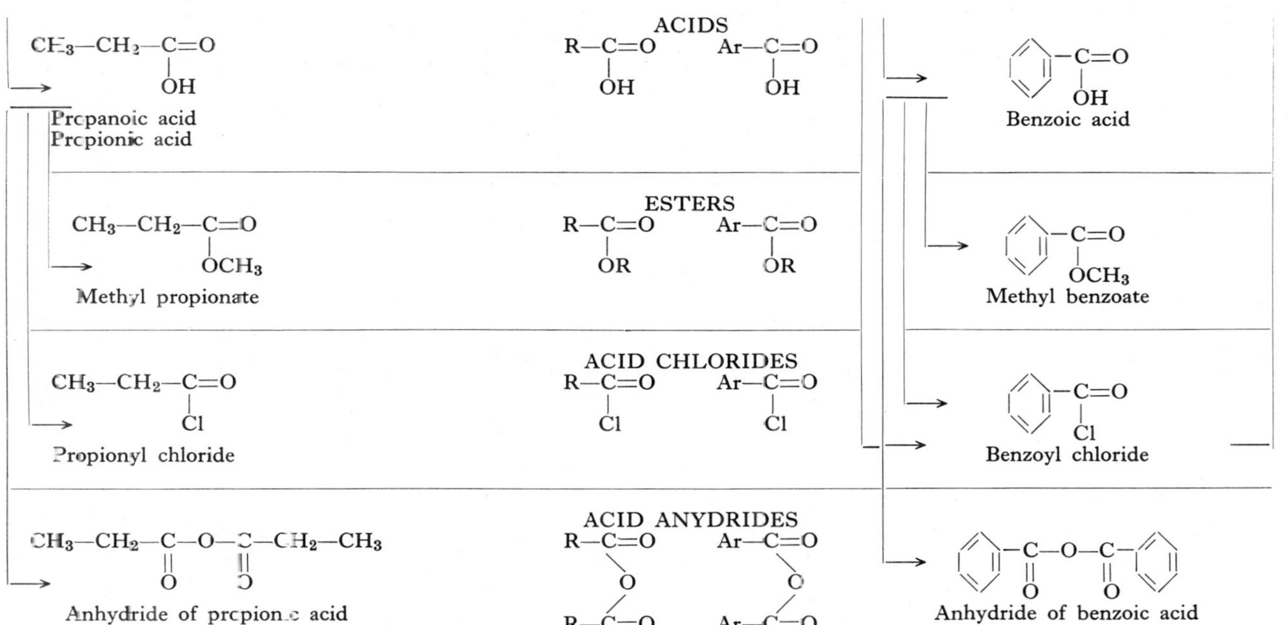

28.1. An outline indicating how the principal characteristic oxygen compounds are related to each other and are obtainable from open-chain and cyclic hydrocarbons. See Sections 23.4 and 23.6 for the meaning of R and Ar.

a mixed compound is seldom found to live up to this rule. This is unfortunate for the textbook writer, because if the additive relation held he would not have to make a separate study of the properties of mixed compounds. It is fortunate for the research worker, however, because he finds that a mixture of groups in a molecule often gives him some unusual and very useful properties to work with. It is this last factor that we shall discuss in this chapter.

28.4. Nomenclature. The systematic method of naming carbon compounds has been described in preceding chapters, and the rules already stated cover most of the types of compounds in which the beginner is interested. However, it is appropriate, at this point, to add a few rules of a summary nature. In Section 10.9, we defined what is meant by a functional group and stated that the name of this group appears in the main part of the name of a compound. When there are several functional groups in the molecule, the "chief" function is selected for use in the main part of the name. A brief list of functional groups in the order of preference is: carboxyl, aldehyde, ketone, alcohol, phenol, ether, etc. As an example, CH_2OH-CH_2-COOH is 3-hydroxypropanoic acid, the acid function being the chief function with $-OH$ as a lesser function, or substituent. The chain is numbered so as to give the chief function the smaller number.

Theoretically, the simplest of the mixed compounds are the olefins and acetylenes studied in Chapter 13. At that time we showed how the chain of carbon atoms can be built up of $C=C$ units and $C\equiv C$ units as well as $C-C$ bonds, to give three families of compounds of which the following are representative:

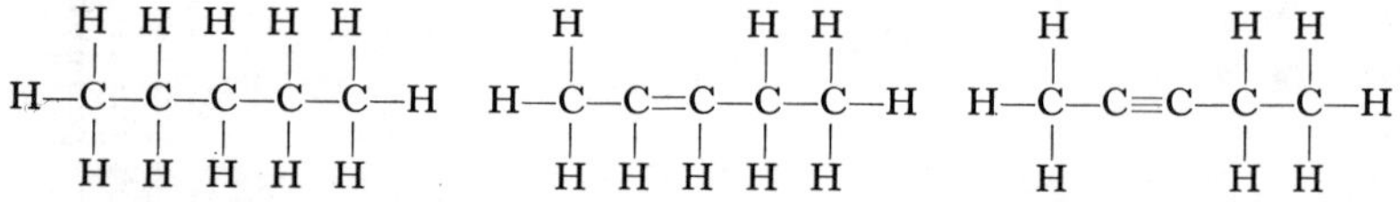

$$
\begin{array}{ccc}
\underset{\substack{\text{Pentane} \\ \text{A paraffin}}}{\text{H-}\overset{\displaystyle H}{\underset{\displaystyle H}{C}}-\overset{\displaystyle H}{\underset{\displaystyle H}{C}}-\overset{\displaystyle H}{\underset{\displaystyle H}{C}}-\overset{\displaystyle H}{\underset{\displaystyle H}{C}}-\overset{\displaystyle H}{\underset{\displaystyle H}{C}}\text{-H}}
&
\underset{\substack{\text{2-Pentene} \\ \text{An olefin}}}{\text{H-}\overset{\displaystyle H}{\underset{\displaystyle H}{C}}-\overset{\displaystyle \,}{\underset{\displaystyle H}{C}}=\overset{\displaystyle \,}{\underset{\displaystyle H}{C}}-\overset{\displaystyle H}{\underset{\displaystyle H}{C}}-\overset{\displaystyle H}{\underset{\displaystyle H}{C}}\text{-H}}
&
\underset{\substack{\text{2-Pentyne} \\ \text{An acetylene}}}{\text{H-}\overset{\displaystyle H}{\underset{\displaystyle H}{C}}-C\equiv C-\overset{\displaystyle H}{\underset{\displaystyle H}{C}}-\overset{\displaystyle H}{\underset{\displaystyle H}{C}}\text{-H}}
\end{array}
$$

The olefins and acetylenes can be viewed as mixed compounds. The properties of the 2-pentene molecule, for example, are the simple sum of the properties of the *olefin* $-C=C-$ group as given in Section 13.4, and of the *paraffin* grouping $-C-C-$ as listed in Section 10.6. In common practice, however, the unsaturated hydrocarbons are not considered to be mixed compounds.

28.5.   Olefin-Halogen Compounds.   A simple family of mixed compounds was mentioned in Section 17.3.   From that discussion it will be apparent that compounds of the following structures

$$CH_2{=}CH{-}CH_2 \quad\quad\quad\quad CH_2{=}C{-}CH_3$$
$$\quad\quad\quad\;\; | \quad\quad\quad\quad\quad\quad\quad\quad\quad\quad |$$
$$\quad\quad\quad\; Cl \quad\quad\quad\quad\quad\quad\quad\quad\quad Cl$$

3-Chloro-1-propene   2-Chloro-1-propene

have markedly different properties because of the different relative positions of the –C=C– group and the Cl atom.   When the chlorine atom is attached to the –C=C– group it becomes inert and does not show the pronounced activity of the chlorine atom in the $-CH_2Cl$ group.   We see, then, that the properties of an organic grouping cannot be tabulated with exactness.   In preceding chapters, we have endeavored to give the student a general idea of the nature of these groups, but the exact nature of the group depends on its relative position in the molecule with respect to other groups.

28.6.   Alcohol-Acids.   Molecules that are both alcohols and acids constitute a family of mixed compounds very often used for demonstration, because they lend themselves so readily to an explanation of this type of chemistry.   The alcohol-acids in general show the usual properties of acids and of alcohols, but in addition they behave in unusual ways, depending on the relative positions of the –OH and –COOH groups:

$$CH_3{-}CH{-}C{=}O + O{=}C{-}CH{-}CH_3 \xrightarrow{\text{heat}}$$

with the OH groups (boxed) above and OH / HO below, yielding

$$CH_3{-}CH{-}C{=}O$$
$$\diagup \quad\quad \diagdown$$
$$O \quad\quad\quad O \;\; + 2H_2O$$
$$\diagdown \quad\quad \diagup$$
$$O{=}C{-}CH{-}CH_3$$

a-Hydroxypropionic acid   Lactide
Lactic acid
(2 molecules)

$$CH_3{-}CH{-}CH\;\; C{=}O \xrightarrow[\text{dehydrating agent}]{\text{heat +}} CH_3{-}CH{=}CH{-}C{=}O + H_2O$$
$$\quad\quad |\quad\; |\quad\quad |\quad\quad\quad\quad\quad\quad\quad\quad\quad\quad\quad\quad\quad\quad\quad\quad\quad |$$
$$\quad\quad OH\;\; H\;\; OH\quad\quad\quad\quad\quad\quad\quad\quad\quad\quad\quad\quad\quad OH$$

β-Hydroxybutyric acid   Crotonic acid

$$CH_2{-}CH_2{-}CH_2{-}C{=}O \xrightarrow[\text{(or spontaneous)}]{\text{heat}} CH_2{-}CH_2{-}CH_2{-}C{=}O + H_2O$$
$$|\quad\quad\quad\quad\quad\quad\quad\quad |\quad\quad\quad\quad\quad\quad\quad\quad\quad\quad | \quad\quad\quad\quad\quad\quad\quad |$$
$$OH\quad\quad\quad\quad\quad\;\; OH\quad\quad\quad\quad\quad\quad\quad\quad\quad\quad\quad{-}{-}O{-}{-}$$

γ-Hydroxybutyric acid   Butyrolactone

From the formula of lactic acid it is clear that one molecule of the compound *acting as an acid* can react with another molecule of the same compound, *acting as an alcohol*, to form an ester (see Section 27.11). Heating lactic acid will produce this ester, which is known as *lactide*. Other α-hydroxy acids behave like lactic acid to form esters of this closed-chain or ring type, which are known in general as the lactides. It will be recalled that rings containing four to six atoms are the easiest to form.

The β-hydroxy acids do not form esters when heated; instead, they lose water as shown in the preceding reactions, to form an olefin-acid. The student should notice that if the β-acid were to form an ester like that of the α-hydroxy acid, it would have to make an eight-membered ring, which is quite difficult.

The γ-hydroxy acids, when heated, form esters *within the same molecule*. The acid –COOH group and the alcoholic –OH group, by rotation

$$
\begin{array}{ccc}
\text{CH}_2\text{—CH}_2 & & \text{CH}_2\text{—CH}_2 \\
\text{CH}_2 \qquad \text{C=O} & \xrightarrow{\text{—H}_2\text{O}} & \text{CH}_2 \qquad \text{C=O} \\
\text{OH HO} & & \text{O}
\end{array}
$$

γ-Hydroxybutyric acid   Butyrolactone

about the C–C bonds, may come quite close to each other. We have, then, ideal conditions for the formation of what is known as an *inner ester*, or more commonly called a *lactone*. This is another ring structure, but it should be observed that it is different from a lactide.

The lactone obtained from the γ-hydroxy acid is a five-membered ring. Lactones can also be obtained from δ-hydroxy acids:

$$
\ldots -\text{CH}_2-\underset{\underset{\text{OH}}{|}}{\text{CH}}-\text{CH}_2-\text{CH}_2-\text{CH}_2-\underset{\underset{\text{OH}}{|}}{\text{C}}=\text{O} \longrightarrow
$$

A δ-hydroxy acid

$$
\ldots -\text{CH}_2-\text{CH}-\text{CH}_2-\text{CH}_2-\text{CH}_2-\text{C}=\text{O} + \text{H}_2\text{O}
$$

A δ-lactone

This lactone has a six-membered ring. The five-membered lactone ring is usually distinguished from the six-membered ring by calling one a γ-lactone and the other a δ-lactone. Lactones cannot be obtained easily from hydroxy acids in which the –OH group is more distant from the –COOH group than the δ-carbon atom. The lactone ring is of great

importance in the study of carbon compounds (Section 32.1); ring compounds consisting of C atoms and O atoms are very common in nature.

28.7. Some Important Alcohol-Acids. Table 28.1 lists the names
and formulas of some other hydroxy acids that are important, but cannot
be considered in detail in a book of this size and kind. In this table,
those carbon atoms to which four different groups are attached and can,
therefore, give the molecule optical activity are marked with an asterisk
(*).

Lactic acid is found in sour milk (see lactose in Section 34.17), in sauer-
kraut, and in an extract of muscle tissue. It is optically active and can
exist in D, L, and DL (that is, *racemic*) modifications. In aqueous solution, the D and L forms rotate plane-polarized light in directions oppo-

D (levo)-Lactic acid   L (dextro)-Lactic acid

site to those of their family heads shown in Figure 22.9. The type of projection formulas used here for the lactic acids should be reviewed in
Section 22.9. On the basis of the Sequence Rule (Section 22.16) the
names of these compounds are (R)-*levo*-lactic acid and (S)-*dextro*-lactic
acid. It is of interest to observe that the *dextro*-lactic acid is produced
in flesh by biological processes, whereas the *levo*-lactic acid is produced
by bacterial action on milk sugar. The *dextro* compound is also known
as *sarcolactic* acid, from the Greek word for flesh.

Tartaric acid has two asymmetric carbon atoms and, because of the
symmetry of the molecule, can exist in a *meso* form as well as the D, L,

D (levo)-Tartaric acid   L (dextro)-Tartaric acid   meso-Tartaric acid and *racemic* modifications. The L form occurs in nature in various fruits, especially in the grape. The isomeric forms of tartaric acid were discussed in Section 22.14 and Table 22.1, which should be reviewed. The projection formulas for tartaric acid should be compared with those for 2,3-butanediol in Figure 22.11.

If the Sequence Rule is applied as in Section 22.16, the names of the optically active compounds in this group become (2R, 3R)-*dextro*-tartaric acid and (2S, 3S)-*levo*-tartaric acid.

Malic acid is found in many fruits and plant products. It was first isolated from unripe apples (Latin *malum*, apple). The optically active

$$
\begin{array}{ccc}
\text{D (dextro)-Malic acid} &
\begin{array}{c}
COOH \\
H \leftarrow\!\!\!-\!\!\!\rightarrow OH \\
CH_2 \\
COOH
\end{array} &
\begin{array}{c}
COOH \\
HO \leftarrow\!\!\!-\!\!\!\rightarrow H \\
CH_2 \\
COOH
\end{array} &
\text{L (levo)-Malic acid}
\end{array}
$$

isomers possible are D and L modifications, with an optically inactive *racemic* mixture. The configuration of the D-compound has been proved to be the same as that of D-lactic acid, but the rotation is in the opposite direction. The L isomer, of course, is the mirror image of the D isomer, with OH to the left of the observer when the formula is *projected* in the conventional manner. The L compound is the one that occurs in nature. This isomer in dilute solution in water rotates polarized light to the left (*levo*), but the extent of rotation decreases until, at a concentration of 34%, it changes to *dextro* and at higher concentrations it is increasingly *dextro*. Malic acid has been used as a reference compound to determine whether other optically active compounds are members of the D or L families (see Section 22.9). The names of these compounds according to the Sequence Rule are (R)-*dextro*-malic acid and (S)-*levo*-malic acid.

Glyceric acid, the formula of which is in Table 28.1, has been used as a reference compound for D and L families, as just described for malic acid. When the structure of D-glyceric acid is projected by the conventional procedure the OH on the asymmetric carbon atom will be on the right. The racemic compound (DL-glyceric acid) can be made by careful oxidation of glycerol. (See *glycerose* in Section 34.1.)

Saccharic acid (glycaric acid) can be obtained by oxidation of glucose, and its relation to this sugar will be discussed in Section 34.7. It has four asymmetric carbon atoms; a comment on the calculation of the number of possible optical isomers will be found in Section 34.3

TABLE 28.1. Common Names of Some Important Hydroxy Acids

Carbonic acid (Hydroxyformic acid)	$HO-COOH$
Glycollic acid (Hydroxyacetic acid)	CH_2-COOH | OH

Lactic acid
(a-Hydroxypropionic acid)
	$CH_3-*CH-COOH$
	     |
	     OH

Hydracrylic acid (β-Hydroxypropionic acid)	CH_2-CH_2-COOH | OH
Glyceric acid	$CH_2-*CH-COOH$ | | OH OH

Tartaric acid	$HOOC-*CH-*CH-COOH$
	       | |
	     OH OH

Malic acid	$HOOC-CH_2-*CH-COOH$
	           |
	         OH

Saccharic acid	$HOOC-(*CHOH)_4-COOH$

Citric acid	OH
	       |
	$HOOC-CH_2-C-CH_2-COOH$
	       |
	      $COOH$

Carbonic acid is listed in the table as the hydroxy derivative of formic acid. The solution called carbonic acid is really mostly a solution of carbon dioxide in water: $H_2CO_3 \leftrightarrows CO_2+H_2O$. There are relatively few molecules of H_2CO_3 or $HO\text{–}C\text{–}OH$ in such a solution because of its instability. The rule of instability of two $-OH$ groups on the same carbon atom holds true in this case as in others that have been mentioned (Section 25.8).

28.8. Salicylic Acid. It is also of interest to examine a family of compounds in the benzene series:

o-Hydroxybenzoic acid   m-Hydroxybenzoic acid   p-Hydroxybenzoic acid
(Salicylic acid)

Of these the most important is the *ortho* compound, commonly known as salicylic acid. Salicylic acid can be separated from the other two

isomers by boiling it with calcium chloride and ammonia, which precipitates only the *o-* compound as a calcium salt, $C_6H_4-C=O$. It is
$$O-Ca-O$$
probable that the –COOH and –OH groups are too far apart in the other two compounds to be able to form such a salt. Salicylic acid is a phenol-acid. The H atoms in both the –COOH and –OH groups are therefore acidic, which explains why Ca can replace both these H atoms in salicylic acid.

Salicylic acid is used in the preparation of many important substances, such as these:

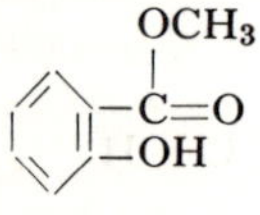

OCH₃   OH   ONa

Oil of wintergreen 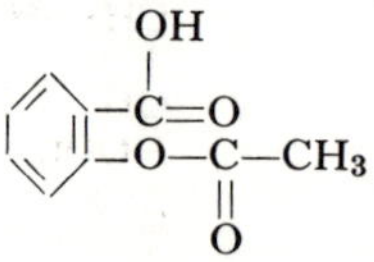   Aspirin 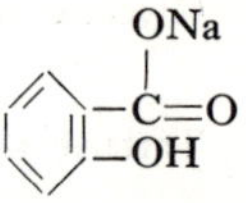   Sodium salicylate

Oil of wintergreen is the methyl alcohol ester of salicylic acid; it is formed by reaction of the –COOH group of salicylic acid with CH_3OH. Aspirin is the acetic acid ester of salicylic acid; it is formed by reaction of the –OH group of salicylic acid with CH_3–COOH.

28.9. Chelation. Salicylic acid, in which the OH and COOH groups are *ortho* to each other, is appreciably different from its *meta* and *para* isomers. This is attributed to hydrogen bonding. It was shown in Section 27.8 that the carboxyl group has a strong tendency to form these bonds *between* molecules. In salicylic acid the bond can form *inside* the molecule to yield a relatively stable six-membered ring. This ring formation also smothers the ability of the OH group to hydrogen bond with other molecules.

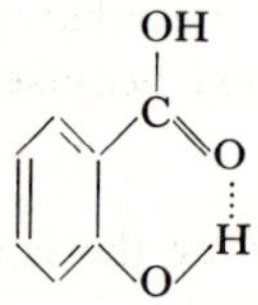

Salicylic acid
o-Hydroxybenzoic acid

Salicylic acid therefore differs from its isomers in its lack of attraction for neighboring molecules. This affects solubility, melting point, and other physical characteristics. The acidity is somewhat enhanced because an inductive effect (Section 17.2) of the hydrogen bond lowers the electron density of the COOH group and permits increased ionization of the H atom in that group.

Ring structures of the type shown here, in which a bond is formed by the sharing of an unemployed electron pair, are called chelate rings and the process of ring formation, which will be further described in Section 28.13, is called *chelation*.

28.10. Steric Hindrance. In the preceding section we observed that a ring structure with groups *ortho* to each other has unusual properties due to interaction of the groups. This is one more illustration of the effect of the space relationships of groups on their activity. We have shown how a double bond in a molecule can produce *cis* or *trans* isomers with different properties (Section 22.3). We also showed how the different space arrangements of the same four groups around a carbon atom can produce *dextro* and *levo* isomers (Section 22.8). These are a few of many examples of the great importance of space relations of the various parts of molecules.

A type of molecule which has interested many research chemists is that in which the steric effect is one of hindrance—that is, where certain groups in the molecule interfere with each other due to space limitations. Two stereoisomeric forms of certain compounds of the structures shown in Figure 28.2 can be prepared. These are diphenyl derivatives (Section 24.4), containing two different atoms or groups represented by A and B in *ortho* positions in the molecule. Molecules of this type can be obtained in one form which is right-handed and another form which is left-handed toward plane-polarized light. These are good illustrations of substances that are optically active, yet have no asymmetric carbon atoms in the molecule, of the type described in Section 22.7. The molecule is asymmetric because A and B prevent free rotation of the two rings, and these rings can therefore be "frozen" into definite positions.

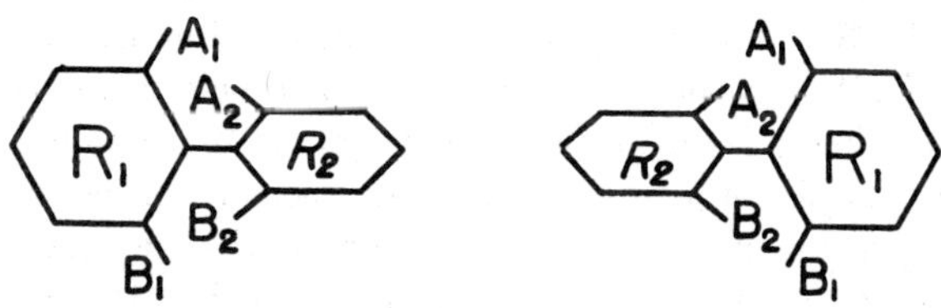

28.2. Mirror-image isomerism due to steric hindrance.

These two modifications are mirror images, and cannot be superimposed on each other. The R_2 ring should be imagined as horizontal, with the R_1 ring vertical; a little study will then show that in the R_2 ring, A_2 can be either on the left or the right with respect to R_1. Much depends on the size of A and B. If they are small, like the fluorine atom, they do not prevent free rotation of the rings and two isomeric forms are then not possible. At least three of the four A and B positions must be occupied in order to produce stereoisomeric forms.

28.11. Tautomerism. We shall now describe compounds that behave as if more *functional* groups are present in the molecule than can be accounted for by a single formula. The phenomenon will be illustrated with *acetylacetone* (2, 4-pentanedione), one of the family of diketones, which behaves like an olefin-alcohol as well as a ketone.

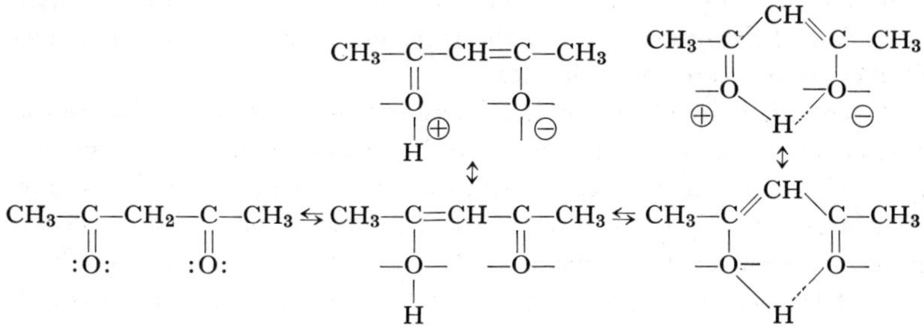

$$CH_3-C-CH=C-CH_3$$

$$CH_3-C-CH_2-C-CH_3 \leftrightharpoons CH_3-C=CH-C-CH_3 \leftrightharpoons CH_3-C \cdots C-CH_3$$

<table>
<tr><td>Acetylacetone
(Keto form)</td><td>Acetylacetone
(Enol form)</td><td>Acetylacetone
(Cyclic enol form)</td></tr>
</table>

The olefin-alcohol structure is designated an *enol*, from *ene* meaning a compound with the C=C bond, and *ol* for the alcohol group, OH. The phenomenon of the same compound acting in two different ways is called tautomerism (Greek *tauto*, the same). The pure liquid is about 80% in the enol form, and this is probably all cyclic because of hydrogen bonding. The high proportion of the *enol* modification is due not only to the stability introduced by the ring structure, but also to the stability brought about by resonance which is possible only in the *enol*. Notice that enolization is brought about by migration of an atom (actually, in this case, transfer of a proton), whereas resonance forms are the result of rearrangement of electrons. Resonance forms are shown connected by double-headed arrows. In the equation, the *keto* structure is shown with pairs of dots for the unshared electrons on the O atoms. In the *enol* formulas, straight lines are used for electron pairs in order to facilitate the electron rearrangements by the procedure described in Section 15.7. The requirement for enolization is that C–H must be adjacent to a C=O group. The H atom in the OH group of an enol is acidic, and is replaceable by active metals such as sodium (see Section 26.5).

In the case of *ethyl acetoacetate*, the ethyl ester of acetoacetic acid, both the *keto* and *enol* forms have been isolated, but they are unstable when alone and in a relatively short time they revert to an equilibrium mixture.

$$CH_3-\underset{\underset{O}{\parallel}}{C}-CH_2-\underset{\underset{O}{\parallel}}{C}-OC_2H_5 \;\rightleftharpoons$$

$$CH_3-\underset{\underset{OH}{\mid}}{C}=CH-\underset{\underset{O}{\parallel}}{C}-OC_2H_5 \;\rightleftharpoons\; CH_3-C\diagup\overset{CH}{\diagdown}C-OC_2H_5$$

Ethyl acetoacetate

Tautomerism may be defined as the dynamic equilibrium of a pair of isomers.

28.12. Acetoacetic Ester Synthesis. Although in this book we do not stress the laboratory procedures of organic chemistry, it is of interest to follow the outline of a typical synthesis involving the ethyl ester of acetoacetic acid.

$$CH_3-\overset{H}{\underset{\underset{O}{\parallel}}{C}}-\overset{\ddot{}}{\underset{\underset{H}{\ddot{}}}{C}}-\underset{\underset{O}{\parallel}}{C}-OC_2H_5 \xrightarrow[-H_2O]{NaOH} \left[CH_3-\underset{\underset{O}{\parallel}}{C}-\underset{\ddot{}}{C}-\underset{\underset{O}{\parallel}}{C}-OC_2H_5\right]^{-} Na^+ \xrightarrow{\overset{\delta^+\ \delta^-}{R-Cl}}$$

$$\text{I} \qquad\qquad\qquad \text{II}$$

$$CH_3-\underset{\underset{O}{\parallel}}{C}-\underset{\ddot{R}}{C}-\underset{\underset{O}{\parallel}}{C}-OC_2H_5 \xrightarrow{NaOH} CH_3-\underset{\underset{O}{\parallel}}{C}-\underset{\ddot{R}}{C}-\underset{\underset{O}{\parallel}}{C}-OH \xrightarrow{-CO_2}$$

$$\text{III} \qquad\qquad\qquad \text{IV}$$

$$CH_3-\underset{\underset{O}{\parallel}}{C}-CH_2-R$$

$$\text{V}$$

The H atoms in the $-CH_2-$ group of I are mobile and "acidic." In II, one of the wandering H^+ ions from the $-CH_2-$ group has been replaced by Na^+ ion to yield a salt. When this salt solution is treated with an alkyl halide (such as the polar CH_3-Cl molecule), the alkyl radical forms a C–C union with the ester (III). This is synthesis as defined in Section 10.2. The Na^+ and Cl^- also become partners. Now the ester is hydrolyzed by means of a weak NaOH solution to IV, to free the carboxyl group, which can then be decarboxylated (Section 27.16) to produce V.

Since R in this outline can be any one of a large number of organic radicals, it is obvious that this synthesis is a useful step in the development of a wide variety of compounds. The ketone in V can be reacted in many ways as described in Chapter 26. Both of the H atoms in the methylene group of acetoacetic ester can be replaced by R groups by the procedure illustrated. Synthesis with malonic ester (Section 27.16) is similar to that outlined in this section.

28.13. Chelate Rings. The properties of the metal derivatives of acetylacetone (section 28.11), because they do not behave like salts, puzzled chemists for a long time, but their structure and behavior were clarified by the electron-valence theory. In Structure I is shown the customary formula of the beryllium compound, in which the divalent beryllium atom exercises its normal valence of two (see Figure 6.1). It can fill up its octet, however, by accepting two pairs of electrons from

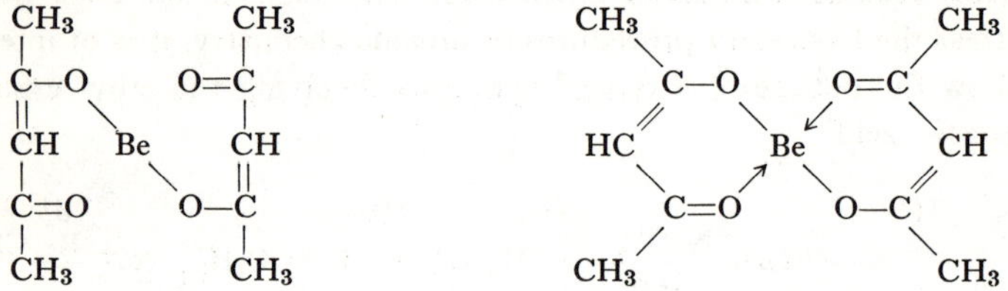

$$
\begin{array}{cccc}
CH_3 & & & CH_3 \\
| & & & | \\
C-O & O{=}C & & \\
\| & & | & \\
CH & Be & CH & \\
| & & \| & \\
C{=}O & O-C & & \\
| & & | & \\
CH_3 & & CH_3 &
\end{array}
$$

I   II

neighboring oxygen atoms by formation of two dative bonds (Section 8.5), as shown in Structure II. This is facilitated by the fact that a six-membered ring is formed, which is generally almost free from strain.

Ring structures of this type, in which a metallic compound with at least one ionic valence bond gives ring closure through dative bonds, are called *chelate* rings. This term is derived from the Greek (*chēlē*, claw) implying the action of a crab's claw, or forceps, and is pronounced *kee-late*. The term is (perhaps unfortunately) also employed to describe other types of ring structures, such as those in which hydrogen bonding causes ring closure (Section 28.9). The ring structure of the *enol* tautomer of ethyl acetoacetate in Section 28.11 is an example of hydrogen bonding rather than true chelation. The use and formation of chelate compounds is being widely exploited in analytical identification, commercial cleansing agents, dye chemistry, pharmacology, etc. Compounds like the beryllium derivative of acetylacetone are soluble in organic solvents and usually can be melted and distilled without difficulty.

28.14. Co-ordination Number. The Be atom discussed in the preceding section has a principal valence of two as indicated by the two electrons in its valence shell in Figure 6.1. Its normal tendency is to acquire stability by losing two electrons and forming two electrovalence bonds in the formation of salts, such as beryllium chloride, $BeCl_2$ (i.e., $Cl^-Be^{++}Cl^-$). As we have just seen, it can also form two covalent bonds by the sharing of electrons.

In addition to its principal valence of two, beryllium is said to have a coordination valence of four. This is also called coordination number and is defined as the maximum valence an element can show in any of its compounds. In the case of the beryllium compound, the coordination number of four was reached by accepting pairs of electrons from other atoms. These two added bonds are usually called dative bonds in this book, but they are also known as coordination bonds (Section 8.5). The meaning of the term coordination number as applied to valence is different from that given in Section 8.1 in the discussion of crystal structure.

29

Nitrogen Compounds

29.1. Introduction. Up to this point we have devoted our attention mostly to families of oxygen compounds—that is, carbon compounds in which there is a C–O linkage. This was reviewed in the last chapter, and the student should once more consult Figure 28.1 to refresh his memory as to the way the various families of chain and ring compounds are related to each other. In this chapter we shall consider the chemistry of compounds characterized by the presence of the C–N bond.

29.2. Amines. These compounds are best regarded as derived from the inorganic substance, ammonia (NH_3), by replacing the H atoms with either alkyl or aryl radicals.

	Aliphatic compounds	*Aromatic compounds*
Primary amines: RNH_2 or $ArNH_2$	CH_3—N—H with H below (Methylamine)	phenyl—N—H with H below (Phenylamine or, Aniline)
Secondary amines: R_2NH or Ar_2NH	CH_3—N—CH_3 with H below (Dimethylamine)	phenyl—N—phenyl with H below (Diphenylamine)
Tertiary amines: R_3N or Ar_3N	CH_3—N—CH_3 with CH_3 below (Trimethylamine)	phenyl—N—phenyl with phenyl below (Triphenylamine)

This classification of amines as primary, secondary, and tertiary is much different from that used in Section 25.3 for alcohols, where the classification depends on the number of carbon atoms attached to a specific C atom. In the amines, it is based on the number of organic groups attached to the N atom. For example, CH_3–CH–CH_3 is 2-aminopropane; it is a primary amine, but a secondary carbon compound. The –NH_2 group is called *amino*.

Mixed amines are possible in which several different radicals are attached to the N atom, as in methylphenylamine, (ring)–N–CH_3. This is also known as N-methylaniline; the prefix N- indicates that the –CH_3 group is on the nitrogen atom and not in the ring.

29.3. Properties of Amines. The aliphatic amines listed in Section 29.2 occur in herring brine and have a characteristic fishy as well as ammoniacal odor. All three amines are gases, very soluble in water, but the higher open-chain amines are liquids insoluble in water, and are practically odorless.

The aromatic amines are liquids or solids, and usually have characteristic but not fishy odors.

The properties of amines are in many respects similar to the properties of ammonia, and we must therefore recall the essential nature of this inorganic substance. Ammonia is a gas lighter than air, easily liquefied and sold commercially in cylinders as liquid ammonia. It is very soluble in water and its chemical properties can be explained by assuming the following equilibria to exist in solution:

$$NH_3 + H_2O \rightleftharpoons NH_4OH \rightleftharpoons NH_4^+ + OH^-$$

Ammonia	Ammonium hydroxide	Ammonium ion	hydroxyl ion

This solution is the familiar ammonium hydroxide, used when a weak alkali is required. Perhaps it would be more accurate to call the solution ammonia water, because there are relatively few NH_4OH molecules in solution, most of the dissolved ammonia remaining as NH_3 molecules. The dissociation of NH_4OH into molecular NH_3 is more pronounced than the dissociation of NH_4OH into NH_4^+ ions.

The amines, when soluble in water, behave like ammonia:

$$CH_3NH_2 + H_2O \rightleftharpoons CH_3NH_3OH \rightleftharpoons CH_3NH_3^+ + OH^-$$

Methylamine	Methylammonium hydroxide	Methylammonium ion	Hydroxyl ion

29.4. The Valence of Nitrogen. Probably the simplest way to explain the chemistry of the amines and of the ammonium compounds is to picture their electronic structures.

The nitrogen atom has five valence electrons, and the hydrogen atom has only one. The combination of these two elements in ammonia is as follows:

$$3H\cdot + \cdot \ddot{N}\cdot \longrightarrow H:\overset{\displaystyle H}{\underset{\displaystyle \cdot\cdot}{\ddot{N}}}:H$$

In ammonia, there is a potential valence bond left unused on the N atom and the ease with which ammonia forms ammonium compounds is due to this *lone pair* of electrons. In the formation of an ammonium compound addition takes place to the lone electron pair, as in the following reaction which shows what happens when ammonia is dissolved in water:

$$H:\ddot{N}:H + H:\ddot{O}:H \longrightarrow \left[H:\overset{H}{\underset{H}{\ddot{N}}}:H \right]^{+} :\ddot{O}:H^{-}$$

Ammonia   Water   Ammonium hydroxide

The nitrogen atom is said to have a valence of five, but it is evident that the fifth valence can only be used by a negative ion, like OH^- or Cl^-, in an electrovalence bond.

Because the exposed electron pair gives them basic properties, ammonia and amines yield salts quite easily. These salts can be prepared by the addition of either an *acid* or an *alkyl halide*:

$$CH_3\overset{\displaystyle H}{\underset{\displaystyle H}{-N}}: + H-Cl \longrightarrow \left[CH_3\overset{\displaystyle H}{\underset{\displaystyle H}{-N}}-H \right]^{+} Cl^- \text{ or, } CH_3NH_2\cdot HCl$$

(An acid)   Methylammonium chloride   Methylamine hydrochloride

$$CH_3\overset{\displaystyle H}{\underset{\displaystyle H}{-N}}: + CH_3-Cl \longrightarrow \left[CH_3\overset{\displaystyle H}{\underset{\displaystyle CH_3}{-N}}-H \right]^{+} Cl^- \text{ or, } (CH_3)_2NH\cdot HCl$$

(An alkyl halide)   Dimethylammonium chloride   Dimethylamine hydrochloride

These "substituted" ammonium salts are represented by the general formula

$$\left[-\overset{|}{\underset{|}{N}}- \right]^{+} X^{-}$$

in which the four valences in the bracket usually occupied by hydrogen atoms, as in NH_4Cl, are now taken up by *one or more* hydrocarbon radicals. Aliphatic radicals, like $-CH_3$, produce more stable salts than the original ammonium salts; aromatic radicals, like the phenyl group, result in less stable salts. Triphenylamine, $(C_6H_5)_3N$, for example, is so weak as a base that it does not add acids to form salts.

It should be observed that the salts derived from amines are sometimes named as "double compounds," as illustrated by the hydrochlorides.

29.5. Quaternary Ammonium Compounds. By the use of methods based principally on the preceding reactions, it is possible to prepare an entire series of amines and substituted ammonium compounds in which one or all of the hydrogen atoms of NH_3 or of the NH_4^+ ion are replaced by radicals:

$$NH_3 \; + H_2O \; \rightleftarrows \; NH_4OH$$
$$RNH_2 + H_2O \; \rightleftarrows \; RNH_3OH$$
$$R_2NH + H_2O \; \rightleftarrows \; R_2NH_2OH$$
$$R_3N \; + H_2O \; \rightleftarrows \; R_3NHOH$$
$$R_3N + R-Cl \longrightarrow R_4N-Cl \xrightarrow{Ag-OH} R_4NOH$$

For the series containing $-CH_3$ groups, the relative basic strength of the hydroxyl derivatives is shown in Table 29.1. The values should be compared with the corresponding ionization constants of organic acids in Table 27.4. These constants are for the *electrolytic* dissociation, such as $NH_4OH \rightleftarrows NH_4^+ + OH^-$, measured under conditions in which the interfering *molecular* dissociation, $NH_4OH \rightleftarrows NH_3 + H_2O$, mentioned in Section 29.3, has presumably been eliminated.

It will be observed that all these bases are weak, except the compound $(CH_3)_4NOH$. Compounds of this type, in which all four hydrogen atoms of the NH_4^+ ion have been replaced by radicals, are known as *quaternary*

TABLE 29.1. Ionization Constants (at 25°C.)

Ammonium hydroxide	NH_4OH	0.000018
Methylammonium hydroxide	CH_3NH_3OH	0.0005
Dimethlammonium hydroxide	$(CH_3)_2NH_2OH$	0.00052
Trimethylammonium hydroxide	$(CH_3)_3NHOH$	0.000074
Tetramethylammonium hydroxide	$(CH_3)_4NOH$	about 1.0
Anilinium hydroxide	$\langle\!=\!\rangle-NH_3OH$	0.00000000046

ammonium compounds, from the Latin word, *quaternarius*, meaning a group of four.

The strongly basic character of quaternary ammonium hydroxides is explained by the absence of a hydrogen bridge, thus permitting formation of an ionic compound.

$$
\begin{array}{c}
\text{H} \\
| \\
\text{R—N:H:OH} \\
| \\
\text{H}
\end{array}
\qquad\qquad\qquad
\left[
\begin{array}{c}
\text{R} \\
| \\
\text{R—N—R} \\
| \\
\text{R}
\end{array}
\right]^{+} \text{OH}^{-}
$$

I   II

It will be seen that in I it is possible for the proton of a water molecule to act as a bridge between pairs of electrons on the highly electronegative elements, oxygen and nitrogen. This phenomenon, described in Section 25.10, accounts for the low extent of ionic dissociation of Structure I as compared with II which does not have this hydrogen bond.

Quaternary ammonium hydroxides are very strong bases; they dissociate in solution to practically the same extent as sodium hydroxide, Na^+ OH^-. Moreover, these hydroxides are hygroscopic solids, similar in appearance to the strong inorganic bases. Analogous arsenic compounds and phosphorus compounds (Section 30.9) have been prepared, and all are *-onium compounds* (Section 23.7).

29.6.   Relative Basicity of Aliphatic and Aromatic Amines. As defined in the electronic theory of acids and bases (Section 27.6) a compound that can share its available electron pair is a base. The relative basicity of the amines shown in Table 29.1 refers to the relative tendency of the lone electron pair on the N atom to be shared. An aliphatic amine such as CH_3NH_2 is a stronger base than NH_3 because of the electron displacement $C+\rightarrow N$. The CH_3 group tends to release electrons (Section 21.5); this increases the tendency of electrons to pile up at the N atom, and consequently enhances the ability of the free electron pair to attract a positively charged particle (such as H^+) or an electron-deficient molecule (such as BF_3). This is an appropriate point at which to review the meaning of the dative bond in Section 8.5.

An aromatic amine such as aniline is a weaker base than ammonia. This is due to resonance possibilities, as a result of which the average state of the molecule is one in which the free electron pair is transferred from the N atom into the ring. Substitution reactions at *ortho* and *para* positions in the ring are facilitated (Section 21.6)), but the electron pair is less available for reactions involving the N atom, and the negative charge

Resonance structures of aniline

on the N atom can be said to be decreased.   In other words, the basic
tendency of aniline should be much lower than that of NH_3, and this is
borne out by the ionization constant given in Table 29.1 for aniline (more
strictly the constant of anilinium hydroxide).

When reactions are carried out with aniline in acid solution, anilinium
salts, such as $C_6H_5NH_3^+Cl^-$, are formed.   Since the anilinium ion does
not have the resonance possibilities shown for the free base, there is no
transfer of electrons into the ring.   Substitution reactions, therefore, do
not take place *ortho* and *para* in acid solution, but practically entirely
meta, and are slower.

Since CH_3NH_2 is more basic than NH_3, one should expect increasing
basicity with a larger number of alkyl radicals on the nitrogen.   From
Table 29.1, however, it is seen that $(CH_3)_3N$ is a weaker base than CH_3
NH_2, although the difference is not very great.   The smaller basicity has
been ascribed to steric hindrance (Section 28.10) which can be understood from Figure 8.6 and from Diagram II in this Section.   Bulky groups
at the back of the molecule may interfere with the ability of the molecule
to share its lone electron pair.   Larger alkyl radicals as substituents in the
amines give a more pronounced steric effect.

Willingness to share an electron pair is the reason that tertiary amines
can be oxidized to amine oxides:

Trimethylamine   Trimethylamine oxide

This type of reaction is also given by aromatic amines.   The reaction
played an important role in the history of the development of the electron
theory of valence.   According to the old classical valence theory, in
order to give the N atom its valence of five the oxides were written
$R_3N=O$, but the compounds had no properties associated with an $N=$
O bond.   The new theory showed that the bond is essentially a single

bond, which is in better accord with the properties of the compounds. The formation of the bond is dative (Section 8.5).

Amine oxides in which the three organic radicals are all different can be resolved into optically active forms (see Section 22.7). It is obvious from Diagram I that the N atom in such a molecule is asymmetric. In Section 30.6 we shall describe sulphur compounds that are optically active and that lead us to look for optical activity in analogous nitrogen compounds

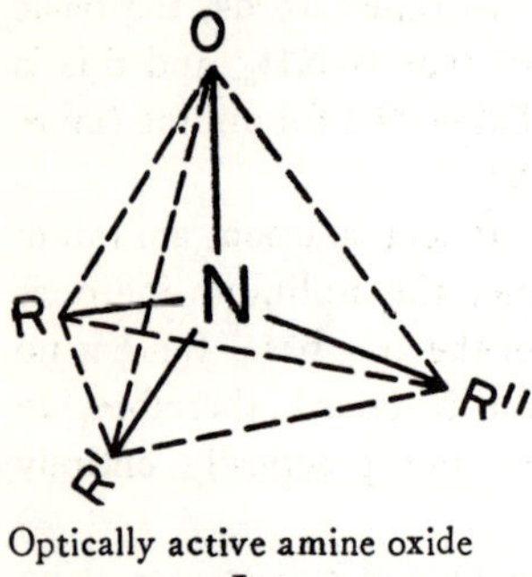

Optically active amine oxide
I

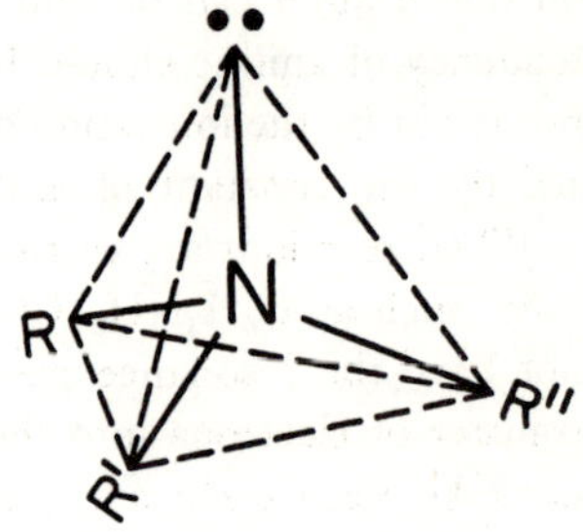

A tertiary amine
II

like that shown in Diagram II. These nitrogen compounds, however, are not optically active. The reason is that these molecules are in a constant oscillatory motion like the constant reversing of an umbrella. This intra-

molecular motion in the case of ammonia, NH_3, has been located in its absorption spectrum. It absorbs energy corresponding to a wavelength of 1.25 cm. Because of this constant oscillation (the frequency, according to Table 14.2 is 2.4×10^{10} Hz for the ammonia molecule), the unshared electron pair is not fixed in position and this precludes the possibility of an asymmetric molecule.

29.7. Properties of Amines Dependent on their Structure. It is obvious from the preceding remarks on the properties of the primary, secondary, and tertiary amines that they all have the general characteristics associated with the lone pair of electrons on the N atom. There are, however, a number of substances with which they react differently. Indeed, the chemistry of amines is so varied and complex we can in this section only cite a few examples, stressing the reactions that can be used to differentiate them as primary, secondary, and tertiary.

Nitrous acid (HO–N=O) is a reagent with a long history in the chemistry of amines. It is unstable; its water solution (from dissolved sodium nitrite) contains a variety of nitrogen compounds including nitrite ion, dinitrogen trioxide (N_2O_3), and nitric oxide (NO). The reaction of nitrous acid with a primary amine is complex, but under the proper conditions in strong acid this summary equation can be written for the products obtained with ethylamine:

$$C_2H_5-NH_2 + HO-N=O \longrightarrow C_2H_5-OH + N_2 + H_2O$$

Ethylamine   Nitrous acid   Ethyl alcohol   Nitrogen   Water

The driving force in this reaction is the tendency to form the very stable N_2 molecule; the yield of nitrogen is quantitative and can be used to estimate the quantity of primary amine in a sample. The other products obtained depend on the nature of the alkyl group in the primary amine and the experimental conditions.

Secondary amines react with nitrous acid to give yellow *nitroso* compounds, which are yellow liquids or solids and are stable:

$$CH_3-\underset{\underset{\displaystyle H}{|}}{N}-CH_3 + HO-N=O \longrightarrow CH_3-\underset{\underset{\displaystyle N=O}{|}}{N}-CH_3 + H_2O$$

Dimethylamine   Dimethylnitrosoamine

Tertiary amines will dissolve in nitrous acid solutions to form salts; other complex reactions may then take place, but absence of an H atom on the N atom in a tertiary amine precludes reactions like those illustrated.

29.8. Azo, Diazo, and Diazonium Compounds. The French word for nitrogen is *azote*, and this is the basis for the names of several groups of important compounds in the organic chemistry of nitrogen.

Azobenzene   Diazomethane   Benzenediazonium chloride

The electronic structures of these compounds will be elaborated on presently; at this point it should be observed that in the *azo* linkage the two N atoms are bonded to two C atoms (a 1:1 ratio), whereas in the *diazo* structures the two N atoms are linked to only one C atom (a 2:1 ratio).

Azomethane, $CH_3-N=N-CH_3$, is a yellow gas, made by oxidation of *sym*-dimethylhydrazine (Section 29.10). The great importance of azo compounds, however, lies among the aromatic compounds in the area of dyes, which the student can readily observe by glancing at the contents of Chapter 36. A point of interest that may be mentioned here is that *azobenzene*, due to the N=N group, exhibits the *cis, trans* type of isomerism described in Section 22.3. The more stable form is *trans*; action of light causes partial transformation to *cis*.

Diazomethane is a toxic, yellow gas. It is so likely to be explosive in the gaseous state that it is generally employed in solution (ether is a good vehicle). Diazomethane is a compound of great importance in the research laboratory. The substance is apparently a resonance hybrid, shown here in accordance with the schemes used in this book for writing electronic structures.

$$\left[\ \overset{H}{\underset{\oplus}{H-C}}::N::\overset{..\ominus}{N}:\ \leftrightarrow\ H-\overset{H}{C}:N:::N:\ \right]\ \text{or}\ \left[\ H-\overset{H}{\underset{\oplus}{C}}=N=\overset{|:\ominus}{N-}\ \leftrightarrow\ H-\overset{H}{C}-N\equiv N-\ \right]$$

Resonance structures of diazomethane

From these formulas it will be seen that the C atom has considerable negative character. The formal charges shown are determined as explained in Section 8.6. An example of the use of the compound in the laboratory is the preparation of *methyl* esters from carboxylic acids. The CH_2 group has sufficient negativity to extract a proton (H^+) from an acid to give a methyldiazonium ion, $CH_3-N\equiv N \leftrightarrow CH_3-N=N$. This splits out the very stable N_2 molecule as the positive diazonium ion reacts with the negative acid ion:

$$CH_3-\overset{}{\underset{O}{C}}-O^- + CH_3N_2{}^+ \longrightarrow CH_3-\overset{}{\underset{O}{C}}-OCH_3 + N_2$$

Acetate ion   Methyldiazonium   Methyl   Nitrogen
ion   acetate

The reaction is run in ether and is complete when evolution of N_2 ceases.

When diazomethane is subjected to decomposition by light, it yields two free radicals that are highly reactive and useful if prepared in contact with the proper reagents. These two radicals are *carbene* ($:CH_2$) and *methylene* ($H-\dot{C}-H$), and they are further described in Section 24.11. At this point attention is simply called to the *insertion·reaction* which is due to methylene. As shown in the methylation reaction just illustrated, a CH_2 unit may be regarded as inserted between O and H in acetic acid,

CH_3COOH; similar insertions take place with other compounds containing H. The following illustrations are for insertions between C–H in the isopropyl group:

$$CH_3\text{--}\underset{\underset{OH}{|}}{\overset{\overset{H}{|}}{C}}\text{--}CH_2\text{--}CH_3 \xleftarrow{\ CH_2\ } CH_3\text{--}\underset{\underset{OH}{|}}{\overset{\overset{H}{|}}{C}}\text{--}CH_2\text{--}H \xrightarrow{\ CH_2\ } CH_3\text{--}\underset{\underset{OH}{|}}{\overset{\overset{CH_3}{|}}{C}}\text{--}CH_3$$

sec-Butyl alcohol   Isopropyl alcohol   *t*-Butyl alcohol

The *aromatic diazonium* compounds, from the point of view of their importance in chemical research operations, rank with the Grignard reagents (Section 30.15) in the esteem of organic chemists. They are prepared from aromatic primary amines by the nitrous acid reaction (see section 29.7) but the procedure is modified in order to prevent the formation and elimination of N_2. The preparation is carried out at $0°$; the sodium nitrite solution is added to the salt of the amine in an excess of a mineral acid (HCl, for example). The process is called *diazotization:*

$$\text{⟨⟩}\text{--}NH_3{}^+Cl^- + HO\text{--}N{=}O \longrightarrow$$

Aniline   Nitrous acid
hydrochloride

$$\left[\text{⟨⟩}\text{--}\overset{|}{N}{=}\underset{+}{\overset{|}{N}} \leftrightarrow \text{⟨⟩}\text{--}N{\equiv}\underset{+}{\overset{|}{N}}\right]Cl^- + 2H_2O$$

Benzenediazonium chloride

The reaction is more complex than shown by this equation, and the formula of the diazonium ion shows it is a resonance hybrid, the structures of which are built up by the method employed in Figure 17.3. By variations of the procedure it is possible to prepare the dry diazonium salts as crystalline solids, many of which are unstable and explosive when shocked.

A diazonium compound has a dual structure. In the strongly acid solution in which the compound is normally made it is an *onium* salt, with the N atoms in the positive ion; but in a solution made strongly alkaline with sodium hydroxide the N atoms appear in the negative ion of a *diazo* type of salt:

$$\text{⟨⟩}\text{--}N{=}N\text{--}OH \qquad\qquad \text{⟨⟩}\text{--}N{=}N\text{--}O^-Na^+$$

Benzenediazo hydroxide   Sodium benzenediazoate
Benzenediazoic acid

The diazoic acids have not been isolated, but the salts are stable and can be handled without difficulty.

Only a few of the many types of organic compounds that can be made through the diazonium reaction will be mentioned here. In acid solution, elimination of the N_2 molecule generally takes place if the solution is allowed to warm up; at about $100°$, water converts a diazonium salt to a phenol:

$$C_6H_5N_2^+Cl^- + H{-}OH \longrightarrow C_6H_5OH + HCl + N_2$$

Benzenediazonium   Phenol
chloride

In neutral or slightly alkaline solution, a reaction called *coupling* takes place to produce azo compounds (see Section 36.10). Although the diazonium ion is only slightly electrophilic, the attack on the aromatic ring is the electrophilic type described in Section 20.2. The coupling

$$\langle \rangle{-}N_2^+Cl^- + H{-}\langle \rangle{-}N(CH_3)_2 \xrightarrow{-HCl}$$

Benzenediazonium Dimethylaniline
chloride

$$\langle \rangle{-}N{=}N{-}\langle \rangle{-}N(CH_3)_2$$

p-Dimethylaminoazobenzene

works well if the substituent already in the ring is one that activates it (Section 21.6). Coupling does not take place if the reaction conditions convert the diazonium compound to the diazo structure.

29.9. Free Radicals from Azo Compounds. The azo compounds, as already stated, tend to eliminate N_2 molecules. Some do so at relatively low temperatures. If they do, the by-product is a free radical ($R{-}N{=}N{-}R{\rightarrow}2R\cdot{+}N_2$). A compound extensively used for generating free radicals when gently heated (about $100°$) is this one:

$$\underset{\underset{CH_3}{|}}{\overset{\overset{CN}{|}}{CH_3{-}C}}{-}N{=}N{-}\underset{\underset{CH_3}{|}}{\overset{\overset{CN}{|}}{C}}{-}CH_3 \longrightarrow 2CH_3{-}\underset{\underset{CH_3}{|}}{\overset{\overset{CN}{|}}{C}}\cdot + N_2$$

a,a′-Azobisisobutyronitrile The free radicals
(Nitriles are defined in (see Section 24.8)
Section 29.14)

Compounds of this nature are used to start chain reactions (Section 17.9), and for initiating polymerizations leading to giant molecules (Section 38.7).

This compound just described is also used as a *foaming agent* in rubber formulations because of the ease with which it gives up its nitrogen and because of its compatibility with the other ingredients.

29.10. Representative Amines. *Ethanolamine*, CH_2–CH_2, is an
$$\overset{\displaystyle |}{OH} \quad \overset{\displaystyle |}{NH_2}$$
alcohol-amine—that is, a mixed compound like those described in the last chapter. Ethanolamine and related compounds have many uses. For example, *triethanolamine*, $(HOCH_2CH_2)_3N$, combines with fatty acids to form detergents, or soapless soaps (Section 27.18). These soaps are almost neutral and are free from injurious effects on textiles or skin. Their structure is similar to that of the sodium soaps (Section 27.17), with the amine in an *onium* ion replacing the sodium ion:

$$R-\underset{\overset{|}{OH}}{C}=O + N(CH_2CH_2OH)_3 \longrightarrow \left[R-\underset{\overset{|}{O}}{C}=O\right]^- \left[\underset{\overset{|}{H}}{N}(CH_2CH_2OH)_3\right]^+$$

Glycine, CH_2–C=O, or aminoacetic acid, is the simplest of the amino-
$$\overset{\displaystyle |}{NH_2} \quad \overset{\displaystyle |}{OH}$$
acids, another class of mixed compounds. Glycine is closely associated with the chemistry of the proteins (Chapter 33).

The family of mixed compounds of which glycine is a simple member presents an unusual characteristic in that the compounds are both acids and bases. There is evidence to show that most compounds of this kind have an uneven distribution of charge due to migration of the hydrogen ion H^+ (which the reader will recall is a proton) from the acid group to the basic group. These are neutral molecules, but are called *dipolar*

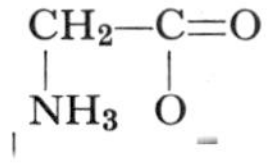

$$CH_2-C=O \qquad\qquad N_3H-\underset{+}{\langle\ \rangle}-S=O$$

Glycine 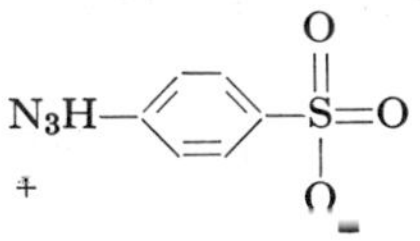   Sulphanilic acid
(see sulphonic acids, Section 20.4)

ions (the German word for this is Zwitterionen). Molecules of this type have also been called inner salts. The crystals of these compounds are probably held together by electrostatic action, as in the case of crystals of ionic compounds (Figure 8.1).

Choline is both an alcohol and a quaternary ammonium base. Through its alcohol group it forms compounds that are of great importance in life processes; an example is *acetylcholine* (Section 35.6).

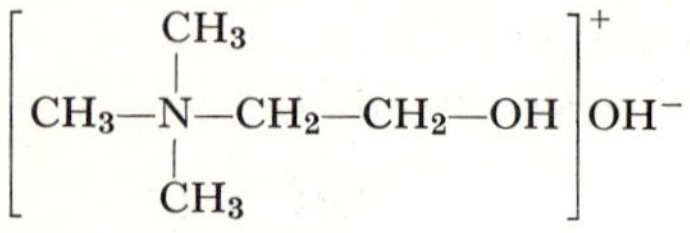

$$\left[\begin{array}{c} CH_3 \\ | \\ CH_3-N-CH_2-CH_2-OH \\ | \\ CH_3 \end{array} \right]^+ OH^- \qquad \left[\begin{array}{c} CH_3 \\ | \\ CH_3-N-CH_2-CH_2-O-\underset{\underset{O}{\|}}{C}-CH_3 \\ | \\ CH_3 \end{array} \right]^+ OH^-$$

Choline 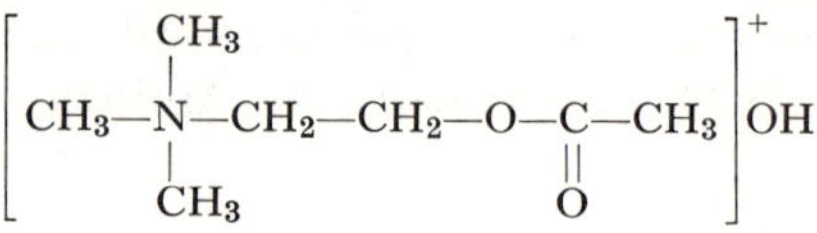   Acetylcholine

Among biochemists, the names choline and acetylcholine are applied to these positively charged ions rather than to complete compounds. In each case the physiological activity is due to the ion, which is associated with whatever negative ions may be present at the site of action in the body.

Hydrazine is an inorganic liquid directly related to ammonia, and has the composition H_2N-NH_2. It forms hydrocarbon derivatives known as *hydrazines*, just as ammonia yields derivatives called *amines*. The best known hydrazine in the laboratory is phenylhydrazine, $C_6H_5NH-NH_2$ (Section 34.9), which is used as a reagent for the identification of aldehydes and ketones (Section 26.8). Important drugs can also be made from hydrazine. *Unsym*-Dimethylhydrazine, $(CH_3)_2N-NH_2$, is used in rocket propulsion systems. With dinitrogen tetroxide, N_2O_4, it forms a *hyper-golic* fuel, one that ignites spontaneously when mixed. The oxidation products in this system are all gases.

29.11. Amides. The amides are a series of nitrogen compounds in which the –OH group in the $-\underset{|}{C}=O$ group of an acid is replaced by

$$\underset{OH}{}$$

the $-NH_2$ group. The relation between the amides and the other families of compounds we have studied can be understood from the series of reactions shown here, starting with a hydrocarbon and ending with an amide.

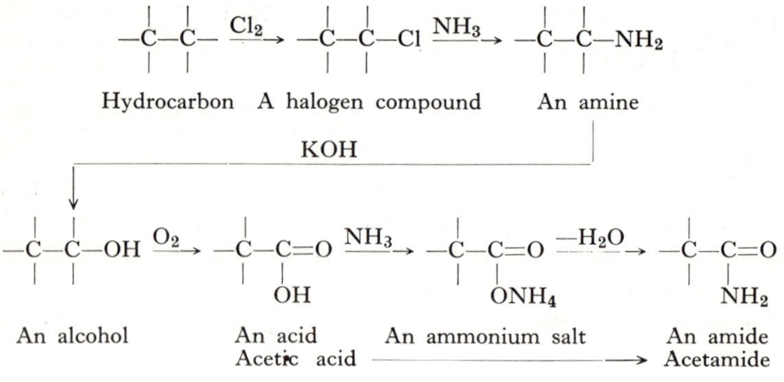

$$-\overset{|}{\underset{|}{C}}-\overset{|}{\underset{|}{C}}- \xrightarrow{Cl_2} -\overset{|}{\underset{|}{C}}-\overset{|}{\underset{|}{C}}-Cl \xrightarrow{NH_3} -\overset{|}{\underset{|}{C}}-\overset{|}{\underset{|}{C}}-NH_2$$

Hydrocarbon   A halogen compound   An amine

KOH

$$-\overset{|}{\underset{|}{C}}-\overset{|}{\underset{|}{C}}-OH \xrightarrow{O_2} -\overset{|}{\underset{|}{C}}-\underset{OH}{C}=O \xrightarrow{NH_3} -\overset{|}{\underset{|}{C}}-\underset{ONH_4}{C}=O \xrightarrow{-H_2O} -\overset{|}{\underset{|}{C}}-\underset{NH_2}{C}=O$$

An alcohol   An acid   An ammonium salt   An amide
     Acetic acid ————————————→ Acetamide

An amide is obtainable from the corresponding acid by heating the ammonium salt of the acid. It is evident from the formulas that the effect of the heat is to remove a molecule of water from the salt.

The amides and amines are similar in that both contain $-NH_2$ groups, but amides are derived from acids, whereas amines are derived from hydrocarbons. Like the amines, it is best to regard the amides as structurally derived from the inorganic substance, ammonia. When an acid chloride and ammonia are brought together, they react rapidly as follows:

$$CH_3-\underset{\underset{Cl}{|}}{C}=O \ + \ H-NH_2 \longrightarrow CH_3-\underset{\underset{NH_2}{|}}{C}=O \ + \ HCl$$

Acetyl chloride   Acetamide

This is an acylation (Section 27.9), analogous to the alkylation reaction by which an amine is formed from an alkyl chloride and ammonia (Section 29.4). It is possible, though difficult, to place two and sometimes even three acyl groups on the N atom, but such compounds are not important. Acetylation of the proper aromatic amines yields compounds useful as drugs. Acetanilid and phenacetin are used as antipyretics for fevers and headaches. The APC type of headache pill contains aspirin, phenacetin, and caffeine.

$$\langle \rangle\!\!-NH-\underset{\underset{O}{\|}}{C}-CH_3 \qquad\qquad CH_3O-\langle \rangle\!\!-NH-\underset{\underset{O}{\|}}{C}-CH_3$$

Acetanilid   Phenacetin

29.12. Properties of Amides. With the exception of the amide of formic acid, $HC\underset{\underset{NH_2}{|}}{=}O$, *formamide*, which is a liquid, the amides are crystalline solids. The lower open-chain members are soluble in water but the higher members and the aromatic amides such as *benzamide*, $\langle \rangle\!\!-\underset{\underset{NH_2}{|}}{C}=O$, can be crystallized from hot water.

Amines are more basic than their parent ammonia; this was explained in Section 29.6 as due to electron-release characteristics of alkyl groups. The *acyl* radicals, such as acetyl $CH_3-\underset{|}{C}=O$, however, are negative in

character and they accordingly tend to neutralize the basic nature of the
$-NH_2$ group.   An amide is therefore practically a neutral compound, but
it does retain some basic properties.

Acetamide, just like the amines and ammonia, forms salts with strong
acids.   It is possible to make *acetamide hydrochloride*,

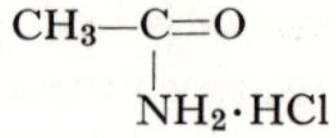

$$CH_3-C=O$$
$$|$$
$$NH_2 \cdot HCl$$

showing that acetamide has basic properties.   The salt formed is not very
stable, however, and is easily hydrolyzed by water.   Acetamide has two
possible structures that are tautomeric (see Section 28.11); the form with
greater stability is *keto*:

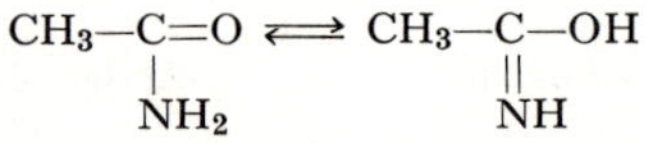

$$CH_3-C=O \rightleftharpoons CH_3-C-OH$$

Keto, or amido form   Imido form

An important property of amides is their hydrolysis with water:

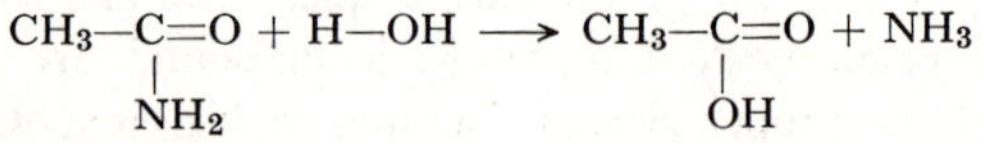

$$CH_3-C=O + H-OH \longrightarrow CH_3-C=O + NH_3$$

Acetamide   Acetic acid

The bond between C and N is broken and the free acid is regenerated.
However, the bond between C and N is very stable in amines, for we
showed that water solutions of amines are stable and are similar to ammonium hydroxide in behavior (Section 29.3).

29.13. Representative Amides.   (*Urea*, NH_2-C-NH_2, is the amide
of carbonic acid.   It is a crystalline solid and is the principal waste
product of the metabolism of foods in the animal body.   It has the
general properties of amides, but has many other individual characteristics and uses.

Barbituric Acid.   Malonic acid (Section 27.16) and urea can be condensed to give barbituric acid, the parent structure of many widely used
soporifics.

Urea   Chloride of   Barbituric acid
         malonic acid
         Malonyl chloride

The two hydrogen atoms in barbituric acid marked by an asterisk can be replaced by radicals. If these H atoms are replaced by ethyl ($-C_2H_5$) groups, the soporific obtained is *veronal*. Related compounds are *luminal, dial, amytal, allonal*, etc., all of which are sleep-producers and have individual characteristics due to the specific radicals used to replace the H atoms in the ring.

Purines. The substance known as purine may be regarded as the parent structure of many compounds important in biochemistry. One is *uric acid*, known since 1776; it is a major component of the excreta of birds and reptiles and is present to a small extent in the urine of mammals. Another is *caffeine*. The formula of purine itself is given here in the usual form for ring compounds.

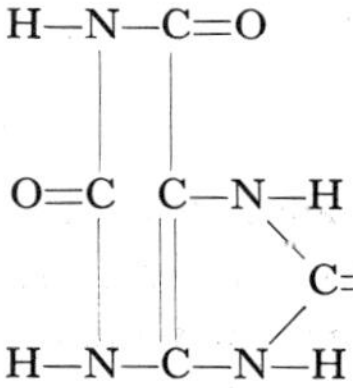

Uric acid 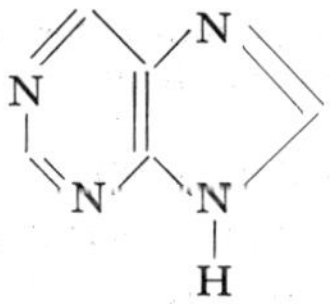　　　　　　　　　　Purine 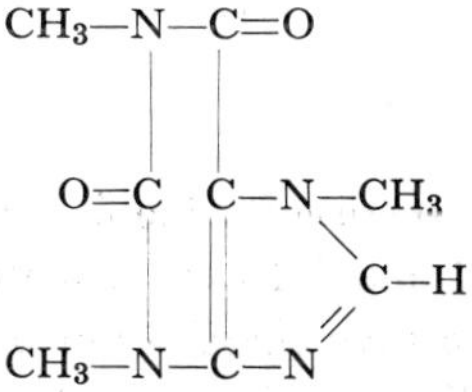　　　　　　　　　　Caffeine

It should be observed that there are two urea skeletons to one of malonic acid in these compounds.

Succinamide, $NH_2-C-CH_2-CH_2-C-NH_2$, is the amide of succinic acid (Section 27.16). It is noteworthy for the ease with which it forms an *imide* when heated. The *imido* H atom of succinimide is replaceable

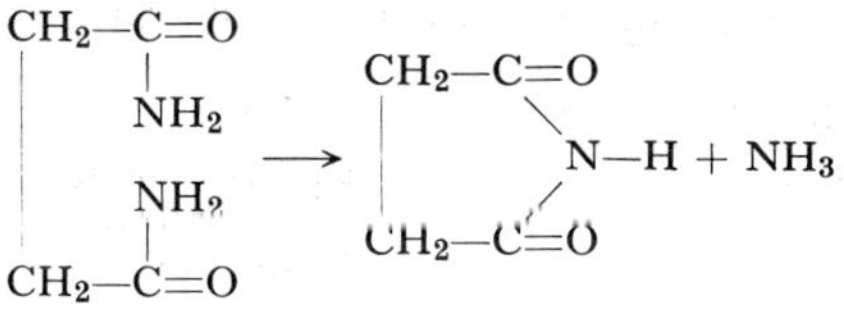

Succinamide　　　　　Succinimide

by sodium or potassium. With such a metal in the molecule, its usefulness in organic syntheses is greatly increased, since it will easily react with many types of halogen compounds.

Guanidine, NH_2-C-NH_2, is the imide of urea. It is a decomposition product of *guanine*, which is a derivative of purine (Section 29.13). Guanine is 2-amino-6-oxypurine; it is present in large quantity in the bird excrement, guano.

Of particular interest in the chemistry of guanidine is the fact that it reacts with only one molecule of an acid such as HCl, instead of the three that its formula leads us to expect. Adding one proton results in

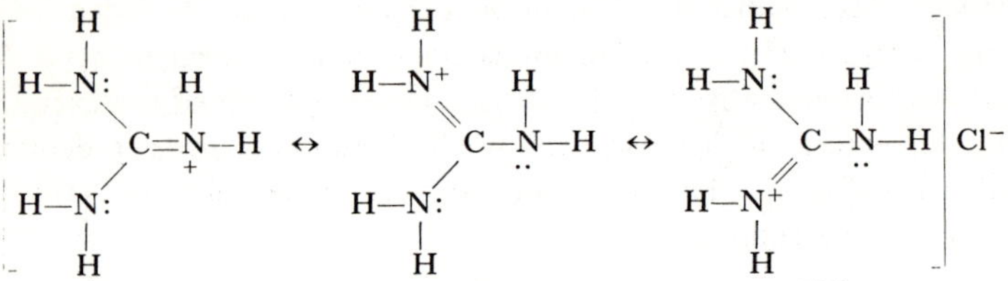

$$(I) \leftrightarrow (II) \leftrightarrow (III)$$

Guanidinium chloride
Guanidine hydrochloride

Structure I, a highly symmetrical guanidinium ion. Resonance Structures II and III can be written by simply rearranging the valence bonds; the positive charge originally acquired by one N atom is distributed among all three N atoms. Since the three structures are equivalent (X-ray data show that the three $-NH_2$ groups in the ion are symmetrically placed about the C atom), the resonance energy is maximum, as explained in Section 16.9c. Ability to form this very stable guanidinium ion is the reason for the strong base activity of guanidine.

29.14. Nitriles. Potassium cyanide, KCN, and sodium cyanide, NaCN, are common reagents in the laboratory of organic chemistry because they are so useful in the preparation of compounds containing nitrogen. It is a simple matter, using the reactive halogen compounds, to attach the $-C \equiv N$ group to a hydrocarbon chain structure:

$$CH_3 - Cl + K^+[:C:::N:]^- \longrightarrow CH_3 - C \equiv N + K^+Cl^-$$

Methyl chloride   Methyl cyanide
                                 Methyl nitrile
                                 Acetonitrile

These organic cyanides are known as *nitriles*. The reaction mechanism in this case is the S_N2 type as described in Figure 22.15, with the CN^- ion attacking from the rear, and Cl the leaving group.

The nitriles are easily reduced to amines:

$$CH_3 - C \equiv N + 2H - H \xrightarrow{\text{hydrogenation}} CH_3 - CH_2 - NH_2$$

Methyl cyanide   Ethylamine

The nitriles are also readily hydrolyzed to amides. The hydrolysis is catalyzed by both acids and bases; a probable mechanism for the acid hydrolysis is given here:

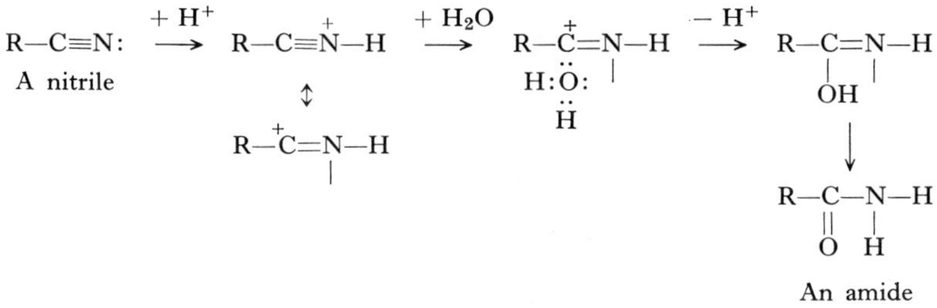

$$R-C\equiv N: \xrightarrow{+\ H^+} R-\overset{+}{C}\equiv N-H \xrightarrow{+\ H_2O} \underset{\underset{\overset{|}{H}}{\overset{\cdots}{H:O:}}}{R-\overset{+}{C}=N-H} \xrightarrow{-\ H^+} \underset{\overset{|}{OH}}{R-\overset{|}{C}=N-H}$$

A nitrile

$$\updownarrow$$

$$\underset{\overset{|}{}}{R-\overset{+}{C}=N-H}$$

$$\downarrow$$

$$\underset{\overset{||\ \ |}{O\ \ H}}{R-C-N-H}$$

An amide

An amide easily hydrolyzes further to an acid (Section 29.12). The common names of the nitriles are obtained from the names of the acids to which they can be hydrolyzed. Methyl cyanide is *acetonitrile* because it hydrolyzes to acetic acid. Similarly, phenyl cyanide is *benzonitrile*, $\langle\!\!\!\bigcirc\!\!\!\rangle$–CN, because it can be hydrolyzed to benzoic acid (Table 27.2).

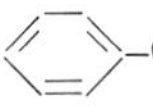

The complex compound described in Section 29.9 is an isobutyronitrile because of this theoretical relation to isobutyric acid.

Isonitriles are isocyanides, and are also called *carbylamines*. They have the RNC structure. It was discovered long ago that this is the main product when silver cyanide, AgCN, is used in the customary preparation of a nitrile from an alkyl halide. The different reaction mechanism with AgCN is apparently due to the greater covalent nature of that compound as compared with salt-like KCN. Other preferred methods for making isonitriles that do not require metallic cyanides are now available.

The structure of an isocyanide is believed to be $R-N\rlap{\;\geqslant}{\equiv}C:$, in which the NC linkage is similar to that in carbon monoxide, $:C\rlap{\;\leqslant}{\equiv}O:$. In both, the C atom is acceptor of a dative bond (Section 8.5). These molecules will also be found written as pairs of resonance structures $(R-N\equiv C: \leftrightarrow R-\overset{\cdots}{N}\overset{\oplus\ \ \ominus}{}$
$=C:)$ and $(:C\equiv O: \leftrightarrow :C=\overset{\cdots}{O}:)\overset{\ominus\ \ \oplus}{}$. The electron structure of carbon monoxide is outlined graphically in Figure 29.1. In these diagrams, the electrons *shared by* the atoms are blacked in, and the electron pair *accepted from* the oxygen atom is designated by crosses. This structure really has all the characteristics of a triple bond, like that shown for acetylene in Section 13.1. In actual practice, the isocyanides and carbon monoxide

29.1.   Electron-sharing in carbon monoxide.

are usually written with double bonds, R–N=C and C=O, for that suffices to indicate their most important chemical properties.

The isonitriles are very reactive because of the strongly dipolar nature of the NC group. They are hydrolyzed faster than nitriles and do not give acids of the same carbon content; an amine is formed and a molecule of formic acid is split off:

$$\overset{\oplus}{R-N} \overset{\ominus}{\equiv C}: + 2H_2O + HCl \longrightarrow R-NH_2 \cdot HCl + H-COOH$$

Isonitrile   Hydrochloride   Formic acid
of a primary amine

Reduction of an isonitrile results in formation of a secondary amine:

$$\overset{\oplus}{R-N} \overset{\ominus}{\equiv C}: + 2H_2 \longrightarrow R-N-CH_3$$
$$\underset{H}{|}$$

A secondary amine

These two reactions are evidence for the RNC structure in these compounds as compared with RCN for normal cyanides. When isocyanides are heated with metallic cyanides they convert to the more stable (less reactive) normal cyanides.

The isocyanides have very offensive odors, whereas the normal cyanides have strong but fairly agreeable odors. The iso compounds are also much more toxic.

Cyanogen, $N\equiv C-C\equiv N$, is a poisonous gas with a characteristic odor resembling that of bitter almonds. It can be called *oxalo*nitrile because it hydrolyzes to oxalic acid (Section 27.15).

Hydrocyanic acid, $H-C\equiv N$, is a volatile liquid with an odor like that of cyanogen, and is very poisonous. It is also known as *formo*nitrile because it can be hydrolyzed to formic acid (Section 27.15).

29.15. Nitro Compounds. The student should review Section 20.3 for the mechanism of nitration by which nitro groups are introduced into organic compounds, and for the difference between nitrates and nitro compounds; and Section 21.7 should be reviewed for the description of the resonance structures of the nitro group.

Nitro compounds have an important role in the chemistry of open-chain compounds as well as their more prominent role in the chemistry of the benzene series. Those nitroparaffins in which C–H is adjacent to the NO_2 group exhibit the phenomenon of *tautomerism*, like that described in Section 28.11 in connection with ketones. *Nitro* and *aci* forms are possible, similar to the *keto, enol* forms mentioned in Chapter 28. This is illustrated by the structures of nitromethane. The *aci* structures in

$$
\begin{array}{ccc}
\underset{\textit{Nitro form}}{H-\overset{\overset{\displaystyle H}{|}}{\underset{\underset{\displaystyle H}{|}}{C}}-\overset{\overset{\displaystyle O}{\|}}{N}\to O} & \text{Nitromethane} & \underset{\textit{Aci form}}{H-\overset{\overset{\displaystyle OH}{|}}{\underset{\underset{\displaystyle H}{|}}{C}}=N\to O}
\end{array}
$$

(See Section 21.7 for this method of writing the NO_2 group.)

this series of compounds are more stable than the *enol* structures among the keto compounds.

All nitroparaffins containing α hydrogen atoms are acidic. Water-soluble salts are produced in dilute alkali solution. When the basic solvent removes a proton (H^+) at the α carbon atom it gives rise to a negative ion which is stabilized by resonance (Structures I and II); this is similar to the behavior of acetone described in Section 28.11. However, when the

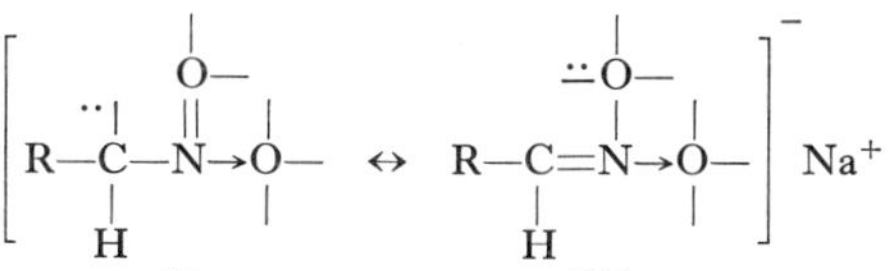

$$
\left[R-\overset{|}{\underset{|}{C}}-\overset{\overset{\displaystyle O-}{\|}}{N}\to\overset{|}{O}-\ \ \leftrightarrow\ \ R-\overset{|}{\underset{|}{C}}=\overset{\overset{\displaystyle \cdot\cdot O-}{|}}{N}\to\overset{|}{O}- \right]^{-} Na^+
$$
$$
\text{(I)}\qquad\qquad\qquad\text{(II)}
$$

Resonance of the negative ion of a nitroparaffin. The electron pairs are shown as in Section 17.3.

solution is acidified, the proton does not return to the α carbon atom (Structure I) but becomes attached to the oxygen atom (Structure II). The acid solution therefore at first contains the *aci* form of the nitro compound; this slowly reverts to the *nitro* form as the proton migrates to the more stable position on the carbon atom.

The acidity of nitroparaffins is thus explained by the same reasoning as employed for carboxylic acids (Section 27.5) and phenols (Section 25.12). The resonance energy of the negative ion is greater than that of the unionized molecule, and this results in acidity—that is, tendency for the H^+ to remain as such in a solution of the substance.

The nitro group in a nitroparaffin is said to have an activating influence on the α hydrogen atoms. Such an activating effect is illustrated by a reaction of nitroparaffins which resembles that of aldehydes (Section 26.9); when catalyzed by dilute alkali they undergo the *aldol* type of addition reactions if an α H atom is present in the molecule:

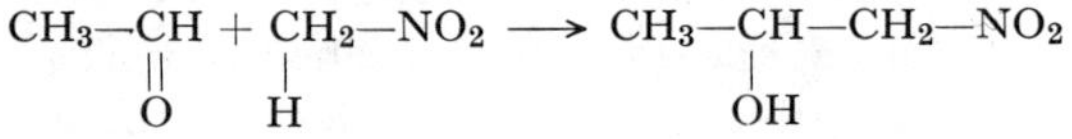

$$CH_3—CH + CH_2—NO_2 \longrightarrow CH_3—CH—CH_2—NO_2$$

$$\text{Acetaldehyde} \quad \text{Nitromethane} \qquad \text{1-Nitro-2-propanol}$$

Acid properties of certain aromatic nitro compounds are also explainable on the basis of resonance stabilization of the negative ions which result when a proton is removed by the solvent medium. This is illustrated by the effect of a base ($:B^-$) on p-nitrotoluene, the formulas being written according to the scheme shown in Section 17.3:

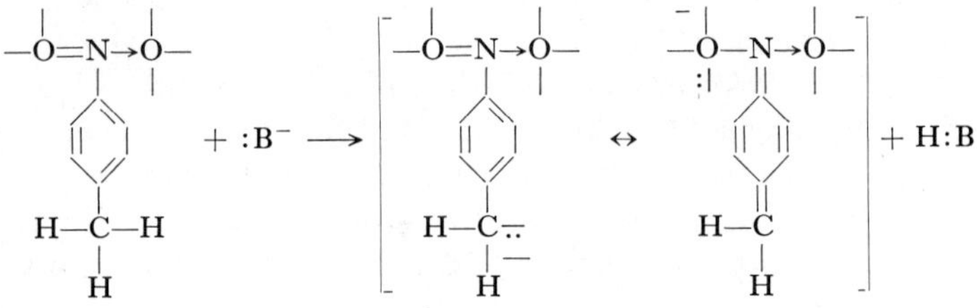

p-Nitrotoluene　　Resonance structures of the negative ion

30

Compounds with Sulphur, Phosphorus, and Other Elements

30.1. Introduction. Sulphur is just below oxygen in the periodic system of the elements and the two elements form several structurally analogous series of compounds. The *simpler* sulphur compounds in organic chemistry are studied as being derived from hydrogen sulphide, H_2S, just as the oxygen compounds are related to water, H_2O, on a

H—O—H	Water	H—S—H	Hydrogen sulphide
C_2H_5—OH	Ethyl alcohol Ethanol	C_2H_5—SH	Ethyl mercaptan Ethanethiol
C_2H_5—O—C_2H_5	Diethyl ether Diethyl oxide	C_2H_5—S—C_2H_5	Diethyl thioether Diethyl sulphide

structural basis. The compounds containing sulphur are often given the names of the analogous oxygen compounds, adding the prefix *thio* from the Greek word for sulphur. There are thioalcohols, thioethers thioaldehydes, thioketones, etc. However, alcohol analogs are more frequently called *thiols*, or *mercaptans*.

The nomenclature system mentioned for oxygen compounds in Section 25.16, in which diethyl ether is called 3-oxapentane, is sometimes also used for sulphur compounds. According to this scheme, diethyl sulphide is 3-thiapentane, and mustard gas (Section 30.5) is 1,5-dichloro-3-thiapentane.

30.2. Electron Bonding in Sulphur Compounds. Although the compounds listed in the preceding section show a marked resemblance on a formula basis between oxygen compounds and sulphur compounds, their chemistry is significantly different. The reason for this was not well understood until the electron-valence theory was rather advanced. The

347

reader should turn to Figure 6.1 and observe that the valence electrons of sulphur are screened from the positively charged nucleus by 10 electrons, whereas the valence electrons of oxygen are screened by only 2. Because of the resulting weaker hold on its valence electrons, the sulphur atom has a lower ionization potential (Figure 5.1) and a lower electronegativity (Section 8.8).

The differences in chemistry between carbon compounds containing oxygen and carbon compounds containing sulphur can generally be accounted for by the lower electronegativity of the S atom, and by its larger size which permits the additional orbitals illustrated in Figure 30.1. An element like oxygen, in the first period of the periodic table, has four orbitals which are described in Figure 7.1 and can accommodate eight electrons; it satisfies the octet rule of Chapter 6. A second-period element, like sulphur, with more orbitals available, can accommodate more electrons and has the means with which to violate the octet rule; it forms stable compounds like sulphur hexafluoride, SF_6, in which there are six covalent bonds and therefore twelve electrons in the valence shell. The six covalent bonds in SF_6 are made possible by sp^3d^2 hybridization at the S atom (Section 7.10), which provides six orbitals. Molecules like SF_6 which have six sp^3d^2 orbitals radiating from the central atom have an octahedral structure or shape, whereas we found that molecules constructed from sp^3 orbitals are tetrahedral (Section 8.4).

The sulphur atom (Figure 6.1) has three principal energy levels; there are two electrons in the first, eight in the second, and six in the third. On the basis of the definitions in Section 7.8, in the third energy level there are 3^2 or 9 orbitals. The sulphur atom in its chemical bonding can use these orbitals: one **3s**, three **3p**, and five **3d**, which makes a total of nine. In earlier chapters, with the exception of the H atom we were concerned only with elements in the 2nd energy level, so we did not generally use a **2** when discussing their atomic orbitals. Now we must distinguish between the energy levels by labeling them.

Inequality in size of atoms generally results in lower bond strength between them. We refer to a pair of atoms which have valence shells with different principal quantum numbers, such as H (quantum number **1**) and S (quantum number **3**); there is a lesser effectiveness when their orbitals overlap head-on to form σ bonds, or coalesce sideways to form π bonds. It is illustrated by these sets of data for bond energy in kcal/mole:

$$
\begin{array}{ll}
\text{O—H} \quad 111 & \qquad \text{O—C} \quad 84 \\
\text{S—H} \quad 81 & \qquad \text{S—C} \quad 62
\end{array}
$$

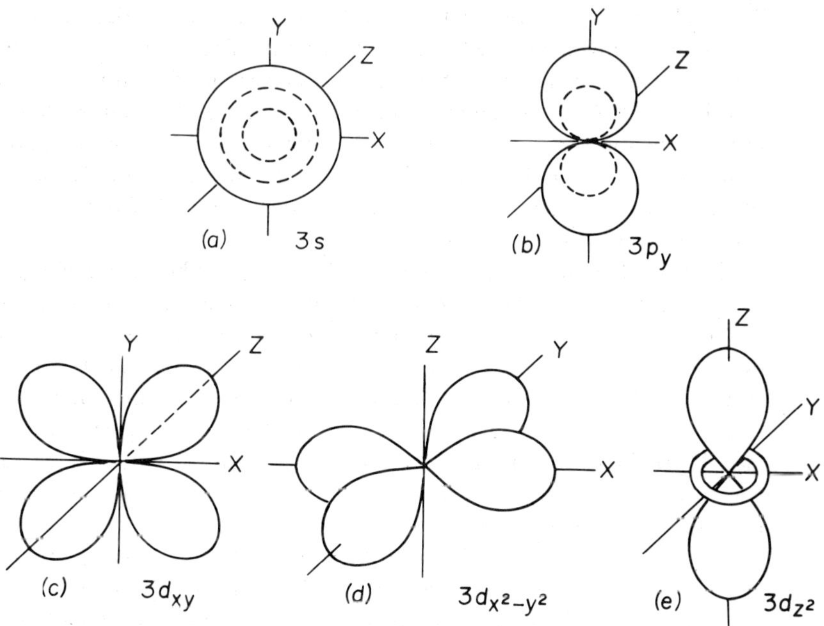

30.1. The nine atomic orbitals in the 3rd energy level of an element in the second period of the periodic system. This applies to sulphur.

   (a)   The **3**s orbital has two spherical nodes, shown by dotted lines. A **2**s orbital has one such node. The **1**s orbital of the hydrogen atom is shown in Fig. 7.2.

   (b)   The **3**p orbital has the planar node shown for a **2**p orbital in Fig. 7.2, and in addition the symmetrical node shown by dotted lines. There are three of these orbitals perpendicular to each other, similar to those shown in Fig. 7.1.

   (c)   The **3**d_{xy} orbital has four lobes, but can hold only the two customary electrons. It has two planar nodes. The lobes in this orbital extend into space between the x and y axes. There are three such orbitals at right angles to each other: in the **3**d_{xz} orbital the lobes are between the x and z axes; and in the **3**d_{yz} orbital the lobes are between the y and z axes.

   (d)   The **3**$d_{x^2-y^2}$ orbital is aligned on the x and y axes. It is conventional to use these axes instead of the other possible combinations. In specialized works on orbital theory the z axis is generally drawn where the y axis is customarily found in this book, and the x axis is often at right angles to the page.

   (e)   The **3**d_{z^2} orbital is shown on a vertical axis in the plane of the page, for the reason given in d.

The orbitals shown in d and e will not be referred to in this chapter.

The OH bond is pictured in Figure 7.2 as the result of overlap of the **1**s orbital of hydrogen and a **2**p orbital of oxygen. This overlap is stronger than that which takes place between the **1**s orbital of hydrogen and a **3**p orbital of the sulphur atom. This means that the SH bond is a looser union than the OH bond; it probably explains why hydrogen sulfide, HSH, is more acidic than water, HOH, the H being more readily ionizable in the sulphur compound. Similarly, ethanethiol, C_2H_5SH, is more acidic than ethanol, C_2H_5OH.

Hydrogen bonding occurs when a proton is trapped between the highly electronegative elements oxygen, nitrogen, and fluorine. An example of the great importance of the H bond among oxygen compounds is that described in Section 27.8; another example is the one mentioned in section 25.10 in connection with alcohols. With sulphur compounds, due to low sulphur electronegativity, the H bond is of little importance. In fact, HSH does not appear to form H bonds at all; it boils at $-61°$ as compared with HOH at $+100°$. The properties of water are known to be highly dependent on its ability to form H bonds.

The electron distribution in the 3rd energy level of the sulphur atom is like that shown in Section 7.9 for the 2nd level of the oxygen atom, and the formation of the H–S–H molecule is similar to that for H–O–H depicted in Figure 7.4 except that the σ bonds are **1**s–**3**p. The bond angle of the molecule should be a right angle (90°) as shown in Figure 7.4 because of the orientation of the p orbitals in space. The measured angle in this molecule is 92.3°. The reader should compare this with the 105° angle of the water molecule discussed in Section 8.4; some of this difference is no doubt due to the greater repulsion by the OH bonds in the smaller water molecule as compared with that of the SH bonds in H–S–H.

Carbon can be linked to sulphur in a C–S bond of the σ type just as in C–C linkages. However, most of the interesting chemistry we have studied so far has been concerned with multiple bonds exemplified by $H_2C{=}CH_2$ and $O{=}C{=}O$, where one bond is the σ variety and the additional bonding is the π type. These molecular π bonds are formed by sideways (parallel) coalescence of **2**p atomic orbitals as described in Figures 15.6 and 15.7. Compounds with C=S bonds are known (see thiourea in Section 30.6) but are not particularly stable, so parallel overlap of a **2**p orbital on carbon with a **3**p orbital of the larger sulphur atom cannot be very efficient.

Carbon compounds containing S=O double bonds are both stable and numerous. In these double bonds the first bond, of course, is the usual

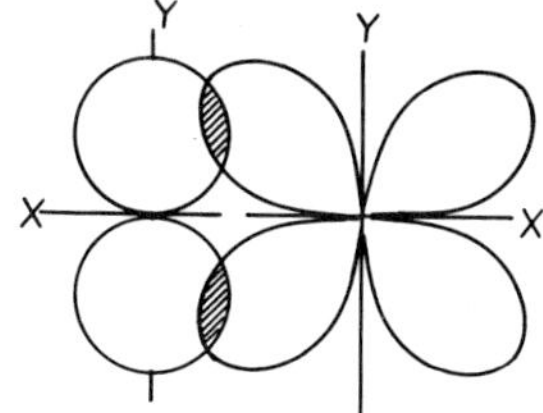

30.2. Formation of a π bond by parallel overlap of a $2p$ oxygen orbital and a $3d$ sulphur orbital. This is a $2p$—$3d$ molecular orbital, and it contains two electrons.

σ bond and it is presumed that the second bond (a π bond) is formed by parallel overlap as described in Figure 30.2, where the S atom makes use of a normally unoccupied d orbital. A compound explainable in this way is dimethyl sulphoxide, for which two structural arrangements can be written as shown in I and II with circles representing the electrons supplied by S and dots for the electrons contributed by carbon and oxygen. Structures III and IV are written with straight lines for electron pairs, as illustrated in Figure 17.2, and the formal charges shown in IV calculated as explained in Section 8.6.

It is believed that dimethyl sulphoxide is a resonance hybrid of I and II.

I   ⟷   II   or   III   ⟷   IV

Most authorities state that Structure I predominates, but this is debated. Some do not agree that the S atom promotes a pair of electrons at all to the higher energy level of a d orbital, which it must do in order to accommodate the 10 electrons shown in the double bond Formula I (or III). There is no doubt that d orbitals can be employed; there is no other explanation for the compound SF_6 mentioned earlier in this chapter. Structure II (or IV) is reminiscent of the dative bond (Section 8.5) which is sometimes found written $(CH_3)_2S{\rightarrow}O$, as in earlier editions of this book, but mostly these compounds are written with double bonds, $(CH_3)_2S{=}O$. The $S{=}O$ double bond does not have the properties associated with the double bond in olefins, aldehydes, ketones, etc., which take part in addition reactions of which the $S{=}O$ double bond is not capable.

30.3. Thiols. These compounds can be prepared by a method that recalls the preparation of the alcohols:

$$C_2H_5{-}Cl + K{-}SH \longrightarrow C_2H_5{-}SH + KCl$$

Potassium   Ethyl mercaptan
hydrosulphide   Ethanethiol

The mercaptans are flammable liquids, with a very unpleasant, garlicky odor. They are not poisonous. A mixture of mercaptans is blended with illuminating gas in some cities in order to help the gas companies detect leaks in the mains.

The mercaptans are similar to alcohols in their fundamental chemical properties. For example, they react with sodium:

$$2C_2H_5{-}SH + 2Na \longrightarrow 2C_2H_5{-}SNa + H_2$$
$$C_2H_5{-}SH + NaOH \longrightarrow C_2H_5{-}SNa + H_2O$$

Sodium ethyl mercaptide

Alcohols do not react with sodium hydroxide to form ethoxides because they are too weakly acidic, but mercaptans react as shown. Furthermore, since mercaptans are more acidic than alcohols, the mercaptides are more saltlike than ethoxides (Section 25.3). The mercaptans get their class name from their ability to form the mercury salt with the composition $(C_2H_5S)_2Hg$, mercury mercaptide. The name mercaptan is derived from the Latin, *mercurium* (mercury) *captans* (seizing).

In Section 30.2 it was stated that sulphur compounds do not hydrogen-bond as readily as the alcohols do; this probably accounts for the fact that ethyl mercaptan boils at the fairly low temperature of 35° as compared with 78.5° for ethyl alcohol.

The mercaptans are easily oxidized to disulphides. This takes place even when the mercaptans are exposed to air, but is usually accomplished in the laboratory by means of hydrogen peroxide or iodine. The reaction is reversible; the disulphide can just as readily be reduced back to the thiol.

$$2C_2H_5{-}SH \underset{reduction\ (+2H)}{\overset{oxidation\ (-2H)}{\rightleftarrows}} C_2H_5{-}S{-}S{-}C_2H_5$$

Ethanethiol   Ethyl disulphide

30.4. Sulphonic Acids. The mercaptans are like the alcohols in that they can be oxidized to acids. Ethyl alcohol yields acetic acid (Section 27.2), and ethyl mercaptan gives ethanesulphonic acid; the reaction mechanisms are much different but are not discussed in detail in this book:

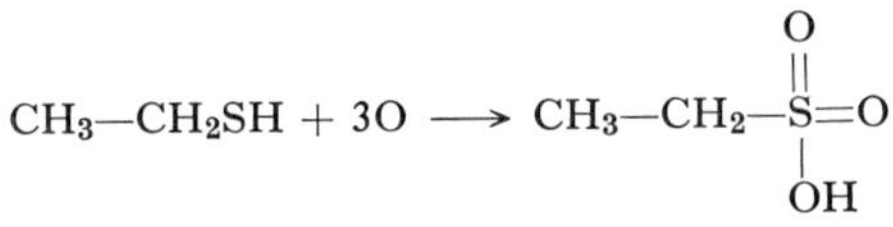

$$CH_3-CH_2SH + 3O \longrightarrow CH_3-CH_2-\underset{\underset{OH}{|}}{\overset{\overset{O}{\|}}{S}}=O$$

Ethyl mercaptan   Ethanesulphonic acid
Ethanethiol

Acetic acid is weak (Section 27.7), but the sulphonic acid is strong, practically as strong as hydrochloric acid. The sulphonic acids have the general properties of the carboxylic acids listed in Chapter 27. They form acid chlorides, salts, esters, and other acid derivatives (see the sulphonamides in Section 30.7).

As already explained in Section 20.4, the sulphonic acids must be distinguished from sulphates. Organic sulphates do not have a C–S bond:

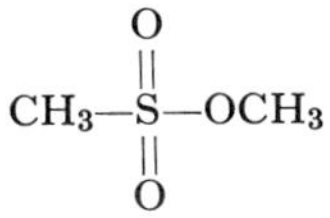

$$CH_3-\underset{\overset{\|}{O}}{\overset{\overset{O}{\|}}{S}}-OCH_3 \qquad\qquad CH_3O-\underset{\overset{\|}{O}}{\overset{\overset{O}{\|}}{S}}-OCH_3$$

Methyl ester of methanesulphonic acid   Dimethl ester of sulphuric acid
Methyl methanesulphonate   Dimethyl sulphate

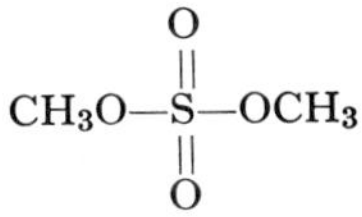

These two compounds are not isomers. The student should, however, refer back to Section 2.2 where related compounds were formulated as derived from sulphurous acid, H_2SO_3. The two compounds in section 2.2 are isomers and are, respectively, diethyl sulphite and the ethyl ester of ethanesulphonic acid.

Sulphonic acids and organic derivatives of sulphuric acid are widely used in commerce, especially in the manufacture of detergents (Section 27.18). In fact, these detergents constitute the major part of those currently on the market. They are good lathering agents, useful in shampoos. These detergents are typical:

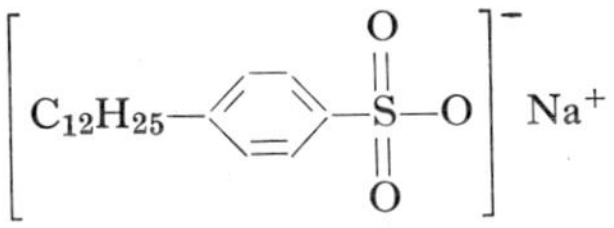

$$\left[C_{12}H_{25}-\!\!\left\langle=\right\rangle\!\!-\underset{\overset{\|}{O}}{\overset{\overset{O}{\|}}{S}}-O \right]^{-} Na^+ \qquad\qquad \left[C_{12}H_{25}O-\underset{\overset{\|}{O}}{\overset{\overset{O}{\|}}{S}}-O \right]^{-} Na^+$$

Sodium dodecylbenzenesulphonate   Lauryl sodium sulphate
(An alkarylsulphonic acid derivative)   Dodecyl sodium sulphate

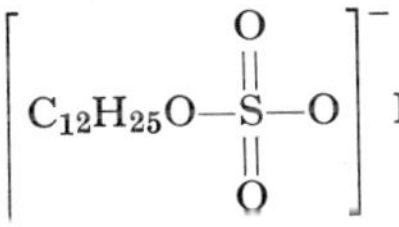

Triethanolamine salts (Section 29.10) are often used instead of the sodium salts.

The reader should refer to Section 20.4 where we described the preparation of sulphonic acids of the aromatic series by direct sulphonation of the hydrocarbon, as in the preparation of benzenesulphonic acid

$\langle\!\!\bigcirc\!\!\rangle$–SO₂–OH. The sulphonic acids in the open-chain series are not easily made by direct sulphonation of the hydrocarbon except in special cases, but can be prepared by oxidation of the mercaptan as described in this chapter.

An intermediate type of compound in the oxidation state of mercaptans is the series of *sulphinic acids*, in which two atoms of oxygen are added to the molecule. These compounds are made by indirect methods rather than by direct oxidation of mercaptans. Their chemistry is relatively unimportant in an elementary study.

$$CH_3-CH_2-\overset{\displaystyle O}{\overset{\displaystyle \|}{S}}-OH$$

Ethanesulphinic acid

30.5. Thioethers. The thioethers are similar to ethers (Chapter 25) in methods of preparation and most of their chemical properties. A representative member of the series is $CH_3–CH_2–S–CH_2–CH_3$, known as diethyl thioether, or diethyl sulfide. Like the mercaptans, the thioethers have piercing, disagreeable odors.

Mustard gas, $Cl–CH_2–CH_2–S–CH_2–CH_2–Cl$, is a symmetrical dichloro-derivative of diethyl sulphide. It is one of the important chemical-warfare agents. It is not a gas, but an oily liquid which vaporizes very slowly and maintains a toxic concentration in the air for a long time. Both the liquid and the vapor cause serious skin burns. The Cl atoms in this molecule are on the β-carbon atoms; when they are on the α-carbon atoms—that is, next to the sulphur—the molecule is less toxic.

The thioethers are more active than their sister oxygen compounds, both chemically and physiologically. For example, the oxygen compound analogous to mustard gas is dichloroethyl ether, $ClCH_2–CH_2–O–CH_2–CH_2Cl$. It is comparatively nontoxic, and is used in large quantities as a commercial solvent and "dry cleaner."

Sulphoxides and *sulphones* are stable substances made by oxidizing thioethers with strong oxidizing agents. By proper choice of oxidizing agent the oxidation can be stopped at the sulphoxide stage. In the early 1960's,

$$CH_3-S-CH_3 \longrightarrow CH_3-\overset{O}{\underset{}{\overset{||}{S}}}-CH_3 \longrightarrow CH_3-\overset{O}{\underset{\overset{||}{O}}{\overset{||}{S}}}-CH_3$$

Dimethyl thioether　　Dimethyl sulphoxide　　Dimethyl sulphone

dimethyl sulphoxide (DMSO) was found to be unique in the ease with which it penetrates the skin. Since it is an excellent solvent, this resulted in extensive research on its ability to carry drugs into the body by that pathway.

Thioethers are alkyl sulphides, R–S–R, analogous in structure to the amines, R–N–R. Accordingly, we find that they form *sulphonium* com-

$$\underset{\overset{|}{R}}{R-N-R}$$

pounds which are analogous to the ammonium compounds (Section 29.5):

$$CH_3\!:\!\overset{..}{\underset{..}{S}}\!:\!CH_3 + .\overset{..}{\underset{..}{I}}\!.CH_3 \longrightarrow \left[\underset{\overset{..}{C}H_3}{CH_3\!:\!\overset{..}{S}\!:\!CH_3}\right]^{+} :\overset{..}{\underset{..}{I}}:^{-} \text{ or } (CH_3)_3S^+I^-$$

Dimethyl　　Methyl　　　　Trimethlsulphonium iodide
sulphide　　iodide

From this product we can prepare trimethylsulphonium hydroxide, $(CH_3)_3S^+OH^-$, a strong base, similar in behavior to sodium hydroxide.

30.6. Asymmetry of Sulphur Compounds. In Section 22.7 we discussed the possibility of optically active compounds in which the asymmetry is not due to the carbon atom. Sulphoxides and sulphonium compounds are of considerable interest in this connection, in that certain of these compounds have been resolved into optically active isomers. It has been accomplished, for example, with several sulphonium salts in which the positive ion has three different groups, represented by R_1, R_2 and R_3, on the sulphur atom. This behavior indicates that the lone electron pair can have the effectiveness of a group at a corner of the tetrahedron. In Section 29.6 it was pointed out that this resolution into optically active isomers is not possible among the corresponding nitrogen compounds with a lone electron pair, because

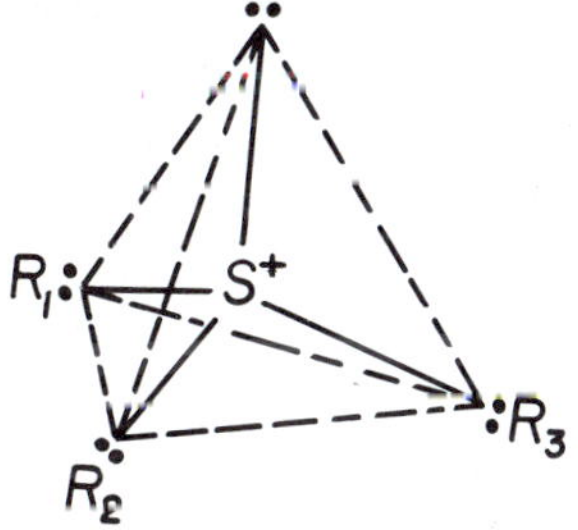

of the rapid inversion of the molecule. This is another illustration of the differences sometimes unexpectedly encountered between elements in the top horizontal row of the periodic system (see Figure 6.1) and those in the next row.

30.7. Other Sulphur Compounds. We have restricted our study of sulphur compounds almost entirely to those that can be visualized as derived from H–S–H. We should, however, mention at least briefly several other important types of compounds.

$$\left[\ C_2H_5O\!-\!\underset{\underset{S}{\|}}{C}\!-\!S\ \right]^{-}K^{+} \qquad\qquad NH_2\!-\!\underset{\underset{S}{\|}}{C}\!-\!NH_2 \qquad\qquad CH_2\!=\!CH\!-\!CH_2\!-\!N\!=\!C\!=\!S$$

$$NH_2\!-\!\underset{\underset{SH}{|}}{C}\!=\!NH$$

Potassium ethyl   Allyl isothiocyanate
xanthate   A "mustard oil"

Thiourea

Xanthic acid, C_2H_5O–CS–SH, is an unstable compound; its salts, however, can be made by the reaction between CS_2 and a solution of KOH in alcohol. The salts are yellow (Greek word for yellow is *xanthos*). There is a series of xanthates, represented by RO–CS–SM, where R is a radical and M is a metal. Xanthates are used in the textile industry as a solvent for cellulose, in the manufacture of rayon, etc. Xanthates are derivatives of dithiocarbonic acid (HO–CS–SH).

Thiourea may be looked on as an amide of thiocarbonic acid (HO–CS–OH). In many of its reactions it apparently tautomerizes (Section 28.11) to the enol type of structure which is known as *isothiourea*. In these compounds the S atom in the C=S group is referred to as *thiono* sulphur, whereas in the —SH group it is *thiol* sulfur.

There is a series of thiocyanates, R–S–C≡N, and a series of isothiocyanates, R–N=C=S. An important example of the second is allyl isothiocyanate. These *iso* compounds are also known as *mustard oils*, a name that originated from the fact that allyl isothiocyanate can be extracted from a glycoside (Section 34.13) in black mustard seed. Allyl isothiocyanate is called oil of mustard; it is mildly vesicant and is used in liniments.

Sulphonamides are commercially important; this is evident from inspection of the formulas of "sulfa" drugs in Section 35.3. In Section 29.12, we mentioned the amide of benzoic acid (benzamide). By sulphonation of the benzene hydrocarbons the corresponding sulphonic acids can be made, and these in turn can be converted to the sulphonamides:

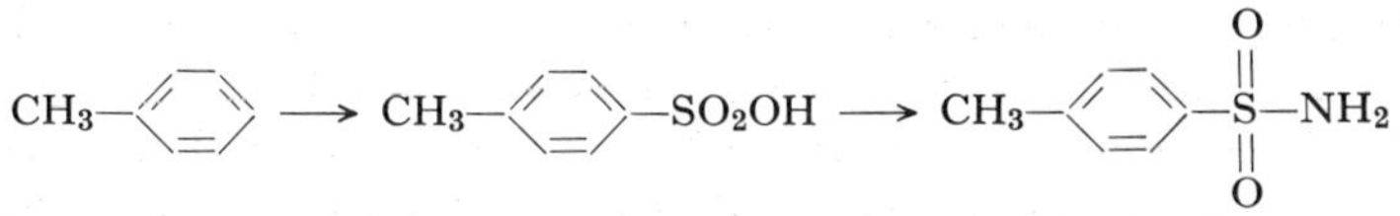

$$CH_3-\langle\rangle \longrightarrow CH_3-\langle\rangle-SO_2OH \longrightarrow CH_3-\langle\rangle-\overset{\displaystyle O}{\underset{\displaystyle O}{\overset{\|}{\underset{\|}{S}}}}-NH_2$$

Toluene   *p*-Toluenesulphonic acid   *p*-Toluenesulphonamide

Sulphonation at $100°C$ gives mostly the *p*- acid shown here, whereas the reaction at room temperature results mainly in the *o*- compound. The *o*- acid can be oxidized (which converts the $-CH_3$ to $-COOH$) and then made into saccharin. This is about 500 times as sweet as sucrose and is used in large quantities as a sweetening agent. The discovery of its sweetness is one of the favorite stories in the history of accidental discovery (about 1880 by Fahlberg and Remsen). Strangely enough, a very similar accidental discovery in 1937 (Sveda and Audrieth) resulted in another sweetener, called *sucaryl*.

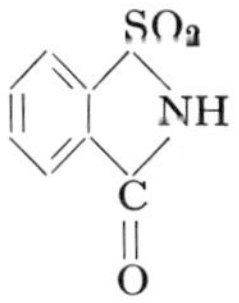

$$HO-SO_2-NH_2$$

Saccharin   Sulphamic acid   Sucaryl

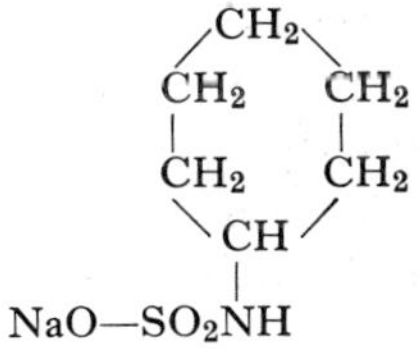

Saccharin and sucaryl are the only compounds approved for use in the sweetening of foods. Sucaryl is about 30 times as sweet as sucrose. It has the advantage of being stable in cookery, even at baking temperatures. Sucaryl is the sodium (or the calcium) salt of the cyclohexyl derivative of sulphamic acid.

PHOSPHORUS

30.8. Electron Bonding in Phosphorus Compounds. This brief presentation of the organic chemistry of phosphorus will be found similar to that on the organic chemistry of sulphur. The P atom is below the N atom in the periodic system (see Figure 6.1) and accordingly forms like compounds, but the fact that the P atom is larger and can make use of the additional orbitals shown in Figure 30.1 introduces significant differences between nitrogen chemistry and phosphorus chemistry. The phosphorus atom can form the useful molecule PCl_5, in which its valence shell accommodates ten electrons through the five hybridized orbitals sp^3d. Because of its smaller size the N atom can only form the trichloride, $:NCl_3$, in which the valence shell of nitrogen has its normal quota of eight electrons (the octet) distributed in four sp^3 hybrid orbitals.

30.9. Phosphines and Phosphine Oxides. The following series of compounds is analogous to the nitrogen compounds described in sections 29.2 to 29.5:

PH_3	CH_3PH_2	$(CH_3)_2PH$	$(CH_3)_3P$	$(CH_3)_4P^+OH^-$
Phosphine	Methyl-phosphine	Dimethyl-phosphine	Trimethyl-phosphine	Tetramethylphos-phonium hydroxide

The phosphines are toxic; they also have a strong, stupefying odor and bitter taste. Whereas ammonia, NH_3, is quite soluble in water due to its hydrogen-bonding ability, the solubility of phosphine is very low. The poor H-bond tendency of the electron pair in $:P\equiv$ molecules is similar to that described for sulphur compounds in Section 30.2. Phosphines absorb oxygen from the air so rapidly to go from a valence of three to a valence of five, that they ignite spontaneously. Representative oxygenated compounds (not made that way!) have these structures:

$$
\begin{array}{ccc}
\text{OH} & \text{OH} & \text{CH}_3 \\
| & | & | \\
\text{CH}_3-\text{P}=\text{O} & \text{CH}_3-\text{P}=\text{O} & \text{CH}_3-\text{P}=\text{O} \\
| & | & | \\
\text{OH} & \text{CH}_3 & \text{CH}_3 \\
\\
\text{Methanephosphonic} & \text{Dimethylphosphinic} & \text{Trimethylphosphine} \\
\text{acid} & \text{acid} & \text{oxide}
\end{array}
$$

The problem of the double bond in those compounds in which phosphorus has a valence of five is exactly like that discussed at length in Section 30.2. Two resonance structures can be written for a compound like trimethylphosphine oxide, as in I and II, where unshared electron

$$\text{(I)} \quad (CH_3)_3P=O- \quad \leftrightarrow \quad (CH_3)_3\overset{\oplus}{P}-\overset{\ominus}{O}- \quad \text{(II)}$$

pairs on the oxygen atom are represented by straight lines; this is similar to the treatment of dimethyl sulfoxide in section 30.2. The double bond in I consists of a σ bond together with a π bond constructed as shown in Figure 30.2. Measured bond distances indicate that compounds of this type have a PO bond length intermediate between that calculated for the shorter double bond of I and the longer single bond of II. The character of the PO bond will vary with the nature of the groups attached to the P atom.   The CH_3 group has electron-release properties (Section 21.5) that promote electron density in the PO bond, so that this compound can well be written $(CH_3)_3P \rightarrow O$; it reacts with BF_3 to form $(CH_3)_3PO:BF_3$ (see Section 8.5).   On the other hand, when there are three electronegative atoms like Cl on the P atom, as in Cl_3PO, the molecule has the properties of one with a double bond, $Cl_3P{=}O$, because of withdrawal of electron density from phosphorus.   These compounds are in general written with double bonds by most authors, except when matters of reaction mechanism are concerned.

30.10. Acids of Phosphorus.   In connection with the organic chemistry of phosphorous acid (Formula I) it is of interest to observe that the acid is dibasic and probably is better represented by formula II;

$$
\begin{array}{cccc}
OH & OH & OCH_3 & OCH_3 \\
| & | & | & | \\
:P-OH \longrightarrow & H-P{=}O & :P-OCH_3 & H-P{=}O \\
| & | & | & | \\
OH & OH & OCH_3 & OCH_3 \\
I & II & III & IV
\end{array}
$$

<table>
<tr><td>Phosphorous acid</td><td>Trimethyl phosphite</td><td>Dimethyl ester of phosphonic acid</td></tr>
</table>

OH on trivalent phosphorus is not stable and rearranges to give the P atom a pentavalent state.   Compounds derived from phosphorous acid can have Structures III or IV.   There is a series of phosphonic acids in which the H atom of IV is replaced by a radical; this is illustrated by methanephosphonic acid in the preceding section.   Phosphoric acid, $(HO)_3P{=}O$, forms esters, but there is no C-P bond in such compounds.

The chemistry of phosphorus took on considerable importance in the 1920's when it was found that phosphorus metabolism is important in the animal body (Section 35.1a).   Phosphorus chemistry has been increasingly important since World War II as a result of German investigations in the fields of insecticides and chemical warfare.

SILICON

30.11.   Electron Bonding in Silicon Compounds.   Just as oxygen is the most plentiful of the electronegative elements in the crust of the earth, silicon is the most plentiful of the electropositive elements.   The two elements together constitute about 76% of the earth's crust. Although the organic chemistry of silicon had its origins in the middle of the last century, this family of compounds remained essentially laboratory curiosities until about 1940. During World War II, it was discovered that certain of the organosilicon compounds have important commercial uses, and they are now being made in large quantities.

Silicon, of all the elements, most resembles carbon. It is just below carbon in the periodic table of the elements. Like carbon, it forms compounds in which it has a valence of four, but unlike carbon, it also exhibits a valence of six, as it does in a fluosilicate, such as Na_2SiF_6, where the negative ion is $(SiF_6)^=$. Like the other elements described in this chapter it has available the d orbitals shown in Figure 30.1, so that by means of sp^3d^2 hybridization it can accommodate six pairs of electrons. We found that it is possible for sulphur and phosphorus to form double bonds with other elements by using their d orbitals in a sideways overlap as in Figure 30.2, but that these double bonds do not have the properties of the double bonds in the chemistry of carbon. The silicon compounds in organic chemistry do not form double bonds at all.

30.12. Silanes. Silicon forms silanes which are analogous to the paraffin hydrocarbons (Table 10.1), and both aliphatic and aromatic derivatives of the silanes can be prepared. The two simplest silanes are

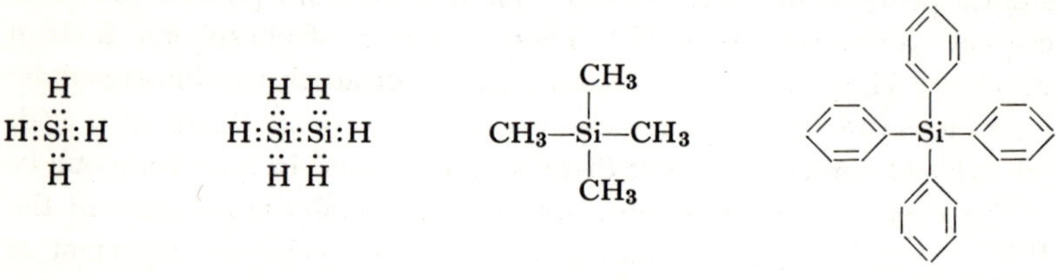

Silane	Disilane	Tetramethylsilane	Tetraphenylsilane

(Silanes (Si_nH_{2n+2}))   Silane derivatives

gases that are stable to heat, but are oxidized so readily that they ignite in air. When the Si-H bond is absent, as in the two derivatives shown here, this easy oxidation is no longer possible. In fact, the C-Si bond is more stable than C–C, and tetraphenylsilicon (tetraphenylsilane) is a solid that when melted can be distilled at 530°C without decomposition. This is unusual stability for an organic compound.

Silanes with a Si–Si chain longer than six atoms have never been prepared, whereas in paraffin hydrocarbons C–C chains as long as 100 atoms are known (Section 10.3). A method of preparation of silanes consists of treating a metallic silicide with an acid, which results in a mixture of products for which no simple reaction can be written.

$$MgSi \xrightarrow{\text{HCl}} SiH_4 + Si_2H_6 + MgCl_2 + SiCl_4, \text{ etc.}$$

Halogen derivatives of the silanes are not prepared by direct action of the halogen because of the violence of the reaction, but can be made by use of the halogen acid if assisted by a catalyst:

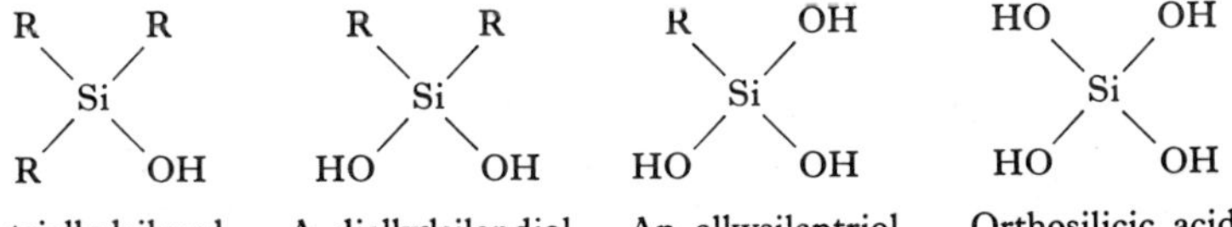

$$\overset{\displaystyle H}{\underset{\displaystyle H}{H-\overset{|}{\underset{|}{Si}}-H}} + HCl \xrightarrow{\text{AlCl}_3} \overset{\displaystyle H}{\underset{\displaystyle H}{H-\overset{|}{\underset{|}{Si}}-Cl}} + H_2$$

Silane   Chlorosilane

By this procedure, the more completely halogenated compounds can also be prepared, such as SiH_2Cl_2 *dichlorosilane*, and $SiHCl_3$ *trichlorosilane*. These compounds are useful in syntheses, just as are the corresponding carbon compounds (Section 9.4). For example, they can be methylated by means of the Grignard reagents (Section 30.15):

$$SiH_3Cl + CH_3MgCl \longrightarrow SiH_3—CH_3 + MgCl_2$$

Chlorosilane   Methylsilane

30.13. Silicones. Compounds obtained by hydrolysis of halogenated silanes are of great commercial importance. Typical halogen compounds are R_3SiCl, R_2SiCl_2, $RSiCl_3$, and $SiCl_4$. These give rise to the following hydroxyl compounds when treated with water:

A trialkylsilanol   A dialkylsilandiol   An alkysilantriol   Orthosilicic acid

Orthosilicic acid forms esters, such as tetraethyl silicate, of some importance, but these compounds contain no C–Si bond and we shall not discuss them. The other types of compounds listed are used in large quantities to make *silicone resins* (Section 38.6) due to the ease with which they condense to form large polymeric molecules. The alkyl compounds are less stable (more easily condensed) than the aryl compounds where R is a phenyl group. For example, *diphenylsilandiol* $(C_6H_5)_2Si(OH)_2$ can be obtained as a simple crystalline compound, whereas the *dimethyl* compound condenses with itself as follows:

$$\begin{array}{ccccccc}
CH_3 & CH_3 & CH_3 & CH_3 & CH_3 & CH_3 & \\
& Si & & Si & & Si & Si \\
HO & OH & HO & OH & OH & OH & HO
\end{array}$$

If a sheet of filter paper is held for a second in the vapor of dimethyl-dichlorosilane, hydrolysis takes place on the surface of the paper due to the water vapor normally present. The hydrolysis product reacts as just illustrated and the product which then coats the paper has the structure

$$\begin{array}{ccccccc}
R & R & R & R & R & R \\
& Si & & Si & & Si \\
& & O & & O & & O
\end{array}$$

A silicone

The paper, which is normally easily wetted, is now waterproof. The oxygen atoms in the chain are attached to polar groups in the paper, so that the paper presents a hydrocarbon surface, analogous to that illustrated in Figure 27.1, which repels water.

ORGANOMETALLIC COMPOUNDS

30.14. Carbon Compounds and the Periodic Table. Those carbon compounds in which there is a direct link between a carbon atom and a metallic element are referred to as organometallic compounds. Such substances do not occur in nature, and must be synthesized. An instructive glance at the possibilities and magnitude of this field of chemistry is furnished by Table 30.1, which lists representative elements linked to methyl groups. The compounds to the right of the dotted line are types already studied in this book—that is, halogen compounds, ethers, amines, and so on. Some of these compounds are found in nature.

Because the elements to the left of the periodic chart are metallic, or electropositive, the bonds they form with hydrocarbon radicals may have considerable ionic character. Methylsodium, for example, is very reac-

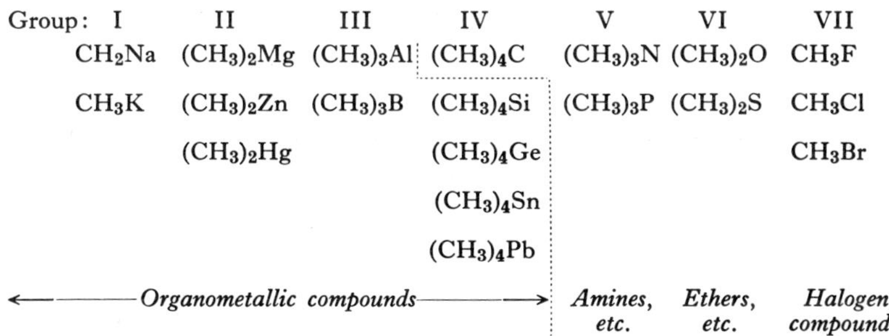

TABLE 30.1. Hydrocarbon Derivatives of the Elements According to the Periodic System

Group: I	II	III	IV	V	VI	VII
CH_2Na	$(CH_3)_2Mg$	$(CH_3)_3Al$	$(CH_3)_4C$	$(CH_3)_3N$	$(CH_3)_2O$	CH_3F
CH_3K	$(CH_3)_2Zn$	$(CH_3)_3B$	$(CH_3)_4Si$	$(CH_3)_3P$	$(CH_3)_2S$	CH_3Cl
	$(CH_3)_2Hg$		$(CH_3)_4Ge$			CH_3Br
			$(CH_3)_4Sn$			
			$(CH_3)_4Pb$			

←————Organometallic compounds————→ : *Amines, etc.*   *Ethers, etc.*   *Halogen compounds*

tive; it burns spontaneously when exposed to air. From the data in Table 8.2 and text in Section 8.9, the ionic character of a carbon-sodium (C–Na) bond may be estimated at close to 50%. The CH_3Na molecule is the sodium methide mentioned in Sections 9.6, 17.12, and 23.7. Dimethylmercury, on the other hand, is mostly covalent in character, is relatively stable, and is more easily handled in the laboratory.

The high reactivity of the type of organometallic compound shown in Table 30.1 is made use of in many industrial processes. Tetraethyllead, and more recently tetramethyllead, are added to gasoline as an "antiknock" to improve its performance as a motor fuel. Organoborons are jet fuels; trialkylaluminums and phenylsodium are catalysts in the synthesis of high polymers; and organotin compounds are stabilizers for vinyl plastics. The list of uses is quite large. More complex examples of useful organometallic compounds are *salvarsan* (Section 35.3), the first commercial organometallic compound (about 1900), and *mercurochrome* (Section 36.11).

30.15. The Grignard Reagents. By far the most valuable of the organometallic compounds is the family of substances known as the Grignard reagents. These were first prepared in 1899 by Barbier, but the credit for investigating their possibilities and uses is due to Grignard. The preparation of a Grignard reagent can be written

$$Mg + R\text{—}X \longrightarrow R\text{—}Mg\text{—}X$$

where R stands for practically any radical, X represents chlorine, bromine or iodine, and Mg is magnesium.

The actual laboratory manipulation is not so simple, because of the great reactivity of these substances. It consists in placing clean magnesium ribbon in carefully dried ether, and then adding the halogen compound.

The reaction takes place as follows:

$$CH_3—CH_2—I + Mg \longrightarrow CH_3—CH_2—Mg—I$$

Ethyl iodide   Ethyl magnesium iodide
(A Grignard reagent)

The ingredients must be kept dry and in an inert atmosphere. With water, the Grignard reagent reacts in this way:

$$CH_3—CH_2—Mg—I + HO—H \longrightarrow CH_3—CH_3 + HO—Mg—I$$

Ethane   Basic
magnesium
iodide

a *pure* hydrocarbon being formed. This is a good general method for preparing hydrocarbons, provided the proper Grignard reagent can be made in the first place. No attempt is made to isolate the Grignard reagent. It is made in the ether solution when required and then mixed with other substances to create the desired products. For example, to prepare a tertiary alcohol (Section 25.3), we can mix a ketone with a Grignard reagent:

$$CH_3—CH_2—\underset{\underset{O}{\|}}{C}—CH_3 + CH_3—Mg—I \longrightarrow CH_3—CH_2—\underset{\underset{OMgI}{|}}{\overset{\overset{CH_3}{|}}{C}}—CH_3$$

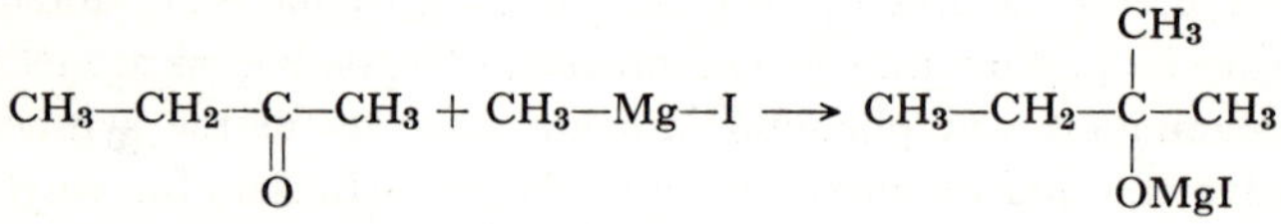

2-Butanone
(A ketone)

and the initial complex substance formed is then decomposed with water,

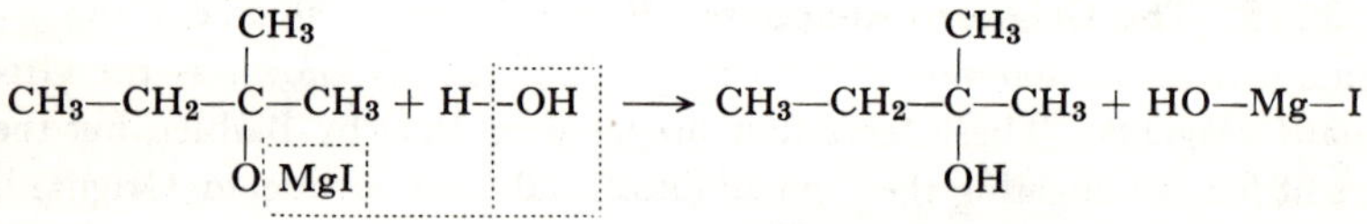

$$CH_3—CH_2—\underset{\underset{O\ MgI}{|}}{\overset{\overset{CH_3}{|}}{C}}—CH_3 + H—OH \longrightarrow CH_3—CH_2—\underset{\underset{OH}{|}}{\overset{\overset{CH_3}{|}}{C}}—CH_3 + HO—Mg—I$$

2-Methyl-2-butanol
(A tertiary alcohol)

Similarly, an aldehyde with a Grignard reagent yields a secondary alcohol.

By proper choice of the Grignard reagent and the molecule to which we add it, nearly all the fundamental types of structures known to organic chemistry can be prepared. The value of the Grignard reagents can be judged from the fact that the Nobel prize in chemistry was awarded to Grignard in 1912.

A great effort has been made to determine the real structure of a Grignard reagent.   There is proof that RMgX does not exist as such in solution.   From a study of dielectric constants in ether, the evidence is good that a solution of C_2H_5-Mg-Br contains a 1:1 mixture of $(C_2H_5)_2Mg$ and $MgBr_2$.   The most likely structure is A, which is complexed with two molecules of solvent $(Et=C_2H_5)$.

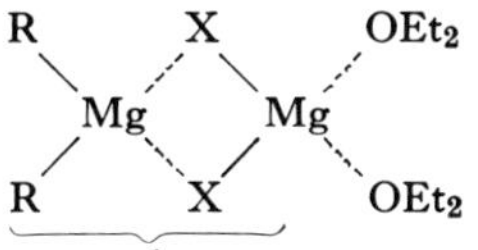

$$
\begin{array}{ccccc}
R & & X & & OEt_2 \\
 & Mg & & Mg & \\
R & & X & & OEt_2
\end{array}
$$

A

Ether plays an important part in the Grignard synthesis, because it can act as a base to the reagent—that is, it supplies electron pairs to form an oxonium compound (Section 25.17).   In general, the Grignard reagent in ether will be attacked by a reactant which is more basic than ether to the reagent and will displace it; the importance of ether is that it is basic enough to dissolve the reagent, but not so basic that it cannot be displaced.   Although the nature of the intermediate complexes formed prior to ion formation is conjectural, on the electronic basis the ketone reactant may be expected to displace the ether and behave as follows:

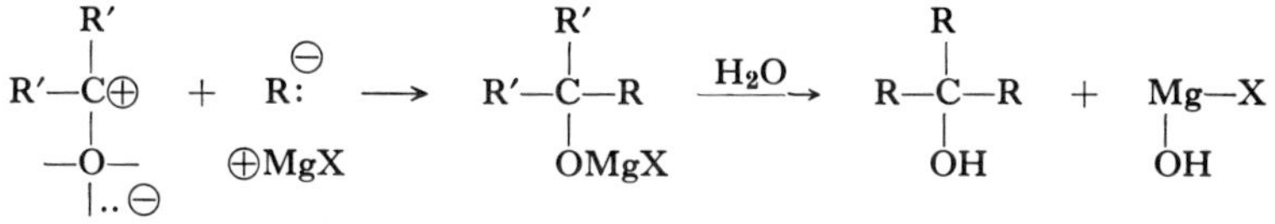

$$
\begin{array}{ccccccc}
R' & & & R' & & & R \\
| & & & | & & & | \\
R'\!-\!C\!\oplus & +\ R\!:^{\ominus} & \longrightarrow & R'\!-\!C\!-\!R & \xrightarrow{\ H_2O\ } & R\!-\!C\!-\!R & +\ Mg\!-\!X \\
| & \oplus MgX & & | & & | & | \\
-\!O\!- & & & OMgX & & OH & OH \\
|..\ominus & & & & & &
\end{array}
$$

If a solvent more basic than ether is used (for example, pyridine), the reaction is slower and may give different products.

Part 4

SPECIAL TOPICS IN ORGANIC CHEMISTRY

In Chapter 1 we showed how organic chemistry originated about 150 years ago as the result of the discovery that the carbon atom is so closely associated with life.

Although it was easy enough to show that "the element carbon stands out as the central luminary in the great mystery of life" (see p. x), it was not so easy to demonstrate *why* it is so important. It took years of persistent endeavor to work out the principles governing the architecture of carbon compounds described in Parts 2 and 3. Now that these principles are well understood, progress is being made rapidly in the study of large classes of compounds that are of interest because they are an essential part of living nature.

In Part 4 we shall point out as briefly as possible the fundamental characteristics of proteins, dyes, sugars, etc., substances of great importance physiologically and industrially. It will be apparent to the student that the structures of these more complex molecules are not at all terrifying, provided we have a thorough knowledge of the various kinds of structural groups described in Part 3.

In a book of this size we cannot go into great detail, but we can, even in a small space, make an instructive survey of these interesting fields.

3I

Structures of Complex
Compounds

31.1. Structure from Chemical Properties. The determination
of the various structural groupings in a molecule requires a good working knowledge of the principles we have met in Part 3. The application
of these chemical principles may be illustrated by the following simple
example.

Assume that a liquid substance is found by a qualitative analysis to contain only the elements carbon, hydrogen, and oxygen and that a quantitative analysis shows the composition to be C_3H_8O. Since carbon nearly
always has a valence of four ($-\overset{\displaystyle |}{\underset{\displaystyle |}{C}}-$), the possibilities narrow down to
these three structures:

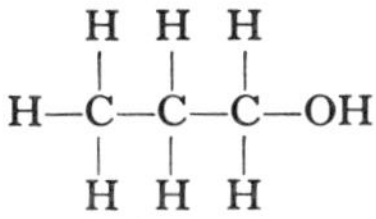

1-Propanol	2-Propanol	Methyl ethyl ether

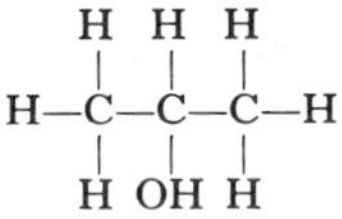

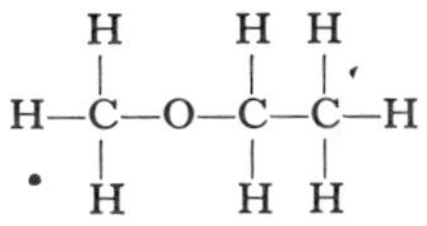

The alcohols and ethers are different in their chemical properties and
only a few simple tests are necessary to prove the structure of the unknown
substance. If metallic sodium dissolves in the liquid with evolution of
hydrogen (Section 25.3) and if the substance reacts with acetyl chloride
to form an ester (Section 27.9), the substance is an alcohol. If, on treatment with concentrated HI, the substance breaks down as shown in
Section 25.17, it is an ether.

If we find the substance is an alcohol, then by oxidizing it we can determine whether it is 1-propanol or 2-propanol. On oxidation, the primary

369

alcohol would yield an aldehyde, whereas the secondary alcohol would result in a ketone (Section 25.7).

31.2. Structure from Physical Properties. To solve the simple problem described in the preceding section, we really need nothing more than a few physical properties. The different structural possibilities are easily worked out on paper, and the physical properties of each of these possible compounds can then be compared with accepted values given in the handbooks of chemistry. The physical properties usually employed for identification are odor, melting point, boiling point, density, refractive index, etc. The two alcohols and the ether we have just discussed can easily be differentiated by their boiling points.

31.3. Refractive Index. Physical properties are used not only for identification, but also for proving the structures of substances. The effect of structure on refractive index is illustrated by these compounds:

$$CH_2{=}CH{-}CH_2{-}CH_2{-}CH{=}CH_2 \qquad CH_3{-}CH{=}CH{-}CH{=}CH{-}CH_3$$

Diallyl   Isodiallyl

The *iso* compound has a higher refractive index, which leads to the conclusion that the double bonds are conjugated (Section 15.7). Conjugated systems of double bonds commonly give rise to such an *optical exaltation*. The student should refer to the allyl group (Section 17.3) for nomenclature of these compounds.

Refractive-index data are often found listed in terms of *molar refraction*, R_M, which is calculated from the refractive index n, the density d, and the molecular weight M by the following expression:

$$R_M = \frac{n^2 - 1}{n^2 + 2} \cdot \frac{M}{d}$$

The value depends on the wavelength of light used (which is usually the D-line of the sodium spectrum).

31.4. Dipole Moment. Measurements of certain physical characteristics have in some cases even reversed the identity of compounds that had long been thought of as definitely known. The following *cis, trans* isomers analogous to those described in Section 22.3 are now known to have

cis-Dichloroethene (b.p. 60.2°)   *trans*-Dichloroethene (b.p. 48.3°)

the boiling points as indicated, but before the theory of dipole moment was developed it was thought that *trans*-dichloroethene boils at 60.2°C and the *cis* compound at 48.3°C.

The error was recognized when the dipole moments (Section 8.3) were measured. The dipole moment of the *trans* compound is zero and that of the *cis* isomer is relatively high according to the theory of dipole moments. The boiling points, shown with the structures, correspond to these respective dipole moments. Not all symmetrical *trans* compounds have a zero dipole moment; in the case of certain groups attached to the double bond the moment is not strictly *axial* (Section 12.6) and the dipole moment may then be significant.

31.5. The Parachor. In its simplest form the equation for the parachor is

$$P = V\gamma^{1/4}$$

where V is the molecular volume and γ is the surface tension of the substance. The molecular volume is the molecular weight divided by the density. The parachor is in general a completely *additive* property. It is a function of (1) the kinds of atoms in the molecule, (2) the number of each kind of atom, and (3) the structural arrangement of the atoms.

The parachor for CH_3-CH_3 is 110.5, whereas for $CH_3-CH_2-CH_3$ it is 150.8. The difference is the value of P for $-CH_2-$ and is 40.3. The value of P for $-CH_2-$ has been obtained from many series of compounds, such as C_3H_6O and C_4H_8O, $C_2H_4O_2$ and $C_3H_6O_2$, etc. The average value of the parachor for $-CH_2-$ from all these determinations is $P=39.0$.

The value of P for the H atom is 17.1, and the same value is obtained from measurement with liquid hydrogen as from difference measurements in organic compounds. Since the parachor of $-CH_2-$ is 39.0 and of two H atoms is 34.2, the parachor of the carbon atom is 4.8.

A few atomic parachors and structural parachors, sufficient to serve as illustrations, are as follows:

Atomic Parachors		*Structural Parachors*	
Carbon	4.8	Double bond	23.2
Hydrogen	17.1	Triple bond	46.6
Nitrogen	12.5	5-Membered ring	8.5
Oxygen	20.0	6-Membered ring	6.1

An interesting application of the parachor is its use in confirming the Kekulé structure of benzene. According to accepted theory, there are the following atoms and structural arrangements in benzene:

6 H atoms	= 102.6
6 C atoms	= 28.8
3 Double bonds	= 69.6
6-Membered ring	= 6.1
	207.1

The actual parachor value for benzene, obtained experimentally, is 206.2, indicating that our conception of the benzene ring as a special case of an alternate ring of single and double bonds is consistent.

31.6. Spectral Fingerprints. In Section 14.17 we discussed at some length the fact that most organic compounds have groups of atoms in a state of vibration which absorbs infrared radiation. Typical absorption spectra are shown in Figure 14.7. Recording these spectra is popularly known as fingerprinting organic molecules. It is a rapid and certain way to identify whole molecules and to determine their structural components.

31.7. Raman Spectra. Infrared absorption spectra are obtained by passing a wide range of infrared radiation through a sample and finding out what has been absorbed. Raman spectra are obtained with light of a single wavelength (usually from a mercury-vapor lamp); this radiation is generally scattered by the molecules of the sample through which it passes, giving rise to lines on the spectrogram representative of new wavelengths in addition to that of the exciting radiation. The positions of the new lines relative to that of the exciting radiation is called the Raman shift. The shifts have been quantitatively related to modes of vibration of the atoms in specific groups, and are easily recognized. For example, the C=C bond produces twice the shift of the C–C bond, and the C≡C shift is three times as great. It is evident that the Raman effect is a structural effect.

31.8. Nuclear Magnetic Resonance. The peaks in a spectrum obtained with visible light or infrared radiation show at what frequencies certain motions in a molecule are in resonance with the radiation frequencies passed through the sample. Nuclear physics has recently given chemists another resonance phenomenon which can be presented pictorially as a mapped spectrum and is useful in studying molecular structures. It is called nuclear magnetic resonance (NMR), or nuclear spin resonance (NSR).

The nucleus of an atom, when in a constant magnetic field, *precesses* like a spinning top (Section 24.10). The precession frequency ω is proportional to the intensity of the magnetic field H according to the equation $\omega = \gamma H$, where γ is a fundamental constant of a nucleus which shows the ratio of its spin quantum number to its magnetic moment. For some nuclei, such as C^{12} and O^{16}, this constant is zero, but for the nuclei of many atoms and of all the hydrogen isotopes (Section 37.3), it is finite and for these the equation can be applied. The precessional rotation frequency ω around a constant magnetic field is in the range of radio frequencies, about 3×10^7 Hz, which is below the range of optical methods for detection (see Table 14.2). The nuclei of ordinary hydrogen, H^1, precess at two angles which represent two energy levels; in the lower level the nuclei are aligned in the direction of the external magnetic field, and in the upper level against the field. The nuclei can be made to flip from one angle to the other by a smaller magnetic field perpendicular to the principal field. This is the resonance effect to be detected.

The sample in a test tube is subjected between the poles of a large magnet to a uniform field. A coil around the sample also subjects it to a radiofrequency excitation of about 3×10^7 Hz, which induces the nuclei to precess in step rather than at random. A coil at right angles to the constant magnetic field then "sweeps" the sample with a slow change in field strength. When the frequency of the sweep forces is in resonance with the natural frequency of the spin/magnetic moment forces of the nuclei in the sample, a small voltage is developed in a coil which is loosely coupled with the coil that supplies the radiofrequency energy. This induced voltage is amplified and presented on an oscilloscope or recorded on a chart. The spectrum obtained resembles a light-absorption spectrum, but shows voltage peaks plotted against magnetic field intensity.

For a compound like ethyl alcohol, CH_3–CH_2–OH, peaks will be obtained for H atoms in the CH_3 group, in the CH_2 group, and in the OH group, and they are well separated on the spectrum. *The height of a peak is proportional to the number of similar nuclei involved; the peak for CH_3 is three times as high as that of OH.* *The position of a peak* on the spectral chart will always be a definite distance from that due to the H nuclei of a reference compound such as $(CH_3)_4Si$; this separation is called the *chemical shift*. This is another fingerprint procedure; the chemical shifts of many organic groups have been tabulated, and in this way it is possible to identify groups of atoms such as CH_3 and OH in a molecule even though the nuclei of C^{12} and O^{16} are inert to this procedure. The different positions of resonance peaks for CH_3 and OH are due to the different environments of the H nuclei in CH_3 and in OH; by environment we refer principally to the diamagnetic shielding effects (Section 24.10) by the electrons in these groups. Compounds containing hydrogen nuclei are those most used in NMR spectroscopy, but F^{19} and P^{31} also give strong signals.

31.9. Electron Paramagnetic Resonance.   The electron spins much faster than a nucleus and for that reason it is a stronger magnet.   In an externally applied magnetic field it precesses faster, so that when the basic principles described in the preceding section are used to find the resonance point a higher frequency must be employed.   A resonance effect in spinning nuclei is found at a precession frequency in the radiofrequency range of about 30 to 100 MHz, but with spinning electrons it is in the microwave range of 10,000 MHz.   To detect the resonance requires the equipment of microwave techniques based on waveguides.

The chemical bond consists of paired electrons in which magnetic effects cancel.   However, as described in Section 24.10, there are compounds with odd electrons, and these are paramagnetic.   An odd electron can be detected with extremely high sensitivity by this electron paramagnetic resonance (EPR), which is also known as electron spin resonance (ESR). Chemists thus have available a method that can hardly be overemphasized for following reactions involving free radicals.

32

Aromatic Character in Heterocycles and Condensed Cycles

32.1. Heterocyclic Compounds. In Part 3 we met a number of ring compounds containing in the ring other atoms besides the usual carbon atoms. As examples we may recall the following compounds:

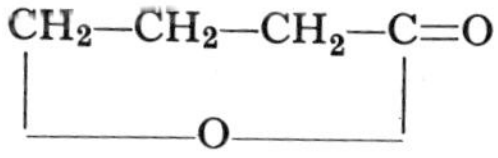

$$CH_2-CH_2-CH_2-C=O$$

Butyrolactone
(Section 28.6)

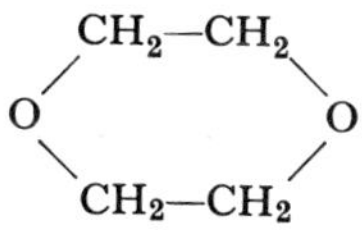

Dioxane
(Section 25.19)

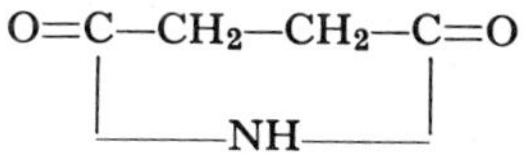

Succinimide
(Section 29.13)

These are called heterocyclic rings because there are several different kinds of atoms in the ring (from the Greek *heteros*, meaning other).

Heterocyclic rings are to be distinguished from the homocyclic rings in which all the atoms are the same (from the Greek *homos*, meaning one and the same). Familiar examples of homocycles are cyclobutane (Section 11.2) and benzene (Section 19.2).

As discussed in Section 12.7, the existence in natural products of rings larger than a 6-atom ring was not known until 1926, when it was shown that the odorous substance *civetone*, obtained from the civet cat and of importance to the perfume industry, is a ring compound with the composition $C_{17}H_{30}O$, and the structure

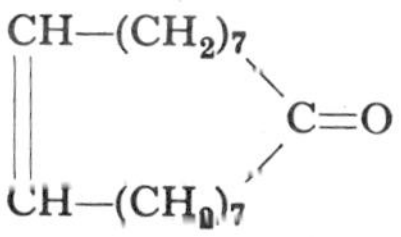

$$CH-(CH_2)_7$$
$$\qquad\qquad\quad C=O$$
$$CH-(CH_2)_7$$

This is a *homo*cyclic compound and is an unsaturated ketone. As a result of this discovery related compounds containing up to about thirty carbon atoms in the ring have been prepared.

The heterocyclic compounds described in the preceding paragraphs are members of the *aliphatic* series of compounds. In other words, they are simply closed chains. We shall devote most of this chapter to a study of compounds that are not only heterocyclic in character but are also *aromatic*. These compounds have the peculiar properties of the benzene ring.

32.2. Five-Membered Heterocycles. The most important heterocyclic compounds with five atoms in the ring are these three:

Thiophene	Furan	Pyrrole

According to these structures, the compounds appear to be unsaturated because of the C=C bonds and should therefore resemble the olefins quite closely. They do undergo addition reactions, but not as readily as the olefins. In their chemical behavior they are more like benzene. For example, thiophene is always present to the extent of about 0.5% in the benzene obtained from coal tar, but its presence was not suspected until 1883, because it behaves so much like benzene. It will be remembered that one name for benzene is phene (Section 19.1), and this sulphur compound was accordingly called *thio*phene when it was discovered.

Pyrrole is by far the most important of the five-membered heterocycles. The pyrrole ring occurs in many complex compounds found in nature. Such substances are the chlorophyll of leaves, the coloring matter of the blood, the nicotine of tobacco, and the decomposition products of the proteins.

The furan ring is present in the structure of certain types of sugars (Section 34.4). An important commercial substance containing the furan

Furfural	Scheme for naming derivatives of five-membered rings	1-Methyl-2-chloro-4-nitropyrrole

ring is furfural or furfuraldehyde. This is a liquid with many uses and is obtained, on a large scale, by the distillation of corncobs and cornstalks with dilute acids. It may also be prepared by the dehydration of pentose sugars (Section 34.1). Its structure is given, together with the general method of naming derivatives of five-membered rings.

32.3. Sextet Theory of Aromatic Character. In 1925 it was suggested that the resemblance in chemical behavior between benzene and the heterocyclic compounds just described, i.e. the fact that they have similar aromatic properties (Section 19.2 and Chapter 20), is a result of their having conjugate systems (Section 15.7) with six π electrons. The student should review the benzene ring in Figure 19.1, where the C atoms are linked by σ bonds in a flat ring, with six atomic p orbitals that can coalesce sideways to form molecular π orbitals: these molecular orbitals provide a circular path for the six *delocalized* electrons.

If we take thiophene as representative of the 5-membered heterocycles, Formula I shows two pairs of unused electrons on the S atom after the σ bonds are formed; one pair consists of s electrons and can be disregarded, but the other pair are in a p orbital like those shown in Figure 19.1. Thiophene can therefore take the shape of a flat molecule in which the p orbitals coalesce to yield a circular path for six π electrons. In a magnetic field the π electrons set up a strong ring current detectable as diamagnetism, as already stated in Section 19.4 in connection with benzene.

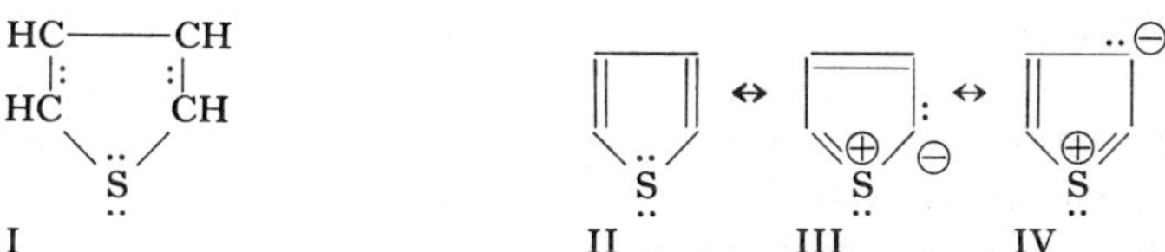

<table>
<tr><td>I</td><td>II</td><td>III</td><td>IV</td></tr>
<tr><td>Thiophene</td><td colspan="3">Some resonance structures of thiophene</td></tr>
</table>

Resonance stabilization, as well as a sextet of electrons, must be considered a necessary factor in "aromatic" behavior. A few of the possible resonance structures that can be written for a heterocycle like thiophene are shown in II, III, and IV, where it will be seen that an electron pair from the *hetero* atom can distribute itself over the entire ring. (The scheme for working out such diagrams was described in Section 21.6.) As a result of this contribution of an electron pair to the aromatic sextet of the ring, thiophene is not capable of forming sulphonium salts (Section 30.5), whereas such salts can easily be made from the saturated compound

$$\begin{matrix} CH_2 & CH_2 \\ | & | \\ CH_2 & —CH_2 \end{matrix} \Big\rangle \ddot{S}:$$

for which resonance structures are not possible.

The ease with which the *hetero* atom contributes to the aromatic ring sextet is a measure of the aromatic stabilization conferred on the ring by resonance. The estimated resonance energies are listed in Table 32.1.

TABLE 32.1. Resonance Energies of Heterocyclic Compounds

	kcal/mole
(Benzene)	36
Thiophene	31
Pyrrole	22.5
Furan	21.5
Pyridine	43

In connection with the aromatic sextet of the five-membered heterocycles, it is of interest to look into the properties of the related homocyclic compound, cyclopentadiene (Formula V). This compound does not have aromatic properties (see Section 32.10) and has but a very small resonance energy. An important characteristic is its acidity; one of the H atoms in the CH_2 group can be replaced by a metal, such as po-

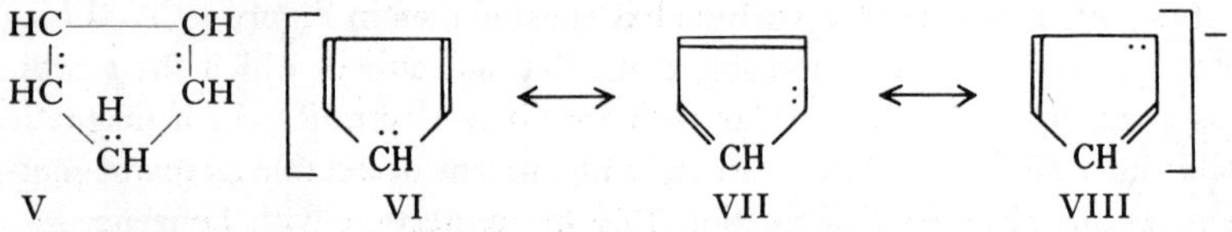

V

Cyclopentadiene

VI   VII   VIII

Some resonance structures of the negative ion of cyclopentadiene

tassium. If an H^+ ion leaves the molecule, an electron pair is exposed; this completes the sextet and the *negative ion* therefore has aromatic stability. A number of resonance structures for the ion can be worked out, but only a few are given here (VI, VII, and VIII). The stability of the ion due to this resonance is responsible for the acidity of the compound, for the same reasons as given under phenol (Section 25.12) and under acetic acid (Section 27.5). Alternatively, acidity of cyclopentadiene (tendency to ionize) can be attributed to the demand set up by the ring structure for a pair of electrons to complete the aromatic sextet.

The term *nonbenzenoid* aromatic compounds has been introduced to describe compounds that do not have a benzene nucleus but have considerable resonance stabilization, as shown in Table 32.1, and consequently aromatic properties.

32.4. Six-Membered Heterocycles. The cyclic compounds shown in the following with six atoms in the ring contain the same *hetero* atoms as the corresponding five-membered rings which we have just described.

The student should work out their electronic formulas and observe that they all have a *sextet* of electrons in the ring. A method of naming the substitution products is shown in the formula of pyridine. Another

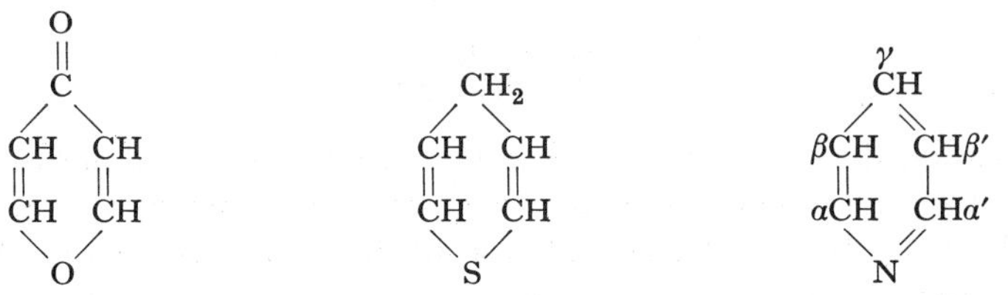

γ-Pyrone   Penthiophene   Pyridine

system is to number the atoms in the ring, starting with the *hetero* atom, as described for the five-membered ring in Section 32.2. Thus, 2-chloro-5-nitropyridine is the same thing as α-chloro-β'-nitropyridine.

Pyridine is a colorless liquid, miscible with water. It has a characteristic, unpleasant odor. When greatly diluted, the odor is that of stale tobacco smoke. It is one of the chief constituents of bone oil. Pyridine is an unusually stable substance, more stable in fact than benzene itself. Substitution in the pyridine ring takes place less readily than in the benzene ring. The pyridine ring occurs in many complex substances in nature, such as nicotine and other alkaloids.

The pyrone ring is found in certain types of sugars (Section 34.4) and in some of the compounds present in the coloring matter of flowers. γ-Pyrone itself is interesting because of its tendency to form salts. This is supposedly due to the ability of the oxygen atom in the $C=O$ group to form *oxonium* compounds, similar to those described in Sections 25.4 and 15.17.

α- and β-Pyrones are also known. Penthiophene is unimportant.

32.5. The $(4n+2)$ Rule of Aromatic Character. In 1931, Hückel made the observation that ring structures should attain aromatic stability when they have $(4n+2)\pi$ electrons, where n can be 0, 1, 2, 3, etc. This conclusion was reached mainly on the basis of orbital theory calculations. For the compounds we have studied in this chapter we can set n equal to 1 and then have 6π electrons; this is the magic number expressed by the term "aromatic sextet" as employed in Section 32.3. The $4n+2$ rule has inspired a large amount of research on aromatic ring structures, beyond the scope of this book. It might be mentioned, however, that 3-carbon rings have been made which have only one double bond and therefore only two π electrons, and fit the rule when n is zero. An example is this positive ion,

$$Pr-C{=\!=}C-Pr$$
$$^+C-Pr$$

Tri-n-propylcyclopropenyl ion
(Pr is $CH_3CH_2CH_2-$)

which is said to have aromatic properties (1962). The $4n+2$ rule is meant to apply only to monocyclic structures, but it does apply to fused rings such as naphthalene (Section 32.6) which has 10π electrons ($n=2$), and anthracene which has 14π electrons from 7 double bonds ($n=3$).

NMR measurements (Section 31.8) are now being accepted as evidence for or against the presence of aromatic character in a ring compound. High diamagnetism is a good indication of delocalized electrons; in a magnetic field perpendicular to the ring, the π electrons circulate freely and set up an opposing (diamagnetic) field which is detectable by NMR through the shielding effect on protons. There are critics who believe that physical properties like this (and ability to satisfy the $4n+2$ rule) are not enough to prove aromatic character; the molecules should also behave like benzene chemically.

32.6. Condensed Cycles. In Section 24.4 we showed how two cyclic structures may be linked in this manner, to form what may be called a *polycylic* compound. Many polycyclic compounds are known, however, in which the rings are joined in a much closer type of union. These are called condensed cycles. An important example is naphthalene; it will be seen that naphthalene is also *homocylic* in that the rings contain only carbon atoms.

Naphthalene

Naphthalene, $C_{10}H_8$, is present in quite large quantities (5–10%) in coal tar. It is a colorless, crystalline solid, and has a characteristic odor resembling that of coal tar itself. It is used in large quantities in making dyes (Section 36.10) and lampblack, and to fight the clothes moth.

The chemistry of naphthalene is similar to that of benzene. There are eight hydrogen atoms in naphthalene which can be replaced by other atoms or groups, and for reference, their positions are designated, as shown, by either numbers or Greek letters. Thus, $\alpha_1\,\alpha_4$-dichloronaphthalene is the same thing as 1, 8-dichloronaphthalene. Ten isomers of such a *di*-substitution product of naphthalene are possible according to our structure theory and for the dichloro compounds they are all known.

When naphthalene is oxidized, one ring is destroyed:

Naphthalene   Phthalic acid   Phthalic anhydride

The phthalic acid formed is a compound with the two COOH groups *ortho* to each other. This has been proved by its chemical behavior. It is also a proof that the two rings in naphthalene are joined to each other through the o- carbon atoms. Phthalic anhydride is an important intermediate in synthetic chemistry (Section 36.11).

Naphthalene can be reduced to a completely saturated compound, just as benzene can be reduced to cyclohexane (Section 19.2):

Naphthalene   Decalin

Decalin is used as a solvent. It has also been the subject of much research on its stereochemical shapes. As for naphthalene itself, the two rings are plane, as in the case of benzene.

Anthracene, $C_{14}H_{10}$, is a three-ring cyclic compound. Like naphthalene, it consists of *condensed* cycles. Anthracene is present to a small extent (0.5%) in coal tar. The name is derived from the Greek word, *anthrax*, for coal. It is a colorless, crystalline compound, with a blue fluorescence.

There are ten replaceable H atoms in anthracene, so that possibility of isomerism of substitution products is extensive. There are three possible isomers of chloroanthracene, in which α-, β-, or γ-hydrogen atoms

Anthracene

are replaced, and all three are known.

In general, anthracene is more reactive than benzene. It is especially sensitive to oxidizing agents, the sensitive positions being the carbon atoms 9 and 10, which are oxidized to $C=O$ groups. The oxidation product is anthraquinone, the structure of which is to be found in Section 36.9 in the formulas of a few common dyes.

An important isomer of anthracene is phenanthrene,

This ring system is found in certain alkaloids such as morphine (Section 32.8).

32.7. Condensed Heterocycles. In the preceding section, we described condensed cycles consisting of two or more benzene rings. In nature, a great many compounds are found consisting of several rings of different kinds. The names and outline structures of some important substances of this type are shown in the following examples. With the name of each of these compounds will be found the names of the two cycles present in the structure. All these examples are compounds in which one of the rings is *heterocyclic*.

Indole is the parent substance of many compounds of physiological interest, especially among the proteins, and is closely related to indigo (Section 36.12). *Chromone*, as its name implies (Greek *chroma*, color), is an ingredient of the coloring matter of flowers in combination with sugars and phloroglucinol (Section 25.15). *Quinoline* and its isomer, *isoquinoline*, occur to a small extent in coal tar and bone oil. They are

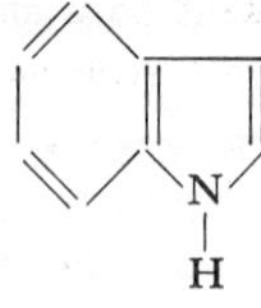

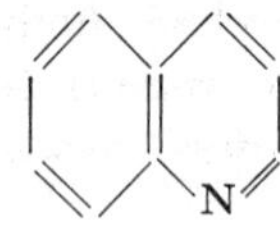

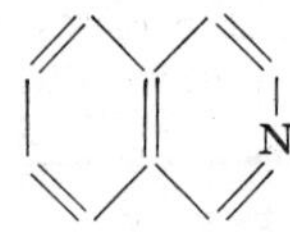

Indole   Quinoline   Isoquinoline

(Benzene + pyrrole)   (Benzene + pyridine)   (Benzene + pyridine)

Chromone
(Benzene + γ-pyrone)

Acridine
(Benzene + pyridine)

the principal constituents of the molecules of many alkaloids (Section 32.8). *Acridine* is present in crude anthracene. Many important dyes are made from it.

32.8. Alkaloids. Formerly, all basic compounds that were found in plants and contained nitrogen atoms were called alkaloids. When the structures of many of these compounds were determined, it was found best to classify them into more or less distinct groups. For example, caffeine and related purine compounds are called *vegetable bases* and certain amines are called *ptomaines*. This leaves the term *alkaloid* reserved for certain ring compounds containing nitrogen that occur almost exclusively in the form of salts in plants. The N atom is in the form of a tertiary amine except in comparatively few cases. Alkaloids are nearly always optically active and they are all poisonous.

The alkaloids can be classified according to their source as the opium alkaloids (the opium poppy), cinchonine alkaloids (the cinchona tree), etc., but chemically it is convenient to group them according to the ring systems they contain. This gives four classes of alkaloids—namely, pyridine, quinoline, isoquinoline, and phenanthrene alkaloids. These ring structures should be reviewed.

32.9. Cyclooctatetraene. This is an 8-membered ring (Figure 32.2) with a conjugate system of single and double bonds, and at one time it

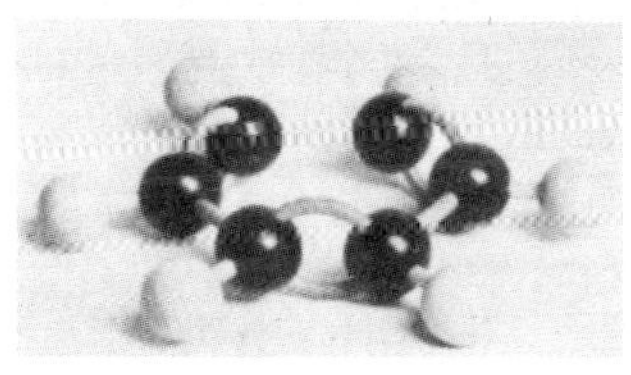

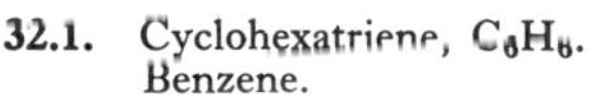

32.1. Cyclohexatriene, C_6H_6.
Benzene.

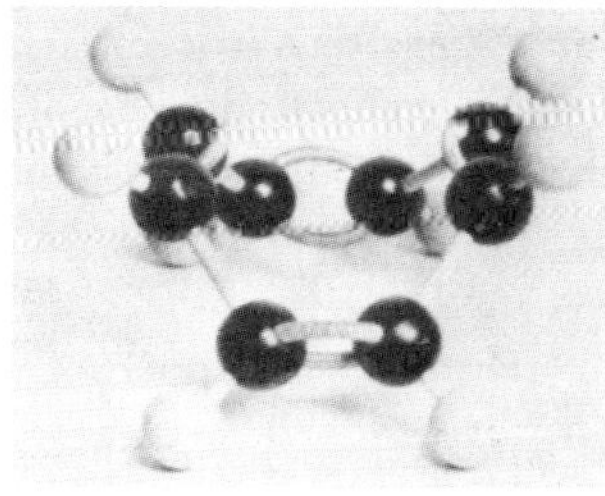

32.2. Cyclooctatetraene, C_8H_8.
"Tub" form.

aroused great interest because of its similarity to the conjugate system in benzene. It was thought that perhaps a cyclic conjugate system might be the key to aromatic character. When the compound finally became available, as the result of the discovery that it could be made from acetylene by polymerization, its properties were found to be those of an olefin and not of a *benzenoid* compound. Since it has four double bonds, cyclooctatetraene has 8π electrons, and this does not agree with the $4n+2$ rule for aromatic character as explained in Section 32.5.

The molecule in Figure 32.2 does not permit the resonance phenomenon responsible for aromatic nature, which requires a *flat* structure like that in Figure 32.1.

32.10. Cyclopentadiene is a compound of commercial importance and in Section 32.3 it was shown to be of much theoretical interest in connection with problems discussed in this chapter. It does not present a sextet of electrons to the ring and is therefore not aromatic. On the contrary, cyclopentadiene is a typical olefin; an example of this be-

1,3-Cyclopentadiene (C_5H_6)
(2 molecules)   →   Dicyclopentadiene
($C_{10}H_{12}$)

havior is the ease with which it combines with itself by the 1,4-addition mechanism (Section 17.6). This is an illustration of the Diels-Alder synthesis, which is widely used for the preparation of polycyclic compounds containing six-membered rings.

The Diels-Alder synthesis is so general for molecules with a conjugate olefinic system that it is often used to determine if such a structure is present in a compound. One of the favorite reagents is maleic anhydride, which adds on conjugate systems very readily. The student should com-

1,3-Butadiene
(Section 15.7)   +   Maleic
anhydride   →   Tetrahydrophthalic
anhydride

pare maleic anhydride with succinic acid (Section 27.16) and with malic acid (Table 28.1). It is also of interest to observe that the presence of an active conjugate system in the middle ring of anthracene (Section 32.6) can be proved by its reaction with maleic anhydride; the student should try to work out the structure of the addition product. Diels and Alder received the Nobel prize in chemistry in 1950. Their names became part of our language when two powerful insecticides known as *dieldrin* and *aldrin*, related chemically to dicyclopentadiene, were introduced commercially.

32.11. Sandwich Compounds. It was pointed out in Section 32.3 that cyclopentadiene has one H atom that is acidic; this atom is replaceable by an active metal to give a salt. Such a salt reacts with ferrous chloride to yield a compound that is representative of a fascinating series of substances called sandwich compounds. The nature of the bonding between the aromatic rings and the metal atom is still being debated.

$$2\ \text{(cyclopentadienyl)}{-}\text{Na} \ + \ Fe^{II}Cl_2 \longrightarrow \ Fe\text{(sandwich)} \ + \ 2\ NaCl$$

Cyclopentadienylsodium   Ferrous chloride   Bis-(cyclopentadienyl)-iron   Sodium chloride
Ferrocene

The theory that follows is modeled on that used to explain complex coordination compounds usually studied in inorganic chemistry. The student should refer to the position of iron in the periodic table on the inside cover, and to the cubic pictures of the atoms in Figure 6.1. The "noble" elements in Group 0 have all their electron shells filled and have great stability. Other elements, like chlorine, have vacant orbitals in the outer (valence) shell. We have not studied elements like iron, in Group 8 of the periodic table, which has four principal quantum levels and has vacant orbitals not only in the outer valence shell but also in the energy level just beneath it. Group 8 elements are called "transition" elements, and can take part in chemical reactions in which *both* outer energy levels are involved.

Iron has an atomic number of 26, which means it has 26 electrons. A driving force in its chemistry is a tendency to reach greater stability by assuming the electron structure of krypton, which is the next higher noble element in the periodic table and has 36 electrons with all its shells filled. In the reaction above, iron has lost two electrons to the Cl atoms, which leaves 24; it can accommodate 12 electrons from the two $C_5H_5^-$ ions, each of which has an aromatic sextet of electrons as described in Section 32.3. In that way the iron atom fills all its empty orbitals with electrons and looks like the noble element, krypton. Ferrocene has the iron atom sandwiched between two parallel ring structures and is extremely stable, so stable in fact that it can be boiled without decomposition at 249°. It is an orange, crystalline compound and has the aromatic properties of benzene; the rings have most of the properties described in Chapter 20. It does not add H_2, nor will it add to maleic acid in the type of reaction shown in Section 32.9, which is good evidence that the π electrons are in a firm bond between the carbon rings and the metal atom. The two rings appear to have freedom of rotation about the bonds that link them. Ferrocene was discovered accidentally in 1951. Many similar compounds have been made but ferrocene is the most stable. They are given the class name of *metallocenes*.

33

Proteins

33.1. Introduction. Animal and plant cells contain three important types of compounds—namely, fats, carbohydrates, and proteins. Of these, the most important from a physiological point of view and certainly the most complex type from the chemical standpoint are the proteins.

The great importance of the proteins lies in the fact that they are indispensable to animal life, whereas animals can live for a long time without fats and carbohydrates in the diet. The word *protein*, from the Greek *proteios*, meaning "to be the first," indicates its first importance to the structure of living cells.

33.2. The Amino Acids in Proteins. Our knowledge of the structure of proteins has been obtained largely by breaking them down to recognizable fragments. The breaking-down method is either a hydrolysis in dilute acid or dilute alkali, or a hydrolysis by means of enzymes. The decomposition products are almost entirely α-amino acids and these are regarded as the building blocks of the protein molecule.

Amino acids are mixed compounds (Section 28.1) in which the characteristic groups on the carbon skeleton are $-NH_2$ (or $=NH$) and $-\overset{|}{\underset{OH}{C}}=O$. Some representative members of this family of compounds are shown in Table 33.1. These examples are obtainable by the hydrolysis of naturally occurring proteins.

The method of naming the amino acids is similar to that of the other classes of mixed compounds which we have studied. The position of the $-NH_2$ group is indicated by a Greek letter showing how far it is from the $-COOH$ group. The important fact about proteins is that on hydrolysis they yield α-amino acids almost exclusively; that is, the $-NH_2$ group is on the carbon atom next to the COOH group. Lysine, one of the amino acids listed in the table, contains two NH_2 groups, one on the α carbon

TABLE 33.1.   Amino Acids Obtained from Proteins*

Glycine	Aminoacetic acid	$H-CH-COOH$ with NH_2
Alanine	α-Aminopropionic acid	$CH_3-CH-COOH$ with NH_2
Leucine	α-Aminoisocaproic acid	$CH_3-CHCH_3-CH_2-CH-COOH$ with NH_2
Proline	α-Pyrrolidinecarboxylic acid	CH_2-CH_2 / $CH_2-CH-COOH$ joined by NH
Methionine	α-Amino-γ-methylthiol-butyric acid	$CH_3-S-CH_2-CH_2-CH-COOH$ with NH_2
Cystine (see Section 33.6)	Di-(α-amino-β-thiopropionic acid)	$HOOC-CH-CH_2-S-S-CH_2-CH-COOH$ with NH_2 and NH_2
Lysine	α,ε-Diaminocaproic acid	$CH_2-CH_2-CH_2-CH_2-CH-COOH$ with NH_2 and NH_2
Histidine	α-Amino-β-imidazole-propionic acid	$N-C-CH_2-CH-COOH$; HC CH NH_2; NH
Tyrosine	α-Amino-β-(p-hydroxy-phenyl)-propionic acid	$HO-C_6H_4-CH_2-CH-COOH$ with NH_2
Aspartic acid	Aminosuccinic acid	$HOOC-CH_2-CH-COOH$ with NH_2
Glutamic acid	α-Aminoglutaric acid	$HOOC-CH_2-CH_2-CH-COOH$ with NH_2
Serine**	α-Amino-β-hydroxy-propionic acid	$HO-CH_2-CH-COOH$ with NH_2

---

* About twenty such acids are obtainable from proteins (Section 33.3).

** This compound serves as a basis for configuration in the D- and L-families of α-amino acids (Section 22.9 and Section 33.8).

atom and the other on the ε carbon atom.

The chemistry of the amino acids is analogous to the chemistry of the hydroxy acids as given in Section 28.6. The hydroxy acids should be reviewed before proceeding to the following discussion. Like the hydroxy acids, the α, β, and γ amino acids behave differently when heated, in accordance with the position of the NH_2 group in the molecule.

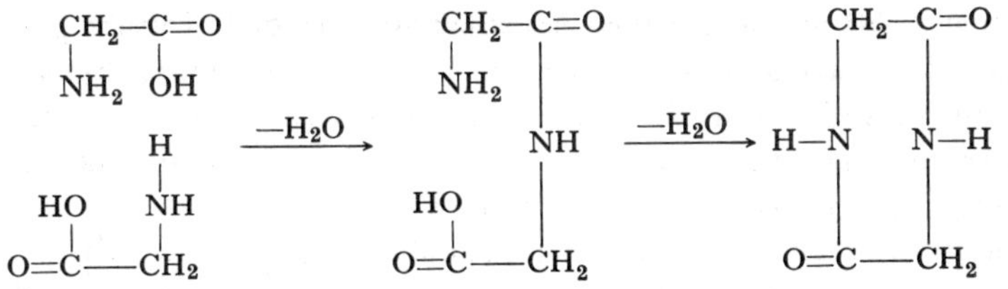

$$CH_2-C=O \qquad CH_2-C=O \qquad CH_2-C=O$$

Glycine
(Two molecules)

Glycylglycine
A "dipeptide"

Anhydride of glycine
(Compare with lactide in Section 28.6)

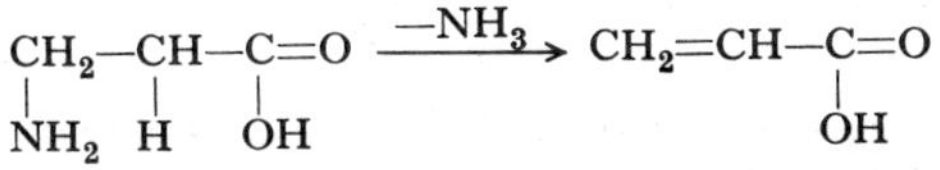

$$CH_2-CH-C=O \xrightarrow{-NH_3} CH_2=CH-C=O$$

β-Aminopropionic acid

Propenoic acid

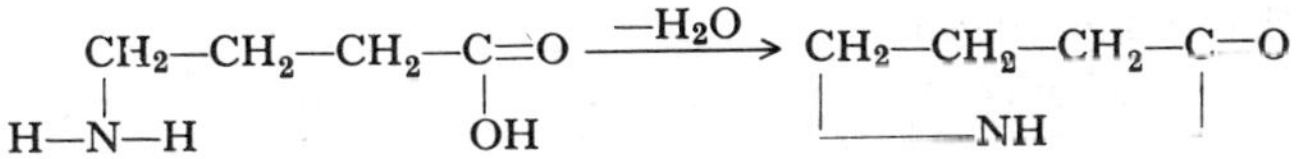

$$CH_2-CH_2-CH_2-C=O \xrightarrow{-H_2O} CH_2-CH_2-CH_2-C-O$$

γ-Aminobutyric acid

Butyrolactam (compare with the lactones in Section 28.6)

33.3. Polypeptides. From the standpoint of the subject matter of this chapter—that is, the structure of the proteins—our principal interest is in the behavior of the α-amino acids as illustrated by glycine. Molecules of this type, composed of a large number of amino acids, are known in general as *polypeptides*. The general form of a polypeptide is

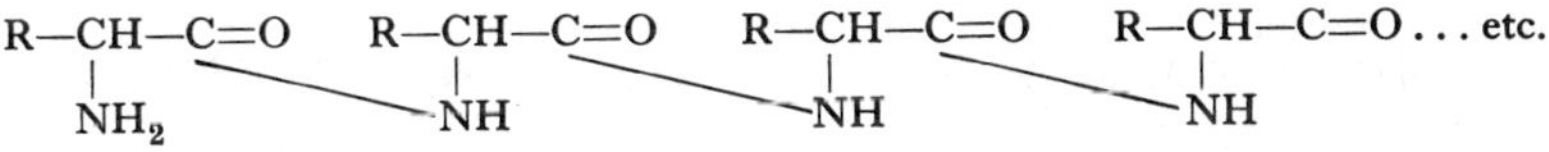

$$R-CH-C=O \quad R-CH-C=O \quad R-CH-C=O \quad R-CH-C=O \ldots \text{etc.}$$

Note that two adjacent amino acids condense by elimination of H_2O from the $-NH_2$ and $-COOH$.

The same Emil Fischer who laid the foundation for a systematic study of carbohydrates performed a similar task in the field of proteins. He made a polypeptide (about 1905) that contained fifteen glycine units and three of leucine, with a molecular weight of more than 1,000. Compounds of this size resemble the *proteoses* and *peptones* which are among the decomposition products of proteins. The molecular weights of proteins cover a wide range from about 10,000 up to many millions.

There are 20 α-amino acids in proteins, assembled by nature in polypeptide chains in an infinite number of combinations. It has been found that ten of the amino acids are essential in the food of healthy animals.

Actually, this means that they are essential in the diet of young growing rats. For example, lysine is found in most proteins, but is absent from the proteins of corn meal. Animals fed exclusively on corn-meal protein ration soon have their eyesight impaired, if the rest of the diet is not changed. It is reasonable to assume that animals require lysine for building up their tissue and cannot manufacture lysine themselves. The 20 amino acids are listed in Table 33.2, with the 10 indispensable items marked by an asterisk. It is claimed that eight of these ten are also indispensable for man; the two exceptions are arginine and histidine.

The compounds in Table 33.2 are listed roughly in the order in which their presence in proteins was recognized or proved by isolation through hydrolysis. Leucine was isolated in 1818, and threonine in 1935. The lists of the 20 amino acids are not always in agreement. Some omit either cystine or cysteine, for reasons that will be obvious from their formulas and relationship shown in Section 33.6. Hydroxyproline may also be found missing since it is closely related to proline. Asparagine and glutamine may be included, although they are simply amides of aspartic acid and glutamic acid. These selections are based on the opinions of experts as to which compounds are "prime" and become modified in the protein.

**TABLE 33.2.   The Amino Acids in Proteins
(and Their Abbreviations).**

Leucine*	Leu		Histidine*	His
Glycine	Gly		Cystine	Cys
Tyrosine	Tyr		Tryptophan*	Tryp
Serine	Ser		Proline	Pro
Aspartic acid	Asp		Cysteine	CysH
Glutamic acid	Glu		Hydroxyproline	Hypro
Alanine	Ala		Isoleucine*	Ileu
Phenylalanine*	Phe		Valine*	Val
Lysine*	Lys		Methionine*	Met
Arginine*	Arg		Threonine*	Thr

* Indispensable, as explained in the text.

33.4. Classification and Composition of Proteins. The elements that may be found in a protein are C, H, O, N, S, P, and perhaps a metal. The composition of the proteins obtained from various sources is approximately the same; the nitrogen content, for example, varies over the small range of 14–19%.

Most proteins are soluble only in water, but they form colloidal solutions rather than true solutions. They are all optically active in solution; this is also true of the polypeptides and amino acids, with the exception of glycine.

A few proteins, such as the albumin from the white of an egg, have been prepared in crystalline form. In general, however, they are amorphous substances with no definite melting point, and are easily charred when heated. When in solution, they are easily coagulated by heat, as in the familiar case of poached eggs.

The proteins were originally classified according to some of their physical properties, such as reaction to heat, or solubility in certain salt solutions. An attempt at a chemical classification of the proteins has been made by the Society of Biological Chemists and their classification is the one followed at present. The complete classification can be found in the larger textbooks.

From the general structure of a polypeptide shown in Section 33.3 it will be seen that they all have the *same backbone*, and that the variations among polypeptides (and proteins) are due to the nature of the R groups in the molecule. Let us assume a hypothetical molecule that has three

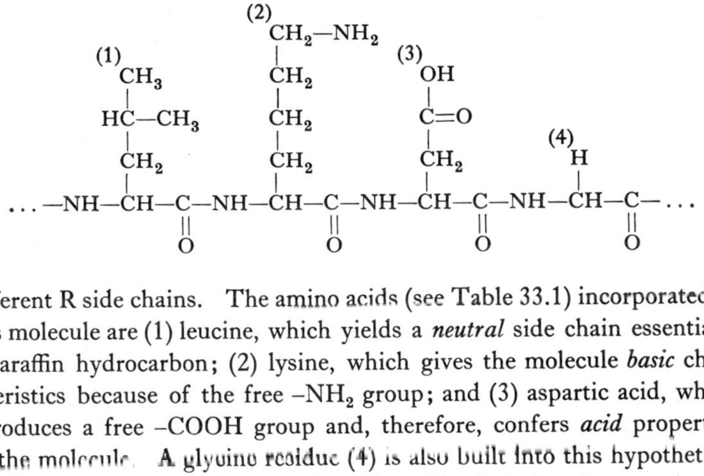

$$
\begin{array}{cccc}
& (2)\ CH_2{-}NH_2 & & \\
& | & & \\
(1)\ CH_3 & CH_2 & (3)\ OH & \\
| & | & | & \\
HC{-}CH_3 & CH_2 & C{=}O & \\
| & | & | & (4)\ H \\
CH_2 & CH_2 & CH_2 & | \\
| & | & | & \\
\cdots{-}NH{-}CH{-}\underset{\parallel}{C}{-}NH{-}CH{-}\underset{\parallel}{C}{-}NH{-}CH{-}\underset{\parallel}{C}{-}NH{-}CH{-}\underset{\parallel}{C}{-}\cdots \\
O \qquad\quad O \qquad\quad O \qquad\quad O
\end{array}
$$

different R side chains. The amino acids (see Table 33.1) incorporated in this molecule are (1) leucine, which yields a *neutral* side chain essentially a paraffin hydrocarbon; (2) lysine, which gives the molecule *basic* characteristics because of the free $-NH_2$ group; and (3) aspartic acid, which introduces a free $-COOH$ group and, therefore, confers *acid* properties on the molecule. A glycine residue (4) is also built into this hypothetical molecule; obviously it does not furnish a side chain but just another H atom.

33.5. The Isoelectric Point. This is one of the more important properties of amino acids and proteins. It can be understood by referring back to the discussion of glycine as a *dipolar ion* or *zwitterion* (Section 29.10). Glycine is amphoteric; it can react with either acids or bases. It may be appropriate to review Sections 27.6 and 27.7 if the meaning of acids and bases is not clear.

$$\overset{\oplus}{H}-NH_2 \;\; \begin{matrix} CH_2-C=O \\ | \qquad | \\ O_{\ominus} \end{matrix} + H:\overset{..}{\underset{..}{O}}:H \longrightarrow \left[\begin{matrix} CH_2-C=O \\ | \qquad | \\ NH_2 \quad O \end{matrix} \right]^{-} + \left[H:\overset{..}{\underset{..}{O}}:H \right]^{+} \quad (1)$$

Glycine reacts as an acid with a base.

$$\overset{\oplus}{H}-NH_2 \;\; \begin{matrix} CH_2-C=O \\ | \qquad | \\ O_{\ominus} \end{matrix} + H_2O \cdot H^{+} \longrightarrow \left[\begin{matrix} CH_2-C=O \\ | \qquad | \\ H-NH_2 \quad OH \end{matrix} \right]^{+} + H_2O \quad (2)$$

Glycine reacts as a base with an acid.

A zwitterion does not have equal activities as an acid and as a base. In water solution, glycine reacts more by (1) than by (2)—that is, it gives up protons more readily than it accepts them. Consequently, the pH of a glycine solution in pure water is a little below 7, slightly acid. (It will be recalled that the neutral point for water is pH 7; when the pH is below 7 it is acid, and above 7 it is alkaline.) If an acid, such as HCl is added to glycine solution in water, the formation of negative ions by reaction (1) will be repressed, and at pH 6 reactions (1) and (2) will be balanced in that equal numbers of positively and negatively charged ions will be present. This is the isoelectric point for glycine. At this pH not only are the positive and negative charges equal in number, but the total number of charged particles is at a minimum.

On the alkaline side of the isoelectric point (above pH 6) glycine forms negative ions which migrate to the positive pole in an electrolysis cell; on the acid side (below pH 6) it forms positive ions which migrate to the negative pole. At the isoelectric point there is no migration to either pole because the *net* charge is zero.

On the alkaline side of the isoelectric point, glycine, *and proteins*, react with bases; on the acid side they react with acids.

At the isoelectric point, the molecules of a protein in solution are in a favorable condition to coagulate and precipitate, since at that pH the number of positive charges on a protein surface (that is, in the R groups shown in Section 33.4) is equal to the number of negative charges and electrostatic repulsion between molecules is consequently minimized.

Proteins can often be separated from each other and purified by this principle of adjusting their solutions to the pH values corresponding to their characteristic isoelectric points. Thus, lysine in Table 33.1 is seen to have an extra NH_2 group in its molecule, which gives it stronger basic than acidic properties; it has an isoelectric point of pH 9.7. Its normal tendency in pure water is to add H^+ ions to form lysine ions with a positive charge. In order to repress this tendency and balance it with the formation of negative lysine ions equal in number to the positive ions, it is necessary to make the solution alkaline to the extent of pH 9.7. Aspartic acid, with its additional –COOH group, is more acidic than basic and the pH must be lowered to 2.8 to reach the isoelectric point. Proteins containing these amino acids have roughly analogous isoelectric points.

Isoelectric precipitation is favored by the presence in a protein molecule of amino acids like alanine or leucine (Table 33.1) which furnish purely hydrocarbon groups with little polarity to the R side chains shown in the typical structure in Section 33.4. The precipitation is more difficult when the protein molecule contains amino acids which are basic, like lysine and histidine or acidic, like aspartic acid and tyrosine.

Precipitation at the isoelectric point is the basis for the commercial isolation and purification of insulin.

33.6. Structures of Some Proteins. The remarkable advances in protein chemistry in recent years are the result of progress in many different fields of science. Enzymes have been found to be highly specific in their ability to break up a peptide bond. *Trypsin*, for example, hydrolyzes only a peptide bond in which the $C=O$ group is part of an amino acid with a positively charged side group (see lysine in Table 33.1, with its second NH_2 group); *chymotrypsin* acts preferably on a bond where $C=O$ is part of an amino acid with a 6-carbon ring in a side chain (see tyrosine); other enzymes act on peptide bonds connected to terminal amino acid units, and only if they have certain structures. In addition to these degradation processes, new synthetic methods have been devised, and new instrumentation for identifying small and large fragments has been developed.

The first polypeptide hormone synthesized (in 1953) is *oxytocin*, built up from eight different amino acids. Oxytocin is an extremely powerful pituitary hormone, one of the functions of which is to promote flow of milk in a pregnant female. In 1954 a description of the beef *insulin* molecule was published which gives the exact arrangement of all of its atoms; this is one of the simpler proteins, and is a hormone. It consists of two chains connected by S-S links; one of the chains has 21 amino acid units and the other, 30. By 1960, the chemical structures of the enzyme, *ribonuclease*, containing 124 amino acid units, and of the protein of *tobacco mosaic virus*, with 158 amino acid units, were established.

It is customary to label the components in a polypeptide chain with the first three letters of the names of the amino acids. The illustration in the last paragraph of section 33.4 can be written -Leu-Lys-Asp-Gly-. More letters are needed in a few instances (see Table 33.2).

Particular attention should be given to the unique character of the cystine molecule with its S–S bond. This bond is very easily broken by reduction (Section 30.3), and very easily regenerated by oxidation. When the link is broken the product is two molecules of cysteine:

$$
\begin{array}{c}
\overset{\displaystyle NH_2}{\underset{|}{}} \\
HOOC\text{—}CH\text{—}CH_2\text{—}S \\
| \\
HOOC\text{—}CH\text{—}CH_2\text{—}S \\
\underset{\displaystyle NH_2}{\overset{|}{}}
\end{array}
\qquad
\begin{array}{c}
2[H] \\
\rightleftharpoons \\
[O]
\end{array}
\qquad
\begin{array}{c}
\overset{\displaystyle NH_2}{\underset{|}{}} \\
2HOOC\text{—}CH\text{—}CH_2\text{—}SH
\end{array}
$$

Cystine (Cys)   Cysteine (CysH)

Cysteine has not been isolated from proteins in the free state, presumably because it is so easily oxidized to cystine in non-living tissue. It is thought that the cystine-cysteine equilibrium is of importance as a catalyst in metabolisms involving oxidation and reduction.

We mentioned that the insulin molecule has two long chains; these are connected in two places by cystine units—that is, by two of these S–S links. Cyclization, or folding of the chain is also caused by this type of bond. The most common bond in polypeptide chains, however, is the hydrogen bond. A third factor important in holding proteins in their specific shapes is an electrostatic attraction the nature of which can be visualized from the discussion of glycine in Section 33.5.

The backbone of the polypeptide example in the last paragraph of Section 33.4 is a straight line. A more realistic representation is in Figure 33.1. From a study of the dimensions of the amide group, it is known that whereas the normal CN link has a length of 1.47 Å in the amide the length is 1.32 Å. The CN link, therefore, has considerable double-bond character (see Figure 16.4), which is illustrated graphically in part (b) of Figure 33.1. The shortened bond is due to resonance. The double-bond character of the bond forces *all the atoms* in (b) *into the same plane*. These atoms are also shown in part (a) of the figure,

33.1.   Resonance in a peptide bond.

indicated by the asterisks. It is believed that in most polypeptides the NH is *trans* to CO across the CN double bond.

Some proteins (such as muscle fibers) are elastic, showing ability to contract as much as 55%. To account for the ease with which these fibers can be contracted and stretched, structural models of a helical type have been suggested. Figure 33.2 shows one view of such a model. The reader should study the top line of atoms in this model and pick out those that are indicated by the asterisks in part (a) of Figure 33.1, in order to observe that these atoms are *in one plane*.

A side view of this same helical model is in Figure 33.3. The model has 3.7 amino acid residues per turn, the helix being held together by hydrogen bonds between C=O and H–N groups (see the dotted bonds in the model). This is the contracted structure; stretching can be explained by the breaking of a certain number of the hydrogen bonds. The

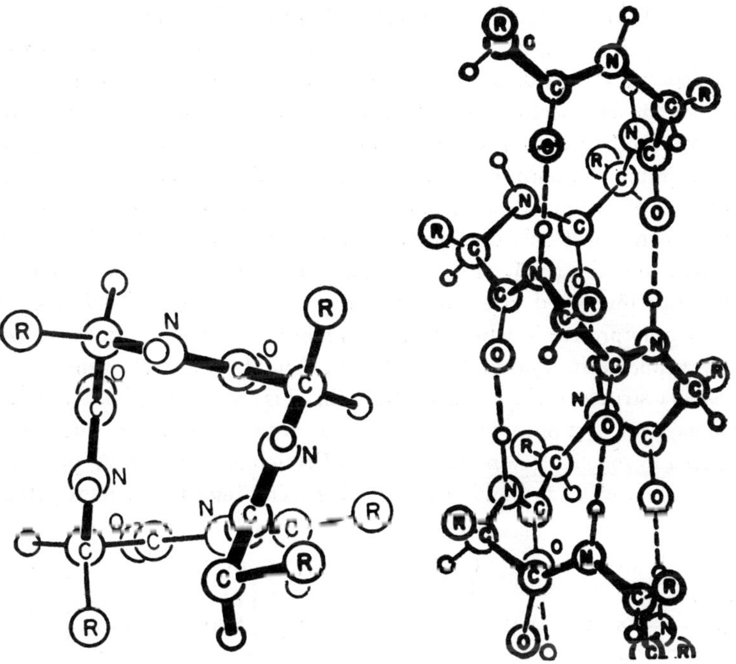

33.2. Polypeptide helix, looking through the cylindrical hole from one end [Adapted by permission from L. Pauling, R. B. Corey, and H. R. Branson, *Proc. Nat. Acad. Sciences*, **37**, 208 (1951)].

33.3. Polypeptide helix, side view [Adapted by permission from L. Pauling, R. B. Corey, and H. R. Branson, *Proc. Nat. Acad. Sciences*, **37**, 207 (1951)].

model described here is the α-helix, and is left-handed. These models illustrate only a portion of the protein problem. Certain properties of proteins demand structural models built of chains, sheets, and groups of chains and helixes in "rope" formation. In all these theories, however, the hydrogen bond between C=O and H–N is an important consideration.

33.7. Enzymes. The proteins that catalyze digestive and synthetic reactions in living things are called enzymes. Although enzymes are made by living cells, they may retain their activity outside the cell. Some enzymes are linked (conjugated) with relatively simple inorganic or organic molecules referred to as coenzymes. The same coenzyme may function with several different enzymes. Enzymes that have been named systematically have names ending in *-ase*, the rest of the name indicating the *substrate* on which it acts; for example, maltase catalyzes the hydrolysis of maltose (its substrate) to glucose (Section 34.17). Most enzymes are remarkably specific; urease will hydrolyze nothing but urea. Some, however, are not so particular; peroxidase in the presence of peroxides will catalyze the oxidation of several types of organic compounds.

Since the chemical reactions that take place in the living cell are carried out by essentially specific enzymes, a vast amount of research is being done to determine the nature of this specificity. The *primary* structure of a protein is the sequence of the amino acids in the polypeptide chain, and it is known that if only one amino acid unit in a chain of hundreds is not the right one the result on the organism is profound. For example, the disease of the blood called sickle-cell anemia is due to a defect of the hemoglobin protein in which a single glutamic acid unit has been replaced by a valine unit, caused by malfunction in the patient's genes. The *secondary* structure of a protein is the helix (Figures 33.2 and 33.3) or some related configuration. The research goal is to discover how the genetic material in the cell (see nucleic acids in Section 35.1) enables a protein to reproduce itself or to build other proteins according to a constant pattern.

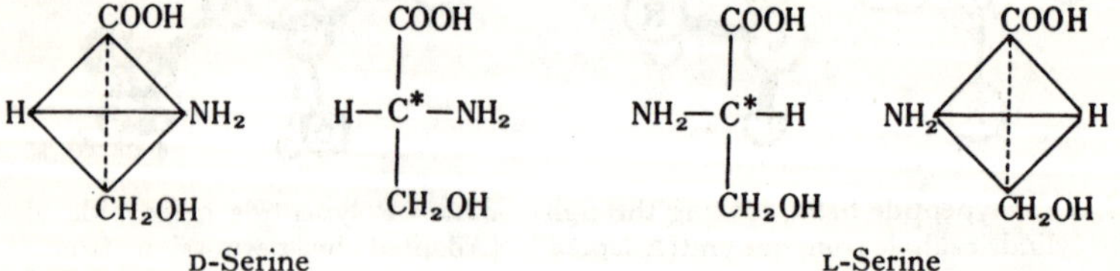

D-Serine   L-Serine

33.4. Projection formula of serine (C* is the asymmetric α-carbon atom).

Early in 1969 it was announced that the enzyme ribonuclease (Section 35. 1b) had been synthesized and that its properties are identical with those of the natural substance. This is a major advance in the understanding and control of life processes.

33.8. Optical Activity of Amino Acids. The α carbon atom in all of the α-amino acids (except glycine) is asymmetric, which means, of course, that all these compounds are optically active. An α-amino acid is allocated to a D or L family according to the configuration at the α-carbon atom, regardless of the direction of the optical activity. The reference standard for the amino acids is serine (Table 33.1), the projection formula of which is worked out in Figure 33.4. Nearly all of the α-amino acids in natural proteins have the L configuration, possibly indicating a uniform method of synthesis in nature. (See the first paragraph in Section 22.16.)

34

Carbohydrates

SIMPLE SUGARS

34.1. General Nature of Sugars. Sugars are the simple types of carbohydrates. They are the building blocks used by plant cells for the manufacture of more complex carbohydrates, like starch and cellulose. To understand what is meant by a simple sugar it is necessary to look at a more complex one, like sucrose in Section 34.15. Sucrose is called a complex sugar because by hydrolysis it can be broken into two smaller sugars. These smaller units, that do not hydrolyze further, are the simple sugars.

A sugar is a mixed compound, either an aldehyde-alcohol or a ketone-alcohol. A substance satisfying this definition is a mixture called *glycerose*, obtained by oxidation of glycerol (Section 25.8), but the properties that will be described in this chapter as characteristic of sugars are really

$$
\begin{array}{ccc}
\underset{\displaystyle \overset{|}{OH}}{CH_2} - \underset{\displaystyle \overset{|}{OH}}{CH} - \underset{\displaystyle \overset{|}{OH}}{CH_2} & \xrightarrow{O_2} & \underset{\displaystyle \overset{|}{OH}}{CH_2} - \underset{\displaystyle \overset{|}{OH}}{CH} - \underset{\displaystyle \overset{|}{H}}{C} = O \quad \text{and} \quad \underset{\displaystyle \overset{|}{OH}}{CH_2} - \underset{\displaystyle \overset{\|}{O}}{C} - \underset{\displaystyle \overset{|}{OH}}{CH_2}
\end{array}
$$

Glycerol	Glyceraldehyde	Dihydroxyacetone
	An aldose	A ketose

not encountered until the carbon chain is five atoms long. There are many sugars, but it is a relatively easy matter to keep track of them.

The names of all sugars end in *ose*, and there are two series—the aldehyde sugars (aldoses) and the ketone sugars (ketoses). The general formula of a sugar is $(CH_2O)_n$; the simple sugars of both theoretical and commercial importance are those where $n=5$ and $n=6$. These are pentoses and hexoses, $C_5H_{10}O_5$ and $C_6H_{12}O_6$, illustrated in Figure 34.1.

CHO	CHO	CHO	CHO	CH$_2$OH
HO—C—H	H—C—OH	H—C—OH	H—C—OH	C=O
H—C—OH	H—C—OH	HO—C—H	HO—C—H	HO—C—H
H—C—OH	H—C—OH	H—C—OH	H—C—OH	H—C—OH
CH$_2$OH	CH$_2$OH	CH$_2$OH	H—C—OH	H—C—OH
			CH$_2$OH	CH$_2$OH
D-Arabinose	D-Ribose	D-Xylose	D-Glucose	D-Fructose
	Aldo-pentoses		Dextrose	Levulose
			An aldo-hexose	A keto-hexose

(For the meaning of the prefix D- in these names, see Section 34.3)

34.1.   Representative pentoses and hexoses.

An exception to the $(CH_2O)_n$ formula will be mentioned in Section 34.22.

D-*Arabinose* is obtained by boiling gum arabic with dilute acids (Section 34.21), which accounts for its name. D-*Xylose*, named from the Greek word *xylon* (meaning *wood*) may be obtained by boiling wood gum, straw, or jute with dilute acids. It is used in nature to build carbohydrates which form structural parts of plants, just as glucose is employed to make the structural parts called cellulose (section 38.2). Ribose and deoxyribose are important in the chemistry of the nucleic acids (Section 35.1).

D-*Glucose* occurs widely distributed in nature in fruits and in honey. It is known as grape sugar; it is also called dextrose, because it is a right-handed molecule with respect to plane-polarized light. On a commercial scale it is made by hydrolysis of starch or cellulose. It is the sugar of the blood.

D-*Fructose* is usually found associated with glucose, in fruits and in honey. It has long been known as fruit sugar; it is also called levulose because it is a left-handed molecule with respect to plane-polarized light. It is obtainable in large quantities from certain complex carbohydrates (Section 34.21).

34.2. Open-Chain Isomerism. Before we can hope to understand the chemistry of the sugars we must be in position to understand their space arrangements. We shall first describe the model invented by Emil Fischer (1891) who did the classical pioneer work in the determination of the structures of the sugars. The Fischer system consists of so-called *projection formulas* shown in (*a*) and (*c*) of Figure 34.2. In (*a*), four of the carbon atoms of the sugar molecule are tetrahedra (Section 12.2), with their back edges in line and resting on the printed page. This is an abnormal way for us to look at a carbon-to-carbon link; as indicated in (*b*), which is the same structure but showing the tetrahedral valence angles, the carbon-carbon bonds are joined at an angle instead of in the customary straight line which is illustrated in Figure 12.5. The projection formula is, however, purely a *convention*, which is useful for tracing the relations among the sugars. A more orthodox convention will be described in Section 34.4.

The Fischer type of formula commonly used is that in (*c*) of Figure 34.2. The functional group (Section 10.9) in an *aldose* like D-glucose is always written at the top and assigned the 1 position in the molecule. The terminal –CH$_2$OH group, containing the 6 carbon atom, is at the bottom. If the sugar is a *ketose*, like D-fructose depicted in Figure 34.1, the C atoms in the chain are numbered from that end that gives the functional C=O group the smallest possible number. In D-fructose, the functional group is in the 2 position.

With the aid of Figure 34.2 (*c*), we can more easily perceive why there are so many different sugars. If we arrange H atoms and –OH groups

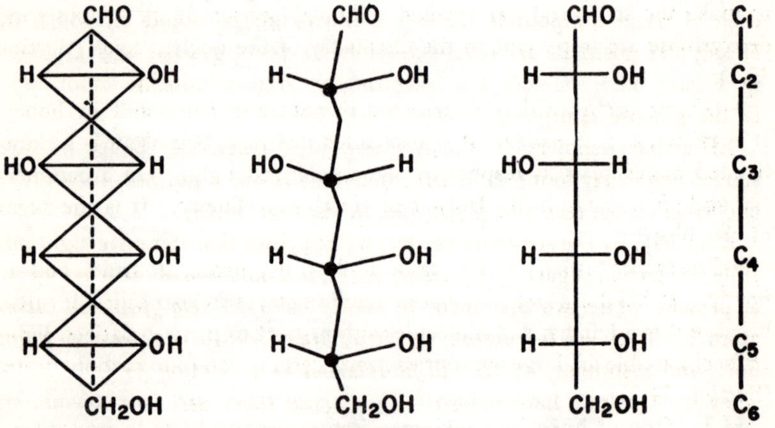

CHO	CHO	CHO	C$_1$
H⟷OH	H—OH	H——OH	C$_2$
HO⟷H	HO—H	HO——H	C$_3$
H⟷OH	H—OH	H——OH	C$_4$
H⟷OH	H—OH	H——OH	C$_5$
CH$_2$OH	CH$_2$OH	CH$_2$OH	C$_6$
(*a*)	(*b*)	(*c*)	(*d*)
Tetrahedra with back edges in line.	*Showing C—C bonds joined at an angle.*	*The final projection formula.*	*The number system.*

34.2. Projection formula of D-glucose.

CHO	CHO	CHO	CHO
H | OH	HO | H	H | OH	H | OH
H | OH	H | OH	HO | H	H | OH
H | OH	H | OH	H | OH	HO | H
H | OH	H | OH	H | OH	H | OH
CH$_2$OH	CH$_2$OH	CH$_2$OH	CH$_2$OH
D-Allose	D-Altrose	D-Glucose	D-Gulose

CHO	CHO	CHO	CHO
H | OH	HO | H	H | OH	HO | H
H | OH	HO | H	HO | H	H | OH
H | OH	H | OH	HO | H	HO | H
HO | H	H | OH	H | OH	H | OH
CH$_2$OH	CH$_2$OH	CH$_2$OH	CH$_2$OH
L-Talose	D-Mannose	D-Galactose	D-Idose

In the space models of these sugars, shown in model II of Figure 34.6, the atoms on the *left* of the chain are on *top* of the ring.

34.3.   Scheme for working out the possible *aldo*hexoses.

on the skeleton in all possible ways, sixteen different isomeric arrangements of the *aldehyde* hexoses will be found possible. The structures of eight of these hexoses are indicated in Figure 34.3, where we successively change the positions of the –OH groups to the left-hand side of the chain, and progressively increase the number of –OH groups on that side. For each of these compounds there is a mirror-image isomer in which the H and OH positions are reversed, as shown for D-glucose and L-glucose in Figure 34.4. There are, accordingly, sixteen possible aldohexoses in the glucose family.

All sixteen isomers of glucose have been identified. Three of these sixteen sugars are found in nature; their names are D-glucose, D-mannose, and D-galactose (see the glycosides in Section 34.13).

Similarly, by the structure theory, we can show that there are eight possible six-carbon sugars in the *ketone* series, but only five of these are known at present. The two that occur in nature have the *keto* group on carbon atom 2. They are D-fructose, the structure of which is shown in Figure 34.1, and L-sorbose, which will be referred to in Section 34.7.

34.3.   Optical Isomerism.   In glucose there are four asymmetric carbon atoms, marked with an asterisk, thus: HOCH$_2$–C*–C*–C*–C*–

CHO. An inspection of the formula of glucose will reveal there are four different groups attached to each of these carbon atoms. To calculate the number of possible isomers we make use of the expression 2^n, where n is the number of asymmetric carbon atoms. For glucose, $n=4$, so that

$$
\begin{array}{ccc}
\text{CHO} & \text{CHO} & \text{CH}_2\text{OH} \\
| & | & | \\
\text{H--C--OH} & \text{HO--C--H} & \text{C}=\text{O} \\
| & | & | \\
\text{HO--C--H} & \text{H--C--OH} & \text{HO--C--H} \\
| & | & | \\
\text{H--C--OH} & \text{HO--C--H} & \text{H--C--OH} \\
| & | & | \\
\text{H--C--OH} & \text{HO--C--H} & \text{H--C--OH} \\
| & | & | \\
\text{CH}_2\text{OH} & \text{CH}_2\text{OH} & \text{CH}_2\text{OH}
\end{array}
$$

D-Glucose	L-Glucose	D-Fructose
(Dextro rotation)	(Levo rotation)	(Levo rotation)

34.4. Projection formulas of three hexoses.

there are 2^4 or 16 isomers. This agrees with the number of isomers of glucose which we have already determined should be possible by assuming that the structure of a carbon chain is that shown in Figure 34.2, and by working out all the possible arrangements of H and OH on the chain. The formula we used for calculating the number of possible optical isomers holds only when the two end groups are different. If the terminal groups are the same, as in saccharic acid (Section 34.7), the possible number of isomers will be less than 2^n, but the correction factor can be calculated.

It will be recalled that mirror-image isomers are said to be enantiomorphous; D-glucose and L-glucose (Figure 34.4) are therefore called *enantiomorphs* or *enantiomers*. All the sugars in Figure 34.3 are isomeric and are optically active, but they are not mirror images. Thus, D-glucose is not the enantiomer of D-galactose; such optically active isomers were defined in Section 22.14 as *diastereoisomers*.

A system initiated by Fischer for allocating a sugar to a D or L family turned out to be troublesome and was modified (M. A. Rosanoff, 1906) by the selection of D-glyceraldehyde and L-glyceraldehyde as the *reference standards* of the two families. The glyceraldehyde structure was

$$
\begin{array}{cccc}
\text{CHO} & \text{CHO} & \text{CHO} & \text{CHO} \\
H\longleftrightarrow\text{OH} & \text{H--C}^*\text{--OH} & \text{HO}\longleftrightarrow\text{H} & \text{HO--C}^*\text{--H} \\
\text{CH}_2\text{OH} & \text{CH}_2\text{OH} & \text{CH}_2\text{OH} & \text{CH}_2\text{OH} \\
(a) & (b) & (c) & (d)
\end{array}
$$

D-Glyceraldehyde		L-Glyceraldehyde	

34.5. Projection formula of glyceraldehyde (C* is the asymmetric carbon atom).

mentioned briefly in the introduction to this chapter and should be reviewed. Its projection formula is written in (*b*) of Figure 34.5, where it is arbitrarily stated that when the asymmetric carbon atom is placed with its back edge on the page as shown in (*a*), with the –OH group to the right of the observer, the molecule is to be called D-glyceraldehyde (it is also a *dextro* molecule). Similarly, L-glyceraldehyde, which is a *levo* molecule, arbitrarily has the structure shown in (*c*) of Figure 34.5, with the projection formula (*d*).

The current practice in naming sugars is to use the prefix D or L to indicate whether its *configurational family* is that of D-glyceraldehyde or L-glyceraldehyde. This is done by projecting the formula of the sugar in the conventional Fischer manner (as in Figures 34.1 to 34.5) and examining the *asymmetric* carbon atom farthest removed from the –CHO group (in other words, the asymmetric carbon atom with the highest number according to the scheme in figure 34.2 (*d*), where it is the C atom next to the –CH$_2$OH group). If that carbon atom has the same configuration as the C* atom in Figure 34.5 (*b*), with –OH on the right, the molecule belongs to the D family. If –OH is on the left, as in Figure 34.5 (*d*), it belongs to the L family; an example is L-talose in Figure 34.3.

The prefixes D and L are pronounced "dee" and "ell," not "dextro" and "levo." If desired, the fact that the compound is right-handed or left-handed with respect to plane-polarized light can be shown by the supplementary use of *dextro* and *levo*. This may be illustrated by the name D (*levo*)-fructose for the compound in Figure 34.4. Similarly, the arabinose in Figure 34.1 is a D sugar because of its family relations, but it is levorotatory and may be written D (*levo*)-arabinose.

The student should review Section 22.9, where it was shown that these D and L family configurations are now on an *absolute* basis. In Section 22.16 there is a discussion of the Sequence Rule, according to which each asymmetric carbon atom can be designated either (R) or (S). When a sugar molecule is portrayed as indicated in these paragraphs, any carbon atom with OH on the right is (R) in the Sequence Rule, and with OH on the left is (S).

34.4. Ring Isomerism. The projection formula described in Section 34.2 and Figure 34.2 is an artificial way of picturing an open chain of carbon atoms. We found in Section 12.3 that a carbon chain zigzags, and that the ends may meet if the chain is long enough.

We shall now permit the chain in Figure 34.2 (*b*) to coil on itself, as it should to give C–C bonds of customary angles. In Figure 34.6, Model I shows the same chain as depicted in Figure 34.2 (*b*), and Model II is

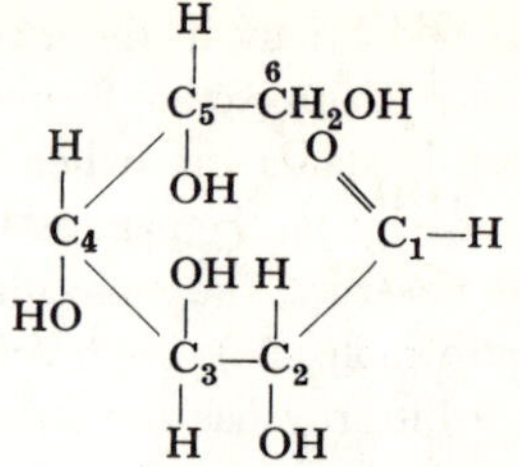

I. Straight-chain formula of the hexose, D-glucose. (The C atoms are numbered for reference in the discussion.)

II. Space model of the hexose, D-glucose.

34.6. Transition from projection formula to coiled structure of D-glucose.

the ring version. (A look ahead at Figure 34.9 will help the student with the physical picture.)

In Model II the partly closed chain is to be visualized as resting on a table. Certain –H and –OH groups then point up, while others point down and are in actual contact with the table. When we read the flattened (straight-chain) formulas of sugars on a printed page as in Model I, we should picture them as uncoiled versions of the partially closed ring shown in Model II. The atoms on the *left* of the chain are on *top* of the ring.

The cyclic structures of the sugars pictured in the rest of this chapter are referred to as *perspective* formulas (Haworth and Peat, 1929). Also see the way deoxyribose is pictured in Figure 35.1.

It is not surprising that the sugars exist in cyclic forms, in view of the great reactivity of the $C=O$ group and the probability of its proximity to the –OH on a supposedly distant carbon atom. The result of the closure is shown by the transition from Model II to Models III and IV in Figure 34.7.

The six-membered ring in these α and β sugars is known as the *pyranose* ring. It is a modification of the pyrone cycle described in Section 32.4. Structure III is more specifically α-D-glucopyranose, and IV is β-D-glucopyranose, but they are more generally known by the names given to them in Figure 34.7. A water solution of glucose is almost entirely in the two cyclic forms.

The ring closure in glucose can be pictured as addition by C–OH across the $O=C$ bond, in the manner described in Section 26.7; α or β rings will form depending on which of the two links in the double bond is broken. These cyclic forms of the sugars can be regarded as examples of

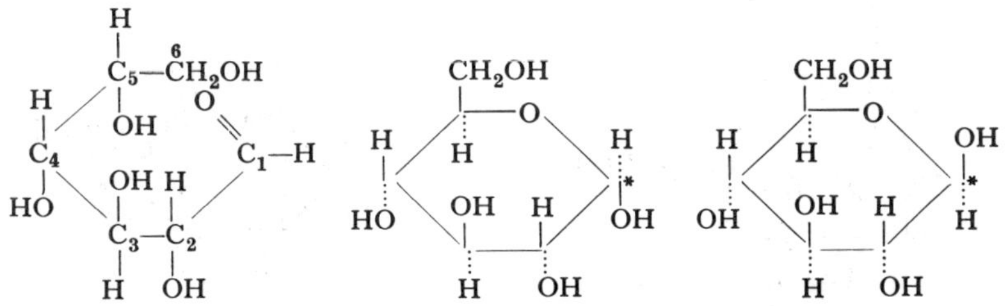

II. D-Glucose, open-chain form   III. α-D-Glucose   IV. β-D-Glucose

34.7.   Ring closure in glucose;   * marks the anomeric C atom.

the *cis, trans* (geometric) isomers described in Section 22.3. For example, α-D-glucopyranose has a *cis* and β-D-glucopyranose has a *trans* pair of –OH groups on the C_1 and C_2 atoms. However, the important thing is that the ring closure gives rise to a new asymmetric carbon atom. In glucose it is the C_1 atom, and in fructose, the C_2 atom. The α- and β-isomers are optically active, and are said to be anomeric forms, or *anomers*. The C atom which is involved is called the anomeric carbon atom, as indicated in Figure 34.7.

The two anomers of D-glucose are not mirror images; they are, therefore, *diastereoisomers* as discussed in this chapter in Section 34.3. By means of molecular models, however, it can be shown that mirror-image pairs in this sugar family do exist. Examples are α-D-glucopyranose and α-L-glucopyranose. The models are illustrated in Figures 34.9 and 34.10.

Fructose, the *keto* sugar, forms compounds in which there are 6-membered rings (Section 34.12) similar to those of glucose. The α anomer of such a pyranose ring is shown in Model VI of Figure 34.8; in the β anomer the OH on the anomeric carbon atom would be on the same side of the ring as the OH on the C_3 atom. Ordinary commercial fructose, or levulose, is the β variety. Fructose also yields compounds in which it has a 5-membered ring structure resembling furan (Section 32.2), consequently known as the *furanose* form. This modification of fructose is present in the molecule of sucrose (Section 34.15). Model VII in Figure 34.8 shows the β anomer of the 5-membered ring, with the anomeric carbon atom at the right in the β standard formula notation. In VIII this *same* formula is printed with the anomeric C atom at the left, as it appears in the formula of sucrose. The α anomer of VII would resemble VI with respect to the positions of the OH groups on the C_2 and C_3 atoms. (The student should try to construct VIII from VII on paper without

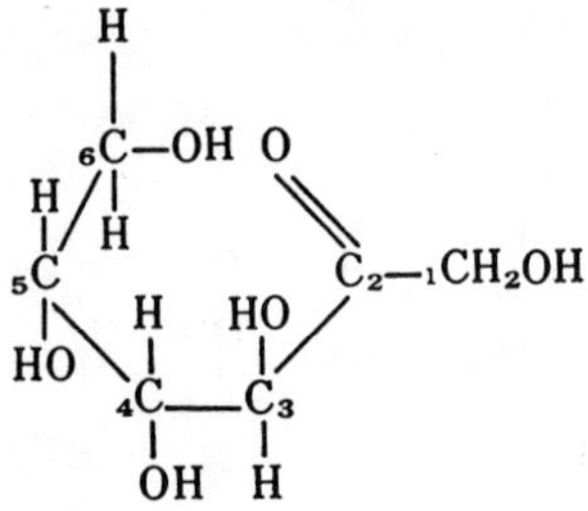

V. D-Fructose, open-chain form

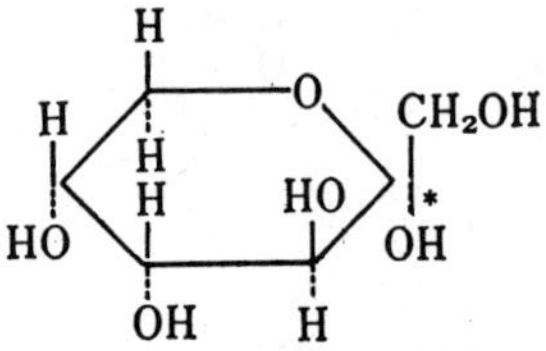

VI. α-D-Fructose

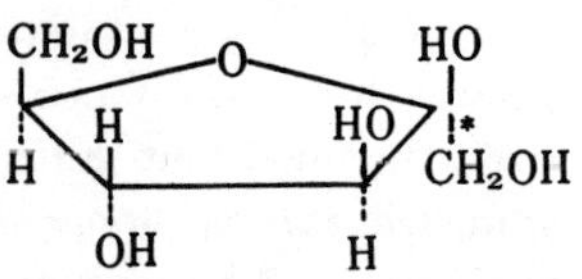

VII. β-D-Fructofuranose

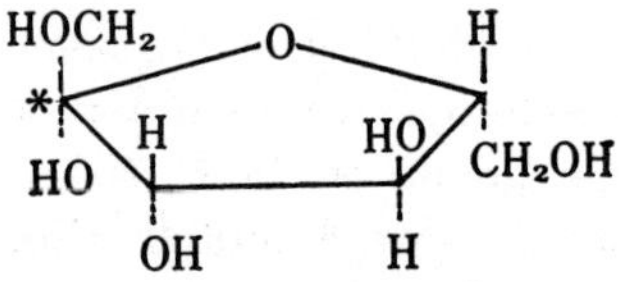

VIII. β-D-Fructofuranose

34.8. Ring closure in fructose; * marks the anomeric carbon atom. VII and VIII are identical (see text).

recourse to this printed page. A set of atom models will help.)

34.5. Conformation of Ring Structures of Sugars. The student should review Sections 12.5 and 12.6 on conformation analysis of cyclic carbon compounds. The sugar rings are puckered as shown by the models in Figures 34.9 and 34.10. In carbohydrate literature the *anti* conformation illustrated by Figure 12.16 is generally called the *chair* form, and the less stable conformation of Figure 12.15 is the *boat* form. The energy barrier between the two conformations is rather small, but in only a few instances is the *boat* form believed to be present among the sugars in stable compounds.

The important difference between α-glucose and β-glucose is in the position of OH on the C_1 atom (the anomeric C atom); in the α sugar it is axial, perpendicular to the OH bond on the C_4 atom, whereas in the β sugar it is equatorial, or parallel to the OH bond on C_4. (See the models of cellulose and starch in Section 38.2.)

34.6. The Possible Isomers of Glucose. In Figure 34.3 there are eight different compounds with the composition $C_6H_{12}O_6$. Each of these has a mirror image, which makes 16 isomers. Each of the D and L isomers exists as α or β cyclic anomers; this makes $32+16=48$ possible isomers. And this does not take into consideration the keto sugars and

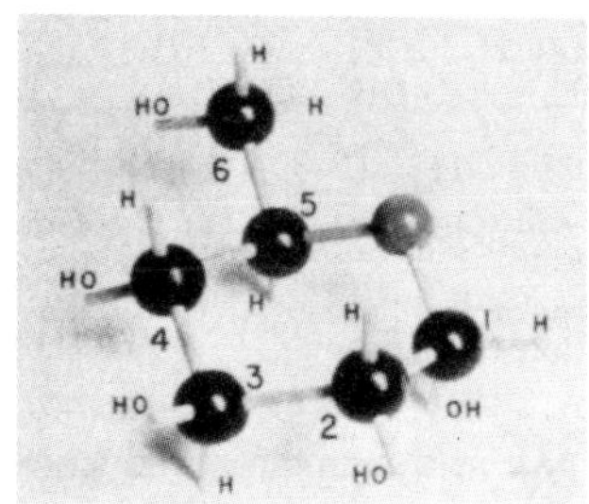

34.9a.   Model of α-D-glucose.

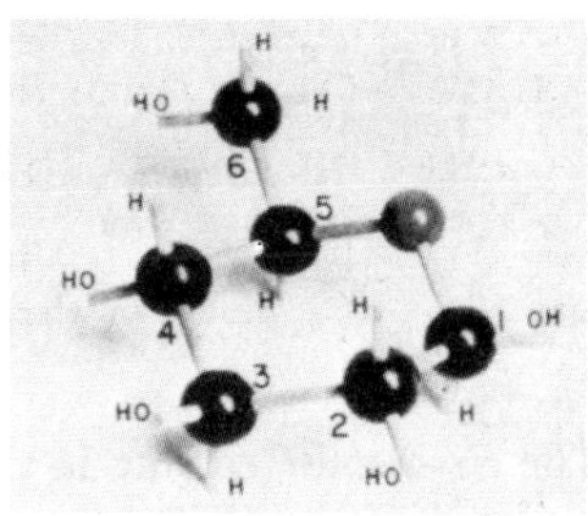

34.10a.   Model of β-D-glucose.

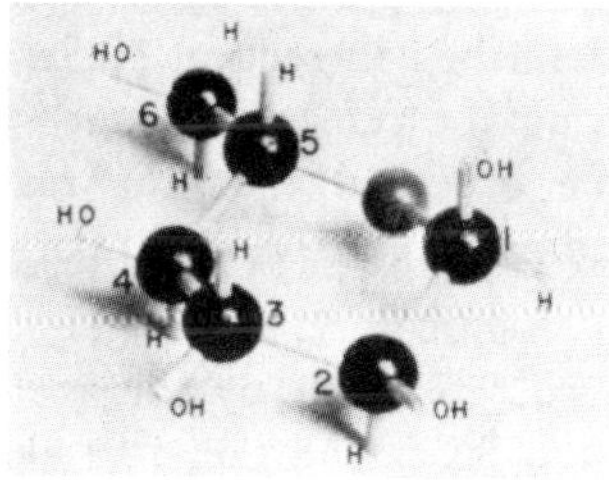

34.9b.   Model of α-L-glucose.

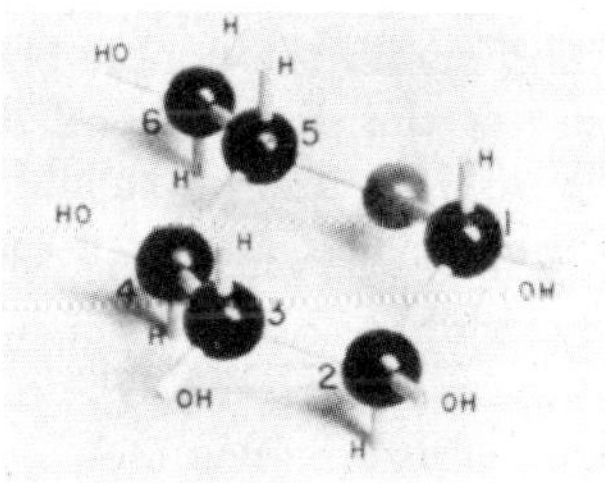

34.10b.   Model of β-L-glucose.

the possible cyclic isomers that have 5-membered rings. Some of the 6-carbon sugars can also exist as 7-membered rings, in which the CHO group at one end of the molecule reacts with CH_2OH at the other end.

34.7. Chemical Relatives of Glucose. The compounds in Figure 34.11 are a few representative of those which are related to glucose (Figure 34.1) and are often mentioned in sugar chemistry. The literature on sugars may currently be a little confusing because of changes in nomenclature. The common occurrence of the prefix *gluc-* or *glyc-* is due to derivation from the Greek word *glykys*, meaning sweet. Classes of compounds have had name changes in which the prefix is altered from gluc to glyc, as in gluconic acids to glyconic acids; and the saccharic acids, always somewhat a misfit in nomenclature, are changed to glycaric acids.

D-Sorbitol is widespread in plants, being found in the *fresh* juice of mountain ash berries, in certain seaweeds, and in fruits, such as pear, cherry, etc. L-Sorbose is a keto sugar, found originally in the *fermented* juice of the mountain ash berry; its presence is due to a bacterial oxidation of D-sorbitol, and it is made that way commercially for use as an intermediate in the synthesis of vitamin C (Section 35.8). The D and L

$$
\begin{array}{ccccc}
& & \text{OH} & \text{H} & \text{OH} \\
& & | & | & | \\
\text{CH}_2\text{OH} & \text{CH}_2\text{OH} & \text{C=O} & \text{C=O} & \text{C=O} \\
| & | & | & | & | \\
\text{H—C—OH} & \text{H—C—OH} & \text{H—C—OH} & \text{H—C—OH} & \text{H—C—OH} \\
| & | & | & | & | \\
\text{HO—C—H} & \text{HO—C—H} & \text{HO—C—H} & \text{HO—C—H} & \text{HO—C—H} \\
| & | & | & | & | \\
\text{H—C—OH} & \text{H—C—OH} & \text{H—C—OH} & \text{H—C—OH} & \text{H—C—OH} \\
| & | & | & | & | \\
\text{H—C—OH} & \text{C=O} & \text{H—C—OH} & \text{H—C—OH} & \text{H—C—OH} \\
| & | & | & | & | \\
\text{CH}_2\text{OH} & \text{CH}_2\text{OH} & \text{CH}_2\text{OH} & \text{C=O} & \text{C=O} \\
& & & | & | \\
& & & \text{OH} & \text{OH}
\end{array}
$$

D-Sorbitol	L-Sorbose	D-Glyconic	D-Glycuronic	D-Saccharic
(A polyatomic alcohol)	(A keto sugar)	acid (An acid-alcohol)	acid (An aldehyde- acid-alcohol)	acid (An acid-alcohol)

34.11.   Chemical relatives of glucose.

family relations are interesting. D-Fructose (Figure 34.1) can be regarded as the product of oxidation of D-sorbitol (Figure 34.11) at the C_2 atom in the chain, reading from the top. L-Sorbose is made by the bacterial oxidation of D-sorbitol at the C_5 atom in the chain, and therefore should have the structure shown in Figure 34.11. However, it is conventional to use the smallest possible number for the C=O group and, to regard it as a 2-*ketose*, we rotate the formula 180° in the plane of the paper. In this aspect, L-sorbose is shown to be related to L-glucose (Figure 34.4), since now the C_5 atom has the OH on the left.

CHEMISTRY of the SIMPLE SUGARS

34.8.   The Potential C=O Group.   There is so much evidence to show that the simple sugars exist in a cyclic rather than a straight-chain form that the question arises as to how we explain their "aldehyde" or "ketone" properties. The answer is that the cyclic modifications exist *in solution* in equilibrium with the open-chain type. Glucose in solution is cyclic, about two-thirds of it in the β form, one-third in the α form, and only very small amounts in the open-chain, aldehyde form. The situation is more complicated in solutions of fructose, the keto sugar, because of its tendency to form 5-membered as well as 6-membered rings. Since the sugars are so often written in the cyclic form, the beginner is often puzzled in trying to pick out the $C=O$ group. Glucose, for example, has the properties of an aldehyde, but there is no $-\overset{\displaystyle |}{\underset{\displaystyle H}{C}}=O$ group appearent in the cyclic structure. In this book, for convenience, we have put an asterisk (*) on that C–OH group in the cyclic structure that forms the $C=O$ group when the ring opens to the straight chain (review Figure 34.7). This particular C–OH is the "potential $C=O$ group" that gives the molecule typical sugar properties, and the carbon atom in that group is the anomeric carbon atom.

34.9. Reactions of the $C=O$ Group. The reactions of glucose we shall consider in this section are characteristic of all aldoses and, in general, of all the reducing sugars. These are the properties usually associated with sugar chemistry. It should be observed that the sugar properties are essentially aldehyde properties, altered more or less by the presence of –OH groups in the molecule. Fructose, even though it is a keto sugar, gives most of the reactions of glucose. This is due to the influence of the neighboring –OH groups in the molecule on the $C=O$ group.

1. Glucose shows most of the reactions used as qualitative tests for aldehydes. It reduces a solution of silver nitrate in ammonium hydroxide ($AgNO_3/NH_4OH$), depositing a film of metallic silver. It also reduces the copper in Fehling's solution (Section 26.6) to cuprous oxide, Cu_2O. This test with Fehling's solution differentiates sugars like glucose, which have a free or potential $C=O$ group in the molecule, from those like sucrose (Section 34.15) that do not. Glucose is called a *reducing* sugar, whereas sucrose is a *nonreducing* sugar.

2. A glucose solution forms a brown resin when boiled with dilute alkalies, as is characteristic of aldehydes.

3. Reactions in which the O atom of the $C=O$ group is replaced are listed in Figure 26.1. It was shown that aldehydes and ketones react with hydrazine to yield hydrazones with the characteristic structural grouping $C=N\ NH_2$. Among sugars, the *hydrazone* reacts further to give an *osazone*; in glucose the first two carbon atoms would look like this when the reagent used is phenylhydrazine:

$$H-C=N-NHPh$$
$$|$$
$$C=N-NHPh$$
$$|$$
$$R$$

An osazone (Ph is the phenyl group)

The osazones are more important than the phenylhydrazones formed at first, because they are less soluble and more easily crystallized. The melting point and solubility of their osazones are among the important constants used in the identification of sugars.

The reader should examine the structure of D-fructose in Figure 34.4 and observe that its osazone must be identical with that of D-glucose. In fact, this is one of the reactions that proved that outside of the first and second carbon atoms, D-glucose and D-fructose are identical.

34.10. Specific Rotation. One of the constants employed in the identification of the sugars is the specific rotation. In Section 22.5 we discussed optically active substances, which rotate the plane of polarized light, and we mentioned the polarimeter, an instrument by which the angle of rotation can be measured. The sugars are optically active, and each sugar rotates plane-polarized light through a certain number of degrees and in a certain direction, that is, right ($+$) or left ($-$). The specific rotation data for sugars are usually obtained at 20°C, with the D line of the spectrum (yellow sodium light). The specific rotation of glucose (dextrose) is $+52.7°$ and of fructose (levulose), $-92.4°$.

Specific rotation is calculated from the equations

$$\left[a\right]_{\text{D}}^{20°} = \frac{a}{ld} \times 100 \qquad\qquad \left[a\right]_{\text{D}}^{20°} = \frac{a}{lgd} \times 100$$

(for pure liquids)   (for solutions)

where a is the measured rotation in degrees, l is the length of the sample in decimeters, d is the density of the pure liquid or of the solution, and g is the concentration of solute in g/100 g solution. The result is "specific" in that the rotation is calculated for unit quantity of solute, unit volume, and unit length of sample.

34.11. Mutarotation. When a-glucose is dissolved in water, the solution has a specific rotation of $+112.2°$, but the value is not constant. It gradually decreases to a lower limit of $+52.7°$, at which it does remain constant. The phenomenon is known as mutarotation (from the Latin *mutare*, to change). When β-glucose is dissolved in water its initial rotatory power is $+18.7°$, but this value gradually *increases* until it reaches an upper limit of $+52.7°$. The same specific rotation, therefore, is reached, no matter which of the two glucose isomers we start with, because of the phenomenon of mutarotation. One form changes into the other until equilibrium is reached. An ordinary glucose solution is an equilibrium mixture of the a- and β-cyclic forms, in which the β- form predominates.

The nomenclature of the a- and β-anomers follows rules suggested by

Hudson. In the D series, the more dextrorotatory of a pair of anomers is named α and the other is then β. Examples are α-D-glucopyranose (rotation, $+112.2°$) and β-D-glucopyranose (rotation, $+18.7°$). In the L series the more levorotatory anomer is given the name α and the other β.

34.12. Glucosides. When an aldehyde and an alcohol react, the initial unstable product is a hemiacetal, which in the presence of acids reacts with more of the alcohol to yield a stable acetal:

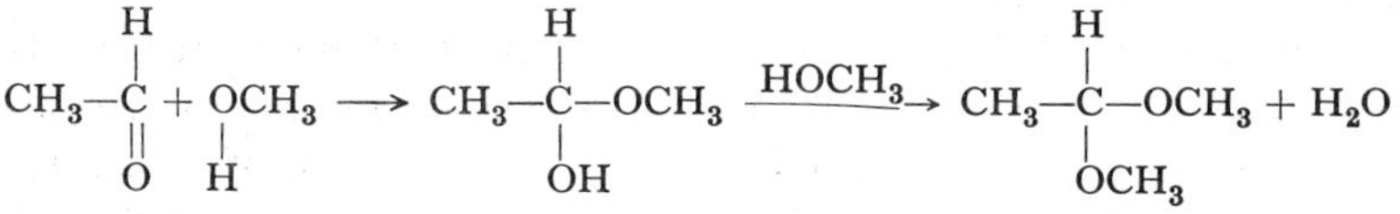

$$CH_3-\underset{\underset{O}{\|}}{\overset{\overset{H}{|}}{C}} + \underset{\underset{H}{|}}{OCH_3} \longrightarrow CH_3-\underset{\underset{OH}{|}}{\overset{\overset{H}{|}}{C}}-OCH_3 \xrightarrow{HOCH_3} CH_3-\underset{\underset{OCH_3}{|}}{\overset{\overset{H}{|}}{C}}-OCH_3 + H_2O$$

A hemiacetal   An acetal
Methylal

It will be observed that hemiacetals and acetals have an ether type of linkage in the molecule.

The formation of a hemiacetal consists in "addition across the double bond," explained in section 26.7 and further illustrated in the ring closure of sugars in Figure 34.7. The cyclic forms of the sugars shown in Figures 34.7 and 34.8 are therefore referred to as hemiacetals by sugar chemists. The cyclic form of α-glucose behaves just like a hemiacetal in that it can be made to react with an alcohol to yield an acetal, but the product is generally called a glucoside. With methyl alcohol the product is α-methyl glucoside. From β-glucose, we can obtain by this procedure the β-methyl glucoside.

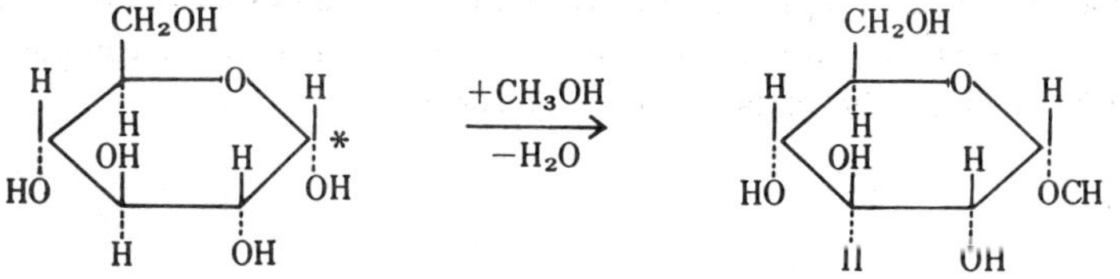

α-Glucose
α-D-Glucopyranose
(A hemiacetal) α-Methyl glucoside
Methyl α-D-glucopyranoside
(An acetal)

The α- and β-glucosides have markedly different properties:

	α-Methyl glucoside	*β-Methyl glucoside*
Specific rotation	159°	−34.2°
Melting point	165°C	104°C
Hydrolyzed by	Maltase	Emulsin

The interesting fact is that the two compounds cannot be broken down by the same enzyme. Maltase acts only on the α-compound, whereas emulsin affects only the β- compound. Immediately after hydrolysis, α-glucose or β-glucose shows its characteristic initial specific rotation, which, later, by mutarotation, goes to the equilibrium value of $+52.7°$. Fructose forms analogous compounds in which both the 5-membered and 6-membered rings shown in Figure 34.8 are utilized. The methyl fructosides with the furanose ring are somewhat less stable to hydrolysis than those with the pyranose ring.

Glucosides are important constituents of most vegetable matter. In most plants, substances needed as reserve food or for emergencies are stored up by combining them with glucose in the form of glucosides. Some natural glucosides and their component parts obtained by hydrolysis or fermentation are the following:

Amygdalin (in bitter almonds) yields glucose + benzaldehyde + HCN.
Salicin (in willow bark) yields glucose + saligenin. When administered to man, the saligenin is changed internally to salicylic acid.
Arbutin (in bearberry leaves) yields glucose + hydroquinone. The hydroquinone is an antiseptic.

Glucosides are used by plants to combat diseases and for other purposes; they were used by man long before the reason for their powers was known. It will be shown that the more complex sugars, like cane sugar and milk sugar, are also formed in the manner of the glucosides. Most natural glucosides are derivatives of β-D-glucose.

34.13. Glycosides. All compounds of the aldoses and ketoses similar in structure to the glucosides that we have described in the preceding section are known as *glyco*sides. Galactosides, mannosides, and fructosides, for example, are known and for convenience these compounds have been given the class name of glycosides.

COMPLEX CARBOHYDRATES

34.14. Complex Sugars. The sugars we have discussed up to this point are called the *simple* sugars because they cannot be broken down by hydrolysis into any types of sugars that are less complex. Instead, these simple sugars are apparently used in nature for the building up of more complex sugars, and of those highly complex carbohydrates known as starch and cellulose. The most important complex sugars are those formed by the union of two simple six-carbon sugars, thus:

$$C_6H_{12}O_6 + C_6H_{12}O_6 \rightleftharpoons H_2O + C_{12}H_{22}O_{11}$$

Two molecules of same, A complex sugar
or different, sugars

It will be observed that a molecule of water is eliminated when the complex sugar is formed. This is important, *even though nature may not build the complex sugars in exactly this way*. The process can be reversed in the laboratory; water can be added to the complex sugar in such a way as to decompose it into the two simple sugars.

34.15. Sucrose. By far the most important of all the complex sugars is sucrose. It occupies among the sugars the same position as alcohol among the "alcohols," and ether among the "ethers." It is, therefore, usually called simply "sugar." Other names for it are cane sugar, beet sugar, and saccharose. These names indicate where sucrose comes from, and that it is sweet. Sucrose is the α anomer of D-glucose in the six-membered ring form (Figure 34.7, Model III), combined with the β anomer of D-fructose in the five-membered ring form (Figure 34.8, Model VIII).

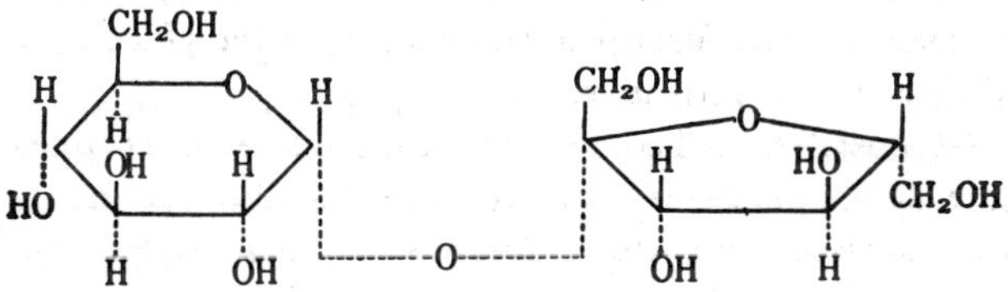

Sucrose, $C_{12}H_{22}O_{11}$

1-α-D-Glucopyranosyl-β-D-fructofuranoside

Sucrose is interesting because it is stable toward several chemical reagents with which other sugars react easily. For example, it does not have reducing properties toward Fehling's solution, and it does not react with phenylhydrazine (Section 34.9). Also, it does not show mutarotation (Section 34.11), but has a constant specific rotation of $+66.4°$. These characteristics show that the union between glucose and fructose takes place through the OH groups on the anomeric carbon atoms in Figures 34.7 and 34.8. These are the carbon atoms in the "potential carbonyl groups." Since sucrose does not have a potential C=O group it is a *nonreducing* sugar. We can regard sucrose as either a glucose fructoside or a fructose glucoside.

34.16. Inversion. When sucrose is boiled with dilute acids, it is hydrolyzed to its constituent sugars, glucose and fructose. Sucrose has a positive rotation of $+66.4°$. The mixture of glucose and fructose has a negative rotation of $-39°$. Since sucrose is dextrorotatory and the mixture of sugars obtained from it by hydrolysis is levorotatory, the hydrolysis is called inversion. The mixture of glucose and fructose is called *invert sugar*. A certain enzyme from yeast breaks down sucrose into glucose and fructose just as easily as boiling with acids. This enzyme is called *invertase*.

34.17. Other Disaccharides. Sucrose, described above, is a disaccharide, which means that it can be hydrolyzed to two simpler sugars. Three other disaccharides, to be described in this section, are maltose, cellobiose, and lactose. It is obvious from their formulas that in each molecule the ring structure printed on the right has its anomeric carbon atom (potential $C=O$ group) intact. In other words, these sugars can exist in α and β forms, and can exhibit the phenomenon of mutarotation. They react with phenylhydrazine and reduce Fehling's solution, the sugar properties that can result when the ring structure can open up to the chain structure with its aldehyde behavior. These disaccharides are generally termed *reducing* sugars to contrast them with sucrose, which is *nonreducing*.

Maltose is found in malt, which is the sprouted seeds of cereal grains, especially barley. Maltose is also the product of hydrolysis of starch (Section 38.2) when it is acted on by the enzyme *diastase*. Both hexose molecules in maltose are the same, namely, D-glucose, and maltose is known to be an α-glucoside because it can be broken down to its constituent sugars by the enzyme *maltase*, which splits only the α-glycosidic linkage (section 34.12).

Cellobiose can be obtained by degradation of cellulose (section 38.2). This sugar is exactly like maltose (built up from two D-glucose units) but the linkage is of the β-glycosidic type since it is hydrolyzed to its constituent sugars only by the enzyme *emulsin*.

Lactose is the principal sugar in the milk of suckling animals, and is known as milk sugar. It is obtained as a by-product of the manufacture of cheese. *Rennet*, an enzyme, precipitates the casein and fat (the cheesy materials) from the milk, and leaves a clear liquid called *whey*. Crystalline lactose is obtained from the whey by evaporation, after neutralizing with a mild alkali. The two hexoses in lactose are D-glucose and D-galactose. The fact that it can be split by the enzyme *emulsin* and not by *maltase* shows that it has the β-glycoside type of union: it is a glucose-β-galactoside.

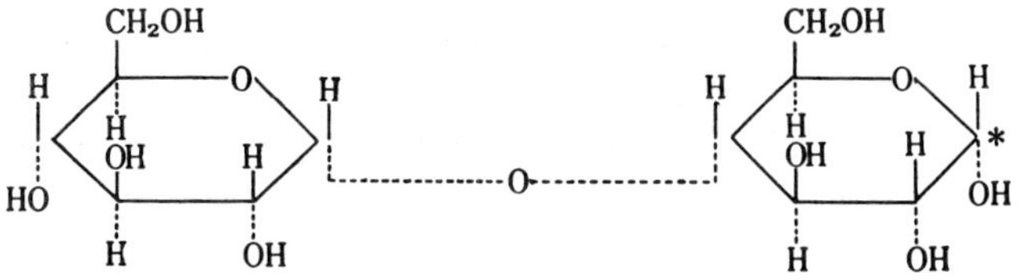

α-Maltose
4-(α-D-Glucopyranosyl)α-D-glucopyranose
Structural unit of starch (Section 38.2).

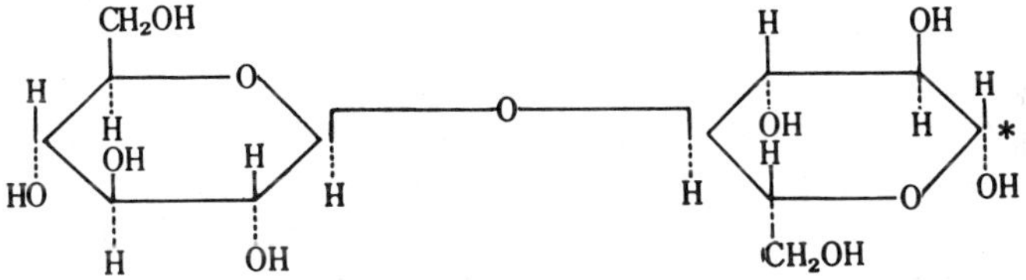

β-Cellobiose
4-(β-D-Glucopyranosyl)-β-D-glucopyranose
Structural unit of cellulose (Section 38.2).

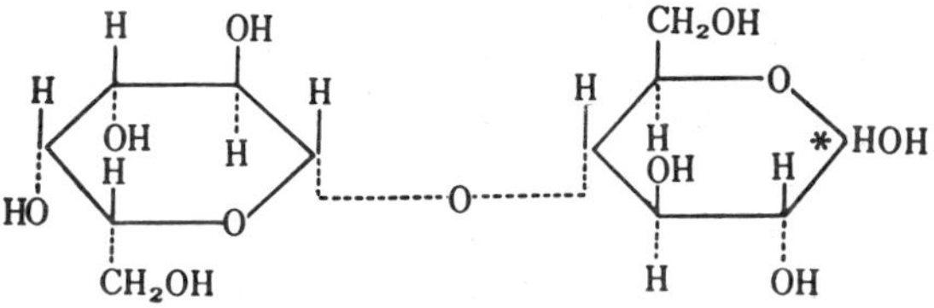

Lactose
4-(β-D-Galactopyranosyl)-D-glucopyranose
In the formula as printed, α-lactose would have the OH at the (*) position
below the ring.

34.18   Raffinose. This is a complex sugar built up from *three* simple sugars and has the composition $C_{18}H_{32}O_{16} \cdot 5H_2O$. When hydrolyzed, it breaks down to glucose, fructose, and galactose. It is practically tasteless. Raffinose is found in beet-sugar molasses, and in seeds, such as barley and cottonseed. It is a nonreducing sugar and consequently has no potential C=O group in its structure.

34.19. Starch. The starches are like the complex sugars in that they are anhydrides of simple sugars; on complete hydrolysis they take up water to form simple sugars. Since the starches yield C_6 sugars (hexoses) when hydrolyzed, we call them *hexosans*; this means they are anhydrides of hexoses.

The starches are the reserve food material of most plants. The various starches are classified, according to their source, as potato starch, wheat starch, corn starch, etc. When a starch is boiled with dilute acids, or acted on by certgain enzymes, it is gradually broken down according to the following scheme:

$$\text{Starch} \longrightarrow \text{Series of dextrins} \longrightarrow \text{Maltose} \longrightarrow \text{D-Glucose}$$

Although starch is accordingly an anhydride of glucose (a *glucosan*) it is probably better considered as built up of maltose units. This will be discussed further in Section 38.2 where it will be pointed out that starches from various sources contain variable amounts of an *amylose* fraction and an *amylopectin* fraction, which have different properties.

The dextrins obtained by breaking down the starch molecule are similar to starch in appearance, but are more readily soluble and are used as adhesives. The enzyme *diastase*, present in leaves and seedlings of plants, hydrolyzes starch to dextrins and finally to maltose. Another enzyme, *maltase*, completes the process by hydrolyzing maltose to glucose, in which form the plant uses carbohydrates as food.

Animals lay by a reserve food material called "animal starch" or *glycogen*. This substance is manufactured by mammals when there is an excess of sugar in the system, and is stored in the liver as reserve. When hydrolyzed, glycogen yields the same products as obtained from starch—namely, maltose and glucose.

34.20. Cellulose. $(C_6H_{10}O_5)_n$ is a hexosan and is more particularly a glucosan, because it can be hydrolyzed to glucose. It is therefore closely related to starch, which is also a glucosan and yields glucose when hydrolyzed. However, just as starch is to be regarded as built up from units of maltose, cellulose is to be considered as constructed from cellobiose. Cellulose is a constituent of the cell walls of plants; it is always found associated with other substances. It is insoluble and very inert; it can be easily obtained from all kinds of plant tissue by first removing other substances by appropriate extraction methods. Cotton is nearly pure cellulose.

34.21. Other Complex Carbohydrates. The complexes described in the preceding pages are anhydrides of glucose. In this section, we shall mention briefly certain substances in nature which are built up of other units of simple sugars. *Inulin* is a substance which resembles starch, but is not now classified as a starch. However, just as starch is a glucosan, inulin is a fructosan that yields D-fructose (and a small amount of glucose) when hydrolyzed. It occurs in Jerusalem artichokes and dahlia tubers; in fact, D-fructose is obtained commercially from the second source.

Xylan and *araban* are examples of naturally occurring pentosans which yield pentoses when hydrolyzed, as mentioned earlier in Section 34.1. Xylan (Section 38.2) is found widely distributed in such substances as straw, wood, algae, and peanut shells and araban is present in complexes in peanuts. Gum arabic contains araban and the calcium salt of arabic acid (the gums, in general, are complex carbohydrates which are partially oxidized to acids).

34.22. Nomenclature and Classification of Carbohydrates. The name "carbohydrate" is an old one, but it is not considered so good a name now as it was in the early days of organic chemistry. When the early chemists isolated certain sugars from nature, they found that the sugars appeared to be hydrates of carbon (carbohydrates), as in the case of cane sugar, which may be written $C_{12}H_{22}O_{11}$ or $C_{12}(H_2O)_{11}$. This apparent water of hydration may be eliminated easily, as is familiar in the charring of burnt sugar and the charring caused by dehydration of sugar with concentrated sulphuric acid. In both cases, pure carbon remains.

Some sugars are known, however, in which the ratio of hydrogen to oxygen is not the simple 2:1 ratio of H_2O. An example is rhamnose, $C_6H_{12}O_5$, the structure of which is CH_3–CHOH–CHOH–CHOH–CHOH–CHO. Moreover, many compounds are known which cannot possibly be classed as carbohydrates, yet have the same molecular composition as a carbohydrate. Two examples are formaldehyde, CH_2O, and acetic acid, $C_2H_4O_2$.

The classification of carbohydrates, in brief, is as follows:

		Examples
I.	Monosaccharides, or simple sugars	
	Trioses, $C_3H_6O_3$	Glycerose
	Pentoses, $C_5H_{10}O_5$	Arabinose
	Hexoses, $C_6H_{12}O_6$	Glucose and Fructose
	(Simple sugars from trioses up to decoses, $C_{10}H_{20}O_{10}$, are known.)	
II.	Polysaccharides, or complex sugars	
	Disaccharides, $C_{12}H_{22}O_{11}$	Sucrose
	Trisaccharides, $C_{18}H_{32}O_{16}$	Raffinose
	(These are dihexoses and trihexoses. Complex sugars derived from pentoses are also known.)	
III.	Colloidal polysaccharides	
	Pentosans, $(C_5H_8O_4)_n$	Araban
	Hexosans, $(C_6H_{10}O_5)_n$	Starch and Cellulose
	(These are not true sugars. Their molecular weights are of the order of 50,000 to more than 1,000,000 and, in modern terminology, they are referred to as colloids.)	

35

Chemistry in Plant and Animal Life

35.1. Nucleic Acids and the Genetic Code. This is a subject of prime interest to biochemistry, and its study is often referred to as *molecular biology*, a term now in general use to describe the fundamental processes that take place in living things on a molecular level.

In Section 33.4 we found that all proteins are built up on the same backbone; proteins differ in respect to the specific amino-acid side chains attached to the backbone (20 kinds of amino acids are available), and they also differ in the sequence in which the amino acids are aligned (a practically infinite number of different sequences is possible). The information for building proteins, and for replicating themselves, is stored in giant molecules called nucleic acids; these are present in the chromosomes in the nucleus of the living cell. The nucleic acids are built up of nucleotides (Figure 35.1) on a common backbone, just as the proteins are made of amino acids on a backbone common to all of them.

The *n*ucleic *a*cids inside the nucleus are built up from nucleotides in which the sugar component is *d*eoxyribose, and is therefore called DNA (note the italicized letters in this sentence); whereas the nucleic acids that operate outside the nucleus in the cell plasma are constructed from nucleotides in which the sugar component is *r*ibose, and is called RNA.

35.1a. *The DNA Molecule.* The composition of a nucleotide present in DNA is shown in Figure 35.1. Of these components, the H_3PO_4 molecule should need no introduction to the reader. Deoxyribose is a pentose (5-carbon) sugar in its furanose ring form (Figure 34.1 and Section 34.4); it differs from ribose in that it does not have an OH group at the C_2 atom. Adenine is a nitrogen base; for example, see the purines in Section 29.13. The union of these three molecules results in a nucleotide, where the link at the C_1 atom of deoxyribose is always by way of the β anomer, and the sugar belongs to the D family.   It is a glycosidic attachment (section 34.13) through the N atom of the base.   In a DNA molecule four different bases are found, as illustrated in Figure 35.2.   The backbone is the series of phosphate-sugar links, with bases attached as side chains to the sugar

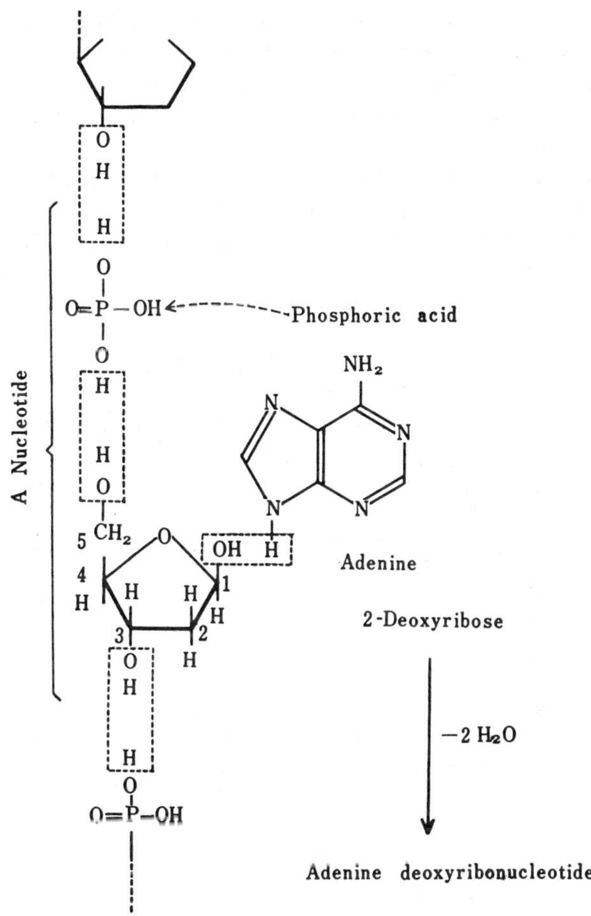

35.1.   Formation of one of the nucleotides which are the structural units of the nucleic acid, DNA.   This nucleotide unit is made from one molecule of phosphoric acid, one molecule of adenine, and one molecule of the sugar, deoxyribose.   In the DNA molecule, thousands of such units join end to end, with certain nitrogen bases as side chains at the C_1 atom of the sugar.   The backbone of the DNA molecule is the long chain of phosphate-sugar radicals.

rings.

The major event in this field of molecular biology was the discovery in 1953 of the structure of the DNA molecule. The α-helix of proteins was published several years earlier (Figure 33.3 is dated 1951). This is a single-strand helix held together by hydrogen bonds lengthwise through the molecule (see the broken lines in Figure 33.3). One of the important clues in arriving at that structure was the *flat* nature of the amide group of proteins as shown in Figures 33.1 and 33.2, which minimized the number of possible arrangements of the atoms in the helix. Soon after the advent of the α-helix it was shown that collagen consists of three helixes assembled as in a rope.

Watson and Crick in 1953 suggested that DNA is a double helix. Instead of being held together by H bonds lengthwise as in proteins, the two strands are held together crosswise as in Figure 35.2 by the H bonds

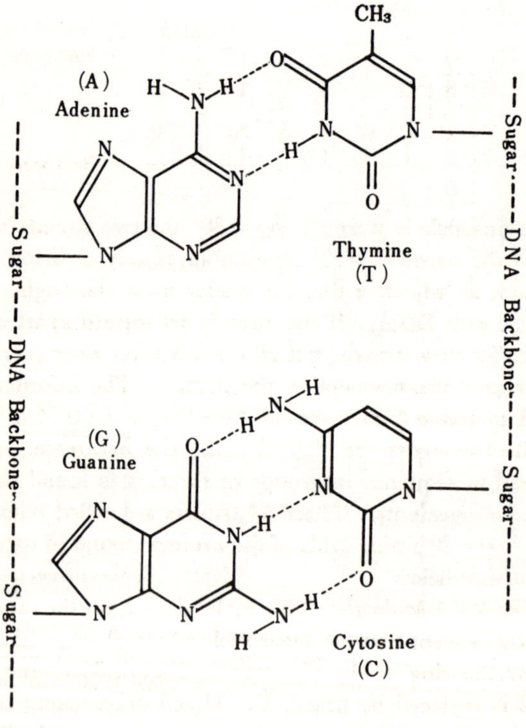

35.2. The two pairs of bases that cross-link the double helix of DNA.

between pairs of nitrogen bases. These were the major clues in working out the DNA structure: (1) when DNA is hydrolyzed only four bases are found, labeled **A, T, G,** and **C** in Figure 35.2, and (2) the amount of **A** in the hydrolysis product is always equal to that of **T,** and the amount of **G** is always equal to that of **C,** whereas the **A/G** and **T/C** ratios vary. This suggested that **A** and **T** are paired, and that **G** and **C** are paired, and models showed that if the pairs are hydrogen-bonded as in Figure 35.2 the distance across each pair, between the backbones, is the same. The cross-links in Figure 35.2 are *flat*. When the two backbones are twisted into a double helix the flat nitrogen base-pairs that hold them together resemble the steps in a spiral staircase. Another appropriate picture is that of a ladder twisted into a helix.

The order in which the base-pairs of Figure 35.2 appear along the DNA double helix may be represented in this way, but this is not intended to represent an actual DNA molecule:

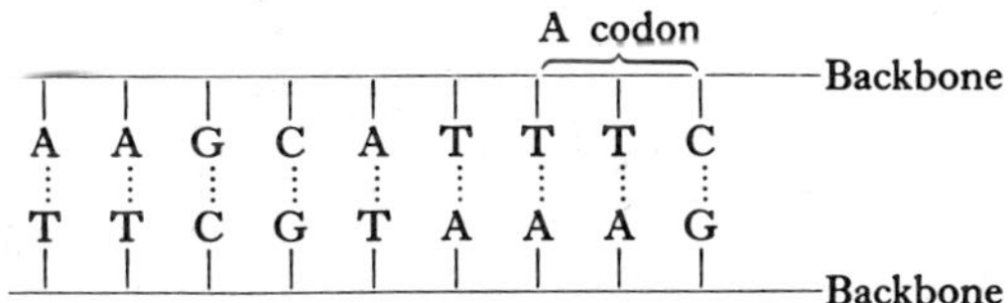

If the DNA molecule is warmed above 80° the two strands separate, but when cooled the partners must reassemble exactly as before. They are *complementary*, an adjective that the reader must thoroughly understand in connection with DNA. If the strands are zipped apart each can act as a template for new strands, but all the offspring when paired will have the same ordered arrangement as the parent. The information packed into a DNA molecule (which we call heredity) is controlled by the order in which the base-pairs are aligned along the backbone. If the base-pairs are read in sequence in groups of three, it is found that there are 64 possible arrangements. These 64 triplets are called *codons*, and they correspond to the 20 amino acids of the proteins found in nature. This is explained further below.

35.1*b*. *The RNA Molecule.* This molecule is built the same way as the DNA structure except that the sugar molecule is ribose, which has an OH group below the ring at the C_0 atom in Figure 35.1, and thymine in Figure 35.2 is replaced by uracil, **U.** Uracil differs from thymine only in the absence of the CH_3 group in its ring structure. The four bases in RNA are therefore **A, U** and **G, C.**

As mentioned before, the DNA molecules operate inside the nucleus of the cell, and bear the genetic code for reproducing proteins that will duplicate the original organism. DNA transfers its information to RNA by an unzipping process that exposes its templates, and complementary RNA's are formed; wherever **A** appeared in DNA there will be **U** in RNA, and where there was **G** in DNA there will be **C** in RNA, and so on. An RNA molecule can operate outside the nucleus, in the cytoplasm, and an RNA is the actual template from which proteins are constructed; it is generally single-stranded.

35.1c. *Messenger RNA and Transfer RNA.* In the cytoplasm (cell plasm) there are minute chemical factories called *ribosomes*, at which synthesis takes place. The RNA that has received its information from DNA inside the nucleus is called messenger RNA, and is sometimes referred to as soluble RNA or simply sRNA. It can get out to the cytoplasm to reach a ribosome. In the cytoplasm there are also transfer RNA's, which conduct amino acids to the ribosomes. There is a specific tRNA for each of the 20 amino acids of proteins. At the ribosomes the amino acids are deposited one at a time in accordance with the triplet code on the messenger RNA backbones, and linked by temporary hydrogen bonds. A tRNA probably has anti-codons on its backbone which guide it to the proper place where it is to lay its amino acid cargo on messenger RNA.

By 1963 the codons for all 20 amino acids had been established. For example, the amino acid, threonine, is deposited where these triplets are found on messenger RNA: **ACU, ACC, ACA, ACG,** which indicates that the first two bases are the important ones in the threonine codon. The codon for phenylalanine has been found to be **UUU.** Certain of the codons are signals to terminate the formation of a protein chain on the RNA template, and to remove the finished portion of a chain (an unzipping process).

35.2. The Body as a Chemical Laboratory. Advances in organic chemistry that take place in the laboratory are always followed with interest by the biochemist in his study of the chemistry that takes place in the body. The introduction of a new drug, or of a new chemical widely used by industry, is accompanied by a survey of its physiological effects. It is well known that nature makes use of defense mechanisms; such defense methods as protective coloring of insects and animals are readily observable, but chemical defense to substances absorbed through the lungs, or skin, or alimentary tract is usually far from obvious. However, just as foods must be metabolized into forms which can be used by the animal, poisons and drugs must be modified to structures that the organs are equipped to assimilate or eliminate.

The body makes use of processes of oxidation, reduction, and synthesis. Certain alcohols, for example, are destroyed in the body by complete oxidation to carbon dioxide and water, through intermediate aldehyde and acid stages. An example of reduction is the transformation of chloral (Section 26.11) to the corresponding alcohol; this metabolite (product of metabolism) is not eliminated as such, but is combined by a synthesis with a compound supplied by the body:

$$CCl_3CHO \longrightarrow CCl_3CH_2OH \xrightarrow[\text{acid}]{\text{Glycuronic}} \text{Urochloralic acid}$$

Chloral   Trichloroethyl
alcohol

Glycuronic acid (Section 34.7) is one of the defense compounds used by the body for conjugation with a toxic substance or its metabolite and subsequent elimination in the urine. The term *conjugation* is extensively used in biochemical literature, but should not be confused with conjugation phenomena described elsewhere in this book (Section 15.7). In this example, glycuronic acid forms a glycoside type of compound (Section 34.13), but it also forms esters through the carboxyl group, and ethers through the hydroxyl groups.

A simpler compound, used by the animal body for detoxication, is glycine (Section 33.5). In fact, the first such mechanism discovered

Benzoic acid   Glycine   Hippuric acid

(1842) was the conjugation of glycine with benzoic acid to give hippuric acid, eliminated in the urine. When studying the effects of drugs on experimental animals, pharmacologists must establish that the defense mechanism of the animal is similar to that of man. For example, glycine is one of the twenty amino acids required by the body for normal growth. In young rats, the glycine supply is more limited than in man, so that use of the compound by a rat to destroy a toxic compound may interfere with his growth, whereas this situation may not prevail for other animals and man. Similarly, nearly all animals detoxify drugs of the aniline type (Section 29.11) by introducing an acetyl group into the $-NH_2$ group attached to the benzene ring, but the dog cannot.

35.3. Chemotherapy. This word was coined by Paul Ehrlich, who received the Nobel prize in 1908 for his research on the synthesis of compounds that can destroy specific toxic organisms in the body. Protozoal, bacterial, and spirochete parasites are difficult to combat in the blood stream without also killing the patient. The "therapeutic index" of a medicinal substance is the ratio of the amount that will kill the host to that required for a curative dose. Medical practitioners seldom use an index lower than ten. Ehrlich's pioneer work was concerned with protozoa and spirochetes and the treatment of syphilis. Starting with the clue that the organism that causes this disease can be stained with a certain dye, as observed with the microscope, Ehrlich made many compounds to find one that would affix itself strongly enough to immobilize or kill the parasite. He met with success at the 606th compound, which he named Salvarsan.

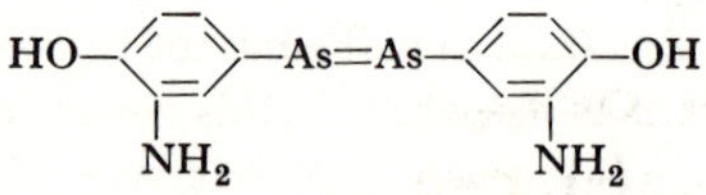

$$HO-\underset{\underset{NH_2}{|}}{\left\langle\bigcirc\right\rangle}-As=As-\underset{\underset{NH_2}{|}}{\left\langle\bigcirc\right\rangle}-OH$$

Salvarsan (arsphenamine)

Since Ehrlich's time, several striking advances have been made in the chemical attack on disease organisms. In 1932, in a screening program in which many compounds were tested in animals, it was found that the red dye *prontosil* was bactericidal. It was the first *antibacterial* drug that could be used systemically. Investigation of its fate in the body showed that prontosil is converted to sulfanilamide, which is essentially one-half the dye molecule and is both simpler and more effective. This led to the synthesis of many *sulfa* drugs, in which organic radicals replace an H in the SO_2NH_2 group of sulfanilamide. An important example is sulfapyridine, the first specific pneumonia drug.

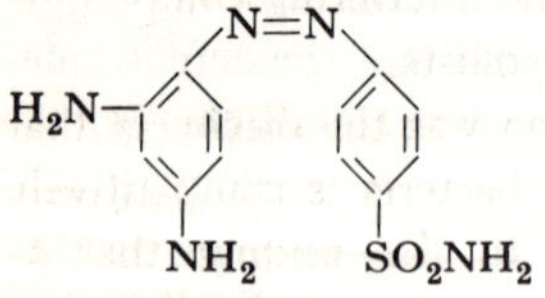

Prontosil 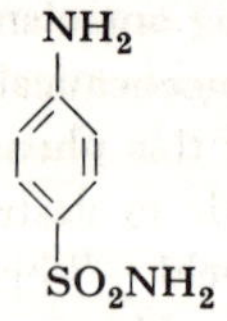   Sulfanilamide 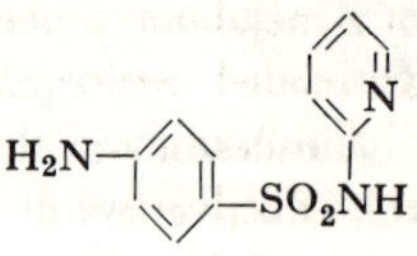   Sulfapyridine

Although the sulfa drugs were relegated to a secondary position with the advent of the *antibiotics*, their discovery gave an impetus to chemotherapeutic research similar to that caused by Ehrlich's work. Antibiotics are the chemotherapeutic substances isolated from cultures of bacteria, molds, and other microorganisms. Ever since Pasteur (1877) it was known that some soil bacteria produce substances that will kill other microorganisms in their vicinity, and it was even suggested that these antagonisms might be used to combat infectious diseases. In 1929, Fleming isolated what he called a *penicillin* from a Penicillium mold, and showed it was not toxic to animal tissues although it destroyed pathogenic bacteria. Ten years later several other antibiotics were discovered, and the penicillin era was born, aided by a helpful fermentation industry. Many antibiotics are now used in medicine, and some have been synthesized. The best known is penicillin G; its salts are water-soluble and practically nontoxic and are effective in treatment of gonorrhea, syphilis, preumococcal pneumonia, and other infectious diseases. There are several penicillins, which vary in the R group and in their effect on microorganisms.

$$R-\underset{\underset{O}{\|}}{C}-NH-CH-\underset{\underset{|}{}}{CH}\overset{S}{\diagup\diagdown}C(CH_3)_2$$

$$O=C\underline{}N\underline{}CH-C=O$$

$$OH$$

Penicillin G (benzyl penicillin)	R is $C_6H_5-CH_2-$
Penicillin F (2-pentenyl penicillin)	R is $CH_3-CH_2-CH=CH-CH_2-$

Modern successes with sulfa drugs and antibiotics should not let us forget that chemotherapeutic treatment for malaria was found empirically hundreds of years ago in cinchona bark, the source of quinine.

35.4. Biochemical Antagonism. One of the important developments in science in the period since 1940 is the realization that compounds of similar chemical structure often interfere with each other in body processes. This phenomenon has been studied rather extensively in connection with metabolites—that is, substances that occur as a result of a metabolic process in a living organism. The interfering compounds are called *antimetabolites*, or biochemical antagonists.

A milestone in the history of this phenomenon was the discovery that the effectiveness of sulfanilamide in destroying bacteria is competitively reversed by *p*-aminobenzoic acid. This led to the finding that *p*-aminobenzoic acid is essential to the growth of the bacterial cell (in the nature of a vitamin) and that it is important generally as a biochemical. Because of its similar size and shape, sulfanilamide can take up the same position in the cell or on an enzyme structure which is normally

$$\text{H}_2\text{N}-\langle\ \rangle-\overset{\overset{\text{O}}{\|}}{\text{C}}-\text{OH}$$

$$\text{H}_2\text{N}-\langle\ \rangle-\overset{\overset{\text{O}}{\|}}{\underset{\underset{\text{O}}{\|}}{\text{S}}}-\text{NH}_2$$

p-Aminobenzoic acid   Sulfanilamide

occupied by *p*-aminobenzoic acid; but the ultimate effects are different and the organism dies. Many analogous antagonistic systems have been found in the enzyme reactions of animals and man.

35.5. Pharmacophores. In Chapter 36 on dyes we shall find that the dye-maker, with his knowledge of the relation between color and constitution, can often work out prospective dyes on paper and obtain the expected results in the laboratory. The drug-maker is not that fortunate, although he does have some correlations that are valuable. In other words, the maker of drugs has found in certain classes of compounds structural groups that can be labeled "pharmacophores," analogous to the chromophores (Section 36.2) of the maker of dyes, but the search for new drugs is still largely empirical. When the pharmacophore in a class of compounds has been found, it is then possible by trial and error to make many related compounds in an effort to attain more desirable properties in the drug. For an example we need go no further than the sulfa compounds. The pharmacophore in prontosil is the sulfanilamide structure. Thousands of derivatives of sulfanilamide have been made in the search for drugs with superior toxicity to a specific microorganism, fewer side effects on the patient, etc. As a result, many sulfa drugs are now available in medicine for treatment of such diverse ailments as leprosy, urinary infection, and diabetes.

One of the most extensive studies of the drug activity of compounds of similar structure has been done on derivatives of β-phenylethylamine,

$\langle\ \rangle-\text{CH}_2-\text{CH}_2-\text{NH}_2$. Probably the most important of this series of

compounds is the hormone, adrenaline.

Adrenaline   Ephedrine   Benzedrine

These substances all have a styptic effect. They contract the blood vessels and raise the blood pressure (the "pressor" effect). They are also referred to as *sympathomimetic* amines in that they produce effects that mimic those obtained when the sympathetic nervous system is excited by other agencies.

35.6. Animal Hormones. Those glands in the animal body with which man has been most familiar in the past carry their secretions to some body surface by means of channels or ducts and discharge their secretions at the surface. For examples, we may cite the salivary glands in the mouth, the sweat glands on the skin, and the milk glands of the female. There are some glands in the body, however, which have no ducts to carry their secretions from one place to another. The secretions of these "ductless" glands are called "internal" secretions. Substances called hormones are secreted by these glands under some body conditions, are drawn directly into the blood stream through the veins of the gland, and arouse certain functional activities of the body. The name hormone comes from the Greek *hormaein,* meaning "to arouse." The great physiological activity of the hormones is shown by the fact that adrenaline exhibits a definite activity when in a dilution of 1 part in 250,000,000. The nature of the hormones will be illustrated with just a few examples.

Acetylcholine (Section 29.10) and *adrenaline* are the hormones of the autonomic nervous system. This portion of the nervous system controls the involuntary movements of muscles in the body— that is, movements not requiring a mental effort, such as contraction of the pupil of the eye, motility of the intestinal tract, beat of the heart, etc. This automatic control of muscles is accomplished through a sympathetic system of nerves and a parasympathetic system, which are essentially antagonistic (when their nerve fibers reach the *same* muscle). Thus a stimulus through the sympathetic nerves will open the eye pupil, but a parasympathetic-nerve stimulus will close it; the sympathetic stimulus will accelerate heart beat, whereas the parasympathetic will slow it down.

A nerve consists of a series of segments called neurons, each of which has a cylindrical core (the axon) and fine, branchlike appendages (the dendrites). The surface contact between the axon of a neuron and the dendrites of the next neuron is called a synapse. Transmission of a stimulus past these synapses through a system of neurons is made possible by acetylcholine and is apparently an electrochemical process. At the nerve-muscle junctions of the sympathetic system, the hormone liberated is *adrenaline* or a closely related substance; similarly, in the parasympathetic system, the hormone liberated is *acetylcholine.* Ultimate control of these competing systems is up to the enzymes present, which destroy the hormones as soon as they have served their purposes (aminoxidase for adrenaline, and choline esterase for acetylcholine).

When the hormones are injected into the animal body in small amounts, they cause effects that *mimic* the stimulations of their corresponding nervous systems, by overpowering the enzyme concentration normally present in the body. Many drugs are known which similarly have a sympathomimetic effect (Section 35.5) and others are known that are parasympathomimetic.

Cortisone is a hormone from the cortex (outer layer) of the adrenal gland, and is remarkable for the effectiveness with which it counteracts inflammations that range from arthritis to iritis. Cortisone belongs to a large family of compounds known as *steroids*; these include *sterols* such as cholesterol and ergosterol, vitamin D, cholic acid of the bile, cardiac glycosides, and sex hormones.

Stereochemistry is an important feature of the steroids. The fused ring system is not aromatic, but aliphatic like that shown in Figure 12.17. The dotted bonds in these formulas indicate whether the OH group is above or below the plane of the ring system. If above, the compound is β, and if below it is α. An example of a male hormone, testosterone, is shown here.

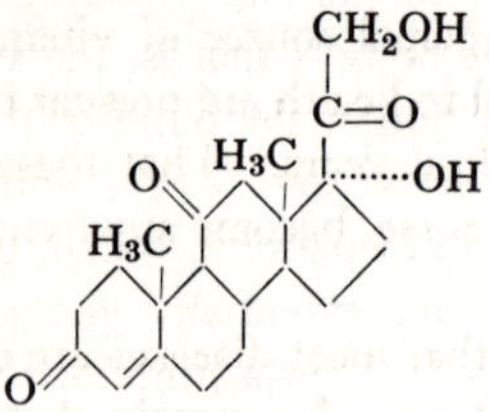

Cortisone 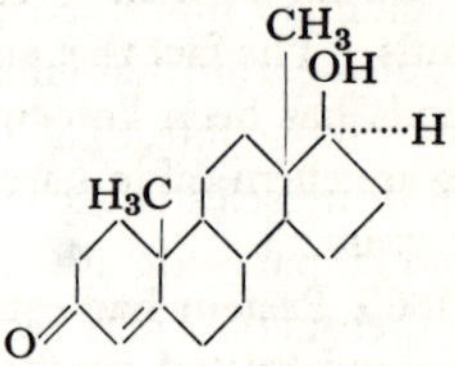   Testosterone

The male hormones as a class are known as androgens (from the Greek *andros*, man). The female hormones are estrogens (from the Latin *oestrus*, the female sex cycle); some have been developed for use in birth control.

35.7. Plant Hormones. Substances have been extracted from bark, seeds, etc., which, when applied to growing plants, bring about responses that resemble activity of hormones in animals and man. The first proof of this plant growth promotion was provided in 1926 by Went. Many relatively simple organic compounds have been found to be far superior to these crude extracts in their hormonelike behavior.

Perhaps the simplest of plant hormones is ethene, $CH_2{=}CH_2$, which is used extensively to accelerate ripening of citrus fruits and bananas in warehouses, after being picked green. When treated with 1-naphthalene-acetic acid, potatoes can be stored almost indefinitely without shrinkage and the budding tendency is delayed. The same compound is used in orchard sprays to prevent apples from dropping before they are harvested.

A result of the study of plant growth regulators is the discovery that they can be used as *herbicides*. At the extremely low concentration of 10 parts per million, or less, the hormones have a beneficial effect on a plant, but if the concentration is raised to between 100 and 1,000 parts per million, the abnormal growth results in death. This action as a plant killer is quite different from that of herbicides used earlier, which had a caustic action on plants that showed up in such symptoms as "burning" of the leaves, etc. These new herbicides have a selective action which often permits eradication of pestiferous plant life without injury to crop plants. However, there are herbicides that will kill cereal crops in preference to weeds.

35.8. Vitamins. The vitamins are extremely active substances physiologically, and seem to be necessary for the welfare of all living things. There is no essential difference between vitamins and hormones. They both serve to regulate body functions. Hormones are manufactured by certain glands of the body, whereas vitamins are produced in other various ways. So far as man is concerned, the principal source of vitamins is foodstuffs. The fact that substances essential to health are present in certain foods has been known for several hundred years. That these substances are chemical in nature could not, of course, become apparent until recent years.

By 1860, Pasteur had established the fact that most diseases are due to bacteria and related organisms. By 1887, it was also certain that some diseases, like beriberi, are not due to bacteria, but to absence of certain foods in the diet. By 1912, it was shown that foodstuffs contain at least two kinds of "essential substances"; one of these, which is necessary for growth of young animals, was called fat-soluble A and the other, which is necessary for prevention of beriberi, was called water-soluble B.

These names indicated the solubility of these potent substances and also the method of extracting them from certain foods in more concentrated form. They were generally called *vitamines*, because it was *proved* that they are essential to life (vital), and because it was *thought* that they are amine compounds, analogous to amino acids (Chapter 33).

By 1920, it was recognized that a third in this series of substances, water-soluble C, is present in certain foods and prevents scurvy. It was also clear by this time that these vitamines are not amines, and it was suggested to drop the "e" from the end of their name, using the alphabet to record the order of their discovery, thus: vitamin A, vitamin B, vitamin C, etc. Today scientists prefer to call vitamins by their chemical names.

The potency of these substances is extremely great. Vitamin D, for example, shows a detectable physiological effect on a rat when diluted 2,240,000,000 times. The daily human requirement is said to be only 0.000,025 g.

The chemical nature of the vitamins was determined so rapidly that ideas concerning vitamins often changed overnight. In 1930, an expert in the field of vitamin chemistry declared that "vitamin A and vitamin C isolation seems, at present, quite remote," but in the following two years these same vitamins were thoroughly investigated and their chemical nature was well established.

Space is not available in this book for a discussion of all the known vitamins. Most are relatively simple. The D vitamins are related to the steroids (Section 35.6), and vitamin A is one-half the molecule of β-carotene (Section 38.3). We shall give some details on vitamin C as representative.

Vitamin C is the antiscorbutic (antiscurvy) vitamin. It is found in citrus fruits, especially in lemons and oranges; in tomatoes and most fresh vegetables; in green and red peppers; and it is also present in small amounts in the animal body in the adrenal gland. It is related to the hexose sugars. It is a lactone (Section 28.6) derived from a six-carbon sugar acid (see sorbose, in Section 34.7). Vitamin C has been named ascorbic acid, from its antiscorbutic properties. It is optically active and

$$\text{HO—CH}_2\text{—}\overset{\displaystyle |}{\underset{\displaystyle |}{\text{C}}}\text{—}\overset{\displaystyle\overset{\text{OH}}{|}}{\underset{\displaystyle\underset{\text{H}}{|}}{\text{C}}}\text{—}\overset{\displaystyle\overset{\text{H}}{|}}{\text{C}}\text{=}\overset{\displaystyle\overset{\text{OH}}{|}}{\text{C}}\text{—}\overset{\displaystyle\overset{\text{OH}}{|}}{\text{C}}\text{=O}$$

Ascorbic acid, or vitamin C

occurs in the form of *dextro* and *levo* molecules. Vitamin C is the *levo*-ascorbic acid; this molecule is more than forty times as active physiologically as the *dextro* compound. In fact, the *dextro* molecule is practically useless for the prevention of scurvy.

35.9. Insect Hormones. Of the internal hormones of insects, four were known by 1966. Two are hormones of the brain and gonads; two that control moulting are *ecdysone* and *juvenile hormone*. Ecdysone is a steroid that has not only been identified but synthesized. Man, by the use of a juvenile hormone that controls the moulting schedule, will have an excellent means for control of insect populations.

Some compounds that behave like hormones, but are secreted externally, are called *pheromones*. They are used to give alarms, to mark trails toward food supplies, to attract the opposite sex, etc., and are essentially for communication in the same species. One of the first of these identified is the sex attractant of the gypsy moth.

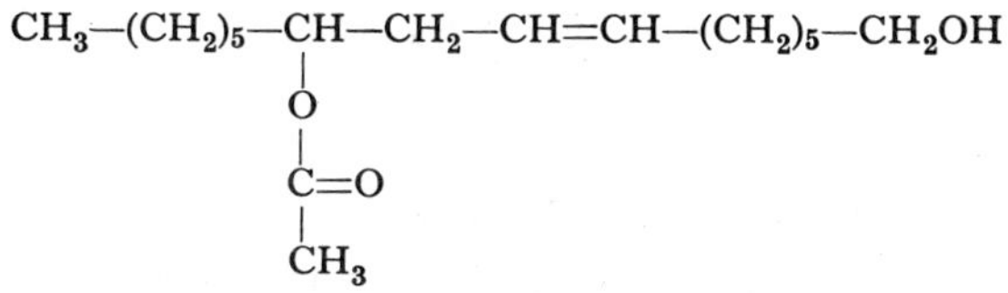

$$CH_3-(CH_2)_5-CH-CH_2-CH=CH-(CH_2)_5-CH_2OH$$

$$\begin{array}{c} | \\ O \\ | \\ C=O \\ | \\ CH_3 \end{array}$$

Gypsy moth attractant (gyplure)

36

Dyes

36.1. Color and Structure. Dyes are substances that have color and can be firmly fixed to textiles by the dyeing process. The principal problems of dyeing are illustrated by the following two compounds:

$$\text{Nitrobenzene} \quad\quad\quad \text{p-Nitroaniline}$$

Nitrobenzene (a colored substance)   p-Nitroaniline (a dye)

Nitrobenzene is yellow, but it does not dye fabrics; the color is easily washed away. *p*-Nitroaniline, however, is not only yellow but also sticks to fabrics and is, therefore, a dye. There are two facts about a dye molecule that are of interest: the reason for its color and the reason for its ability to stick to a textile fiber. Let us consider first the reason for color.

The reader will recall that ordinary white light consists of "all the colors

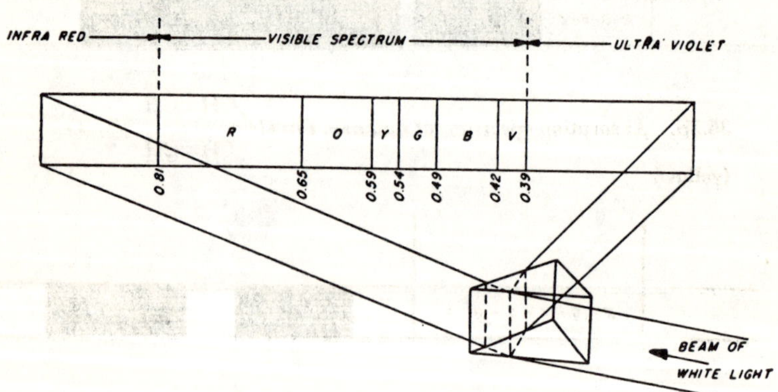

36.1. The spectrum. (Wave lengths are in microns. A micron is 0.001 mm).

of the rainbow," and that these colors are committed to memory in the order VBGYOR—that is, violet, blue, green, yellow, orange, and red. When a beam of light is passed through a prism and allowed to fall on a white surface, these colors are arranged side by side in what is called a spectrum, as shown in Figure 36.1.

Our eyes do not see all the colors in the spectrum, but only those in the "visible spectrum." When the visible spectrum is intact it appears to us as white light. Although the very rapid vibrations (ultraviolet) and the relatively slow vibrations (infrared) make no impression on our eyes, they do make an impression on a photographic plate, and that is the way spectra are usually mapped.

Now suppose that the beam of light in Figure 36.1 is made to pass through a sample of benzene before it passes through the prism and is spread out into its spectrum. The result is shown in Figure 36.2a. A

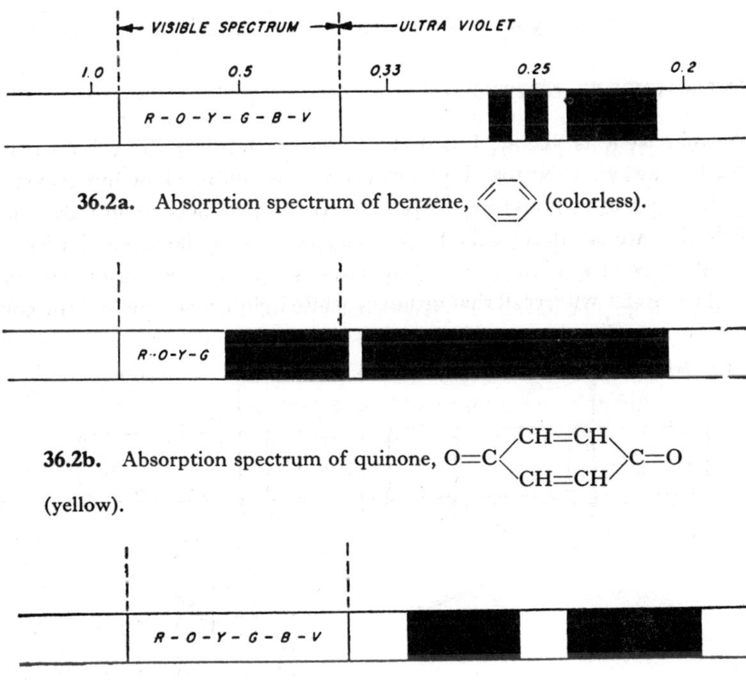

36.2a. Absorption spectrum of benzene, (colorless).

36.2b. Absorption spectrum of quinone, $O=C\langle \substack{CH=CH \\ CH=CH} \rangle C=O$

(yellow).

36.2c. Absorption spectrum of quinol, $HO-\langle \rangle -OH$

(colorless).

spectrum of this nature, obtained after a beam of light passes through a solution or a liquid, is called an *absorption spectrum*. That portion of the spectrum that is absorbed is called an *absorption band*.

There may be several bands, but for a layer of definite thickness and definite concentration, these bands always occur in the same positions and are characteristic of the substance being examined. It will be seen from Figure 36.2a that when a beam of light passes through benzene the visible spectrum is not disturbed. That is why benzene has no color. Light transmitted through benzene, or reflected from it, has no part of its visible spectrum absorbed, and benzene is therefore said to be color-less.

When a light beam passes through a quinone solution (Figure 36.2b), part of the visible spectrum is removed. The light that emerges has had its violet and blue colors absorbed and what we see then is a mixture of colors that is yellow. Light reflected from quinone crystals or transmit-ted through quinone in solution is yellow because the quinone molecule disturbs the visible spectrum. Coal is black because it absorbs all the visible spectrum when light falls on it.

In contrast with quinone there is the closely related compound, quinol, the absorption spectrum of which is given in Figure 36.2c. Quinol is color-less because the absorption bands occur in the ultraviolet spectrum and our eyes are not adjusted to changes in that part of the spectrum.

If we examine the structures of benzene, quinol, and quinone, it will be apparent that quinone is much different from the other two compo-unds; it does not have the typical structure of the benzene ring. Com-pounds with the quinone (or quinoid) structure are generally colored. In other words, the arrangement of the atoms in the quinoid structure tends to shift the absorption bands into the visible spectrum to produce color. We have much evidence of a similar nature proving that in ge-neral the color of an organic compound is a function of its structure.

36.2. Chromophore Groups. Typical structures that frequently give color to the common dyes are the following unsaturated groups :

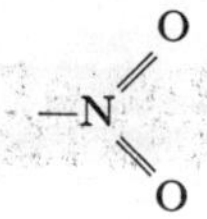

<table>
<tr><td>

O
//
—N
\\
O

Nitro

</td><td>

CH=CH
/   \
=C   C=
\   /
CH=CH

Quinoid structure
(Section 26.11)

</td><td>

=C—C=C—C=

Conjugate system
(Section 15.7)

</td></tr>
</table>

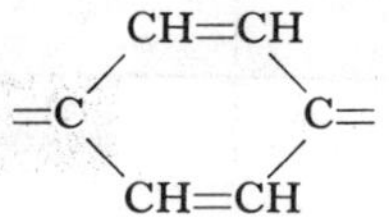

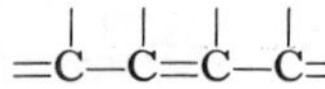

Such groups are called chromophores, meaning color-bearer in Greek.

Typical dyes in which the chromophores are quite evident will be found described in the following pages.

36.3. Auxochrome Groups. At the beginning of this chapter we stated that nitrobenzene, $C_6H_5-NO_2$, is simply a colored substance, whereas the amino derivative, $H_2N-C_6H_4-NO_2$, is a dye. The color, in both cases, is due to the chromophore, $-NO_2$. The dye qualities of the second compound must therefore be due to the $-NH_2$ group, which is an anchoring group that fixes the molecule to the fabric. Groupings that can be used for attaching dye molecules to textiles are either acidic or basic in character:

$$-NH_2 \qquad -OH \qquad \underset{\text{OH}}{-\overset{|}{\underset{|}{C}}=O} \qquad -\overset{\overset{\text{O}}{\|}}{\underset{\text{OH}}{\underset{|}{S}}}=O$$

(also $-NHR$ or $-NR_2$)
Amino · Hydroxyl · Carboxyl · Sulphonic acid

In basic dyes · In acid dyes

These anchoring groups are sometimes called auxochrome groups (Greek, meaning aid to color) because of their fixative effect, but this name is at present more often used in connection with another characteristic that will be described in Section 36.15.

Wool and silk are *animal* fibers, whereas cotton is a *vegetable* fiber. Wool and silk are essentially proteins in composition (Section 33.5) and they therefore contain both acid groups and basic groups. They are accordingly easily dyed, since they are capable of reacting with both acid dyes and basic dyes. Cotton is nearly pure cellulose (Section 38.11*b*) and is rather inert to chemical reagents. It is therefore more difficult to dye cotton than to dye wool or silk.

Rayon is made from cotton or other cellulose materials, such as wood, but the cellulose is so modified in the process that rayon is more reactive and easier to dye than cotton. Nylon resembles a protein in composition (Section 38.10), and thus its behavior to dyes resembles that of wool and silk.

36.4. Mordants and Lakes. In order to dye cotton it is often necessary to use mordants. These are substances that easily penetrate cotton fibers and for some reason possibly because of physical adsorption, are capable of sticking firmly to the fibers. After the fabric has been mordanted (impregnated with the mordant), a dye is added which forms an insoluble precipitate with the mordant that is firmly attached to the cotton.

The precipitate obtained from the reaction between the mordant and the dye is called a *lake*. The dye is said to be "laked out." When a basic dye is applied, an acid mordant, such as tannic acid, is used. For an acid dye, basic mordants, like aluminum salts or iron salts, are employed. The ability of some dyes to form lakes with certain salts is probably due to a structural configuration that permits chelate compounds to form like those described in Section 28.13.

CLASSIFICATION OF DYES ACCORDING TO THEIR METHOD OF APPLICATION

Dyes are sometimes classified very broadly as acid dyes or basic dyes, depending on whether the auxochrome group is acid or basic. This division of dyes is not common in commerce. The usual classification is based on the methods of application to the fabric.

36.5. Substantive Dyes. These are also known as direct dyes, since they are capable of dyeing fabrics directly. This is usually a simple matter in the case of silk and wool, but for cotton there are comparatively few direct dyes. A common example of direct dyes for cotton is Congo red (Section 36.10).

36.6. Adjective Dyes. These are also called indirect dyes, because they do not form a stable union with fabrics but need the aid of mordants to obtain firm combinations. An example of an indirect dye is alizarin (Section 36.9). Adjective dyes, incapable of dyeing cotton directly, are often capable of dyeing silk and wool without a mordant.

36.7. Vat Dyes. The vat dyes are formed by chemical reaction while in contact with the fiber. The dyes of the other two classes that have just been defined are all ready-made by the manufacturer, and dissolved as needed. The vat dyes, however, are not bought completely made. Most vat dyes are put into the dye bath in a leuco, or colorless state (Section 36.11) and the cloth is thus impregnated with a practically colorless substance. When the cloth is removed from the vat, oxidation by the air converts this leuco substance to the colored dye, right on the fibers. An example of a vat dye is indigo blue (Section 36.12). As a rule, vat dyeing yields very stable colors.

CLASSIFICATION OF DYES ACCORDING TO STRUCTURE

In comprehensive books on dyes they are usually classified according to their structures. More than two dozen such classes are known, of which only a few will be described here with specific examples. The structural classification is often based on the chromophores.

36.8. Nitro Dyes. The $-NO_2$ group in a molecule usually tends to impart a yellow color. Most nitro dyes are phenol derivatives, the phenol group acting as the anchoring group. Martius' yellow was at

Martius' yellow　　　　　　　　　　Picric acid

one time used as one of the food colors. These are dyes that the government certifies for use in foods to give them a more appetizing appearance, but Martius' yellow was found to be harmful. However, it was accidentally learned that it has moth-proofing properties, and this discovery has been exploited on a large scale in the manufacture of dyes with ability to repel moths. Picric acid was formerly used extensively as a yellow dye for silk and wool. It is better known as an explosive.

36.9. Anthraquinone Dyes. The most common members of this class of dyes are alizarin and purpurin. The chromophore is the

Alizarin　　　　　　　　　　Purpurin
1,2-Dihydroxyanthraquinone　　　　1,2,4 Trihydroxyanthraquinone

quinoid structure. These dyes occur in nature in the madder root as complex compounds in combination with glucose (see the glucosides in Section 34.12). When mordanted in the proper way, alizarin produces the well-known Turkey red color on cotton. Alizarin was obtained for several thousand years from madder root, but is now made synthetically. The synthetic 1, 8-dihydroxyanthraquinone is a laxative (Modane) used for treatment of chronic constipation; related compounds are found in cascara sagrada and other natural products. The structure of anthracene should be reviewed in Section 32.6.

36.10. Azo Dyes. It is instructive to compare the following two compounds from the standpoint of color:

$$H_3C-N{=}N-CH_3$$

Azomethane (colorless)

Azobenzene (yellow)

Azomethane has an absorption band in the ultraviolet and therefore has no color, whereas azobenzene absorbs in the visible spectrum and is a deep yellow. The color in azobenzene is attributed to the long conjugate system, $C{=}C{-}C{=}C{-}N{=}N{-}C{=}C{-}C{=}C$, formed with $-N{=}N-$ (the azo group) and the two ring structures.

The azo dyes are among the most important of the synthetic dyes. The auxochrome (anchoring group) is usually the basic—NH_2 group or the acidic sulphonic acid group, or both. The simple dyes derived from azobenzene are yellow to orange, but more complex derivatives vary from orange to red and even to black. One well known azo dye is Congo red, and another is methyl orange (see next page).

Congo red (see naphthalene, Section 32.6 and diphenyl, Section 24.4)

The synthesis of azo dyes was briefly mentioned in Section 29.8 in the discussion of diazonium salts. If this is reviewed, the preparation of methyl orange can be easily understood as a reaction in which *diazotized* sulphanilic acid (Section 29.10) is *coupled* with dimethylaniline. The coupling takes place in the *para* position to the dimethylamino group.

36.11. Triphenylmethane Dyes. Many dyes are derived from triphenylmethane, $(C_6H_5)_3CH$. They are rather complex substances, the

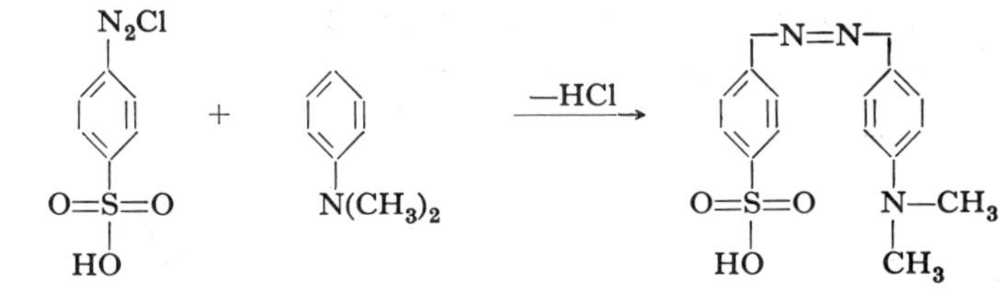

Diazotized   Dimethylaniline
sulphanilic acid

Methyl orange
(See figure 36.7)

nature of which may be illustrated by the dye known as pararosaniline.

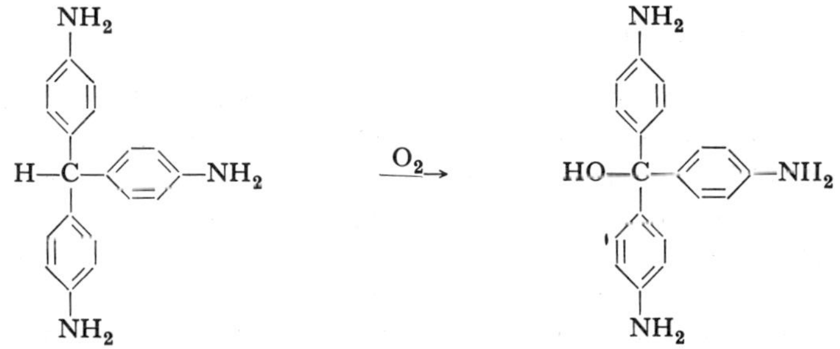

Leuco base of pararosaniline

Color base of pararosaniline

The synthesis of many of the dyes in this class results initially in a color-less compound which is called a *leuco* base, from the Greek *leucos*, meaning white. Mild oxidation converts the leuco base to the *color* base. In some cases, the color base is made directly without an intermediate leuco stage, an example of this procedure being the preparation of pararosaniline itself. The color base is without color, but dissolves readily in acids to form highly colored salts in which the weakly basic—OH group is replaced by acid radicals:

$$(H_2N—C_6H_4)_3C—OH + HCl \longrightarrow HOH + [(H_2N—C_6H_4)_3C]^+Cl^-$$

This will be discussed further at the close of the chapter in connection with the reason for color of these compounds, the chromophore (color-bearing) group not being evident in the structure as written here.

The triphenylmethane dyes fall into two categories. There are the basic dyes of which pararosaniline with its basic –NH$_2$ group is typical. There are also the acid dyes, such as the *phthaleins*, in which acid groups, e.g. phenolic –OH, anchor the dye molecules to fabrics. The phthaleins are made by the action of phenols on phthalic anhydride (Section 27.16).

This accounts for the class name of phthaleins. The most common member of this class of compounds is phenolphthalein.

Phenolphthalein   Mercurochrome

Phenolphthalein is a colorless solid that dissolves readily in alkaline solutions with development of a deep-red color. This is to be compared with pararosaniline which is a base and dissolves in acids. Phenolphthalein is used extensively in patent medicines as a laxative. It is also widely used in analytical chemistry as an indicator, the theory of which will be given in Section 36.16.

Fluorescein is a phthalein dye made from resorcinol (Section 25.14) and phthalic anhydride. A bromine derivative of fluorescein known as *eosin* is used in red inks and is employed extensively as a dye. *Mercurochrome* is a derivative of fluorescein. It is a red dye which has been widely used as a local antiseptic in place of tincture of iodine.

36.12. Indigoid Dyes. Indigo and a number of its derivatives constitute a family of dyes called indigoids. Most of the indigo used now is made synthetically and is much purer than the natural product obtained from the blue coloring matter in the leaves of the indigo plant. Indigo is one of the vat dyes. It is insoluble in water and in strong alkalies and therefore cannot be directly applied to fabrics. It must be converted to the leuco compound, which is soluble. In the dyeing process the fabric

Indigo white (soluble)   Indigo blue (insoluble)

is dipped into an alkaline solution of indigo white. When it is removed from the vat and exposed to air, the colorless compound on the fibers is oxidized to indigo blue.

The indigoids, in general, are fast to light and wear. The purple of Tyre, or Tyrian purple, was highly prized and reserved for the use of the aristocracy in ancient times. It is an indigoid dye, found in certain small marine animals.

36.13. Color Indexes. From the brief account of dyes in this chapter, it is obvious that nomenclature of dyes can be quite confusing. Descriptive names (or even nicknames) are used more commonly than systematic names that would throw light on their chemical structure. The dye chemist usually has at his elbow reference works that catalog the dyes and assign numbers to them. One such catalog is the *Color Index* (the C. I. numbers); another is Schultz's *Farbstofftabellen* (the S. numbers).

36.14. Chromophores. The explanation for the color-producing activity of chromophore groups is closely associated with the resonance theory and with unsaturation (that is, double-bond structures, such as $N=N$ and $C=C$). In connection with the resonance theory, we showed in Section 15.7 and Figures 15.10 and 16.2 how certain molecules can plausibly be assigned structures in *polarized* states as well as in *normal* states. The transformation to a polarized state is essentially one in which there is a shift or oscillation of electrons from one part of the structure to some other part, accompanied by rupture or formation of double bonds. Electrons taking part in such changes are relatively loosely bound, which means that a relatively small amount of energy is associated with their oscillations. In Table 14.2, which shows the relation between wavelength and energy, it is clear that lower energies correspond to longer wavelengths. If the bonding energies associated with the structure of a molecule are such that electron oscillations can take place in it at frequencies corresponding to wavelengths of visible light, the conditions are favorable for color phenomena. Specifically, the absorption frequency is equivalent to the energy separation from the ground level of the molecule (Section 14.13) to the excitation level. It is also essential that the movement of electrons from one part of the molecule to another results in a change in dipole moment and in a *polarized* structure—that is, a structure that can be represented as resulting from a transfer of electric charge. This will be more apparent if we review the resonance structures of benzene (Figure 19.1) which can be pictured as having pairs of π electrons chasing each other around a closed ring. Such structures, which are not polarized, are not conducive to color production.

The function of a chromophore in a colored molecule is to offer a path by which electronic oscillations can take place to give polarized structures. The conjugate system (see especially *glyoxal* in Section 26.11, and see Section 15.7) is useful for this function, and is a highly important chromophore structure. The carotenes (Section 38.3) are very long conjugate systems; they are all intensely colored.

36.15. Auxochromes. Some groups are highly effective in increasing the color when inserted into structures that already have color and are therefore called auxochrome groups. They were defined in another sense in Section 36.3. The ability of auxochrome groups to increase the color is due to their ability to act as electron donors; electron donors were discussed in Section 17.3, where the descriptive term "electron source" was employed. The electrons supplied by the auxochrome groups require, of course, a chromophore path through which to travel. The chromophore, on this basis, is an electron acceptor, or "electron sink."

36.3. A resonance structure of nitrobenzene (see Structure IX in Figure 21.2).

For a first illustration, consider the fact that nitrobenzene, $C_6H_5-NO_2$, is yellow, but *p*-nitroaniline, $H_2N-C_6H_4-NO_2$, is much more intensely colored. In Section 21.7 we discussed the resonance structures of nitrobenzene and showed that the $-NO_2$ group tends to draw an electron from the ring, leaving the ring positively charged as shown in Figure 36.3. The stability of this resonance structure (which is a quinoid, color-bearing type) is considerably increased when an $-NH_2$ group is attached to the ring,

36.4. In *p*-nitroaniline, the auxochrome ($-NH_2$) increases the magnitude of the conjugate system over that in nitrobenzene.

because the N atom acquires the positive charge more easily by losing an electron. This is illustrated in Figure 36.4.

The student should review the text accompanying Figure 15.10 to find out why the shift of one electron in these formulas looks like the shift of an electron pair.

In Section 36.3 several types of groups, acidic and basic, were referred to as auxochromes that serve to attach dye molecules to fibres. The –OH group is useful not only in that respect, but also as a color intensifier according to the mechanism just described for the $-NH_2$ group. The –COOH group and $-SO_3H$ group, however, are not auxochromes in the sense discussed in this section; they are not electron donors.

36.16. Acid-Base Indicators. The dye p-nitrophenol is an acid, just like phenol. It is colorless in acid solution, but intensely yellow when the solution is made alkaline. The reason for this is that the original unionized molecule is colorless, but its corresponding negative ion is colored. In alkaline solution, excess base removes H^+ ion from the

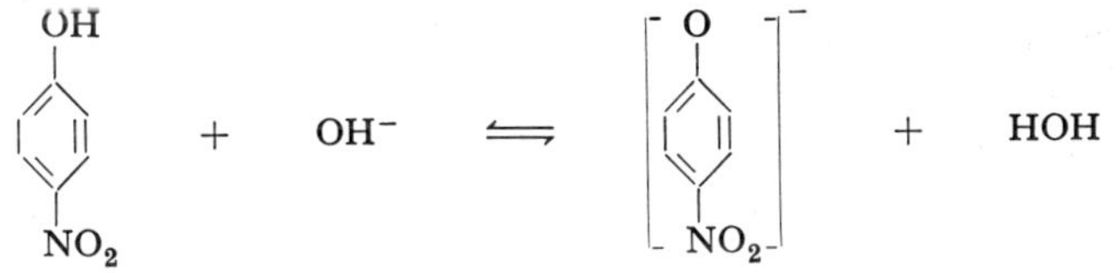

OH
$+$ OH^- $\rightleftharpoons$ O^- $+$ HOH

Colorless molecule in acid   Yellow ion in alkalies

p-Nitrophenol

molecule as shown here and the solution becomes yellow because of the freed negative ion. The salts of p-nitrophenol, like $[O_2N-C_6H_4-O]^-$ Na^+, are colored since they contain the negative ion.

The presence of color in the ion of p-nitrophenol as compared with lack of color in the acid itself is explained by the resonance theory. The explanation is similar to that employed in discussing the acidity of phenol in Section 25.12 (which the reader should review), where it was shown that phenol, C_6H_5OH, has resonance possibilities, but that resonance energy of the phenate ion, $C_6H_5O^-$, is even greater because of absence of the proton (H^+) from the molecule. Similarly, resonance of the negative ion of p-nitrophenol is more effective than that of the unionized molecule and tends to stabilize those polarized structures which are important in light absorption. Using the method of writing formulas as shown in Figure 21.2, a few resonance possibilities indicated in Figure 36.5 can be formulated for this negative ion. In the resonance resulting from ionization of p-nitrophenol there are quinoid structures which are suf-

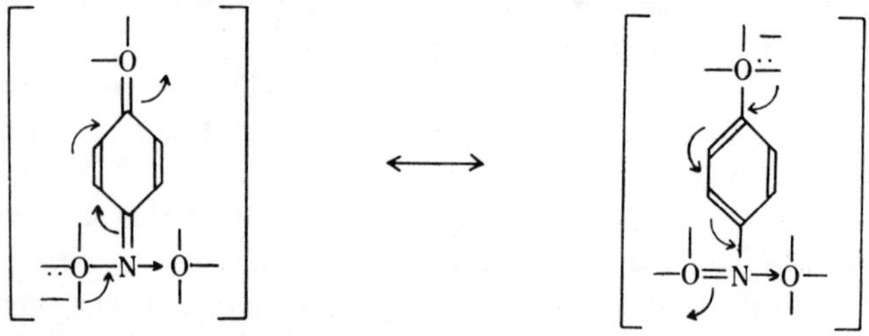

36.5. Some resonance structures of the negative ion of p-nitrophenol.

ficiently active toward the visible spectrum to produce color.

As illustrated by the preceding simple example, an indicator may be defined as an acid (or base) the ionization of which is accompanied by a change in the absorption of light in the visible spectrum. The ionization of different indicators takes place at different pH values—that is, at different concentrations of the H^+ and OH^- ions in solution. Indicators can, therefore, be used in analytical chemistry to determine the degree of acidity or alkalinity of solutions.

A more complex example of an indicator that is a weak acid is that of phenolphthalein. It is colorless in acid or neutral solution but at a certain concentration of OH^- ions in a weak alkali the two H^+ ions shown in its structure (Section 36.11) are removed. The resulting negative ion of phenolphthalein is doubly charged. When freed from the two H^+ ions, highly symmetrical structures of the quinoid type become possible that are in resonance and shift the absorption of light into the visible spectrum. This ion is red, whereas the original molecule is colorless. The nature of the two quinoid rings in the red negative ion of phenolphthalein in alkaline solution can be pictured by referring to the one quinoid ring shown in the structure of mercurochrome. Mercurochrome is red.

Although pararosaniline is not useful as an indicator, its use as a dye is similarly explained as the result of ionization. The color base of pararosaniline (Section 36.11) differs from the two phenol derivatives we have just described in that it is a weak base. It becomes highly colored when dissolved in acid solution, the color in this case being due to ionization in which the OH^- ion is lost and the colored positive ion is liberated. The structures in Figure 36.6 are written with the electron pairs represented by the scheme shown in Figure 21.2. When the molecule is freed from the blocking effect of the OH^- (refer to the structure in Section 36.11), resonance can take place as shown, the resonance structures being highly

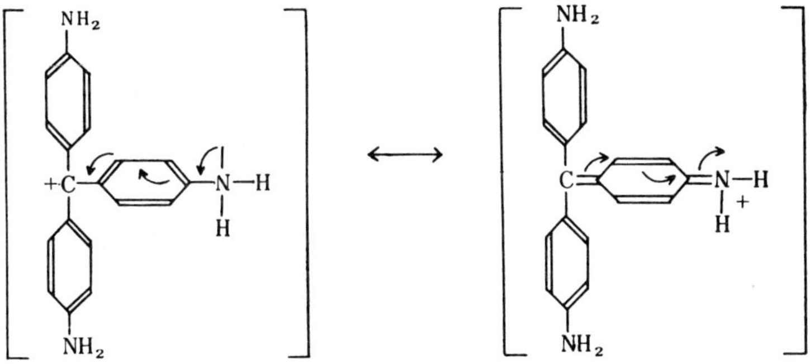

36.6. Some resonance structures of the positive ion of pararosaniline.

absorbent in the visible spectrum.

Methyl orange is an indicator which is both a weak base and a strong acid; its structure is probably not that shown earlier in the chapter in Section 36.10, but it is more likely an inner salt (dipolar ion) of the type described in Section 33.5, which the reader should review. On this basis, the unionized molecule of methyl orange, pictured in Section 36.10, can be rewritten as resonance structures that are present in acid solution and give it a red color (Figure 36.7). The reader should not be confused by the $\oplus$ and $\ominus$ charges in the molecule; it is an electrically neutral molecule, not an ion.

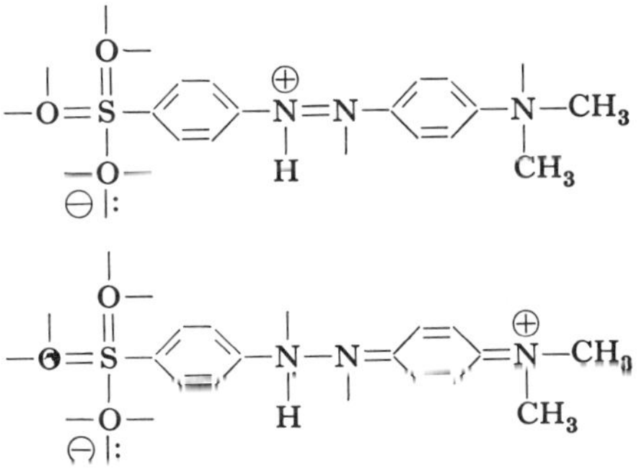

36.7. Resonance structures of methyl orange. Electron pairs are represented as in Fig. 21.2.

The need for the single positive charge on certain of the N atoms in Figure 36.7 can be shown by summing up the number of valence electrons actually on the atoms as compared with the *five* on the neutral atom of nitrogen. For example, in the upper structure of Figure 36.7 the N atom shown with a $\oplus$ charge is linked to other atoms by *four* covalence bonds, to each of which one electron is contributed by nitrogen; the equivalent of the fifth electron is on an oxygen atom at the end of the molecule. The other N atoms in the molecule have their normal quota of *five* (*three* in covalence bonds by which they are linked to other atoms, and *two* as shown by an unshared bond). Calculation of these *formal charges* is explained in Section 8.6.

When the methyl orange molecule does ionize (lose H^+ ion by reaction with a base, such as OH^-) the color turns yellow, the color in alkaline solution being due to:

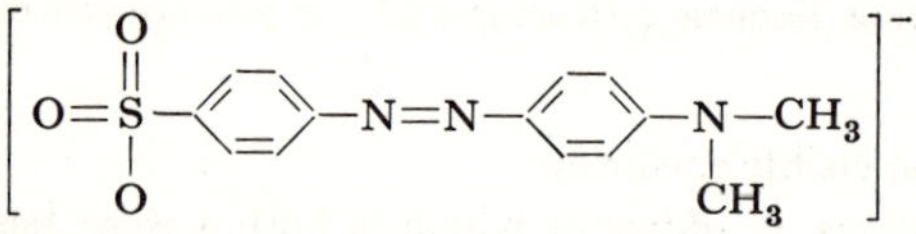

$$\left[O=\overset{\overset{\displaystyle O}{\|}}{\underset{\underset{\displaystyle O}{|}}{S}} - \underset{}{\bigcirc} - N=N - \bigcirc - \overset{}{\underset{\underset{\displaystyle CH_3}{|}}{N}} - CH_3 \right]^{-}$$

This negative ion has the simple azo structure as the molecule in Section 36.10, and we have written the structure in the normal manner, without indicating the shared or unshared electron pairs. Methyl orange is generally sold as the sodium salt, $O_3S\text{–}C_6H_4\text{–}N=N\text{–}C_6H_4\text{–}N(CH_3)_2{}^-$ Na^+.

37

Isotopic Chemistry

37.1. The Hydrogen Atom. Organic chemistry has been properly called the chemistry of carbon, but the observing reader will have noticed that it is almost as truly the chemistry of hydrogen. The hydrogen atom is found associated with carbon in most of the carbon compounds.

When the H atom (turn to Figure 6.1) is employed in the building of molecules, it appears in several modifications. Most important of these is the covalent bond, the normal valence bond, as in the molecule H:H. We have also seen how the hydrogen atom is used as a bridge to link strongly electronegative elements like O, N, and F (see, for example, Section 25.10), due to its ability to hold two electron pairs in what superficially looks like doubly bonded hydrogen.

The hydrogen atom can give rise to the H^+ ion as described in Section 27.6, and this ion can be used to form a hydronium ion (Section 25.3). The hydrogen atom also gives rise to the negative H^- ion, in the salt, lithium hydride, the formula of which is $Li^+:H^-$. The size of the H *atom*, relative to other atoms, is shown in Figure 6.2. The H^+ *ion* is a proton and is negligibly small. The H^- *ion*, however, has an effective diameter even larger than those of the F^- and Cl^- ions shown in black in Figure 6.2. In fact, the H atom may be considered not only as the first alkali metal in group 1 of the periodic table (inside cover), but also as the first halogen of group 7. This is shown by its tendency to add *one* electron to complete a valence shell (containing two electrons) in H:H and in :H^-. The halogens, also, add *one* electron to acquire stability by filling a valence shell (containing eight electrons).

The preceding paragraph sums up the chemistry of the H atom. As to the H:H molecule, it can exist in *ortho* and *para* forms, depending on the directions in which the two protons spin. In *ortho* hydrogen, the protons spin in opposing directions, but in *para* hydrogen, the spins are in the same direction. The *para* modification is produced when hydrogen is in contact with a charcoal surface at very low temperatures.

In 1932, a type of hydrogen was discovered with atomic weight 2, as compared with atomic weight 1 of ordinary hydrogen atoms. Later, a hydrogen with an atomic weight of 3 was found. Similarly, in addition to carbon with atomic weight 12, there are also carbon atoms with weights of 10, 11, 13, and 14. The rest of this chapter will be devoted to the role played in organic chemistry by these recently identified substances. However, we must first consider the meaning of the terms *atomic number* and *isotope*.

37.2. Atomic Number. The reader should refer once again to the schematic pictures of the atoms in Figure 6.1, where each element contains one more planetary electron than the preceding element. This means that each element contains one more proton in the nucleus than the preceding element. Hydrogen has one such proton, helium has two, lithium three, beryllium four, and so on.

The number of protons, or planetary electrons, is the atomic number, or element number. This number gives the sequence of the elements in the periodic system.

On the inside cover of this book will be found a chart of the elements which includes their atomic numbers as well as their atomic weights. According to the periodic law as originally stated, the chemical properties of the elements are a periodic function of their atomic weights. This law does not always hold, as becomes evident by examining the positions in the table occupied by argon and potassium with respect to their weights. Potassium is lighter than argon and according to its weight it should precede argon in the table, but that would put it in the wrong column of the table with respect to its chemical properties.

A more exact statement of the periodic law is that the properties of the elements are a periodic function of their atomic numbers. There are no exceptions to this rule; it will be observed that argon and potassium are placed in their proper positions in a table arranged according to atomic numbers.

37.3 The Hydrogen Isotopes. If we refer again to the pictures of the elements in Figure 6.1, we notice a gap between hydrogen and helium, with respect to atoms that may have atomic weights of 2 and 3. For some time, it was suspected that atoms may exist that contain the combinations of 1 proton + 1 neutron and 1 proton + 2 neutrons in the nucleus and occupy intermediate positions between hydrogen (atomic weight = 1) and helium (atomic weight = 4). These atoms can be represented diagrammatically as follows, using the same scheme as in Figure 6.1; the hydrogen structure is included for reference:

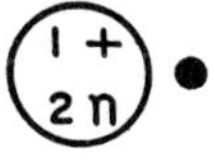

Protium, or Hydrogen (H) Atomic wt. 1 Atomic No. 1	Deuterium (D) Atomic wt. 2 Atomic No. 1	Tritium (T) Atomic wt. 3 Atomic No. 1

Since protons ($+$) and neutrons (n) have nearly the same weight, atoms D and T have weights of 2 and 3, respectively. However, they have only one planetary electron (atomic number 1) and they therefore belong *in the same place* in the periodic system of elements as hydrogen, which is atom number 1. The three kinds of hydrogen are called *isotopes*, from the Greek words *iso*, meaning the same, and *topos*, meaning place. In each family of isotopes, the atomic weights vary, but the number of planetary electrons, which controls the chemical properties, is the same.

The symbols used most generally for representing isotopes of an element can be illustrated by $_1H^1$, $_1H^2$, and $_1H^3$ for hydrogen, deuterium, and tritium, respectively. The lower numeral indicates the atomic number and the upper numeral is the atomic weight. On this basis, the symbol for a neutron is $_0n^1$ and for a proton is $_1p^1$. Often, the lower numeral is omitted, where obvious, as in H^3 for tritium.

The proportion of deuterium in ordinary hydrogen is 0.02%. Tritium is a radioactive isotope (Section 37.7); its concentration in ordinary hydrogen has been estimated at about one part in 10^{18} parts of hydrogen. The term *heavy hydrogen* applies to deuterium. When isotopes were first recognized, they were defined as atoms with different atomic weights but *identical* chemical properties. This is nearly true for most families of isotopes but is far less true for the hydrogen isotopes, which differ markedly in properties because of the great difference in weight. Since deuterium is twice as heavy as hydrogen there is a considerable variation in physical properties and chemical reactivity, not only of the elements but also of their compounds. With respect to physical properties, for example, hydrogen (H_2) boils at 20.38°K, whereas deuterium (D_2) boils at 23.50°K. Similarly, water (H_2O) has a freezing point of 0.00°C and boiling point of 100.00°C, whereas heavy water (D_2O) freezes at 3.82°C and boils at 101.42°C.

The difference in chemical reactivity is not of *kind* but of *degree*. For example, both water and heavy water react with aluminum carbide,

$$12H_2O + Al_4C_3 \longrightarrow 4Al(OH)_3 + 3CH_4 \text{ (methane)}$$

$$12D_2O + Al_4C_3 \longrightarrow 4Al(OD)_3 + 3CD_4 \text{ (deutero-methane)}$$

but ordinary water reacts twenty-three times as fast as heavy water. (There is one molecule of heavy water in about five thousand molecules of ordinary water.)

It has been found that in water solution, the removal of D^+ ions from acids like deuteroacetic acid, CH_3—COOD, is more difficult than separation of H^+ ions from hydrogen acids like acetic acid, CH_3–COOH. This is proved by the smaller ionization constant of the deutero acid. In other words, the D atom is less mobile than the H atom.

37.4. Isotope Exchange Reactions. When acetylene is passed into heavy water containing an alkaline catalyst it is found that the H atoms are replaced by D atoms:

$$H{-}C{\equiv}C{-}H + D_2O \longrightarrow D{-}C{\equiv}C{-}D + H_2O$$

This is called an exchange reaction. Only those compounds in which the H atoms tend to split off as ions go through this type of exchange. The fact that acetylene is capable of ionization was stated in Section 13.13, where it was observed that acetylene is capable of forming salts. The exchange mechanism is to be explained by the fact that the water solution contains both H^+ ions and D^+ ions, both of which can attach to the carbon atom. Since the molecule C_2D_2 forms in preference to the C_2H_2 molecule it is evident that the C–D bond is chemically more stable than the C–H bond. The H atoms in compounds like methane (CH_4) and benzene (C_6H_6) are not ionizable and are not replaceable by heavy hydrogen atoms in such exchange reactions.

Ketones, also, give this exchange reaction, In Section 28.11 it was explained that H atoms on carbon adjacent to a C=O group are acidic, or labile. In an alkaline heavy water solution all six H atoms in acetone are replaced by D atoms. The related compound, diethyl ketone, under the

$$\underset{\displaystyle O}{\overset{\displaystyle ||}{CH_3{-}C{-}CH_3}} \qquad\qquad \underset{\displaystyle O}{\overset{\displaystyle ||}{CH_3{-}CH_2{-}C{-}CH_2{-}CH_3}}$$

Acetone   Diethyl ketone

same conditions, will exchange only four of its ten hydrogen atoms, since only four are on carbon atoms *adjacent* to C=O. The product is CH_3–CD_2–CO–CD_2–CH_3.

37.5. Kinetic Isotope Effect. If sugars are dissolved in heavy water with appropriate catalysts, the H atoms in C–OH but not in C–H can be exchanged for D atoms. In Section 34.4, some of the possible structures of the glucose molecule in solution were pictured and it was explained that the straight-chain form can change into a cyclic form as the result of migration of the H atom from a certain –OH group from one position to another. It has been observed that the velocity of mutarotation of α-glucose (Section 34.11) is considerably lower in heavy water than in ordinary water. The mutarotation of glucose is a process slow enough to be followed; when H is replaced by D, its isotope, the process is even slower. This is called the isotope effect, or kinetic isotope effect, because it is evident in the changed kinetics of a chemical reaction. An illustration of the use of this principle is furnished by the substitution reactions in benzene summarized in Section 20.5. In the nitration of benzene, C_6H_6, the rate at which the intermediate σ complex breaks up is the same as in the nitration of deuterobenzene, C_6D_6. It is not, therefore, a rate-determining, slow step because if it were slow with C_6H_6 it would be much slower with C_6D_6 and easily detectible. On the other hand, in the sulphonation reaction, the intermediate σ complex decomposes slower with C_6D_6 than with C_6H_6. Since the removal of a D^+ ion from the σ complex by a base is slower than the removal of an H^+ ion, this must be the slow step in the sulphonation reaction. Through this isotope effect it is possible to determine which step in a reaction is slow enough to be called rate-determining.

37.6. Isotope Indicators. An atom like heavy hydrogen is often referred to as a "tagged" atom in that we can replace ordinary hydrogen with it, and since it is quite different from ordinary hydrogen, we can follow it in the course of a reaction. It is also called a "tracer" or "indicator" atom.

The problem in photosynthesis is to determine how CO_2 and H_2O are employed in biochemical reactions to build up complex carbon compounds with the aid of the energy supplied by light. The theory of photosynthesis with most adherents in the past was based on the assumption that plants synthesize carbohydrates by means of an intermediate synthesis of formaldehyde:

$$CO_2 + H_2O \longrightarrow H\text{—}CHO + O_2$$

This is a reduction process, in which it was thought that the liberated O_2 comes from the CO_2. This theory has been tested by means of "tracer" oxygen atoms with an atomic weight of 18. Although systematically we should differentiate between the light and heavy oxygen atoms by the symbols $_8O^{16}$ and $_8\mathbf{O}^{18}$, in this discussion we shall simply indicate them by O and **O**, respectively.

Experiments have been conducted in which plants were fed CO_2 and H_2O containing light or heavy oxygen atoms, with results as indicated below:

$$CO_2 + H_2\mathbf{O} \longrightarrow \text{some plant product} + \mathbf{O}_2$$
$$CO_2 + H_2O \longrightarrow \text{some plant product} + O_2$$

(Note: these are not balanced equations)

It was thus proved that in photosynthesis characterized by liberation of oxygen the oxygen liberated originates from the water and not from the carbon dioxide as previously assumed. The oxygen from the carbon dioxide is built up into the plant structure.

37.7. Radioactive Isotopes. In Section 37.3, we described three isotopes that correspond to element number 1. Of these, two are *stable*, namely, hydrogen and deuterium. The third isotope, tritium, is

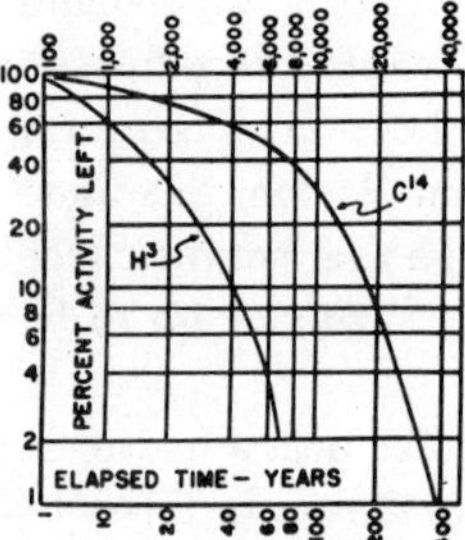

37.1. Decay curves for tritium H^3 and for C^{14}. The scale is logarithmic. *Half-life* is time at which 50% of activity is left. This is 12.4 years for H^3 and 5,400 years for C^{14}. Activity is insignificant after about $5 \times$ half-life.

unstable and is called the *radioactive* isotope; when the nucleus of this atom breaks down, it eliminates a charged particle which is an electron (*e*), also referred to as a *beta particle* or *beta ray* (β^-). If a certain amount of tritium is available at a given time, half of this amount will have decomposed in 12.4 years; this is called the *half-life* of the isotope.

The reaction for the decomposition of tritium is

$$_1H^3 \longrightarrow {_2He^3} + {_{-1}e^0}$$

and can be written down readily when we know the nature of the particle thrown out. The symbol $_{-1}e^0$ for an electron distinguishes it from a positron, which is $_{+1}e^0$. In the symbol $_1H^3$ the lower numeral—the element number—indicates the number of protons in the nucleus. The upper numeral is the weight of the atom (protons and neutrons in the nucleus).

Elimination of an electron causes no change in weight, but the decomposition of a neutron that results in elimination of the electron sets free a proton $(_0n^1 \rightarrow {_1}p^1 + {_{-1}}e^0)$. This results in an element $_2/^3$ with a nuclear charge of two protons and an atomic weight of 3, and from Figure 6.1 this must be an isotope of helium; its nucleus has two protons and one neutron, and its symbol is $_2He^3$.

The trace amounts of tritium present in ordinary hydrogen probably originate from nuclear processes occurring in the atmosphere. For use in chemical experiments tritium can be made in weighable amounts from lithium. The shorthand expression for this reaction is Li^6 $(n,\alpha)H^3$, which means that when lithium of atomic weight 6 is subjected to bombardment by neutrons (n) it eliminates helium ions (α) and is transformed into tritium. The more conventional equation is

$$_3Li^6 + {_0}n^1 \longrightarrow {_2}He^4 + {_1}H^3$$

It should be observed that the upper numerals (weights) balance on the two sides of this reaction, and that this is also true of the lower numerals (positive charges). The *alpha particle* (α) is the helium ion He^{++} rather than the helium atom, but this is not evident in the *nuclear* reaction written here, which is not concerned with loss of *planetary* electrons resulting in ionization.

Radioactive atoms can be used like ordinary atoms because their chemical reactivities are practically identical. Like the deuterium atoms and the heavy oxygen atoms discussed earlier in the chapter, the radio atoms are useful as "tracer" atoms in the study of reaction mechanisms.

Since it is possible to detect an atom through its radioactivity about a million times more easily than by chemical analysis, the ability to make and use artificially radioactive elements is one of the most important advances in recent years. To illustrate, it is possible to feed a compound containing radioactive atoms to animals, insects, or plants and to determine into what organs the compound penetrates and exactly to what extent. Its travel can be followed by the charged particles which it emits.

37.8. The Isotopes of Carbon. There are five members of the carbon family of isotopes, as mentioned in Section 37.1. Ordinary carbon contains 98.9% of the stable isotope C^{12} and 1.1% of the stable isotope C^{13}. The second is known as *heavy carbon*. The presence of heavy carbon in ordinary carbon accounts for its atomic weight of 12.01 as listed in the chart on the cover of this book.

There are three radioactive isotopes of carbon—namely, C^{10} with a half-life of 8.8 seconds, C^{11} with a half-life of 20.5 minutes; and C^{14}, the half-life of which is 5,400 years. It is obvious that only this last isotope can have extensive use as a radioactive tracer atom, although C^{11} has been employed by investigators near the source of a supply of the material. The decomposition of C^{14} atoms takes place with elimination of beta rays. The use of C^{14} as a tracer atom is more convenient than that of C^{13}; the first requires a comparatively simple detection instrument, such as the Geiger-Müller counter, whereas the second requires the more complicated mass spectrometer.

The advent of chain-reacting piles in World War II, with their abundant neutron supply, makes the manufacture of C^{14} rather simple. One procedure is the nuclear reaction $_7N^{14}(n, p)_6C^{14}$. Ammonium nitrate (NH_4NO_3) solution in a suitable container is placed in the pile and the neutron bombardment causes a nuclear transformation in which protons are ejected and nitrogen atoms are changed into carbon. The product can be worked up by the usual syntheses to yield organic compounds with the labeled carbon atom.

There are many problems in reaction mechanism that can be solved with C^{14} tracer atoms. We showed how CO_2 containing heavy oxygen was used in the study of the fate of oxygen in photosynthesis by plants. Similarly, the use of $C^{14}O_2$ will indicate the ultimate disposition of the carbon atom. Very wide use is being made of C^{14} in studying metabolism in the animal body, according to principles described in Section 35.2. If a labeled substance is fed to an animal, the identification of the excreted compounds containing the tracer atom furnishes a clue as to the chemical reactions involved during its progress through the body.

It is probable that C^{14} is produced by nuclear reactions in the upper atmosphere and subsequently finds its way into plant and animal life. It is apparently present in all living things. Since the half-life is about 5,400 years it would be difficult to detect the radioactivity from samples more than about 40,000 years old, but the age of geologic deposits involving carbonaceous materials can be estimated from the radioactive carbon content if they are not older than that.

38

Giant Molecules

38.1. Introduction. The compounds of carbon can vary in size between wide limits from small molecules like methane (CH_4) to the very large diamond molecule. The student should review Figure 19.2 which shows how the carbon atoms in a diamond crystal are linked in a three-dimensional network, the diamond crystal being one giant molecule. We have also discussed the structure of graphite (Figure 19.3) which consists of parallel, giant sheet molecules; the distance between these parallel sheets is such that they are nearly out of range of intermolecular attraction forces. In Section 10.3 it was said that a straight-chain paraffin hydrocarbon has been made with 100 carbon atoms in the chain; its formula is $C_{100}H_{202}$ and its molecular weight is about 1402. It is a very large molecule indeed, but small in comparison with most of those described in this chapter, which fall in the category of giant molecules. Two families of giant molecules have already been considered in detail; these are the proteins (Section 33.6) and the nucleic acids (Section 35.1), which were sufficiently described in the earlier chapters.

38.2. Giant Carbohydrates. These substances have molecular weights of the order of 50,000 to more than 1,000,000, and are built up of many units of a simple sugar.

38.2a. Cellulose is a hexosan (Section 34.20). In cellulose, the C_1 atom of a β-glucose ring is joined to the C_4 atom of the next ring through an oxygen bridge. Alternate β-glucose rings are reversed, thus,

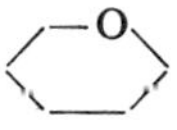

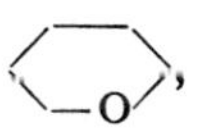

and it should be observed in Figure 38.1 that this yields a pattern in which the β-glucose units zigzag along an essentially straight line. The numbering of the atoms in the ring should be compared with that in Figure 34.10a, where it was indicated that the C–OH bonds on C_1 and C_4 are *parallel*, or *equatorial*; this is an important factor in the ability to build a *linear* cellulose model.

455

38.2b.   *Starch*, like cellulose, has a C_1 atom of a glucose ring joined to the C_4 atom of the next ring through an oxygen bridge. However, the α-glucose ring is the structural unit, and it was shown in Figure 34.9*a* that the C_1–OH bond is *axial*, or *perpendicular* to the C_4–OH bond. When α-glucose units are joined end to end by elimination of HOH this perpendicularity of the two C–OH bonds forces the straight chain into a constant curvature, as illustrated by Figure 38.2 and by the lower model of Figure 38.3.

A starch contains two types of structures; one is a supposedly linear molecule, known as *amylose*, which has a molecular weight range of 10,000 to 100,000 and does not form a paste with water. The other structure in starch is called *amylopectin*; it is present in greater amount than amylose, has a molecular weight of more than 300,000, is highly branched, and forms a paste with water. The linear starch structure is also known as the *A-fraction*, and the branched structure as the *B-fraction*. The A-fraction is soluble in hot water, but precipitates from solution at ordinary

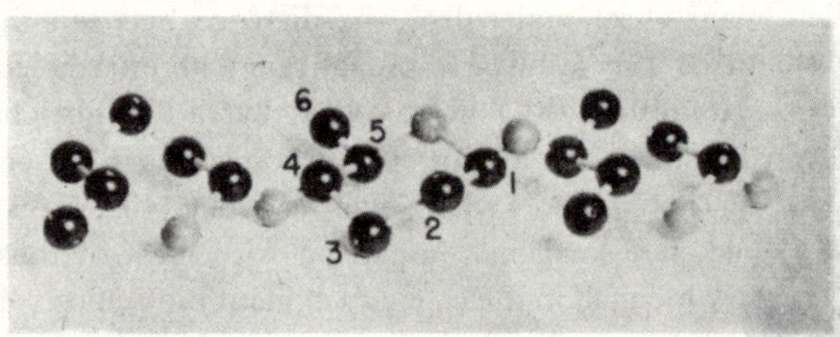

38.1.   Model of cellulose, showing three β-glucose rings as in Fig. 34.10*a*. Atoms and groups attached to the rings are not shown. The carbon atoms are numbered as before, and Ox is oxygen. The chain is tilted slightly toward the reader. Note the positions of the 1,4 oxygen bridges between rings.

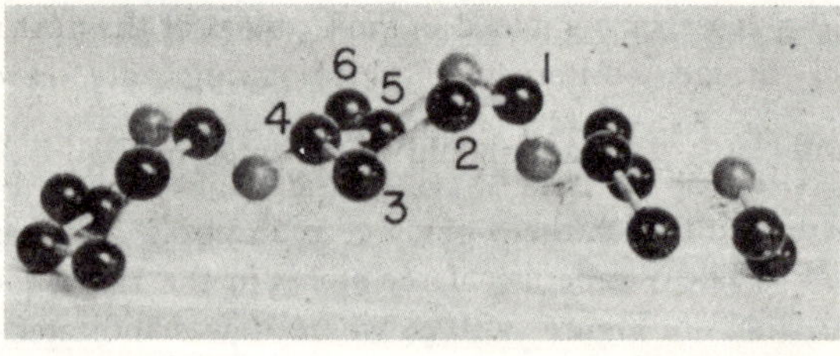

38.2.   Model of the A-fraction of starch, showing three α-glucose rings as in Fig. 34.9*a*. (See comments under Fig. 38.1.)

38.3. The upper model is the probable cellulose structure. The lower model is a suggested structure for the A-fraction of starch. [From G. V. Caesar and M. L. Cushing, *J. Physical Chemistry*, **45**, 776 (1941).]

temperatures. As already stated, this chain molecule does twist on its long axis.

The extensive branching of the molecules in the B-fraction of starch results in a three-dimensional spatial configuration. In this structure a chain of α-glucose units sends out branches which, in turn, branch out just like a tree. The branch chains probably originate through oxygen links at the C_6 atom. The primary difference between *glycogen*, the animal starch (Section 34.19), and the B-fraction of starch is believed to be one of chain length. In glycogen each branch contains about ten glucose units, whereas in the B-fraction of starch they appear to be about twenty-five to thirty glucose units in length. The properties of starch paste are currently ascribed to this interlacing network of branch chains. The ratio of the A-fraction to the B-fraction in a starch depends on its source.

38.2c. Xylan is a pentosan (Section 34.21) built up from many units of xylose. It will be seen from Figure 34.1 that xylose is exactly like glucose except for the absence of the 6th carbon atom. A model of xylan would be similar to that of starch in Figure 38.2 except for two things: the C_6 atom is missing, and the linkage between xylose units is 1, 3 rather than 1, 4 as it is in starch. The 1, 3 linkage is a very important difference; the molecule becomes helical, but three of these helices mesh to form a rope-like straight fibril, the molecules being held together by hydrogen bonds, and the bonding being favored by wetness. This 1, 3 link in xylan was discovered in the early 1960's. Xylan with its 1,3 link is the principal structural carbohydrate in the cell wall of warm water algae, corresponding in importance to cellulose (with its 1,4 link) in land vegetation.

38.3. The Isoprene Skeleton. Isoprene is related to 1,3-butadiene, a compound to which we have already given some attention because of its characteristic arrangement of single and double bonds (section 15.7). The composition of isoprene is C_5H_8, its structure is $CH_2\!=\!C\!-\!CH\!=\!CH_2$, CH_3

and its systematic name is 2-methyl-1, 3-butadiene. It is a liquid which boils at 34°C, and it may be prepared by the dry distillation of India rubber or by passing turpentine through a red-hot tube. Nature has used the isoprene skeleton, C–C–C–C, in the design of an enormous number of compounds of vital importance to man. These include the giant molecule, rubber; red and yellow pigments of vegetables and egg yolks; essential oils like those in oil of turpentine; and vitamin A.

The union of isoprene molecules to form polymers can be visualized as analogous to the 1,4-addition reaction described in Section 17.6, although nature definitely does not perform the job as simply as this:

$$CH_2\!=\!C\!-\!CH\!=\!CH_2 + CH_2\!=\!C\!-\!CH\!=\!CH_2 \longrightarrow$$
$$\dots CH_2\!-\!C\!=\!CH\!-\!CH_2\!-\!CH_2\!-\!C\!=\!CH\!-\!CH_2 \dots$$

The meaning of the word *polymer* can be understood from the definition of *isomer* in Section 10.4. A molecule exactly twice as big as isoprene is a dimer of that compound, one three times as big is a trimer, and so on. A few of the structural arrangements built up from two isoprene skeletons are shown in Figure 38.4. The first compound is open-chain, the second is cyclic, and the third is bicyclic. All have two isoprene (C_5H_8) units, which the student should try to pick out for himself. The isoprene dimers are called *terpenes*, and on that basis isoprene is a hemiterpene, and compounds with three isoprene units are sesquiterpenes (sesqui is Latin for one and one-half, and a C_{15} molecule is that much bigger than C_{10}). In many cyclic terpenes a C–H group is found oxidized to C–OH, and in some cases a C=O group appears; one such compound is *camphor*.

Among important polyterpenes are the *carotenes* (yellow pigments of

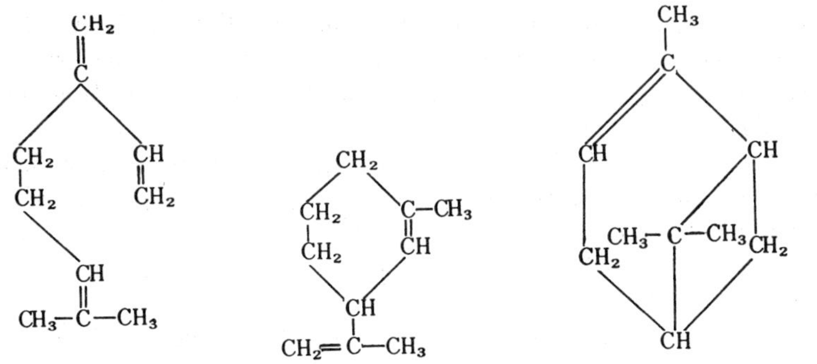

Myrcene, $C_{10}H_{16}$
Occurs in oil of bay

Sylvestrene, $C_{10}H_{16}$
Chief constituent of
Swedish and Russian
oil of turpentine.

Pinene, $C_{10}H_{16}$
Chief constituent of
most varieties of oil
of turpentine.

38.4.   Some terpenes built from two isoprene skeletons.

carrots and other vegetables), and many related compounds built up from 8 isoprene skeletons. There is ample evidence that animals convert a molecule of β-carotene into two molecules of vitamin A; the carotene molecule is split exactly in the center, and at the break the CH groups are oxidized to CH_2OH, so that vitamin A is an alcohol with the composition $C_{20}H_{29}OH$.

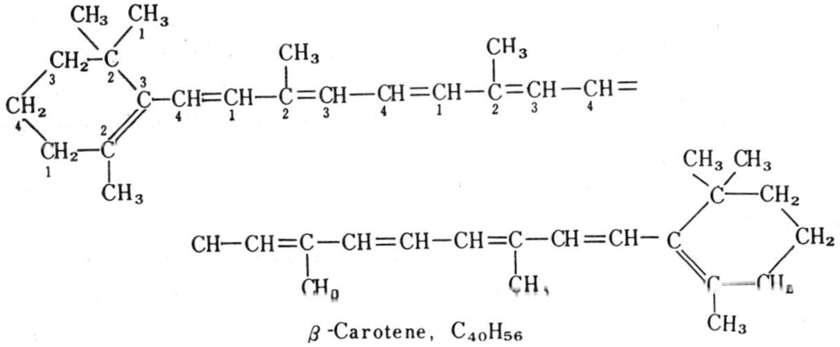

β-Carotene, $C_{40}H_{56}$

38.4. Rubber. The bark of some tropical plants, when cut, exudes a white milk called latex. When the latex is coagulated, crude rubber is obtained. This crude rubber used to be known as caoutchouc, meaning in Indian a "tree which weeps." The latex of certain of these plants contains a substance which differs somewhat from rubber and is known as gutta-percha. Both rubber and gutta-percha have the empirical compo-

sition $(C_5H_8)_n$, where *n* is a very large number. They are both polyterpenes, but whereas the isoprene polymers described in the preceding
section have two to eight isoprene units, these are built up from about
10,000.

Gutta-percha (which is only slightly elastic) has the *trans* configuration shown in the accompanying formulas, whereas rubber (which is quite
elastic) has the *cis* configuration. The *cis, trans* isomerism is based on the

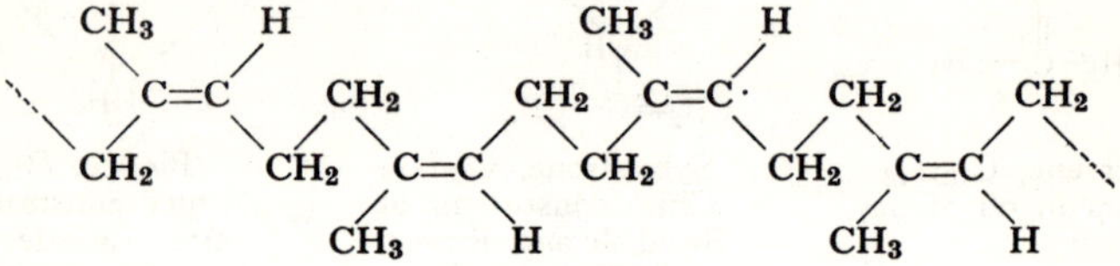

Rubber (cis)

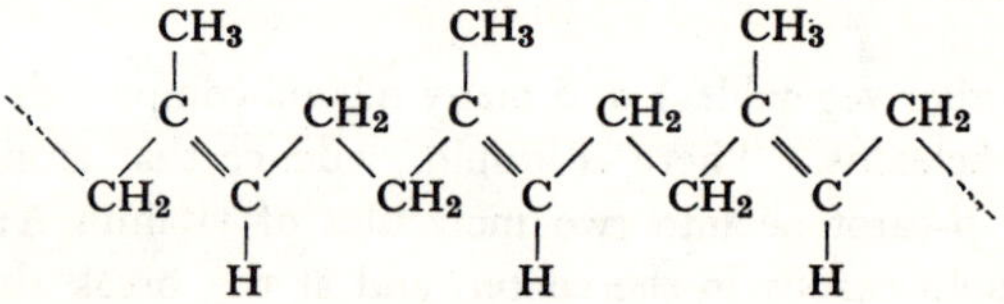

Gutta-percha (trans)

positions of the CH_3 and H units; in rubber they are on the same side of
the double bond (*cis*), and in gutta-percha they are on opposite sides
(*trans*). Freedom of rotation around the single bonds is much greater in
the rubber molecule, resulting in greater flexibility, and the chain is somewhat crumpled. There is a greater regularity in an array of gutta-percha
molecules, similar to that in crystals. Gutta-percha in its α form is hard
and tough, but undergoes a transition at about 70° to a softer β form
which resembles rubber in its pliability. Both forms of gutta-percha are
trans.

The commercial substance we know as rubber is the product obtained
by heating it with sulfur. This process is vulcanization (Section 38.9).
When gutta-percha is vulcanized it is hard enough to be used for golf
ball covers. Efforts to polymerize isoprene in the stereoregularity that
marks the substances in nature did not succeed until 1954, when a true
synthetic rubber was made in this country and a true gutta-percha in Italy.

38.5. Structural Units. We have now described several important
substances that are found in nature and are built up of smaller molecules
discussed in considerable detail in this book. The giant molecules are
often referred to as "high polymers" or "high polymeric substances"

because they contain a great many of the same structural units. Structural units are also called "recurring" units or "repeating" units. The simple molecule that yields the recurring unit of the giant molecule, by rearrangement of the valence bonds or by other methods to be described in the following sections, is referred to as the "monomer," meaning one part.

Addition polymers. The monomer of the rubber molecule has exactly the same composition as the structural unit—that is, C_5H_8. For that reason, the ratio of atoms in the final polymer is the same as in the monomer, although the value of n may vary as represented by $(C_5H_8)_n$. When a high polymer is made by the simple addition of molecules to each other, either by a shift in the double bonds as in the case of isoprene or by any other valence rearrangement, the phenomenon is called polymerization by association or addition.

Condensation polymers. The synthesis in nature of giant molecules like starches and proteins may be regarded as one in which two molecules are joined by elimination of one molecule of water, as shown in Sections 33.3 and 34.14. This is known as polymerization by condensation. The structural unit in the resulting polymer does not have the same composition as the monomer:

$$nC_6H_{12}O_6 \longrightarrow nH_2O + (C_6H_{10}O_5)_n$$

Glucose   Starch

(monomer)   (polymer)

That this brief equation is not entirely correct is evident from Figure 38.5, which shows that the end-groups must be different from the repeating units inside the molecule. This is not a significant error, however, when there are hundreds or thousands of repeating groups. · It should be observed that the terminal units in a starch molecule are not identical. The one on the right still has an anomeric carbon atom (*), which is a potential aldehyde group, but the unit on the other end of the molecule does not. An important point that the student should understand is that the addi-

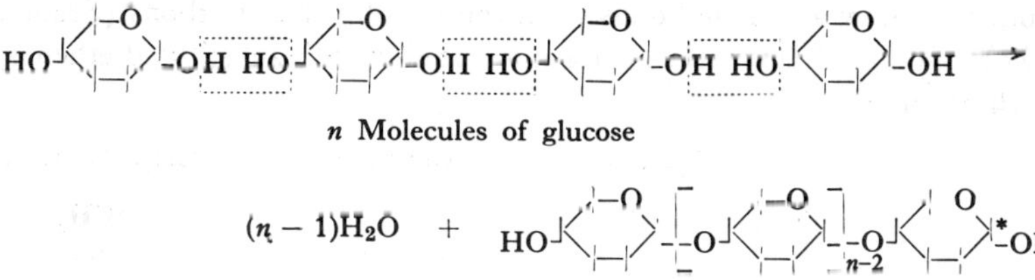

38.5. Hypothetical formation of starch from α-glucose. Only those atoms are shown which pertain to the discussion.

tion polymers are the true polymers and that the term polymer is used loosely in connection with condensation polymers.

38.6. Polymerization by Condensation. Although the chemist is not able to make starch, he has succeeded in making many other polymers on a large scale by the condensation process. A simple example of a reaction of this type is the following:

$$HOCH_2CH_2OH + HOCH_2CH_2OH + HOCH_2CH_2OH \longrightarrow$$
$$2H_2O + HOCH_2CH_2OCH_2CH_2OCH_2CH_2OH$$

Ethylene glycol　　　　　　　　　　　　　　　　Triethylene glycol
Ethene glycol
Ethanediol

The reaction can be made to proceed to the formation of *polyethylene glycols* in the molecular weight range 200 to 1,000, which are liquids, or to the range 1,000 to 7,000 in which they are solids known as *Carbowax* compounds. They are used in ointments, hair dressings, etc. Because of its hygroscopic nature, triethylene glycol is used in air-conditioning systems; in the form of an aerosol it is an atmosphere disinfectant.

A related reaction was described in Section 30.13, in connection with the *silicones*, which also are condensation polymers. The silicones can be varied considerably in character by altering the R groups and by other means. They are generally more stable, especially to high temperature, than the carbon polymers and are used for gasket materials as well as for the manufacture of oils and elastomers. Other examples of condensation polymers are the proteins (Section 33.3), Dacron and the nylons (Section 38.10), and Bakelite (Section 38.11b).

38.7. Polymerization by Addition. The important polymers made by this procedure result from the tendency of the olefin double bond to add to itself:

$$nCH_2{=}CH_2 \longrightarrow \text{end group}{-}(CH_2{-}CH_2)_n{-}\text{end group}$$

Ethene (Ethylene)　　　　　　　　　　　Polyethylene

The product is generally inert chemically, because the carbon-carbon bonds are now all single bonds as in the paraffin hydrocarbons (Section 10.6). Polymerization takes place more readily with substituted ethenes, such as these:

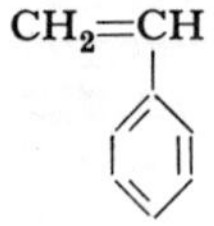

CH$_2$=CH	CH$_2$=CH	CH$_2$=CH	CH$_2$=C—CH$_3$
(phenyl)	CN	Cl	CH$_3$
Styrene	Vinyl cyanide Acrylonitrile	Vinyl chloride	Isobutene Isobutylene

Styrene, in fact, forms polymers simply on standing in its storage container. However, catalysis is usually resorted to for polymerization, and two general mechanisms will be described in this section.

38.7a. Free-Radical Chain Reaction. This type of reaction is broken down into initiation, propagation, and termination as discussed in Sections 17.9 to 17.11. Hydrogen peroxide may be used as a source of free radicals, to act as initiators, by its simple decomposition: $HO{:}OH \rightarrow 2HO\cdot$; to illustrate this mechanism we shall write the olefin double bond, I, in the dot-dot, diradical form (Section 17.10), as shown in II. A free radical fragment unites with II to give III, which now in turn is a

$$\underset{\text{I}}{CH_2{\cdots}\overset{\overset{\displaystyle R}{|}}{C}H} \longrightarrow \underset{\text{II}}{\underset{\bullet}{CH_2{-}}\overset{\overset{\displaystyle R}{|}}{C}H} \xrightarrow{+\ HO\cdot} \underset{\text{III}}{HO{-}CH_2{-}\underset{\bullet}{\overset{\overset{\displaystyle R}{|}}{C}H}} \xrightarrow{+\ II}$$

$$\underset{\text{IV}}{HO{-}CH_2{-}\overset{\overset{\displaystyle R}{|}}{C}H{-}CH_2{-}\underset{\bullet}{\overset{\overset{\displaystyle R}{|}}{C}H}}$$

$$\underset{\text{V}}{HO{-}CH_2{-}\overset{\overset{\displaystyle R}{|}}{C}H{-}CH_2{-}\overset{\overset{\displaystyle R}{|}}{C}H{-}\overset{\overset{\displaystyle R}{|}}{C}H{-}CH_2{-}\overset{\overset{\displaystyle R}{|}}{C}H{-}CH_2{-}OH}$$

$$\underset{\text{VI}}{HO{-}CH_2{-}\overset{\overset{\displaystyle R}{|}}{C}H{-}CH_2{-}\overset{\overset{\displaystyle R}{|}}{C}H_2}$$

free radical and reacts with more of II to yield IV. This lengthening of the chain is the propagation phase. Termination can occur in two ways. There is a *mutual* termination when a propagating chain unites head-to-head with another propagating chain, as exemplified by V, which consists of two moles of IV. There is a *chain transfer* termination when the propagating free radical chain extracts an H atom from some other molecule in its vicinity to give VI. The "other" molecule may be another chain or a chain-transfer agent added for this special purpose to the reaction mixture as a chain stopper. If the H atom for VI is extracted from another chain, that other chain has a new free-radical active center which can start to propagate; this results in chain branching, a phenomenon which is largely eliminated when using the Ziegler catalysts mentioned in Section 38.8.

The molecular weight of an addition polymer can vary from about 20,000 to more than a million. At the ends of the long chain there may be fragments of the catalyst. Ethylene (ethene) is rather difficult to convert to polyethylene (CH_2–CH_2)$_n$ by this mechanism; it requires a pressure of 1,000 atmospheres and high temperature (200°C) with a little oxygen as catalyst (see the diradical structure of oxygen in Section 24.10). Polyethylene is made with better properties and more easily by means of the Ziegler catalysts, without the need for high pressure and temperature.

38.7b. Ionic Chain Reaction. In Section 17.10 it was said that in certain reactions it is convenient to regard ethene as a dot-dot structure, as just described in *a*, whereas in certain other reactions it is to be considered as a polarized structure in which the π electron pair is drawn to one end of the molecule. Many ethene derivatives can be polymerized ionically by acids, bases, and salts, and in such reactions we write the C=C bond in its polarized version. Isobutylene (isobutene) is one of the olefins that polymerizes readily by an ionic mechanism. The two CH_3 groups in isobutylene have electron-donor properties (Section 21.5) that promote

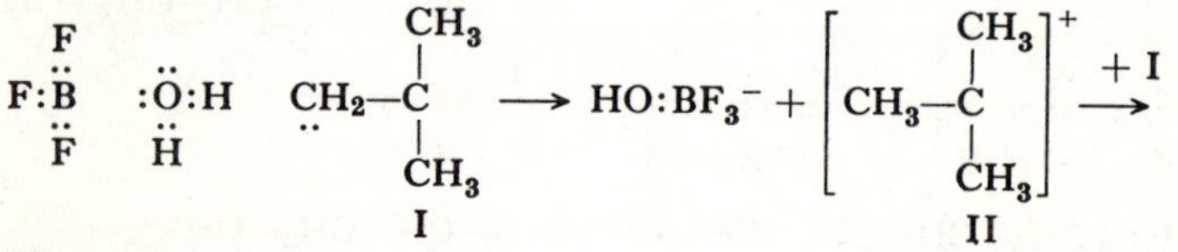

$$\begin{array}{l} F \\ F{:}B \quad {:}\ddot{O}{:}H \quad CH_2{-}\!\!\overset{\displaystyle CH_3}{\underset{\displaystyle CH_3}{C}} \quad \longrightarrow \quad HO{:}BF_3^- \;+\; \left[CH_3{-}\!\!\overset{\displaystyle CH_3}{\underset{\displaystyle CH_3}{C}}\right]^{+} \xrightarrow{\;+\,I\;} \\ F \quad\; H \end{array}$$

$$\text{I} \qquad\qquad\qquad\qquad \text{II}$$

Water
Boron trifluoride
Isobutylene

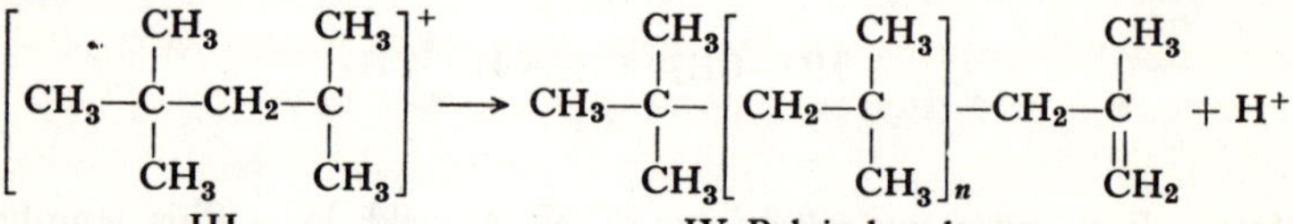

$$\left[CH_3{-}\!\!\overset{\displaystyle CH_3}{\underset{\displaystyle CH_3}{C}}{-}CH_2{-}\!\!\overset{\displaystyle CH_3}{\underset{\displaystyle CH_3}{C}}\right]^{+} \longrightarrow CH_3{-}\!\!\overset{\displaystyle CH_3}{\underset{\displaystyle CH_3}{C}}{-}\!\!\left[CH_2{-}\!\!\overset{\displaystyle CH_3}{\underset{\displaystyle CH_3}{C}}\right]_n\!\!{-}CH_2{-}\!\!\overset{\displaystyle CH_3}{\underset{\displaystyle CH_2}{C}} \;+\; H^+$$

$$\text{III} \qquad\qquad\qquad\qquad \text{IV Polyisobutylene}$$

shift of the π electron pair to the CH_2 group at the other end of the molecule (see I). This is also promoted by a catalyst like BF_3, which is a Lewis acid (Section 27.6) with room for an electron pair on the B atom. The polarizing effect of BF_3 is aided by a trace of water ($H{:}\ddot{O}{:}H$); the overall catalytic effect is similar to that of $AlCl_3$ in the attack of chlorine on the benzene ring described in Section 20.2.

The reactive carbonium ion, II, is the complex at which the chain reaction starts by adding one monomer at a time as in III. No matter how long the chain grows in the propagation phase it is still a reactive carbonium ion. Termination is shown schematically in IV; it probably takes place by elimination of a proton, which renews the catalyst by reacting with the negative ion shown with II.

38.8. Stereoregularity. An important consideration in addition polymerization is the regularity with which monomers add to each other. As just observed, in polyisobutylene synthesized by an ionic mechanism the monomers are joined head-tail-head-tail, etc. Monomers may join in other sequences, and a haphazard arrangement (with branching) is generally likely to take place when polymerization is by a free radical mechanism. Perhaps even more important than this type of regularity is the regularity with which the side chains appear on the backbone of carbon atoms in the polymer chain. This space regularity is known as *tacticity* (from a Greek word meaning arrangement), and is illustrated in Figure 38.6, where the molecules of the polymers are stretched out so that the carbon atoms in the backbone are in the plane of the paper. The models in Figure 38.6 may be taken to represent polypropylene,

$$
n \; \begin{array}{c} H \quad CH_3 \\ | \quad\; | \\ C = C \\ | \quad\; | \\ H \quad H \end{array} \;\longrightarrow\; \text{end group} - \begin{bmatrix} H \quad CH_3 \\ | \quad\; | \\ C - C \\ | \quad\; | \\ H \quad H \end{bmatrix}_n -\text{end group}
$$

with the CH_3 groups indicated by R.

If polypropylene is allowed to polymerize by the mechanisms already described the R groups will occupy random positions along the chain; this lack of ordered arrangement is called *atactic*. The atactic polymer chains do not pack well together and have low strength. In 1954-1955 catalysts were discovered that yield the ordered arrangements shown in Figure 38.6, with R groups all on the same side (isotactic), or alternately on opposite sides (syndiotactic) of the stretched-out chain. These polymers have greater crystallinity and have greater strength.

In Section 12.3 (Figures 12.10 and 12.11) it was said that a chain molecule like butane, CH_3–CH_2–CH_2–CH_3, has the lowest energy and greatest stability in that conformation (*anti*) in which the CH_3 groups are as

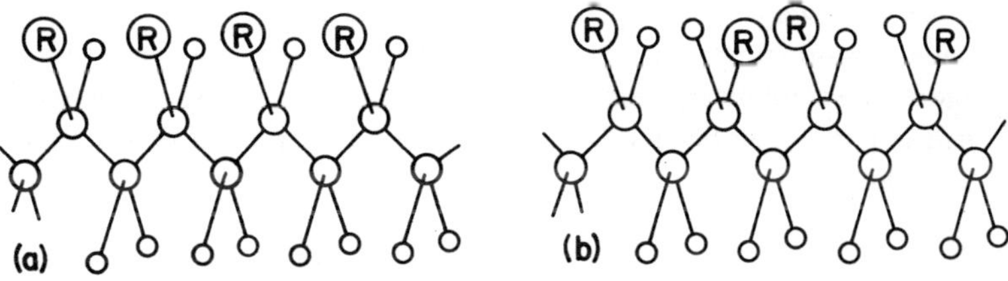

38.6. (a) Isotactic polymer. (b) Syndiotactic polymer.

far apart as they can get. The same effect is found in the isotactic polymer of Figure 38.6, where the efforts of the R groups to reach the most stable conformation twist the chain into a helix similar to those in Figures 33.3 and 38.3. Three monomer units of propylene are in each turn of the polypropylene helix. Karl Ziegler and Giulio Natta shared the 1963 Nobel prize in chemistry for their work on the synthesis of polymers with high tacticity. The original Ziegler catalysts consisted of transition metal salts (e.g. $TiCl_4$) together with aluminum alkyls. The operation of these catalysts is not yet thoroughly understood, but it is believed they form coordination complexes (Section 28.13) with monomers which orient the monomers in a specific space arrangement before they add to each other. The chain grows at its root, the way a hair does on the skin.

38.9. Elastomers. These are elastic polymers that can return rapidly to essentially their original shape and size after having been stretched to many times their original dimensions. Elastomers have properties we normally associate with natural rubber. We have already discussed the chemistry of isoprene (Section 38.3), the building unit of rubber, and have shown that it contains a conjugate system of single and double bonds like that of butadiene (Section 15.7). A property of the conjugate system is the tendency to give 1,4 as well as 1,2 addition; both mechanisms occur in the self-addition which we call polymerization:

$$CH_2{=}CH{-}CH{=}CH_2 \longrightarrow$$

Butadiene

$$[\ldots CH_2{-}CH{=}CH{-}CH_2\ldots]_n \quad \text{and} \quad \left[\begin{array}{c} \ldots CH_2{-}CH \ldots \\ | \\ CH{=}CH_2 \end{array}\right]_n$$

1,4-Polymer 1,2-Polymer

In either case the high polymer still retains double bonds. The presence of these double bonds is responsible for the ability of certain polymers to be *vulcanized*; the properties of the polymers can be modified by partially cross-linking adjacent chains. When sulphur is used as vulcanizing agent, the sulphur atoms probably act as connecting links. The mechanism is not well understood, but in the case of sulphur it is usually represented thus:

$$\begin{array}{l} [\ldots{-}CH_2{-}CH{=}CH{-}CH_2\ldots]_n \quad S \\ [\ldots{-}CH_2{-}CH{=}CH{-}CH_2\ldots]_n \end{array} \longrightarrow \left[\begin{array}{c} \ldots CH_2{-}CH{-}CH{-}CH_2\ldots \\ | \quad\quad | \\ S \quad\quad S \\ | \quad\quad | \\ \ldots CH_2{-}CH{-}CH{-}CH_2\ldots \end{array}\right]_n$$

When the isobutylene described in Section 38.7 is polymerized with about 3% of isoprene the copolymer formed has double bonds and is therefore vulcanizable. This is Butyl rubber. A copolymer of styrene and butadiene was famous in World War II as GR-S rubber (Government rubber type S); it is now known as SBR rubber. The butadiene makes it vulcanizable.

A few other important elastomers are *Vinylite*, a copolymer of vinyl chloride and vinyl acetate; *Koroseal*, a vinyl chloride polymer; *Saran*, a copolymer of vinyl chloride and vinylidene chloride, $CH_2{=}CCl_2$; and *Teflon*, a polymer of tetrafluoroethene, $CF_2{=}CF_2$.

38.10. Fibers. Synthetic fibers can be obtained from certain polymers, generally by dissolving them in suitable solvents and extruding the solutions through openings to give fibers of definite thickness. Saran, which has just been described, is an exception in that the fibers are extruded directly from the molten polymer. In all cases, the fibers are treated chemically and physically immediately after extrusion, to make them suitable for the intended purposes; the fibers may be spun into yarn for manufacture of fabrics or they may be coarse enough for use as bristles.

Vinyon is so inert chemically that it is woven into filter cloths for use in the chemical laboratory. It is a copolymer of vinyl chloride with a small amount of vinyl acetate. *Dynel* is a copolymer of vinyl chloride and acrylonitrile. This fiber can be made into furry materials for liners of winter garments or covers for roller-type paint applicators. *Orlon* is a polymer of vinyl cyanide. Its uses are similar to those of nylon, since the fibers can be processed to yield fabrics which resemble either silk or wool.

Dacron is a polyester, a condensation polymer in which the unit

$$\left[\cdots OCH_2CH_2O-\underset{\underset{O}{\|}}{C}-C_6H_4-\underset{\underset{O}{\|}}{C}\cdots \right]_n \left[\cdots \underset{\underset{O}{\|}}{C}-(CH_2)n-\underset{\underset{O}{\|}}{\underset{H}{\underset{|}{C}}}-\underset{\underset{H}{|}}{N}-(CH_2)n-N\cdots \right]$$

Dacron   Nylon

structure is the ethylene glycol (Section 25.7) ester of terephthalic acid (Section 27.16). *Nylons* have a peptide linkage similar to that in the proteins of silk and wool, and are condensation polymers. An example is the nylon made by heating equimolecular quantities of adipic acid, $HOOC(CH_2)_4COOH$, and hexamethylenediamine, $H_2N(CH_2)_6NH_2$. The condensation takes place with elimination of water.

38.11. Plastic Polymers. The high polymers mentioned in Section

38.9 are classified as *elastic* principally on the basis of their commercial application. In this section we describe a group of polymers said to be *plastic*; this term again is based largely on the end use of the substances. These high polymers prior to formulation for specific use are often called *synthetic resins* from their similarity to natural resins, a familiar example of which is shellac. They are also called synthetic plastics. The first of the synthetic plastics is celluloid (invented in 1869), a nitrocellulose plasticized with camphor. The modern synthetic plastics industry, however, started with Bakelite manufacture in 1907.

38.11a. Thermoplastic Polymers. These polymers have the chain structures described in this chapter, with some possible branching. The chain structures should be distinguished from the three-dimensional network polymers to be described later in Section 38.11b. An example of a thermoplastic polymer is polystyrene. The polystyrenes are soluble in solvents like benzene, the solvent molecules penetrating between the giant molecules and causing first a swelling and then complete solution. The polystyrenes can also be heated above the melting point and then cooled back to the solid state with little change in the chain length. Because polymers built up from linear molecules of this type can be so readily melted (are fusible) without decomposition, they are called *thermoplastics*.

The important plastic known as *Lucite* and as *Plexiglas* is an association polymer made from methyl methacrylate, a derivative of acrylic acid,

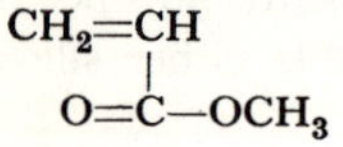

$$
\begin{array}{ccc}
& CH_3 & \left[\quad CH_3 \quad\right] \\
CH_2{=}CH & CH_2{=}C & \left[\,\ldots CH_2{-}C\ldots\,\right] \\
| & | & \left[\quad | \quad\right] \\
O{=}C{-}OCH_3 & O{=}C{-}OCH_3 & \left[\quad O{=}C{-}OCH_3\right]_n
\end{array}
$$

Methyl acrylate 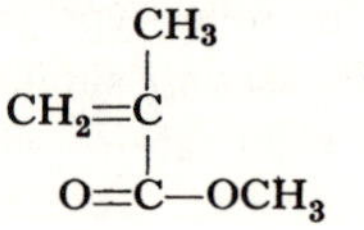   Methyl methacrylate 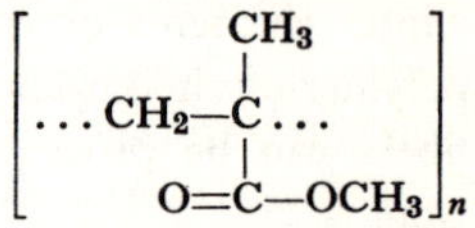   Polymethyl methacrylate

$CH_2{=}CH{-}COOH$. Lucite and Plexiglas, when heated, can be shaped into many useful objects, such as umbrella handles, airplane windows, etc. A copolymer of the acrylate and the methacrylate is known as *Acryloid*, which forms a viscous solution in certain solvents and is used as an adhesive.

38.11b. Thermosetting Polymers. These consist of giant chains *cross-linked* into a three-dimensional network. An example is *Bakelite*—a condensation polymer. The reactions in the formation of Bakelite are complex, but it is believed that the first step is the simple addition of phenol to formaldehyde, and the initial product then condenses to form the giant molecule by elimination of water:

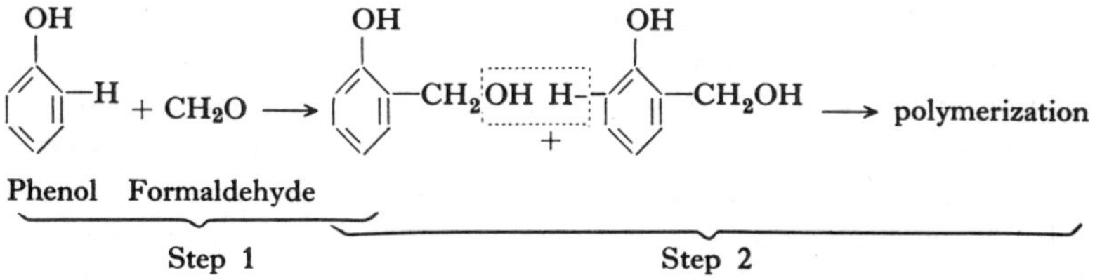

OH | | | | —H + CH_2O ⟶ | —CH_2OH H— —CH_2OH ⟶ polymerization

Phenol   Formaldehyde

Step 1   Step 2

The student should observe that the polymer chains described in Section 37.8 can grow by adding only one unit at a time. They therefore grow at a steady rate. This linear growth is also characteristic of the Bakelite reaction:

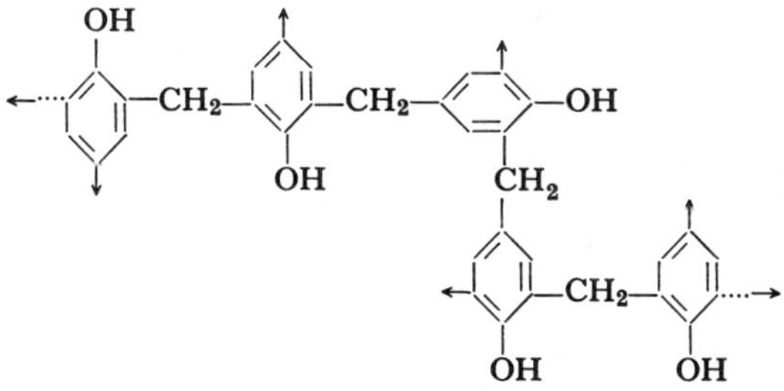

The union between adjacent molecules to form a chain (although a crooked one) is shown here as taking place through positions in the benzene ring which are *ortho* or *para* to the –OH group. Since other *o-* and *p-* positions are still available as reactive points (indicated by arrows), the chain still has possibilities for growth at an angle to this line.

When this linear type of polymer is heated with the proper catalysts, and with proper proportions of formaldehyde to furnish more –CH_2– groups, it becomes a hard, infusible plastic due to cross-linking between adjacent chains. Each molecule that adds to this structure offers new attachment points for other molecules so that the structure grows faster as it gets bigger. The molecular weight of a structure of this kind increases at a geometric rate.

A plastic of the cross-linked type is insoluble in most solvents because the solvent molecules cannot penetrate the solid material to separate the molecules of the polymer. This is due to their tangled, three-dimensional nature, even if in fact the entire solid structure is not one molecule. Three-dimensional polymers are quite hard and infusible. They often decompose if heated to the point where fusion would be expected.

In manufacturing operations, these plastics are usually made directly into the shape of the finished product by placing the incompletely polymerized ingredients in position in a mold for example, and heating until reaction is complete. The time required to finish the operation is the "set time." The mold can then be opened while still hot to remove the solid finished article. Because these polymers are set by heat, they are called *thermosetting* plastics. When making articles from *thermoplastics*, the plastic itself is run into the mold while molten, then must be cooled to the solid state before opening the mold.

The thermosetting plastics, because of their resistance to melting and to solvents, have a very wide use. A typical application is in lacquers to be employed as protective linings in chemical reaction vessels. By contrast, thermoplastics can be used to thicken many liquids, a high viscosity being obtainable with a relatively small concentraction of the plastic in the liquid.

Cellulose was described earlier in this chapter as a high polymer used by nature in the structural parts of plants. It is appropriate to consider it again in this section because of its three-dimensional character. The molecule is an extended chain (Figure 38.1), but each glucose unit has three –OH groups which can form hydrogen bonds with the –OH groups in adjoining chain molecules. Although hydrogen bonds are weaker than primary valence bonds, the cumulative effect of many such hydrogen bonds results in a strong union between adjacent cellulose chains. The chains may be looked on as cross-linked. When cellulose is treated with concentrated alkali solution, the chain molecules are separated and dispersed in solution because of the modification of the –OH groups to –ONa groups with consequent destruction of the hydrogen bonds.

The conversion of cellulose on a commercial scale to thermoplastic plastics makes use of this principle of eliminating the hydrogen bond and making the chain molecules essentially independent of each other. This is done by converting the –OH groups of cellulose chains to ether or ester groups:

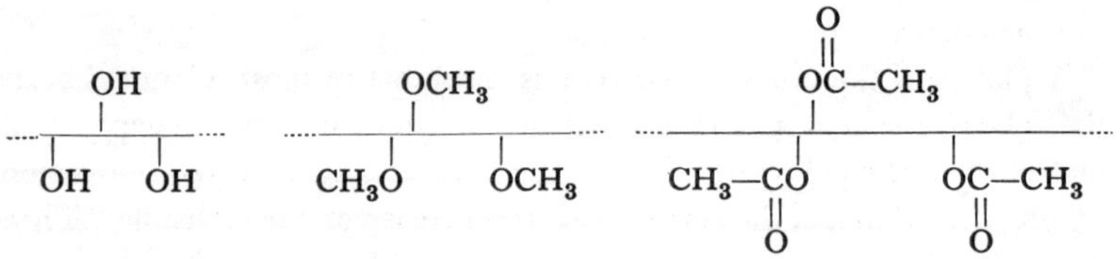

Three —OH groups in each glucose unit of cellulose	Methocel Methyl ether of cellulose	Cellulose acetate Acetic acid ester of cellulose

The properties of the products can be varied by replacing one, two, or all three of the –OH groups.   Methyl cellulose (*Methocel*) is used as a thickening agent, for example, for ice cream.   The plastic known as *Tenite* is a cellulose acetate.   A *rayon* is made from cellulose acetate by dissolving it in acetone and extruding the viscous solution through spinerettes, countercurrent to warm air to evaporate the solvent.

Supplementary Reading

Comprehensive texts are available in most chemical libraries for reading beyond the scope of this book in the areas of general organic chemistry, physical organic chemistry, medicinal chemistry, thermodynamics, and so on. The books and articles in this bibliograpy will be found of interest with respect to special topics. *Books* on special topics are now so numerous a reader cannot hope to find a cited title in any one library, but the *articles* should not be hard to find. The chapters to which these references chiefly apply are shown in parentheses.

N. L. Allinger, "Conformational Analysis in the Elementary Organic Course," J. Chemical Education **41**, 70-72, 1964. *(12)*

G. M. Barrow, *The Structure of Molecules*, W. A. Benjamin, New York, 1963. 168 pp. *(4, 7, 8, 14)*

H. A. Bent, "Distribution of Atomic *s* Character in Molecules and its Chemical Implications", *J. Chemical Education* **37**, 616-624, 1960. *(7, 8)*

R. Breslow, *Organic Reaction Mechanisms*, W. A. Benjamin, New York, 1966. 232 pp. *(16, 17, 20, 24)*

J. A. Campbell, *Why do Chemical Reactions Occur?* Prentice-Hall, New York, 1965. 117 pp. *(16, 17, 18)*

E. Cartmell and G. W. A. Fowles, *Valency and Molecular Structure*, Butterworth, Inc., Washington, D. C. 1961. 295 pp. *(4, 7, 8, 15, 16)*

A. Companion, *Chemical Bonding*, McGraw-Hill, New York, 1964. 155 pp. *(4, 7, 8, 15, 16)*

T. Eisner and J. Meinwald, "Defensive Secretions of Arthropods," Science **153**, 1341-1350, 1966. *(35)*

K. T. Finley, "The Synthesis of Carbocyclic Compounds—A Historical Survey," *J. Chemical Education* **42**, 537-540, 1965. *(12)*

M. St C. Flett, *Physical Aids to the Organic Chemist*, Elsevier Pub. Co., New York, 1962. 388 pp. *(14, 31)*

G. W. A. Fowles, "Lone Pair Electrons," *J. Chemical Education* **34**, 187-190, 1957. *(7, 8)*

C. D. Hurd, "The General Philosophy of Organic Nomenclature," *J. Chemical Education* **38**, 43-47, 1961.

H. N. Hyman, "The Chemistry of the Noble Gases," *J. Chemical Education* **41**, 174-182, 1964; *Ibid*, two articles on pp. 183-186. *(6)*

E. L. King, *How Chemical Reactions Occur*, W. A. Benjamin, New York, 1963. 160 pp. (*17*)

K. Mislow, *Introduction to Stereochemistry*, W. A. Benjamin, New York, 1966. 193 pp. (*22*)

G. Natta, "Macromolecular Chemistry," *Science* **147,** 261-272, 1965. (*38*)

C. R. Noller, "A Physical Picture of Covalent Bonding and Resonance in Chemistry," *J. Chemical Education* **27,** 504-510, 1950. (7, *15, 16*)

L. Pauling, *Nature of the Chemical Bond*, Cornell U. Press, Ithaca, New York, 1960. 644 pp. (7, *16*)

C. C. Price, "Unraveling Sulfur Bonds," *Chemical and Engineering News*, 30 Nov 1964, pp. 58-63. (*30*)

R. D. Preston, "Plants without Cellulose," Scientific American **218** (6) 102-108, 1968. (*34, 38*)

W. A. Pryor, "Organic Free Radicals," *Chemical and Engineering News*, 15 Jan 1968, pp. 70-89. (*17, 20, 24, 38*)

T. J. Schoch, "Cellulose, Glycogen, and Starch," *J. Chemical Education* **25,** 626-631, 1948. (*38*)

J. C. P. Schwarz, *Physical Methods in Organic Chemistry*, Holden-Day, San Francisco, 1964. 350 pp. (*14, 31*)

J. E. Spice, *Chemical Binding and Structure*, Pergamon Press (Macmillan), New York, 1964. 395 pp. (*4, 7, 8, 14*)

J. van Overbeek, "Plant Hormones and Regulators," *Science* **152,** 721-731, 1966. (*35*)

J. J. Vollmer and K. L. Servis, "Woodward-Hoffmann Rules: Electrocyclic Reactions," *J. Chemical Education* **45,** 214-220, 1968. (Somewhat beyond the level of this book, but a major advance in use of orbital theory by chemists.)

R. Waack, "The Stability of the Aromatic Sextet," *J. Chemical Education* **39,** 469-472, 1962.

K. Whetsel, "Infrared Spectroscopy," *Chemical and Engineering News*, 5 Feb 1968, pp. 82-96. (*14, 31*),

C. M. Williams, "Third Generation Pesticides," *Scientific American* **217**(1) 13-17, 1967. (*35*)

R. T. Williams, *Detoxication Mechanisms*, John Wiley, New York, 1949. 288 pp. (*35*)

C. Yanofsky, "Gene Structure and Protein Structure," *Scientific American* **216**(5), 80-85, 1967. This is just one of a series of articles on proteins, enzymes, and the genetic code. A few others: A. Kornberg,

219(4), 64-79, 1968; F. H. C. Crick, **215**(4), 55-62, 1966; M. W. Nirenberg, **208**(3), 80-94, 1963; J. Hurwitz and J. J. Furth, **206**(2), 41-49, 1962. (*35*)

INDEX .

(All topics discussed in this book should be found in this index. However, the book is not intended to be used as a dictionary of chemical compounds. Some compounds are not listed if they simply make tables more complete or if they are not likely to be searched for in an elementary study. A compound like butyrolactone may be found in the page listings under *lactones*. Names of persons are in boldface type.)

α, β Sugars, 404, 411
Acetal, 284, 411
Acetaldehyde, 277, 279, 285, 291
Acetamide, 339
Acetanilid, 339
Acetic acid, 8, 290-94, 305
  dimer of, 298, 417
Acetic anhydride, 303
Acetoacetic ester synthesis, 323
Acetone, 274, 279, 285, 288, 312, 450
Acetonitrile, 342
Acetophenone, 277
Acetoxime, 284
Acetylacetone, 322
Acetylation, 299, 339
Acetyl chloride, 298, 339
Acetylcholine, 338, 427
Acetylene, 97
  model, 90, 129
  orbitals, 133
  properties, 102, 133, 149, 450
Acetylene hydrocarbons, 96, 238, 314
Acetylides, 102
Acid anhydrides, 303-04, 313
Acid-base indicator, 443
Acid chloride, 298-99, 313
Acids/bases, theories of, 295
Acids, carboxylic, 290ff, 313
Acridine, 383
Acryloid, 468
Acrylonitrile, 462
Actinometer, 179
Activated complex, 176
Activation energy, 174ff
Acyl group, 298
Adamantane, 191
1,4-Addition, 158, 381, 458, 466
Addition polymers, 461
Adenine, 419

Adjective dyes, 436
Adrenaline, 226, 426
Alcoholates, 262
Alcohols, 5, 239, 244, 259ff, 282, 313, 364
  1°, 2°, 3° structures, 262
  polyhydric, 265
Aldehyde group, 274
Aldehydes, 274ff, 313
Aldol addition, 285, 346
Aldose, 398
Aldoximes, 284
Aldrin, 385
Alicyclic compounds, 237, 238
Aliphatic compounds, 237
Alizarin, 437
Alkaloids, 383
Alk*anes*, *-enes*, *-ynes*, 240
Alkoxides, 262
Alkoxy radical, 262
Alkylate, 166
Alkylation, 198, 241
Alkyl halides, 243ff, 250
Alkyl radical, 239
Allene, 218
Allyl chloride, 100, 155
Aluminum chloride catalyst, 193
Amides, 338ff, 343
Amido/imido forms, 340
Amine oxides, 331
Amines, 326ff
  1°, 2°, 3° differentiation, 332
Amino acids (table), 386
  optical activity, 397
p-Aminobenzoic acid, 426
Amino group, 327, 435
Ammonia, 46, 49, 327, 332
Ammonium compounds, 242, 309, 327ff
Ammonium cyanate, 8

Periodic Table of the Elements

(Atomic weight values in brackets are for most stable known isotopes)

Series	Period	Group 0	Group 1	Group 2	Group 3	Group 4	Group 5	Group 6	Group 7	Group 8		
1			1 H 1.008									
2	1	2 He 4.003	3 Li 6.94	4 Be 9.012	5 B 10.81	6 C 12.011	7 N 14.007	8 O 15.999	9 F 19.00			
3	2	10 Ne 20.183	11 Na 22.990	12 Mg 24.31	13 Al 26.98	14 Si 28.09	15 P 30.974	16 S 32.064	17 Cl 35.453			
4	3	18 A 39.948	19 K 39.102	20 Ca 40.08	21 Sc 44.96	22 Ti 47.90	23 V 50.94	24 Cr 52.00	25 Mn 54.94	26 Fe 55.85	27 Co 58.93	28 Ni 58.71
5			29 Cu 63.54	30 Zn 65.37	31 Ga 69.72	32 Ge 72.59	33 As 74.92	34 Se 78.96	35 Br 79.91			
6	4	36 Kr 83.80	37 Rb 85.47	38 Sr 87.62	39 Y 88.91	40 Zr 91.22	41 Nb 92.91	42 Mo 95.94	43 Tc [99]	44 Ru 101.1	45 Rh 102.91	46 Pd 106.4
7			47 Ag 107.87	48 Cd 112.40	49 In 114.82	50 Sn 118.69	51 Sb 121.75	52 Te 127.60	53 I 126.90			
8	5	54 Xe 131.30	55 Cs 132.91	56 Ba 137.34	57 La 138.91	58-71 See Lanthanide series						
9												
10	6					72 Hf 178.49	73 Ta 180.95	74 W 183.85	75 Re 186.2	76 Os 190.2	77 Ir 192.2	78 Pt 195.09
11			79 Au 196.97	80 Hg 200.59	81 Tl 204.37	82 Pb 207.19	83 Bi 208.98	84 Po [210]	85 At [210]			
12	7	86 Rn [222]	87 Fr [223]	88 Ra [226]	89 Ac [227]	90-103 See Actinide series						

Lanthanide Series													
58 Ce 140.12	59 Pr 140.91	60 Nd 144.24	61 Pm [147]	62 Sm 150.35	63 Eu 152.0	64 Gd 157.25	65 Tb 158.92	66 Dy 162.50	67 Ho 164.93	68 Er 167.26	69 Tm 168.93	70 Yb 173.04	71 Lu 174.97

Actinide Series													
90 Th 232.04	91 Pa [231]	92 U 238.03	93 Np [237]	94 Pu [242]	95 Am [243]	96 Cm [247]	97 Bk [247]	98 Cf [251]	99 Es [254]	100 Fm [253]	101 Md [256]	102 No [254]	103 Lw [257]